The Climatology Handbook

The Climatology Handbook

Edited by **Andrew Hyman**

R CALLISTO REFERENCE

New York

Published by Callisto Reference,
106 Park Avenue, Suite 200,
New York, NY 10016, USA
www.callistoreference.com

The Climatology Handbook
Edited by Andrew Hyman

International Standard Book Number: 978-1-63239-593-1 (Hardback)

Printed in the United States of America.

Contents

Preface

Climatology is described as the study of climate. It has acquired a much greater position than being considered an individual discipline which treats climate as a phenomenon that fluctuates only within specific boundaries defined by historical statistics. It has been observed that climate alters constantly under the influence of biological as well as physical forces. Therefore, climate cannot be studied in isolation. Rather, climatology requires inputs from diverse scientific disciplines that play their role in comprehending the extremely complicated system of Earth's climate. The current state of climatology has been described in this book. While it provides a wide perspective on climatology, it also acknowledges the regional standpoint which affects people's requirements from climatology. Aspects on the topic of climate change have also been described. It is widely evident these days that current work in climatology has unveiled discoveries which carry profound implications for social and economic policy. Keeping this in mind, the topics covered in this book provide acumens on the applications of various techniques learnt till date.

This book is the end result of constructive efforts and intensive research done by experts in this field. The aim of this book is to enlighten the readers with recent information in this area of research. The information provided in this profound book would serve as a valuable reference to students and researchers in this field.

At the end, I would like to thank all the authors for devoting their precious time and providing their valuable contribution to this book. I would also like to express my gratitude to my fellow colleagues who encouraged me throughout the process.

Editor

Part 1

Synoptic Climatology

Indian Monsoon Depression: Climatology and Variability

Jin-Ho Yoon[1,*] and Wan-Ru (Judy) Huang[2]
[1]Pacific Northwest National Laboratory, Richland, WA
[2]Guy Carpenter Asia-Pacific Climate Impact Centre
School of Energy and Environment, City University of Hong Kong, Hong Kong
[1]USA
[2]China

1. Introduction

The monsoon climate is traditionally characterized by large amount of seasonal rainfall and reversal of wind direction (e.g., Krishnamurti 1979). Most importantly this rainfall is the major source of fresh water to various human activities such as agriculture. The Indian subcontinent resides at the core of the Southeast Asian summer monsoon system (Fig.1) with the monsoon trough extended from northern India across Indochina to the Western Tropical Pacific (WTP). Large fraction of annual rainfall occurs during the summer monsoon season, i.e., June – August[1], with two distinct maxima. One is located over the Bay of Bengal with rainfall extending northwestward into eastern and central India, and the other along the west coast of India where the lower level moist wind meets the Western Ghat Mountains (Saha and Bavardeckar 1976). The rest of the Indian subcontinent receives relatively less rainfall.

Various weather systems such as tropical cyclones and weak disturbances contribute to monsoon rainfall (Ramage 1971). Among these systems, the most efficient rain-producing system is known as the Indian monsoon depression[2] (hereafter MD). This MD is critical for monsoon rainfall because: (i) it occurs about six times during each summer monsoon season, (ii) it propagates deeply into the continent and produces large amounts of rainfall along its track, and (iii) about half of the monsoon rainfall is contributed to by the MDs (e.g., Krishnamurti 1979). Therefore, understanding various properties of the MD is a key towards comprehending the veracity of the Indian summer monsoon and especially its hydrological process.

However, it may be noted that earlier research on the formation and the water vapor budget of the MD may be constrained by limited observation over oceans adjacent to India, especially the Bay of Bengal. Because of this reason, many previous MD studies

---

* Corresponding author
[1] It'll be denoted as JJA hereafter.
[2] Indian monsoon depression, monsoon depression, or MD are used exchangeably in this chapter.

(e.g., Nitta and Masuda 1981, Saha and Chang 1983, Saha and Saha 1988, and many others) mainly focused on a few cases during special field experiments (e.g., the Monsoon Experiment, MONEX[3]). It was not too long ago that a couple of more comprehensive studies on the water vapor budget, life cycle, and dynamical–hydrological cycle interaction were published with more modern datasets (e.g., Yoon and Chen 2005, Chen et al. 2000, Chen et al. 2005, and others).

The MD generally forms around Bay of Bengal and propagates westward or northwestward with the typical life span of three to six days (Ramage 1971, Krishnamurti 1979, Daggupaty and Sikka 1977, Krishnamurti et al. 1975, and 1976). Most of its rainfall occurs in the southwest quadrant of a MD with a recorded maximum of 100 – 200 mm over a 24-hour period (Daggupaty and Sikka 1977, Stano et al. 2002, Saha and Saha 1988). A previous study by Mooley (1973) estimated that MDs could contribute about 11 – 16% of total Indian summer monsoon rainfall using data from six stations (Calcutta, Allahabad, Delhi, Goplur, Nagpur, and Ahmadabad). However, more recent studies (e.g., Krishnamurti et al. 1979, Yoon and Chen 2005) found much higher contribution.

The Indian summer monsoon undergoes an active and break periods of rainfall. This change in rainfall is known as the 'life cycle' of the Indian summer monsoon. The typical life cycle consists of onset, break, revival, and withdrawal (e.g., Krishnamurti and Subrahmanyan 1982). This life cycle is regulated by two intraseasonal modes: the 30-60 and the 10-20 day monsoon modes. Following Yoon and Chen (2005), we call the extreme phases of each mode as active (break) phase for the 30-60 day mode and maximum (minimum) phase for the 10-20 day mode. During the active/maximum (break/minimum) phase of these two intraseasonal modes, the monsoon rain intensifies (weakens). Numerous studies have also reported that the monsoon trough located in northern India deepens and weakens as the 30-60 or 10-20 day mode varies (Krishnamurti and Subrahmanyan 1982, Chen and Yen 1986, Murakami 1976, Krishnamurti and Bhalme 1976). Furthermore, it was suggested that the MD is also affected by the two intraseasonal modes in its occurrence frequency (Chen and Weng 1999) and rainfall intensity (Yoon and Chen 2005), due to changes in the large-scale circulation and convergence of atmospheric water.

Our goal in this chapter is to review recent findings of the MD with focus on the following aspects:

1. Precipitation produced by the MD from genesis to demise and its contribution to monsoon rainfall over central India.
2. The atmospheric water budget of the MD regarding what maintains associated rainfall.
3. Coupling between the hydrological and dynamical processes of the MD.
4. Interaction between the MD and slowly-varying large-scale circulation change such as the Madden-Julian Oscillation (MJO; Madden and Julian 1993), the 10-20 day monsoon mode and the El Nino and Southern Oscillation (ENSO), along with its implication on long-lead climate prediction.

---

[3] The Monsoon Experiment was conducted in 1978 – 1979, as an important part of the First GARP (Global Atmospheric Research Program) Global Experiment (FGGE) (e.g., Murakami 1979).

The chapter is structured along with these focal points mentioned above. In Section 2, the atmospheric water budget and data used are introduced. The climatology of the Indian summer monsoon and intraseasonal variability affecting the Indian summer monsoon are discussed in Section 3. Life cycle, water budget, and dynamical-hydrological processes of the MD are described in Section 4. Interaction between the low-frequency variability, such as tropical intraseasonal oscillation or interannual variability and the MD is demonstrated in Section 5. Concluding remarks are given at Section 6.

2. Atmospheric water vapor budget and data

2.1 Atmospheric water vapor budget

The hydrological cycle of the Indian summer monsoon and the MD can be analyzed with the following atmospheric water vapor budget equation:

$$\frac{\partial W}{\partial t} + \nabla \bullet \vec{Q} = E - P ,$$ (1)

where W is atmospheric precipitable water defined as vertical integral of specific humidity ($W = \frac{1}{g}\int_{p_T}^{p_S} q \bullet dp$), $\vec{Q}$ is vertically-integrated water vapor flux ($\vec{Q} = \frac{1}{g}\int_{p_T}^{p_S} \vec{V}q \bullet dp$), E is evaporation, P is precipitation. g is the acceleration due to gravity, and p_S and p_T are surface pressure and pressure at the top of the atmosphere, respectively[4]. All the analysis is performed at 00Z and 12Z, when upper air sounding is launched and assimilated in atmospheric reanalysis product. Using these high-frequency data, daily- and seasonal-means are computed. Computational details in obtaining each term of the water vapor budget are summarized as follows:

- Convergence of vertically-integrated water vapor flux ($\nabla \bullet \vec{Q}$) is computed using atmospheric reanalysis at every 00Z and 12Z.

- Storage term ($\frac{\partial W}{\partial t}$) at 00Z and 12Z is computed by taking difference 6-hour before and after precipitable water from atmospheric reanalysis. For example, storage term at 12Z is computed taking different between precipitable at 18Z and 6Z, i.e., $\frac{\partial W}{\partial t}\Big|_{12Z} \approx \frac{\Delta W}{\Delta t}\Big|_{12Z} = \frac{W_{18Z} - W_{06Z}}{12h}$. A couple of different methods, Such as using daily mean values, were also tested. However, various different methods yield the same result (Yoon 1999).

- Precipitation (P) is only available as daily-mean value. However, outgoing longwave radiation is available twice a day (Liebmann and Smith 1996). In computing water budget, daily mean precipitation is used, while OLR is used to indicate strong convective regions in tracking the MD.

- Evaporation (E) is neither directly observed nor assimilated. Although it is computed by data assimilation system, evaporation is estimated as a residual in area-mean atmospheric water budget at a given domain (denoted as [ ], and defined as $[\ ] = \iint_A da$):

---

[4] Surface pressure is taken as 1000mb, while the top of the atmosphere is 300mb in our computation. It is because most of atmospheric water vapor resides in the lower troposphere (Peixoto and Oort 1992).

$$[E] = [\frac{\partial W}{\partial t}] + [\nabla \bullet \vec{Q}] + [P] \tag{2}$$

This could include biases or analysis incremental terms of the data assimilation system.

The moisture flux can be separated in to rotational and divergent component ($\vec{Q} = \vec{Q}_R + \vec{Q}_D$), where $\vec{Q}_R$ is rotational and $\vec{Q}_D$ is divergent component of the moisture flux and only divergent component is necessary in the water budget analysis because rotational component does not contribute to the divergence of flux. The importance of divergent water vapor flux in maintaining tropical rainfall was discussed in Chen (1985). Thus, Eq.(1) can be rewritten as

$$\frac{\partial W}{\partial t} + \nabla \bullet \vec{Q}_D = E - P \tag{1a}$$

Further, stream function and potential function of the water vapor flux is computed following Chen (1985):

$$\vec{Q}_R = \hat{k} \times \nabla \psi_Q \tag{3}$$

$$\vec{Q}_D = \nabla \chi_Q \tag{4}$$

$$\nabla^2 \chi_Q = \nabla \bullet \vec{Q} = \nabla \bullet \vec{Q}_D \tag{5}$$

$$\nabla^2 \psi_Q = \hat{k} \bullet \nabla \times \vec{Q} = \hat{k} \bullet \nabla \times \vec{Q}_R \tag{6}$$

Eqs.(5) and (6) are computed in terms of the spectral method based on a horizontal resolution of T31, which is approximately 2.5° latitude × 2.5° longitude.

Considering the long-term mean water budget, one can neglect the storage term ($\frac{\partial W}{\partial t}$) because the atmosphere can hold water about 7 days. Eqs.(1) and (1a) can be rewritten as follows:

$$\nabla \bullet \vec{Q} \approx \nabla \bullet \vec{Q}_D \approx E - P . \tag{7}$$

Further, the water vapor budget of the MD and the Indian summer monsoon is presented by area-averaged values of individual terms in Eqs. (1a) or (7). For the MD, we compute the area-averaged budget terms with a 10° × 10° box centered at the center of the MD, which is about 1,000 km x 1,000 km. Previously the maximum radius of influence affected by the MD is roughly 3,000 km (Godbole 1977, Sikka 1977, Krishnamurti et al. 1975, 1976). However, a relatively conservative estimate was used to analyze all 143 cases.

2.2 Data

Various precipitation estimates are used in this study. For the period of 1979 to 1997, rainfall proxies generated with satellite infrared (IR) observations at Goddard Space

Flight Center (hereafter GPI, Susskind et al. 1997) are used. This dataset is based on daily satellite IR observations with $1° × 1°$ horizontal resolution. In addition, outgoing longwave radiation (OLR) observed by NOAA's polar orbiting satellites (obtained from the Climate Diagnostic Center[5], NOAA) is also used as an indicator of deep convection in the tropics for the early period when satellite retrieved rainfall data was unavailable. As for the later years (1998 – 2002), two different sources of rainfall estimates were used: Tropical Rainfall Measuring Mission (TRMM[6], Simpson et al. 1996)-3B42 and Global Precipitation Climatology Project (GPCP[7], Huffman et al. 1997). Both provide daily rainfall with $1° × 1°$ spatial resolution.

It should be noted that one of the problems in satellite rainfall estimate (especially GPI in our case) is that heavy rainfall over land may not be properly detected. To overcome this deficiency, GPI was blended with station reported rainfall. Merging of two rainfall datasets was done by the methods described by Yoon (1999) and Yoon and Chen (2005). In summary, the Cressman scheme (Cressman 1959) was applied to produce station-reported rainfall at uniform $1° × 1°$ grid first. Then, this was merged with GPI with weighting factors depending on distance to the land. For TRMM and GPCP products used in our study, station-reported rainfall was already merged. Other meteorological variables (e.g., wind field and specific humidity) were obtained from the ECMWF reanalysis data (Gibson et al. 1997) for the 1979 – 2002 period.

3. The Indian summer monsoon: Climatology and intraseasonal variability

3.1 Climatology

More than one billion people live in the south Asian monsoon regions. Previous studies have depicted the important features of the Indian summer monsoon (e.g., Fein and Stephens1987, and many others):

- The monsoon trough extends from northern India across Indochina to the western tropical Pacific (Fig.1a).
- South of the monsoon trough, strong westerly winds at the lower troposphere (extending from the Somali Jet) prevail with a maximum value of 20m/sec over the western Arabian Sea, and a significant eastward deceleration (i.e. low-level convergence) over the eastern Arabian Sea, the Bay of Bengal, and the South China Sea (Fig.1b).
- Heavy rainfall occurs at the western coast of India and the Bay of Bengal. Between these two major wet regions, a semi-arid region situates inside the Indian subcontinent (Fig.1b).

Our analysis domain (80°E – 95°E and 15°N – 27°N indicated as a box in Fig.1b) covers part of the Bay of Bengal and northeast India, which used for area-averaged atmospheric water budget (Fig.2b). This selected area is also quite close to that used by Goswami et al (1999) and more importantly covers the majority of MD tracks (e.g., Yoon and Chen 2005).

---

[5] It is obtained from http://www.cdc.noaa.gov
[6] It is obtained from http://trmm.gsfc.nasa.gov
[7] It is obtained from http://precip.gsfc.nasa.gov/

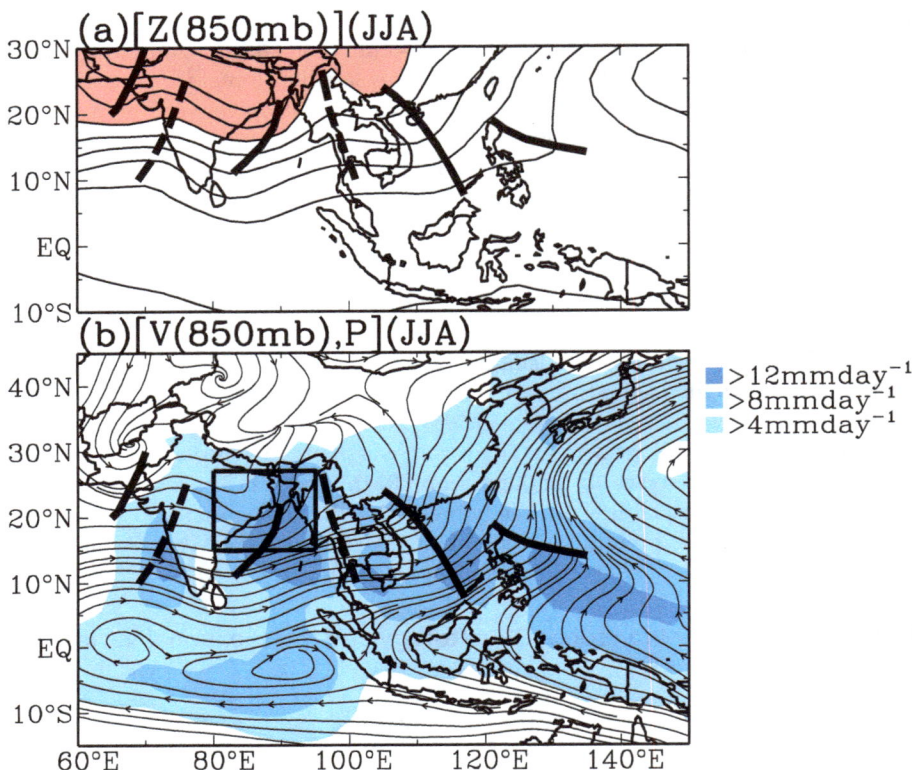

Fig. 1. Summer-mean circulation at the lower troposphere and rainfall. (a) geopotential height at 850mb, and (b) streamline of wind at 850mb and rainfall during northern summer monsoon season (JJA). Monsoon low (lower than 1460m) is colored in (a). Also, trough (ridge) are marked as thick solid (dashed) lines. Analysis domain (80°E – 95°E and 15°N – 27°N) is indicated as a box in (b).

The hydrological processes of the Indian summer monsoon are depicted in Fig.2a with potential function of atmospheric water vapor (χ_Q), divergent water vapor flux ($\vec{Q}_D$) and rainfall (P). Overall the Indian summer monsoon region sits within the vigorous convergent branch of the global hydrological cycle centered at the western tropical Pacific (Chen 1985). Fig. 2 clearly shows that the atmospheric water vapor fluxes converge toward the monsoon trough (located over northern Indian, Indochina, and the WTP; Fig.1). This converging atmospheric moisture maintains the monsoon rainfall centered over the Bay of Bengal, northeast Indian and the western coast of India (Fig. 2a). This feature is further substantiated by area-averaged atmospheric water budget (Fig. 2b) for our analysis domain. The area-averaged rainfall ($[P]$) in Fig.2b is about 12 mmday^{-1} and the convergence of water vapor flux ($-\left[\nabla \bullet \vec{Q}\right]$) is approximately 7 mmday^{-1}, which accounts for about 60% of rainfall. The rest of rainfall is likely maintained by evaporation ($[E]$). This is reasonable because our analysis domain covers part of the Bay of Bengal where evaporation could be substantial during the northern summer season.

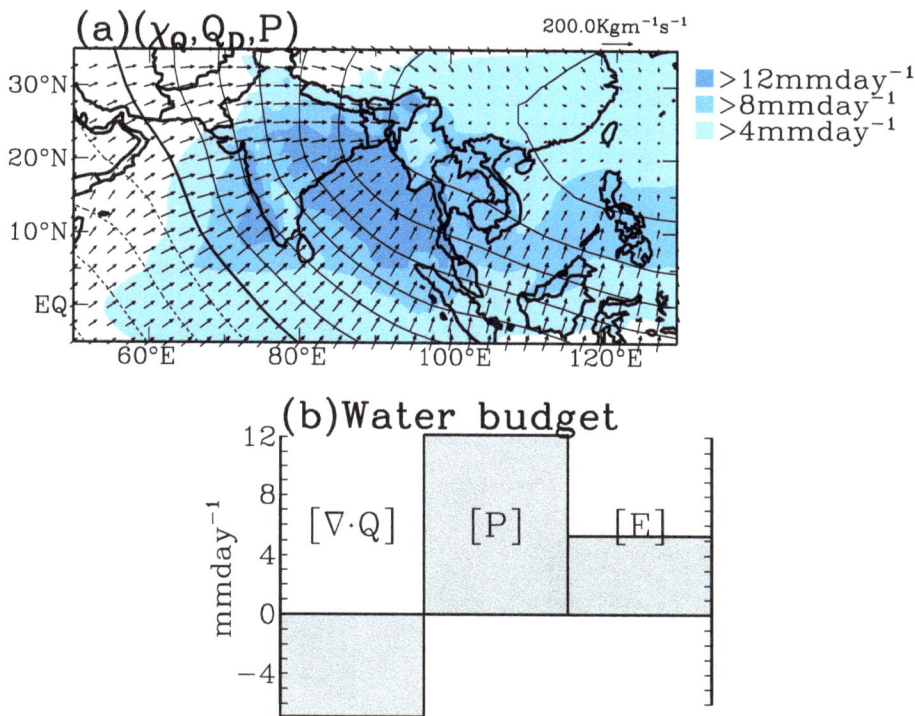

Fig. 2. (a) Summer-mean potential function of water vapor flux (χ_Q), the divergent water vapor flux ($\vec{Q}_D$), and precipitation (P). Contour interval is $2 \times 10^9 \text{m}^2\text{s}^{-1}\text{g} \cdot \text{kg}^{-1}$ for (a). The analysis domain used in (b) was marked as a thick black box in Fig.1(b).

3.2 Intraseasonal variability

The evolution of the Indian monsoon follows a periodical annual cycle but is also affected by quasi-periodic tropical intraseasonal oscillations. For example, the Indian summer monsoon undergoes a 30 – 60 day cycle with active and break phases which are linked to the northward migration of monsoon trough/ridge (Krishnamurti and Subrahmanayam 1982, Joseph and Sijikumar 2004, Krishnamurti and Shukla 2007, Pai et al. 2009) and to the global eastward propagation of the 30-60 day mode (Chen et al. 1988, Madden and Julian 1993). Monsoon trough intensifies (weakens) as monsoon westerlies, the convergence of water vapor flux, and monsoon rainfall intensifies (weakens) during an active (break) phase of the monsoon. To visualize the effect of the 30-60 day mode on the Indian summer monsoon, composite of the active and break phases of the 30-60 day monsoon mode is constructed. To properly isolate this intraseasonal monsoon mode, a band-pass filter designed by Murakami (1979) was used and a 30-60 day filter variable is denoted as (~). The 30 – 60 day band-pass filtered rainfall index over our computational domain (Fig.2a) and the zonal wind ($\tilde{U}(850mb)$) over the Arabian Sea (65°E, 15°N) were used as monsoon indices. Traditionally, the zonal wind index was used to represent the intra-seasonal monsoon life cycle (Krishnamurti and Subrahmanyam 1982). Based on these monsoon indices, all the days above 0.8 (below -0.8) standard deviation of both indices were composited to describe the active (break) phase of the 30-60 day monsoon mode.

Figure 3 exhibits composite filtered anomaly of active/break phases of $\tilde{\chi}_Q, \tilde{\tilde{Q}}_D, \tilde{P}$ (Figs.3a and b) and the area-averaged water vapor budget ($[\nabla \bullet \tilde{Q}_D], [\tilde{E}], [\tilde{P}]$) (Figs.3c and d). Anomalous convergence (divergence) of atmospheric water vapor flux collocates with enhanced (suppressed) monsoon rainfall and trough, consistent with observations of Murakami et al. (1984) and Cadet and Greco (1987). The area-averaged water vapor budget shown in Figs.3c and d reveals that 25% (about 3 mmday^{-1}) increase (decrease) in both precipitation and convergence of water vapor flux during active (break) phase. However, unlike the climatology in Fig.2b the evaporation changes very little by the 30-60 day mode (Figs.3c and d). This may be due to (i) the fact that the 30-60 day mode changes relatively fast so that evaporation cannot respond to this fast-varying mode, or (ii) limitation of residual method used in our study. It is clear that the 30-60 day mode affects the atmospheric water budget through change of the atmospheric water vapor flux.

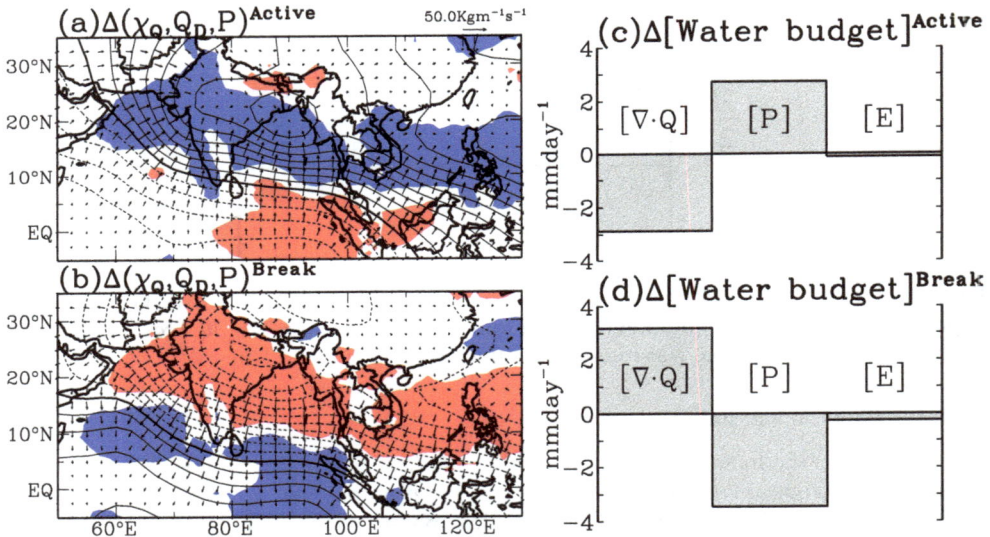

Fig. 3. The 30-60 filtered anomaly of water vapor flux and precipitation ($\tilde{\chi}_Q$, $\tilde{Q}_D$, $\tilde{P}$)active in (a), and ($\tilde{\chi}_Q$, $\tilde{Q}_D$, $\tilde{P}$)break in (b), where (˜) is the 30-60 filtered anomaly. Area-averaged water budget departures from its summer-mean value during phases of (c) active and (d) break phase of the 30-60 day mode. The active and break phases of the 30-60 day monsoon mode are denoted by ( )active, and ( )break, respectively. Contour interval is 2×10⁸m²s⁻¹g·kg⁻¹ for (a) and (b).

In addition to the 30-60 day mode, another important intraseasonal mode affecting the life cycle of the Indian summer monsoon is the 10-20 day mode. In fact, it has been found that the onset of the 1979 summer monsoon was caused by a phase lock of the 30-60 and the 10-20 day modes, not just the 30-60 day mode alone (Krishnamurti et al. 1984). Following similar methodology as used in Fig. 3 with a different frequency (10-20 day), the composite of the 10-20 day monsoon (denoted as (^)) mode was constructed. The synoptic structure of 10-20 day mode and its impact on the Indian summer monsoon were well documented in previous studies (e.g., Murakami 1976, Krishnamirci and Ardanuy 1980, Chen and Chen 1993). Consistent with findings in these previous studies, a dipole structure of potential function of water vapor flux was observed in Fig.4a. In other words, converging center over the Indian subcontinent and diverging center at the equator were observed. It is obvious that the circulation center over the India affects the Indian summer monsoon more directly. On the other hand, the area-averaged water vapor budget indicates that about 17% change in rainfall and convergence of water vapor flux is contributed by 10-20 day mode (Figs.4c and d). This contribution is slightly smaller than that of the 30-60 day mode (Figs.3c and d), but not negligible. Also, change in evaporation is very small as the 30-60 day mode.

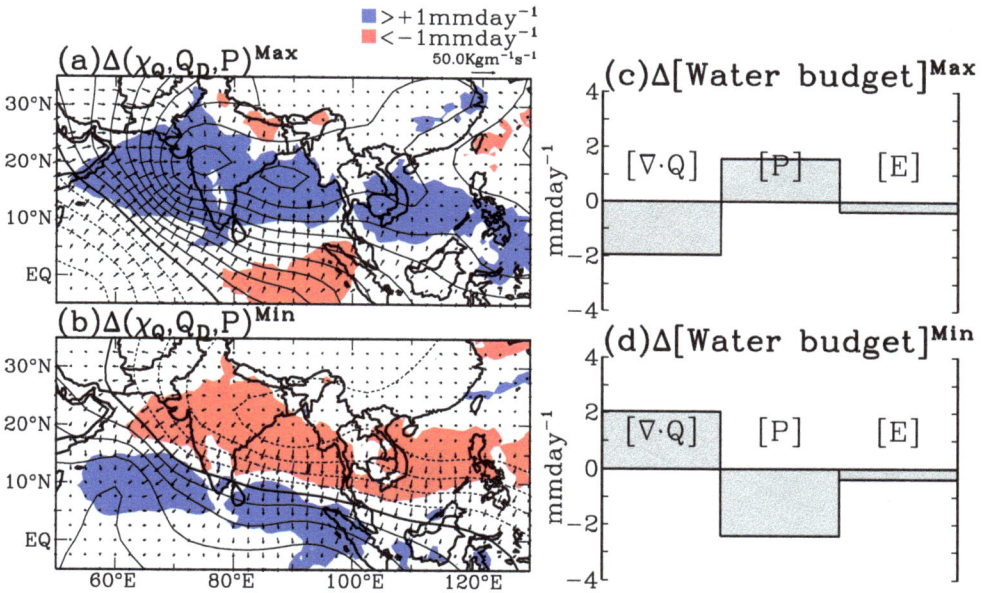

Fig. 4. Same as Figure 3, except for the 10-20 day monsoon mode. Maximum and minimum phases are used instead of active and break phases.

4. Indian monsoon depression

4.1 Life cycle of the MD

Detection of the MD was done with the same method as developed by Chen and Weng (1999) and Yoon and Chen (2005). Tracks of MDs were identified for 1979-1994 by Chen and Weng (1999) and were later expanded to 2002 by Yoon and Chen (2005). One hundred forty three (143) cases were identified over 24 summers (1979 – 2002); the dates were summarized in Fig.1 of Yoon and Chen (2005). Each MD was traced back to its origin position. It is clear from Fig.5 that most of the MDs are formed or intensified over the Bay of Bengal and have their predecessors as residual lows over the Western Tropical Pacific – South China Sea (WTP-SCS) region (Krishnamurti et al., 1977, Saha et al., 1981, Chen and Weng, 1999). There are six genesis mechanisms of the MDs according to Chen and Weng (1999) and Saha et al. (1981). Initial location of the MD with genesis mechanism is summarized in Fig.5.

Population of Monsoon Depressions (1979–2002) and related weather disturbances

(a) First appearance of
- Monsoon Depressions
- 6 Tropical Cyclones
- + Weak disturbances generaged by TCs
- □ Weak disturbances by land–genesis
- * 10–20day monsoon low
- ▲ easterly waves

(b) Last appearance of
1 Tropical Cyclones

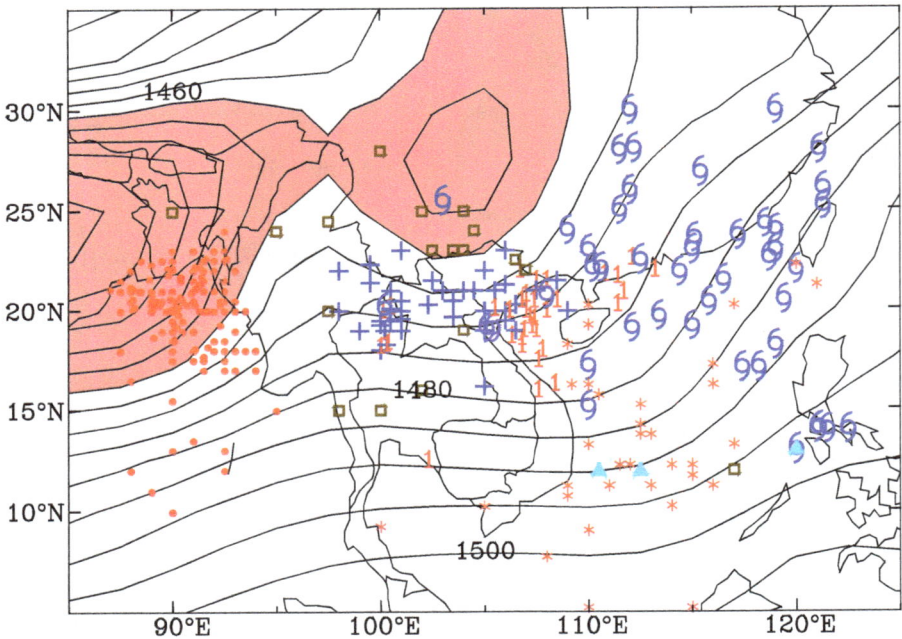

Fig. 5. Genesis locations and mechanisms of the MDs consistent with Fig.5 of Chen and Weng (1999) but for longer record of the MD for 1979 – 2002. First appearance of MDs (red dot), that of weak disturbances apparently linked to tropical cyclones (open tropical cyclone symbol), genesis over land, especially Indochina (brown open square), last appearance of tropical cyclone (marked as 1), 10-20 day monsoon lows (pink diamond), equatorial waves in the WTP-SCS (marked as triangle) are shown on top of the geopotential height at 850mb.

After its formation in the Bay of Bengal, a MD migrates westward or northwestward into the Indian subcontinent with a phase speed of roughly 5° longitude•day^{-1} (e.g., Godbole 1977; Sanders 1984). Because the distance between Bangladesh (~95°E) and northwest India (~70°E) is 25° in longitude, it is estimated that five days are needed for a MD to migrate across the Indian subcontinent. This is consistent with the typical life span of the MD ~ 5 days (e.g., Krishnamurti et al. 1977, Saha et al. 1981, Nitta and Masuda 1981, Chen and Yoon 2000a, and many others). One can estimate that a MD initiated locally or transformed from a residual low over the WTP-SCS gets fully developed over the Bay of Bengal and moves into the Indian subcontinent in 5 days.

To quantitatively illustrate this time evolution of a MD, a composite scheme was developed based on its centered location following Yoon (1999) and Yoon and Chen (2005). First, the average location of all MDs within a 5° – longitudinal zone is obtained and centers of all MDs within this longitudinal zone are adjusted to match this averaged center. For example, all MDs with centers located between 95°E to 90°E, over the Bay of Bengal are averaged to form the MD at Day 1. We repeat this for 2 days prior to and 5 days post its formation over the Bay of Bengal. During the former period, the MD is classified as the prior depression phase (equivalent to the residual low). For the later period, the depression is generally with its rainfall larger than 25 mmday^{-1}, identified as Phase 2 (Day 2 – 4). Note here that some MDs with either slower or faster phase speed than 5° longitude·day^{-1} may exist, which implies our composite method may not ideally fit to these cases. However, phases of these MDs are still decided by their longitudinal location only in our study.

The evolution of MDs prior to and post of phase 2 is classified as phase 1 (Day 1) and Phase 3 (Day 5) when the system exhibits development over the Bay of Bengal and decaying at the western and northwestern part of India following Yoon (1999) and Yoon and Chen (2005). Also, composite of prior depression phase was constructed for two days. All different phases with corresponding location and composite days are listed in Table 1.

Phase	Location	Composite Days
Prior depression	Indochina	Days -2 and -1
Phase 1 or Developing	Bay of Bengal	Day 1
Phase 2 or Mature	Inside the Indian subcontinent	Days 2, 3, and 4
Phase 3 or Decay	Western India	Day 5

Table 1. List of phases, location and days in our composite analysis. More details can be found in Yoon (1999), Yoon and Chen (2005), and Chen et al. (2005).

The life cycle of a MD depicted with wind and divergent circulation in the lower troposphere along with different phases are depicted in Fig.6. For better isolation of this synoptic-scale disturbance from the background flow, zonal-wave filter using Fast Fourier Transform (FFT) is applied. The maximum horizontal scale of the MD is reported to be about 3,000 km (Godbole 1977; Krishnamrti et al. 1975) which is corresponding to zonal waves numbers 12 – 13 at 20°N. Considering variable sizes of individual MD, we applied a Fourier spectral filter with zonal wave number of 6 – 25 (denoted as [ ]^s) following previous studies (Yoon 1999, Chen and Yoon 2000a, Yoon and Chen 2005, Chen et al. 2005).

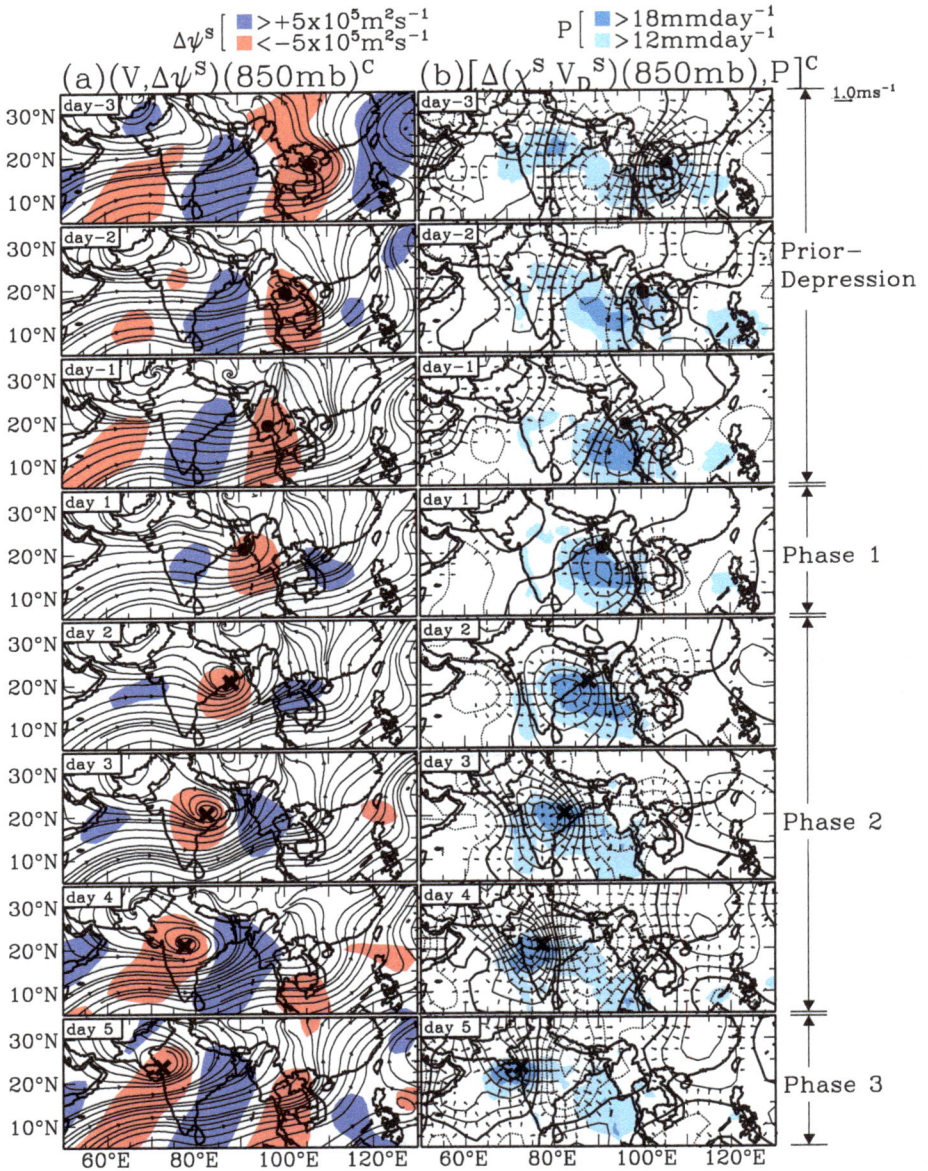

Fig. 6. A composite of MD during twenty four summers (1979−2002) for 7 days (5 (2) days after (before) formation of the MD, listed in Table 1) using (a) wind at 850mb superimposed with departures of the short-wave stream function at 850mb $[\Delta(V,\psi^S)(850mb),P]$, and (b) composite 850-mb velocity potential and divergent wind departures in the short-wave regime superimposed with precipitation, $[\Delta(\chi^S,V_D^S)(850mb),P]$. Contour interval is $2\times10^8 m^2 s^{-1} g \cdot kg^{-1}$. $\Delta(\ )$ is departure from the summer-mean value of $(\ )$.

Here we summarize several important dynamical and physical characteristics based on the composite MD as revealed in Fig.6 and from previous studies (e.g., Krishnamurti et al. 1975, Krishnamurti et al. 1977, Saha et al. 1981, Saha and Saha 1988, Sikka 1977, Daggupaty and Sikka 1977, Godbole 1977, Chen and Yoon 2000a, Douglas 1992a and b, Rajamani and Sikdar 1989, and others):

- The horizontal scale of a MD is in the range of 1,500 to 3,000 km with central pressure down to 990mb. Its propagation speed is 3 – 5 degree per day. Of note, it propagates westward against the strong monsoon westerlies in the lower atmosphere (Fig.6a).
- The vertical extent of a MD is only up to 300mb restrained by the existence of the Tibetan high during Indian summer monsoon season. It implies that the upper level monsoon easterlies may not steer a MD westward (Chen and Yoon 2000a).
- The maximum rainfall and rising motion of a MD were detected in the southwest quadrant of the system (Fig.6b). This is consistent with early observation during MONEX periods (Saha and Saha 1984). Maximum rainfall can be as high as 100mm per day.
- Because the MD does not stay long enough over the Bay of Bengal, energy input from the surface boundary appears unimportant.
- Its evolution and dynamical characteristics have various stages. At its early stage, the barotropic dynamics may play an important role. On the other hand, baroclinic and Conditional Instability of the Second Kind (CISK) dynamics are more important at later stage.
- It is a tropical weather disturbance so that planetary vorticity (f) cannot be larger than relative vorticity (ζ). In other words, one cannot ignore relative vorticity advection (Chen and Yoon 2000a).

4.2 Water vapor budget of the MD

Fig.6b shows that the MD brings a large amount rainfall into central India as it moves westward. To acquire a quantitative measure of how much rain is produced by a MD, we perform area-averaged water vapor budget with a moving window following the MD's position. The research questions addressed here are: *(1) what is the contribution of the MD to the total monsoon rainfall?* and *(2) how this rainfall is maintained within the depression?* Understanding of this hydrological process helps explain the dynamical evolution of a MD, as explained later in Section 4.3. To answer the above questions, composite maps of water vapor fluxes and rainfall are shown with area-average water vapor budget in Fig.7. Important findings obtained from Figs.6 and 7 are summarized here:

1. *Prior depression phase (Days -2 – -1):* A weak low-pressure system moves across Indochina as shown by 850-mb streamline with the short-wave filtered stream function anomalies (colored) in Fig.6a. Although it is a relatively weak system with less-organized circulation structure, a residual low brings some amount of precipitation (Chen and Yoon 2000b) and is maintained by low-level convergence (Fig.6b). As indicated by previous studies (Chen and Weng 1999, Saha et al. 1981), most of the MDs have their predecessors over Indochina. Therefore, it is important to identify or track these systems.
2. *Phase I (over the Bay of Bengal at Day 1):* After reaching the Bay of Bengal, this weak low pressure system transforms into a MD. Our composite captures its transition stage, which signifies the MD's development. Apparently, the cyclonic vortex intensifies (Fig.6a) in accordance with stronger convergence of low-level flow (Fig.6b) as well as water vapor flux and rainfall (Fig.7).

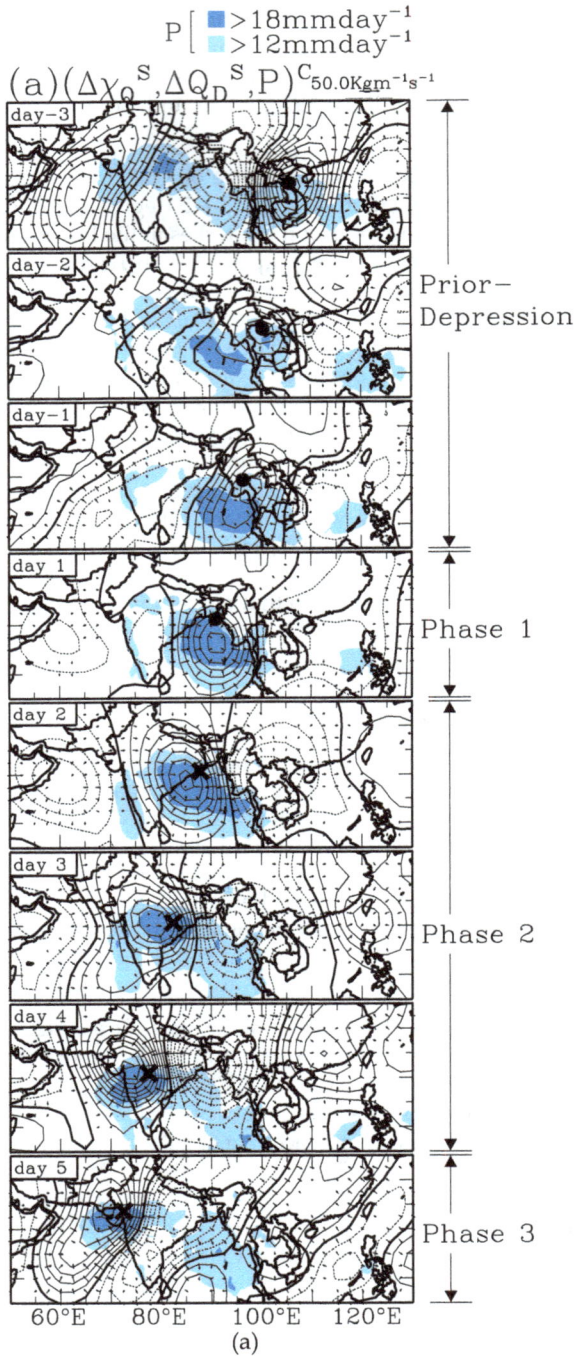

(a)

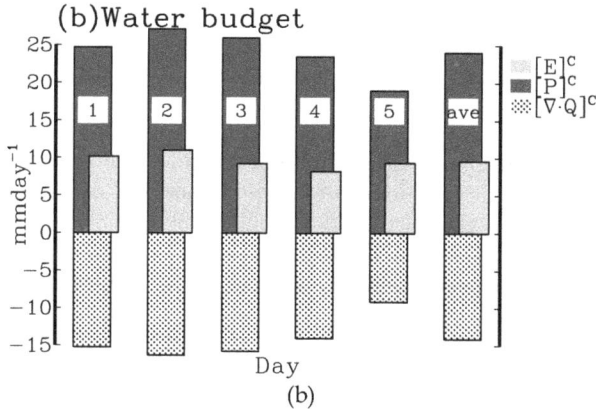

(b)

Fig. 7. (a) same as Fig.6 except potential function of water vapor, divergent water vapor flux, and rainfall $[\Delta(\chi_Q^S,\vec{Q}_D^S),P]$ and (b) area averaged water vapor budget for Day 1 – 5 and average of these five days.

3. *Phase 2 (Days 2 – 4):* At this phase, a MD starts to progress into the Indian subcontinent and produces large amounts of rainfall inland. An important finding here is that as rainfall increases, the convergence of atmospheric water vapor flux intensifies as well (Fig.7a). This implies that the major source of MD rainfall is not from surface source through evaporation but through atmospheric hydrological process. This is further substantiated in terms of area-averaged water vapor flux shown in Fig.7b. At the last day of this phase (Day 4), the entire system starts to weaken.

4. *Phase 3 (Decay at Day 5):* This is the last stage of a MD's lifecycle. However, the low-pressure system can be easily identified and it continues to produce rainfall inside the continent. The demise of the system is clearly seen in the amount of the water vapor flux converging into the system (Fig.7).

It is noted here that during the pre-depression stage, rainfall is less organized but located mainly at the east side of the system. However, as it develops into a MD, major rainfall is concentrated on the western or southwestern corner. Corresponding lower-level divergent circulation is formed across a MD with a convergent (divergent) center located east (west) of the system. Based on area-average water vapor budget, rainfall and convergence of water vapor flux reach their maximum at Day 2 and maintain their strength till Day 4. On the last day (Day 5), convergence of water vapor flux is reduced to about half compared to Day 2.

For most tropical weather systems, water vapor is mainly supplied by evaporation from warm ocean surface (Riehl 1954). However, in the case of the MD, atmospheric water vapor transport and convergence is by far the most important source. In this sense, hydrological processes of the MD are close to those of the mid-latitude cyclones (Chen et al. 1996).

Typically, about six MDs (e.g., Chen and Weng 1999) develop every monsoon season over the Bay of Bengal. It is shown by the composite analysis that a MD could stay about 3 – 4 days (from Day 1 to Day 4) over the Indian subcontinent with rainfall over 25 mmday^{-1}. A simple estimation of the total rainfall by MDs (3 – 4 days × 25mmday^{-1} × 6/season = 450-600 mm/season) is equivalent to about 45 – 55% of the total monsoon rainfall (92 days/season × 12mmday^{-1} = 1104 mm/season) over the computational domain (75°E – 90°E, 15°N – 27°N). This value is higher than that estimated by Mooley (1973), probably due to the different methods and data employed, but is close to the result shown by Dhar et al. (1981). Mooley (1973) used only six stations along the eastern coast of India to estimate the MD contribution to the total rainfall over these stations, while the current study applied composite analysis using rainfall of GPCP, TRMM, and GPI to estimate the contribution by the MD to the total monsoon rainfall. Based on our estimate, about half of the total rainfall over the eastern coast of India is generated by the MDs.

4.3 Coupling between hydrological and dynamical processes of the MD

Our next question is how the aforementioned hydrological processes are linked to dynamical properties of the MD. To answer this question, we'll use the stream function tendency equation, which is another form of the vorticity equation (e.g., Holton 1992). Let's briefly review the vorticity (Eq.(8)) and the stream function tendency equations (Eq.(9)):

$$\frac{\partial \zeta}{\partial t} = -\mathbf{V} \cdot \nabla(\zeta + f) - (\zeta + f)\nabla \cdot \mathbf{V}$$

$$\zeta_t \qquad\qquad \zeta_A \qquad\qquad\qquad \zeta_\chi$$

(8)

First, vorticity tendency (ζ_t) represents whether vorticity at one location becomes more positive (cyclonic circulation) or negative (anti-cyclonic circulation). Second this vorticity tendency is determined by sum of advection (ζ_A) and stretching terms (ζ_χ). Analyzing each term reveals which process is important in a particular system. For example, vorticity advection term by westerly jet in the upper troposphere becomes a dominant process in eastward moving mid-latitude storms (Holton 1992). Vorticity in the tropics is at least one order smaller than that in the mid-latitudes. Thus, it is difficult to apply the vorticity tendency equation to the MD.

To overcome this difficulty, the stream function tendency equation is constructed by applying Inverse-Laplacian to the vorticity equation (Eq.8):

$$\nabla^{-2}(\frac{\partial \zeta}{\partial t}) = \nabla^{-2}[-\mathbf{V} \cdot \nabla(\zeta + f)] + \nabla^{-2}[-(\zeta + f)\nabla \cdot \mathbf{V}]$$

$$\psi_t \qquad\qquad\qquad \psi_A \qquad\qquad\qquad \psi_\chi$$

(9)

This stream function tendency equation (Eq.9) was applied for a case during FGGE-MONEX by Sanders (1984) and Chen and Yoon (2000) to illustrate dynamical properties of the MD. The dynamical implication of each term in Eq.(9) is the same as that of Eq.(8). Name of each term in both equations are summarized in Table 2. Consistent with our analysis earlier, a short-wave filter in zonal direction was applied.

First, let us examine the existence of an east-west asymmetric circulation across a MD proposed by Saha and Saha (1988) based on heat budget analysis. As shown in Fig.6, lower tropospheric rotational and divergent flow has a spatial quadrature relationship. In other words, local maxima or minima of potential function at 850mb ($\Delta\chi^S(850mb)^C$) located at nodal points of the stream function $\Delta\psi^S(850mb)^C$, in between positive and negative centers. On top of this lower tropospheric convergence, upward motion is located in the west and southwest corner of the MD (Fig.4 of Chen et al. 2005). Although its downward branch is less clear, it is conceivable that an east-west divergent circulation exists across the MD.

Vorticity Equation	Stream function tendency equation			
$\dfrac{\partial \zeta}{\partial t}$	$\nabla^{-2}[\dfrac{\partial \zeta}{\partial t}]$	ζ_t	ψ_t	Vorticity (stream function tendency)
$-\mathbf{V} \cdot \nabla(\zeta + f)$	$\nabla^{-2}[-\mathbf{V} \cdot \nabla(\zeta + f)]$	ζ_A	ψ_A	Advection term
$-(\zeta + f)\nabla \cdot \mathbf{V}$	$\nabla^{-2}[-(\zeta + f)\nabla \cdot \mathbf{V}]$	ζ_χ	ψ_χ	Stretching term

Table 2. Vorticity and stream function tendency terms.

Second, each term of Eq.(9) was computed at all the phases of the MD. The result indicates that stretching term is much larger than advection (Chen et al. 2005). Thus, one can approximate this equation as follows:

$$\psi_t^S \sim \psi_\chi^S \qquad (10)$$

Negative stream function tendency is found west and southwest corner of the depression center, which makes the system, i.e., negative stream function center, moves westward (not shown). Convergence at the lower troposphere can generate negative stream function tendency by vortex stretching. Further, total vorticity ($\eta = \zeta + f$) in vortex stretching term can be assumed as a constant and the budget equation can be simplified as follows:

$$\psi_t^S \sim -\eta\chi^S \sim -(\zeta + f)\chi^S . \qquad (11)$$

Our results confirm findings by previous studies (Sanders 1984, Chen and Yoon 2000a, Chen et al. 2005) that advection of total vorticity is negligible in the westward propagation of MDs. Indeed the stream function tendency generated by vortex stretching over the west-southwest sector of a MD is vital to its westward propagation.

Finally, collocation of strong convective rainfall, convergence (divergence) at the lower (upper) troposphere, and upward motion implies that the divergent circulation is closely linked to hydrological processes. In summary, (1) release of latent heat at the center of convection, which is west side of the MD center, produces strong convergence (divergence) at the lower (upper) atmosphere and its counter part to the east of the MD center, and (2) this strong upward branch exerts negative stream function tendency through vortex stretching term (Fig.6 of Chen et al. 2005). This is well depicted by a schematic diagram in Fig.8. Based on the composite $\Delta\psi^S(850mb)$ budget of MDs, it is clear that $\Delta\psi_A^S(850mb)$ (stream function tendency generated by total vorticity advection) is not an effective dynamic process in generating $\Delta\psi_t^S(850mb)$. Instead, $\Delta\psi_\chi^S(850mb)$ coupled with the east-west asymmetric circulation of the MD is the primary dynamic process in propagating a MD westward. The convergent center of the east-west circulation west of a depression center overlaps the negative stream function tendency. Therefore, the depression is propagated westward by this negative tendency.

u(z) at 85°E

mb
100
300
500
700
1000
-30 -15   0   15
m·s^{-1}

Tibetan High

$\chi^s<0$ D $\chi^s>0$ C $\chi^s>0$

$\psi^s<0$

$\psi_t^s\sim\psi_x^s<0$ $\psi_t^s\sim\psi_x^s>0$

$\psi^s<0$

$\chi^s>0$ C   D $\chi^s<0$

Depression

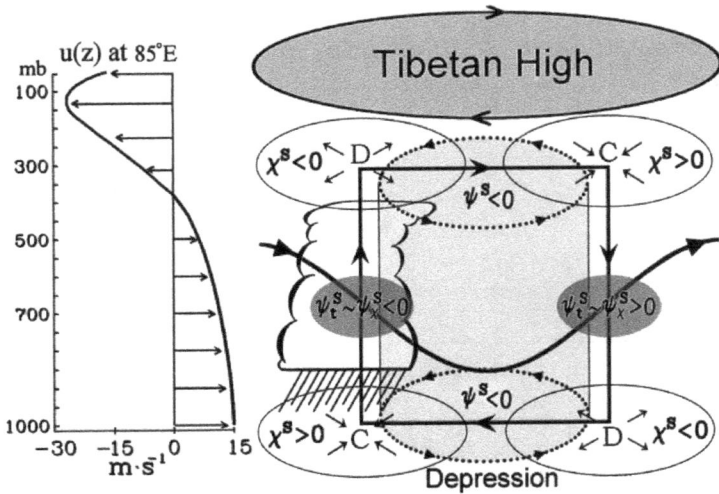

Fig. 8. A schematic diagram of the MD (a stippled cylinder, identified as $\psi^s<0$) embedded in the monsoon westerlies in the lower troposphere (indicated by the zonal wind profile left) and the Tibetan high in the upper troposphere (the elliptic stippled area at the top of the right diagram). The combination of the depression and monsoon westerlies is portrayed by the thick sinusoidal streamline. The east-west asymmetric circulation of the MD is portrayed by solid lines with shafts encircling a cylindrical-shaped depression. The divergent circulation (χ^s , and divergent wind) coupled with the east-west circulation is denoted by thin-line circles centered at Ds (divergent centers) and Cs (convergent centers) at the upper and lower levels. Cumulus convection/rainfall west of the depression center is coincident with the upward branch of the east-west circulation. This depression is moved westward by the negative (positive) stream function tendency (ψ_t^s) generated by the vortex stretching (ψ_χ^s) associated with the upward (downward) branch of the east-west circulation. This was recreated from Fig.1 of Chen et al. (2005) with permission.

5. Interaction between the low-frequency variability and the MD

5.1 Modulation of the MD by the intraseasonal variability

It is shown in Section 3 that the Indian summer monsoon system is modulated by two tropical intraseasonal modes: 30-60 and 10-20 day monsoon modes. Because the MD is the major rainfall contributor to the Indian summer monsoon rainfall, it is conceivable that both the 30-60 day and 10-20 day modes also have a large impact on the MDs and that this impact is achieved by changing the atmospheric water flux.

A previous study by Chen et al. (1999) found that more low-pressure systems propagate into the Bay of Bengal during the active phase of the 30-60 day monsoon mode than the break phase due to changing monsoon westerlies. In this study, we rather focus on changes in the atmospheric hydrological processes. To show the modulation of the MD by the 30-60 day mode, we group all the MD cases into two categories: One with those occurring during the active phase of the 30-60 day mode, and the other during the break phase. The phase of the 30-60 day mode was determined by two monsoon indices mentioned in Section 3.

A MD during active phase of the 30-60 day mode is shown in Fig.9. At its first day over the Bay of Bengal on August 6th 1979, a MD exhibits a clear intensification, moves inland on August 8th, and reaches the other side of the Indian subcontinent on August 10th. Noteworthy here is that even at its last day of propagation, this system still maintains strong rainfall and a well-organized cyclonic flow. Also, another low-pressure system appears over Indochina. In other words, more low-pressure systems can propagate during active phase of the 30-60 day monsoon mode (Chen et al. 1999). In contrast, a case during the break phase is overall weak and less organized (not shown here but refer to Yoon 1999 or Yoon and Chen 2005).

It is found that change in the large-scale circulation, especially the converging atmospheric water vapor flux is responsible for modulation of a MD by the 30-60 day monsoon mode (Fig.9b). The difference in area-averaged atmospheric water budget of the MD during active and break phase of the 30-60 day mode is shown in Fig.10a to further elucidate this process. During the active phase of the 30-60 day monsoon mode, rainfall is about 7mm/day higher than that during the break phase. Relatively large values of atmospheric storage term ($\frac{\partial W}{\partial t}$) and evaporation ($E$) were also observed. This has a physical implication that the surface process is important in tropical intraseasonal variability (Shinoda et al. 1998). However, this large contribution of evaporation in our budget analysis could be due to (1) biases in atmospheric reanalysis used in our study or (2) a sampling bias that numbers of MDs during active are more than break phase so that uncertainty becomes larger in evaporation or storage terms. The former can be fixed using more modern atmospheric reanalysis such as MERRA[8], CFSR[9], or ERA-interim[10], and the latter needs to be tested using more longer analysis period.

The same procedure was repeated for the MD and the 10-20 day monsoon mode. A MD case during maximum phase (June 22nd 1979 – June 26th 1979) is shown in Fig.11. When a MD is collocated with the 10-20 day monsoon low, vigorous and well-organized convection and clear cyclonic flow were observed in Fig.11a during the maximum phase of the 10-20 day mode. During the minimum phase, a clear opposite was observed (not shown here but refer to Yoon 1999 or Yoon and Chen 2005). It is noted here that at last day (June 26th 1979), less intense rainfall and organized cyclonic flow are observed than August 10th 1979 in Fig.11. Although it is only a case, the result implies the strength of the 10-20 day monsoon mode is weaker than that of the 30-60 day mode.

Comparison of the MD water budget between the maximum and the minimum phases of the 10-20 day monsoon mode (Fig.10b) reveals that the hydrological processes of the MD are also affected by the 10-20 day monsoon mode through change of convergence of water vapor flux, just like the 30-60 day mode. Only difference is the weaker intensity. Using average from Day 1 to Day 5, impact of MD by the 10-20 day monsoon mode is about 15% weaker than the 30-60 day mode (Yoon and Chen 2005).

---

[8] MERRA: Modern Era Retrospective-Analysis for Research and Application. More details can be found in http://gmao.gsfc.nasa.gov/merra/
[9] CFSR: The Climate Forecast System Reanalysis. More details can be found in http://cfs.ncep.noaa.gov/cfsr/
[10] ERA-interim: newer version of ECMWF reanalysis. More details can be found in http://www.ecmwf.int/research/era/do/get/era-interim

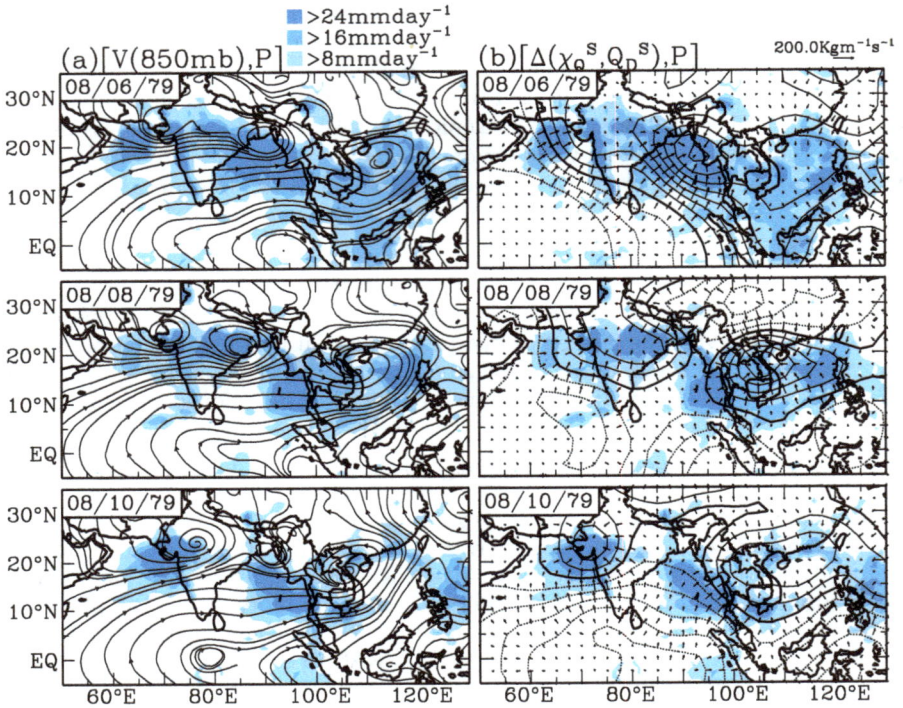

Fig. 9. (a) The 850-mb streamline charts superimposed with P, and (b) $[\,\Delta(\chi_Q, Q_D), P\,]$ of a MD case during the active phase of the 30-60 day mode from August 6th 1979 to August 10th 1979. The contour interval of $\Delta(\chi_Q)$ is $2.0 \times 10^8 \mathrm{m^2 s^{-1} g \cdot kg^{-1}}$, and $\Delta()$ is departure from the summer mean.

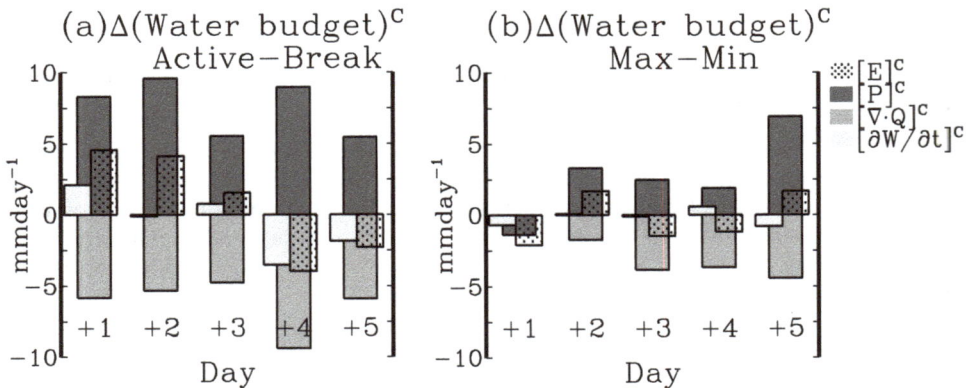

Fig. 10. Difference of the MD water vapor budget between the active and break phases of the 30-60 day monsoon mode (a) and the maximum and minimum phases of the 10-20 day monsoon mode (b).

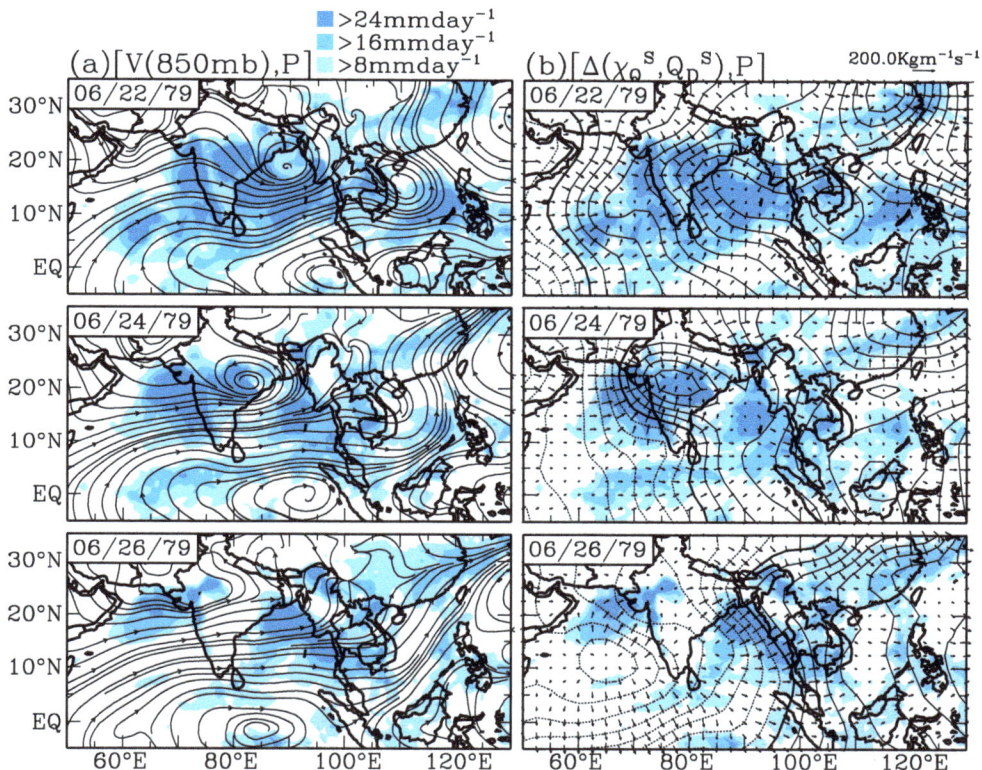

Fig. 11. Same as Figure 9 except the maximum phase of the 10-20 day mode from June 22nd 1979 to June 26th 1979.

5.2 Interannual variation of the MD

A significant year-to-year change of rainfall is found over the Indian summer monsoon region. Most well-known cases are drought occurred in 1987 and flood in 1988 (Krishnamurti et al. 1989). Numerous attempts have been made to explain the mechanisms responsible for this interannual variation of the Indian summer monsoon, such as (i) interannual variation in the sea surface temperature over the tropical Pacific (Palmer et al. 1992, Chen and Yen 1994, Ju and Slingo 1993, Slingo and Annamalai 2000), (ii) land-surface feedback including both long-memory in soil moisture (Meehl 1994) or snow coverage over the central Eurasian continent (e.g., Shukla 1987, Verneka et al. 1995), and (iii) interannual variation of occurrence frequency of MDs (Chen and Weng, 1999). As shown earlier, the MD can be responsible up to 45 – 55% of the total monsoon rainfall. Thus, it is conceivable that any change in the number of the MD could result in significant change in total monsoon rainfall.

To test this hypothesis, total precipitation over the northern part of India (65°E – 90°E and 15°N – 25°N) and occurrence frequency of the MD during summer monsoon season are displayed together in Fig.12 with warm and cold years indicated with red and blue color.

Definition of warm/cold years[11] were adopted from Chen and Weng (1999) which is based on the sea surface temperature (SST) over the NINO3 region (150°W – 90°W and 5°S – 5°N) as an indicator of thermal condition of the eastern tropical Pacific. Drought in 1987 and flood of 1988 are well captured by the total Indian monsoon rainfall index (Fig.12b). On the other hand, more (less) MDs occurred during cold (warm) years of NINO3 (Chen and Weng 1999, Yoon 1999). However, overall correlation between the occurrence frequency and total monsoon rainfall is very low. This clearly indicates that hypothesized mechanism (iii) that occurrence frequency of the MD regulates interannual variation of the Indian summer monsoon rainfall cannot be substantiated.

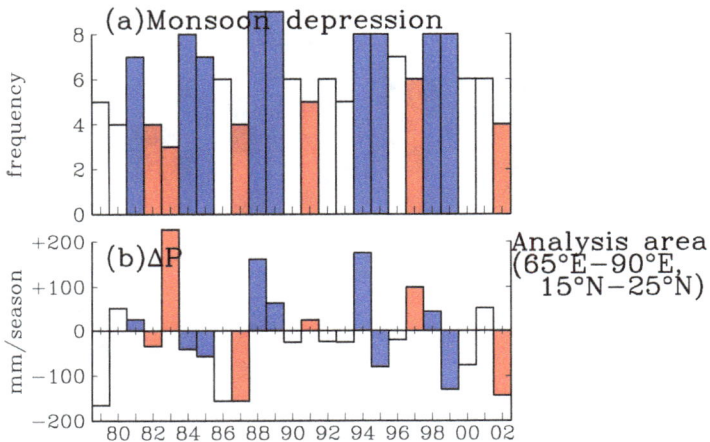

Fig. 12. Histogram of occurrence frequency of the MD at each year from 1979 to 2002 in (a), and total monsoon rainfall over the northern India (65°E – 90°E and 15°N – 25°N) in (b). Warm and cold years of the eastern Tropical Pacific was indicated by red and blue colors. This was based on SST anomaly over the NINO3 region (150°W – 90°W and 5°S – 5°N).

The occurrence frequency of MD (Fig.12a) appears to be well coherent with the NINO3 index, an indicator of the tropical Pacific climate anomaly during summer season. The tropical Pacific serves as a source for the tropical synoptic/meso-scale disturbances that propagate into the Bay of Bengal and then become the MD (Fig.5). In other words, occurrence frequency of the MD is more an indicator of the tropical Pacific weather and climate, but not a direct indicator of the strength of the Indian summer monsoon (Chen et al. 1997, Chen and Weng 1999, Yoon 1999). However, it is noted that the interannual variation of occurrence frequencies of these low-pressure systems has a significant impact on the total rainfall over Indochina (Chen and Yoon 2000b).

In summary, although MDs are responsible for a large fraction of Indian summer monsoon rainfall, the MD itself is not a decisive factor of the total monsoon rainfall change from one year to another. Second, interannual change of the MD occurrence frequency is closely linked to that of the WTP-SCS region which is closely related to the East Asian summer monsoon (Chen and Yoon 2000b, Chen and Weng 1999). Third, the interannual variation of

[11] We have warm summers of 1982, 1983 1987 1991, 1997, and 2002 and cold summers of 1981, 1984, 1985 1988, 1989, 1994, 1995, 1998, and 1999.

the Indian summer monsoon is independent from that of the East Asian monsoon. For example, Lau et al. (1999) and Chen and Yoon (2000b) propose the dynamical and hydrological difference between two major summer monsoon regions: East Asian summer monsoon and South Asian summer monsoon. These studies imply more complicated interaction between two monsoons with global SST on the interannual timescales. Also, there are some prominent hypotheses were not fully discussed or tested here (e.g., Krishnan et al. 2010). Based on these results, more comprehensive research on the interannual variation of the Indian summer monsoon is needed.

6. Concluding remarks

Summer monsoon rainfall is a critical component of human activity over the south Asian regions, where more than one billion people live. Various weather systems contribute this monsoon rainfall. One of the prominent systems is the MD. In this article, we have reviewed various characteristics of the MD. Compared other tropical storms such as hurricane or typhoon, this low-pressure system stands out because it stays mainly over the landmass not over the warm tropical oceans. Because of this special circumstance, its dynamical and hydrological properties are different from other tropical storms. The most important property is that its major water source is not from evaporation from the warm ocean but rather from converging atmospheric water vapor flux.

Several important findings are summarized as follows:

- Our estimation based on the composite of the MD reveals that up to 45% – 55% of total summer monsoon rain is brought by MDs. About 60% of total rainfall is maintained by the converging atmospheric water vapor flux (Fig.5b).
- The life cycle of the Indian summer monsoon (onset – break – revival – withdrawal) is constituted by two intraseasonal modes. As the Indian summer monsoon undergoes its life cycle, convergence of the water vapor flux over India fluctuates coherently.
- The stream function tendency generated by vortex stretching over the west-southwest sector of a MD is vital to its westward propagation, which is located over the upward branch of an east-west circulation across a system. Furthermore, this upward branch is maintained by the latent heat released due to strong convective rainfall. The dynamical and hydrological processes can be established by feedbacks among convective rainfall, diabatic heating, east-west circulation, and divergent circulation.
- The hydrological cycle of the MD is intensified (weakened) by the convergence (divergence) of water vapor flux associated with two intraseasonal modes during active/maximum (break/minimum) monsoons. Also it is revealed that the 30-60 day mode is more effective in modulating the MD than the 10-20 day mode.
- A poor correlation between the MD occurrence frequency and total Indian monsoon rainfall is observed in interannual timescales, despite the fact that MD is responsible for about half of total rain. It is likely that other mechanisms listed in Section 5 is critical in determining the interannual variability of the Indian summer monsoon. Further work is needed in this regard.

Close interaction between the diabatic heating due to strong convection and divergent circulation is similar to the CISK mechanism in the tropics. Difference between the MD and a conventional CISK can be seen in its propagation property (e.g. Hayashi 1970, Lindzen 1974, and many others). This rather complicated dynamical and hydrological processes of the MD provide some challenges to the regional and global climate models. For example, to properly simulate extreme rainfall under global warming, these need to be well simulated by regional or global climate models (e.g., Ratnan and Cox 2006, Vinodkumar et al. 2008 and 2009).

About half of the Indian summer monsoon rainfall is produced by MDs. Thus, the prediction of MDs is important. In order to accurately predict the intensity, propagation, and rainfall of a MD, basic features in various spatial and temporal time scales, such as the Tibetan high, the monsoon trough, monsoon westerlies, thermal contrast, and slowly-varying tropical variabilities should be properly simulated. Also, close interaction between the diabatic heating due to strong convection and divergent circulation is similar to the CISK mechanism in the tropics. Difference between the MD and a conventional CISK can be seen in its propagation property (e.g. Hayashi 1970, Lindzen 1974, and many others). This rather complicated dynamical and hydrological processes of the MD provide some challenges to the regional and global climate models. For example, to properly simulate extreme rainfall under global warming, these need to be well simulated by regional or global climate models (e.g., Ratnan and Cox 2006, Vinodkumar et al. 2008 and 2009). Also, simulation of this complicated interaction will serve as a good test-bed for newly developed physical parameterization.

One of the aspects that have not been discussed in this chapter is the long-term change of the MDs. This is an interesting issue given the fact that (1) our climate has been rapidly changing due to human activities and (2) more importantly that most of the extreme rainfall cases are associated with the MD (Dhar and Nanargi 1995, Sikka 2000). It was found that the number of extreme rainfall events have increased since 1976 (Goswami et al. 2006). At the same time, the number of the MD has been reported decreasing since 1976 (Stowasser et al. 2009, Kumar and Dash 2001, Sikka 2006, Ajayamohan et al. 2010, Rao et al. 2008, Rao et al. 2004, Mani et al. 2009). This contrasting trend – increasing extreme rainfall but reducing total number of depressions – is an interesting area that the community needs to invest resources. Another aspect that needs our attention is the role of aerosol on Indian monsoon, which was pursued as a forcing agent in long-term change of the summer monsoon (e.g., Chung and Ramanathan 2007, Lau and Kim 1999, Bolasina and Nigam 2008). Due to rapid increase of anthropogenic emissions over Asian countries including India and China, roles played by aerosol on the Asian monsoon system need to be well understood.

7. Acknowledgement

Most of the materials were obtained from three published articles at Tellus by the author and M.S. thesis. This work is partially supported by a DOE grant to PNNL. PNNL is operated for the U.S. Department of Energy by Battelle Memorial Institute under contract DE-AC06-76RLP1830. The authors greatly appreciate careful review by Drs. V. Vinoj and Kiranmayi Landu at PNNL and editorial assistance by the editor, Dr. S.-Y.(Simon)Wang.

8. References

Ajayamohan, R. S., William J. Merryfield, Viatcheslav V. Kharin, 2010: Increasing Trend of Synoptic Activity and Its Relationship with Extreme Rain Events over Central India. *J. Climate*, 23, 1004–1013.

Bollasina, M., S. Nigam, and K.-M., Lau, 2008: Absorbing aerosols and summer monsoon evolution over South Asia: An observational portrayal. *J. Climate*, 21, 3221-3239.

Cadet, D. L. and Greco, S., 1987: Water vapor transport over the Indian Ocean during the 1979 summer monsoon. Part I: Water vapor fluxes. *Mon. Wea. Rev.*, 115, 653-663.

Chen, T.-C., 1985: Global water vapor flux and maintenance during FGGE. *Mon. Wea. Rev.*, 113, 1801-1819.

Chen, T.-C., and Yen, M.-C., 1991a: Interaction between intraseasonal oscillations of the midlatitude flow and tropical convection during 1979 northern summer: The Pacific Ocean. *J. Climate*, 4, 653-671.

Chen, T.-C., and Yen, M.-C., 1991b: A study of the diabatic heating associated with the Madden-Julian Oscillation. *J. Geophys. Res.*, 96, 13163-13177.

Chen, T.-C., Tzeng, R.-Y., and Yen, M.-C., 1988: Development and life cycle of the Indian monsoon: Effect of the 30-60 day oscillation. *Mon. Wea. Rev.*, 116, 2183-2199.

Chen, T.-C., and Chen, J.-M, 1993: The 10-20 day mode of the 1979 Indian monsoons: Its relation with the time variation of monsoon rainfall. *Mon. Wea. Rev.*, 121, 2465-2482.

Chen, T.-C., and Yen, M.-C., 1986: The 40-50 day oscillation of the low-level monsoon circulation over the Indian Ocean. *Mon. Wea. Rev.*, 114, 2550-2570.

Chen, T.-C., and Yen, M.-C., 1994: Interannual variation of the Indian monsoon simulated by the NCAR community climate model: Effect of the tropical Pacific SST. *J. Climate*, 7, 1403-1415.

Chen, T.-C., Yen, M.-C., and Schubert, S., 1996: Hydrologic processes associated with Cyclone systems over the United States. *Bull. Amer. Meteor. Soc.*, 77, 1559-1507.

Chen, T.-C., Yen, M.-C., N. Yamazaki, and S. Kienhe, 1997: Interannual variation in the tropical cyclone formation over the western North Pacific. *Mon. Wea. Rev.*, 126, 1080-1090.

Chen, T.-C., and Weng, S.-P., 1999: Interannual and intraseasonal variations in monsoon depressions and their westward-propagating predecessors. *Mon. Wea. Rev.*, 127, 1005-1020.

Chen, T.-C., and Yoon, J.-H., 2000a: Some remarks on the westward propagation of the monsoon depression. *Tellus*, 52A, 487-499.

Chen, T.-C., and Yoon, J.-H., 2000b: Interannual Variation in Indochina Summer Monsoon Rainfall: Possible Mechanism *J. Climate* 13:1979-1986.

Chen T.-C., Yoon, J.-H., and Wang, S.-Y., 2005: Westward Propagation of the Indian monsoon depression. *Tellus, 57A*, 758-769. doi:10.1111/j.1600-0870.2005.00140.x.

Chung, C.E. and V. Ramanathan (2007): Relationship between trends in land precipitation and tropical SST gradient. *Geophys. Res. Lett.*, 34, L16809, doi:10.1029/ 2007GL030491.

Cressman, G. P., 1957: An operational objective analysis system. *Mon. Wea. Rev.*, 87, 367-374.

Daggupaty, S. M., and Sikka, D. R., 1977: On the vorticity budget and vertical velocity distribution associated with the life cycle of a monsoon depression. *J. Atmos. Sci.*, 34, 773-792.

Dhar, O. N., P. R. Rakhecha, and B. N. Mandal, 1981: Influence of tropical disturbances on monthly monsoon rainfall of India. *Mon. Wea. Rev.*, 109, 188-190.

Douglas, M. W., 1992: Structure and Dynamics of Two monsoon depressions. Part I: Observed Structure. *Mon. Wea. Rev.*, 120, 1524–1547

Douglas, M. W., 1992: Structure and Dynamics of Two monsoon depressions. Part II: Vorticity and Heat Budgets. *Mon. Wea. Rev.*, 120, 1548–1564.

Fein, J. and P. L. Stephens (eds), 1987, Monsoons, John Wiley, New York, NY, 632pp.

Gibson, J. K., Kallberg, K., Hernandez, A., Uppala, S., Nomura, A., and Serano, E., 1997: *ERA Description*, ECMWF Re-Analysis Project Report Series, Vol. 1, European Centre for Medium Range Weather Forecasts, 72 pp.

Godbole, R. V., 1977: The composite structure of the monsoon depression. *Tellus, 29A,* 25-40.

Goswami, B. N., Krishnamurthi, V., and Annamalai, H., 1999: A broad-scale circulation index for the interannual variability of the Indian summer monsoon. *Quart. J. Roy. Meteorol. Soc.,* 125, 611-633.

Hayashi, Y., 1970: A theory of large-scale equatorial waves generated by condensation heat and accelerating the zonal wind. *J. Meteor. Soc. Japan,* 48, 140-160.

Holton, J. R., 1992: *Introduction to Dynamic Meteorology.* 3rd ed. Academic Press, 511pp.

Huffman, G. J., R. F. Adler, P. Arkin, A. Chang, R. Ferraro, A. Gruber, J. Janowiak, A. McNab, B. Rudolf, U. Schneider. 1997: The Global Precipitation Climatology Project (GPCP) combined precipitation dataset. *Bull. Amer. Meteor. Soc.,* 78, 5-20.

Joseph, P. V., S. Sijikumar, 2004: Intraseasonal Variability of the Low-Level Jet Stream of the Asian Summer Monsoon. *J. Climate,* 17, 1449–1458.

Ju, J., and J. Slingo, 1995: The Asian summer monsoon and ENSO. *Q. J. R. Meteorol. Soc.,* 121, 1136-1168.

Krishnamurti, T. N., 1979: *Tropical Meteorology.* Compendium of *Meteorology II,* WMO-No. 364, A. Wiin-Nielsen, Ed. World Meteorological Organization, 428 pp.

Krishnamurti, T.-N. and Ardanuy, D., 1980: The 10 to 20 day westward propagating mode and "Breaks in the monsoons". *Tellus,* 32, 15-26.

Krishnamurti, T.-N. and Bhalme, H. H., 1976: Oscillations of a Monsoon System. Part I. observational Aspects. *J. Atmos. Sci.,* 33, 1937-1954.

Krishnamurti, T.-N., and Subrahmanayam, D., 1982: The 30-50day mode at 850mb during MONEX. *J. Atmos. Sci.,* 39, 2088-2095.

Krishnamurti, T.-N., Kanamitsu, M., Godbole, R., Chang, C.-B., Carr, F., and Chow, J. H., 1975: Study of a monsoon depression (I): Synoptic Structure. *J. Meteor. Soc. Japan,* 53, 227-239.

Krishnamurti, T.-N., M. Kanamitsu, R. Godbole, C.-B. Chang, F. Carr, and J. H. Chow, 1976: Study of a monsoon depression (II), dynamical structure. *J. Meteor. Soc. Japan,* 54, 208-226.

Krishnamurti, T.-N., Jayakumar, P. K., Sheng, J., Sugri, N., and Kumar, A., 1984: Divergent circulations on the 30 to 50 day time scale. *J. Atmos. Sci.,* 42, 364-375.

Krishnamurti, T. N., Molinari, J., Pan, H., and Wong, V., 1977: Downstream amplification and formation on monsoon disturbances. *Mon. Wea. Rev.,* 105, 1281-1297.

Krichnamurti, T. N., H. S. Bedi, and M. Subramaniam, 1989: The summer monsoon of 1987. *J. Climate,* 2, 321-340.

Krishnamurthy, V., J. Shukla, 2007: Intraseasonal and Seasonally Persisting Patterns of Indian Monsoon Rainfall. *J. Climate,* 20, 3–20.

Krishnan, R., Ayantika, D. C., Kumar, V. and Pokhrel, S. (2011): The long-lived monsoon depressions of 2006 and their linkage with the Indian Ocean Dipole. *Int. J. Climatol.,* 31, 1334–1352. doi: 10.1002/joc.2156

Kumar, J. R., and S. K. Dash, 2001: Interdecadal variations of characteristics of monsoon disturbances and their epochal relationships with rainfall and other tropical features. *Int. J. Climatol.,* 21, 759–771

Lau, K-M., K-M. Kim, S. Yang, 2000: Dynamical and Boundary Forcing Characteristics of Regional Components of the Asian Summer Monsoon. *J. Climate,* 13, 2461–2482.

Lau, K.-M., and K.-M. Kim, 2006: Observational relationships between aerosol and Asian monsoon rainfall, and circulation. *Geophys. Res. Lett.,* 33, L21810, doi:10.1029/2006GL027546.

Liebmann, B. and C. A. Smith, 1996: Description of a complete (interpolated) outgoing longwave radiation dataset. *Bull. Ameri. Meteor. Soc.*, 77, 1275-1277.

Lindzen, R. S., 1974: Wave-CISK in the tropics. *J. Atmos. Sci.*, 31, 156-179.

Madden, R. A. and Julian, P. R., 1993: Observations of the 40-50 day tropical oscillation – a review. *Mon. Wea. Rev.*, 122, 814-837.

Mani, N. J., E. Suhas, and B. N. Goswami, 2009: Can global warming make Indian monsoon weather less predictable?. *Geophy. Res. Lett.*, 36, L08811, doi:10.1029/2009GL037989.

Meehl, G. A., 1994: Influence of the land surface in the Asian summer monsoon: External condition versus internal feedbacks. *J. Climate*, 7, 1033-1049.

Mooley, D. A., 1973: Some Aspects of Indian monsoon depressions and the Associated Rainfall. *Mon. Wea. Rev.*, 101, 271 -280.

Murakami, M., 1976: Analysis of summer monsoon fluctuations over india. *J. Meteor. Soc. Japan*, 54, 15-31.

Murakami, M., 1979: Large-scale aspects of deep convective activity over the GATE area. *Mon. Wea. Rev.*, 107, 997-1013.

Murakami, T., 1979: Scientific Objectives of the Monsoon Experiments, *GeoJournal*, 3(2), 117-136.

Murakami, T., Nakazawa, T., and He, J., 1984: On the 40-50 day oscillations during the 1979 Northern Hemisphere summer. Part II: Heat and Moisture budget. *J. Meteor. Soc. Japan*, 62, 469-484.

Nitta, T. and K. Masuda, 1981: Observational study of a monsoon depression developed over the Bay of Bengal during the summer MONEX. *J. Meteor. Soc. Japan.*, 59, 672-682.

Pai, D. S., J. Bhate, O. P. Sreejith, and H. R. Hatwar, 2011: Impact of MJO on the intraseasonal variation of summer monsoon rainfall over India. *Clim. Dyn.*, 36, 41-55.

Palmer, C. P., 1952: Tropical Meteorology. *Q. J. R. Meteorol. Soc.*, 78, 126-163.

Palmer, T. N., C. Brankovic, P. Viterbo, and M. J. Miller, 1992: Modeling interanual variation of summer monsoons, *J. Climate*, 5, 379-417.

Peixoto, J. P. and A.-H. Oort, 1992: Physics of Climate, American Institute of Physics, New York, NY, 520pp.

Rajamani, S. and Sikdar, D. N., 1989: Some dynamical characteristics and thermal structure of monsoon depressions over the Bay of Bengal. *Tellus*, 41A,: 255–269. doi: 10.1111/j.1600-0870.1989.tb00380.x

Ramage, C. S., 1971: Monsoon Meteorology. Academic Press, New York, NY, 296 pp.

Rao, Y. P., 1976: *Southwest Monsoon*. Meteor. Monographs, Synoptic Meteorology, Indian Meteorological Department, 367pp.

Rao, V. B., C. C. Ferreira, S. H. Franchito, and S. S. V. S. Ramakrishna (2008): In a changing climate weakening tropical easterly jet induces more violent tropical storms over the north Indian Ocean, *Geophys. Res. Lett.*, 35, L15710, doi:10.1029/2008GL034729.

Rao, B. R. S., D. V. B. Rao, and V. B. Rao (2004): Decreasing trend in the strength of Tropical Easterly Jet during the Asian summer monsoon season and the number of tropical cyclonic systems over Bay of Bengal, *Geophys. Res. Lett.*, 31, L14103, doi:10.1029/ 2004GL019817

Ratnam, J. V., and E. A. Cox., 2006: Simulation of monsoon depressions using MM5: sensitivity to cumulus parameterization schemes. *Meteorology and Atmospheric Physics* 93:1-2, 53-78

Riehl, H. 1954: *Tropical Meteorology*. McGraw-Hill, 392 pp.

Saha, K. and Bavardeckar, S. N., 1976: Moisture flux across the west of India and raindall during the southwest monsoon. *Tellus*, 38, 370-379.

Saha, K. and Chang, C.-P., 1983: The baroclinic processes of monsoon depressions. *Mon. Wea. Rev.*, 111, 1506-1514.

Saha, K., and Saha, S., 1988: Thermal budget of a monsoon depression in the Bay of Bengal during FGGE-MONEX 1971. *Mon. Wea. Rev.*, 116, 242-254.

Saha, K., Sanders, F., and J. Shukla, 1981: Westward propagating predecessors of monsoon depressions. *Mon. Wea. Rev.*, 109, 330-343.

Sanders, F., 1984: Quasi-geostrophic diagnosis of the monsoon depression of 5-8 July 1979. *J. Atmos. Sci.*, 41, 538-552.

Shinoda, T., H. H. Hendon, and J. Glick, 1998: Intraseasonal variability of surface fluxes and sea surface temperature n the tropical western Pacific and Indian oceans. *J. Climate*, 11, 1685-1702.

Shukla, J., 1978: CISK-barotropic-baroclinic instability and the growth of monsoon depressions. *J. Atmos. Sci.*, 35, 495-508.

Shukla, J., 1987: Interannual variability of monsoon, Monsoons, Eds, J. S. Fein, P. L. Stephens, Johns Wiley and Sons, New York, NY, 399-464.

Sikka, D. R., 1977: Some aspects of the life history, structure and movement of monscon drepssions. *Pure and Applied Geophysics*, 115, 1501-1529

Sikka, D. R., 2006: A study on the monsoon low pressure systems over the Indian region and their relationship with drought and excess monsoon seasonal rainfall. Center for Ocean–Land–Atmosphere Studies Rep. 217, 61 pp.

Simpson, J., Adler, R. F., and North, G. R., 1988: A proposed Tropical Rainfall Measuring Mission (TRMM) satellite. *Bull. Amer. Meteor. Soc.* 69, 278–295

Slingo, J. M., H. Annamalai, 2000: 1997: The El Niño of the Century and the Response of the Indian Summer Monsoon. *Mon. Wea. Rev.*, 128, 1778–1797

Stano, G., Krishnamurti, T. N., Vijaya Kumar, T. S. V., and Chakraborty, A., 2002: Hydrometeor structure of a composite monsoon depression using TRMM radar. *Tellus*, 54A, 370-381.

Stowasser, Markus, H. Annamalai, Jan Hafner, 2009: Response of the South Asian Summer Monsoon to Global Warming: Mean and Synoptic Systems. *J. Climate*, 22, 1014–1036.

Susskind, J., Piraino, P., Rokke, L., Iredell, L., and Mehta, A., 1997: Characteristics of the TOVS Pathfinder Path A Datasets. *Bull. Amer. Meteor. Soc.*, 78, 1449-1472.

Vernekar, A. D., J. Zhou, and J. Shukla, 1995: The effect of Eurasian snow on the Indian monsoon, *J. Climate*, 8, 248-266.

Vinodkumar, A. Chandrasekar, K. Alapaty, Dev Niyogi, 2008: The Impacts of Indirect Soil Moisture Assimilation and Direct Surface Temperature and Humidity Assimilation on a Mesoscale Model Simulation of an Indian monsoon depression. *J. Appl. Meteor. Climatol.*, 47, 1393–1412. doi: 10.1175/2007JAMC1599.1

Vinodkumar, A. Chandrasekar, K. Alapaty, Dev Niyogi., 2009: Assessment of Data Assimilation Approaches for the Simulation of monsoon depression over the Indian Monsoon Region. *Boundary-Layer Meteorology* 133:3, 343-366

Yoon, J.-H., and T.-C. Chen. 2005: Water Vapor Budget of the Indian monsoon depression. *Tellus*, 57A, 770-782. doi:10.1111/j.1600-0870.2005.00145.x.

Yoon, J.-H., 1999: Hydrological cycle associated with monsoon depressions, M.S. Thesis, Iowa State University

2

Southern Hemisphere Tropical Cyclone Climatology

Yuriy Kuleshov
National Climate Centre, Australian Bureau of Meteorology, Melbourne
School of Mathematical and Geospatial Sciences
Royal Melbourne Institute of Technology (RMIT) University, Melbourne
Australia

1. Introduction

Each year, around 80 tropical cyclones (TCs) form around the world, with about one-third of them in the Southern Hemisphere (SH) (Gray, 1979). Tropical cyclones within the South Indian Ocean (SIO) and the South Pacific Ocean (SPO) are frequent and intense, and they dramatically affect maritime navigation and the lives of communities in coastal areas. Australia and the island nations are affected each year by TCs. In extreme cases they can have devastating consequences on life, property and the economic well-being of the communities directly affected and the country as a whole, as in the case of one of Australia's most notorious TCs, *Tracy*, which devastated Darwin, the capital of the Northern Territory, on 25 December 1974 (Australian Government, 1977).

A number of high-impact TC events in the SH have occurred in recent years. In December 2002, TC *Zoe* (maximum intensity of 240 km/h) totally devastated the island of Tikopea in the SPO. Cyclone *Heta* developed in the SPO and reached a peak intensity of 235 km/h on 5 January 2004. In Niue, the capital city of Alofi was devastated by the TC and two people left dead. In American Samoa, *Heta* destroyed over 600 homes and damaged 4,000 others; 20 people were injured. In March 2004, TC *Gafilo* affected northeastern Madagascar as the strongest cyclone to ever strike the country. Cyclone *Gafilo* made land-fall at peak intensity of 230 km/h, caused 283 fatalities and over US$250M damage.

It is critical then that we have a full understanding of these disaster-inducing events. There is increasing evidence that the climate is changing on a global scale (IPCC, 2007) and consequently it is important to monitor changes in regional TC frequencies, intensities and tracks (Kuleshov and de Hoedt, 2003; Landsea, 2000a; Trenberth, 2007). Records show that TC frequency and spatial distribution have considerable inter-annual variability, and attempts have been made to find explanations for such variability. The role of El Niño-Southern Oscillation (ENSO) in TC activity has been studied extensively. A significant reduction (*increase*) of TC activity over the Atlantic basin is observed during El Niño (*La Niña*) events (Gray, 1984; Shapiro, 1987). There are significant spatial and temporal variations in TC activity over the western North Pacific associated with El Niño and La Niña events (Chan, 2000). El Niño events cause South Pacific TCs to occur further eastward than normal and also to bring about a general suppression of TC activity in the Coral Sea and north Australia region (Basher and Zheng, 1995; Gray, 1988).

Consequently, attempts have been made to utilise ENSO indices as predictors of TC activity (Chan, Shi and Liu, 2001; Landsea, 2000b; Saunders et al., 2000). However, criteria which different authors chose for defining ENSO phases vary. The lack of a common approach to defining ENSO phases makes it difficult to compare the outcomes of different studies. Recently, a number of key indices and derived lists of ENSO events identified by several studies have been examined by Kuleshov *et al.* (2009b) and a collective list of historical El Niño and La Niña events has been developed which may be recommended for subsequent studies on TC activity in the SH. In this chapter, variability and change in tropical cyclone occurrences across the SH is consistently analysed based on this developed approach.

2. Tropical cyclone archive for the Southern Hemisphere

The World Meteorological Organization (WMO) Tropical Cyclone Programme has established areas of responsibility for TC warning which extend across the regional bodies and also extend across the ocean basins (WMO, 2002a; WMO, 2002b). In the SH, the Regional Specialised Meteorological Centre (RSMC) in La Réunion is responsible for providing cyclone watch over the western SIO. The responsibility for the preparation of marine TC forecasts and warnings in the SPO and the eastern SIO is shared amongst Australia (Brisbane, Darwin and Perth Tropical Cyclone Warning Centres), Fiji (Nadi RSMC), Papua New Guinea (Port Moresby) and New Zealand (Wellington).

The National Meteorological and Hydrological Services (NMHSs) also archive records of TC occurrences in their areas of responsibility. Historical records of TC occurrences in the SH go back a few hundred years. For example, Australian records go back to the late 18th century. However, length of records in TC archives and their quality vary. It is widely accepted that TC records in the Australian region and the SH are complete and reliable after meteorological satellite came in operational use in the late 1960s (Holland, 1981).

Since 1999, considerable efforts were put in preparing consolidated historical TC records under an international initiative "Climate change and Southern Hemisphere tropical cyclones". As the first step, TC archive for the SH (jointly, SHTC) has been prepared at the National Climate Centre, Australian Bureau of Meteorology (ABM) (Kuleshov and de Hoedt, 2003). The archive is a result of multinational efforts of the NMHSs from the SH and has been derived from several different sources. The data for the western SIO (30°E to 90°E) have been provided by Météo-France (La Réunion), for the Australian region (90°E to 160°E) by the ABM, and for the eastern SPO (east of 160°E) by the NMHSs of Fiji and New Zealand. TC tracks from these archives were merged in one archive, ensuring consistency of trajectories when TCs cross regional borders. Since then, data for the recent TC seasons have been regularly added to the SHTC archive and it now consists of TC best track data for the "satellite era" i.e. the TC seasons from 1969/70 to 2009/10, covering name (and/or unique identification number), position and intensity (in terms of central pressure) (Kuleshov et al., 2010a).

The availability of satellite imagery has significantly improved our knowledge of TCs, with satellite remote sensing being vital for accurate estimates of parameters such as TC intensity and position (e.g. the location of minimum atmospheric pressure). Satellite images are used by forecasters for preparing operational and best track data, and a complete digital

Geostationary Meteorological Satellite (GMS) archive for the SH was prepared by the ABM for use in TC reanalysis (Broomhall *et al.* 2010).

Currently, the technique for estimating TC intensity is the Dvorak analogue procedure based on patterns of infra-red brightness temperature (Dvorak, 1984, Velden *et al.*, 2006). The original version of the technique, applied to visible satellite imagery, was published in 1975, with its present form, based on digital infra-red imagery, published in 1984. While polar-orbiting satellite images became available in the Australian region from late 1960s (Holland 1984), geostationary satellite imagery became available only in 1978 and it was only during the mid-1980s that high-resolution multi-channel imagery became available at the Australian Region forecast offices. For the western SIO where the tracks are maintained by Météo-France (La Réunion), significant improvement in data quality and estimation of cyclone intensity occurred with the beginning of geostationary satellite coverage in 1998, as a result of the launch of the MeteoSat-7 satellite (Kossin *et al.* 2007). Within the satellite era, low-resolution geostationary satellite imagery for the SPO (west of about 155°W) became available to forecasters from Fiji and New Zealand from 1980. However, only from the early 1990s, with high-resolution imagery from the GOES-West satellites available, did the Dvorak technique become increasingly used for estimating storm intensity.

This variation between different regions, in the time period for which high resolution satellite imagery is available, influences the accuracy of cyclone intensity estimations and imposes limitations on the analysis of historical data. For example, it restricts the accuracy of long-term trend analysis of changes in cyclone intensity - something required to address the important question of how TC activity is changing over decades and possible relationships to global climate change. Consequently, reanalysis of the historical TCs in order to obtain globally homogeneous records is essential (Kuleshov *et al.* 2010b).

Numerous studies have analysed TC occurrence in the SH (e.g. Kerr, 1976; Revell, 1981; Revel and Goulter, 1986). Nicholls (1985) attributed an increasing trend observed in the number of TCs in the SH, with fewer TCs observed at the beginning of the century, to the improvements in observing systems and networks during this time period. However, strong interannual variability in TC activity occurs around this trend, and Nicholls (1992) showed evidence of a strong statistical link between these inter-seasonal fluctuations of TC numbers and the ENSO phenomenon.

3. Influence of El Niño-Southern Oscillation on tropical cyclone activity

The ENSO is the most important coupled ocean-atmosphere phenomenon affecting inter-annual climate variability on a global scale. ENSO consists of three phases: El Niño (the warm phase), La Niña (the cold phase) and the neutral phase. Despite the fact that ENSO has been the subject of numerous studies over recent decades, there is as yet no international agreement on a definition of ENSO and its phases. Usually, ENSO definitions are based on a quantitative analysis of (i) the intensity of the Southern Oscillation (SO) using indices linked with atmospheric pressure gradients and/or (ii) sea surface temperature (SST) anomalies in the Pacific (equatorial or near-equatorial regions). It is common practice to describe the strength of the SO in terms of a Southern Oscillation Index (SOI), for example the Troup Index (Troup 1965), or mean sea level pressure at certain stations such as Darwin (Australia), Papeete (Tahiti), *etc.* Area-averaged SST anomalies (SSTAs) in various regions such as the NINO3, NINO3.4 and NINO4 regions are commonly used indices to describe oceanic conditions in the Pacific region. However, ENSO is a coupled ocean-atmosphere phenomenon, and combining both atmospheric and oceanic responses in a composite index seems to be a more appropriate index-based approach to a comprehensive description of the phenomenon than using atmospheric and oceanic indices alone. Lack of consensus on a definition of El Niño (La Niña) events typically results in certain warm (cold) episodes being included or excluded from lists of ENSO events compiled by different scientists.

It has been shown that ENSO is a significant contributor to the year-to-year variability in TC activity in most ocean basins (Nicholls 1984, Chan 2000, Landsea 2000b, Saunders et al. 2000, Kuleshov et al., 2008). Once again, the above-mentioned absence of agreement on defining ENSO warm and cold phases leads to investigators employing different criteria in the delineation of El Niño and La Niña events. Thus, NINO3.4 (5°N-5°S, 120°-170°W) SSTA thresholds of above +0.3°C (below −0.3°C) for identifying El Niño (La Niña) were used by Saunders et al. (2000) to investigate ENSO spatial impacts on occurrence and landfall of TCs in the Atlantic and Northwest Pacific basins. On the other hand, while studying TC activity over the Northwest Pacific, Chan (2000) referred to El Niño (La Niña) years if SSTAs in the NINO3.4 region rose above +0.5°C (fell below −0.5°C) sometime during that year. Consequently, even though significant ENSO impact on TC activity has been demonstrated, quantitative comparison of the results from different studies is somewhat difficult.

In this section, some key ENSO indices are examined in an attempt to develop a collective list of historical El Niño and La Niña events which may be recommended for subsequent studies on TC activity in the SH. The NINO3.4 SSTA and the SOI are the two most commonly used indices in defining ENSO phases. The SOI data used in this analysis were obtained from the ABM (www.bom.gov.au/climate/current/) and values for the NINO3.4 SSTA (3-month running mean) - from the Climate Prediction Center, NOAA (ftp.cpc.ncep.noaa.gov/wd52dg/data/indices/).

The Multivariate ENSO Index (MEI) (Wolter 1987) was also used to quantify the start and end of ENSO events (www.cdc.noaa.gov/people/klaus.wolter/MEI/table.html). Another multivariate ENSO index, based on the first principal component of monthly Darwin mean sea level pressure (MSLP), Tahiti MSLP, and the NINO3, NINO3.4 and NINO4 SST indices, was developed and examined. Its base period is 1950-1999. We will denote this (standardised monthly) index the 5VAR index (Kuleshov et al., 2009b). It has a distinct advantage over the MEI in that it can be consistently extended further back in time.

Based on findings from early studies (Trenberth 1997, Reid 2003), the following assumptions have been made to evaluate the indices: the Pacific exhibits El Niño conditions approximately 31% of the time, La Niña conditions 23%, and neutral conditions 46%. Bi-monthly MEI values (in multiples of the standard deviation) for the period from Dec 1949/Jan 1950 to Dec 2005/Jan 2006 were ranked from lowest to highest. Applying the above assumptions to the period 1950 to 2005 (56 years), a warm (cold) ENSO episode was identified if the MEI ranks for at least five consecutive months were greater than or equal to $39 \approx (1-0.31) \times 56$ (less than or equal to $13 \approx 0.23 \times 56$). Another list of ENSO episodes was derived in an analogous fashion using the 5VAR index. The El Niño and La Niña events in these two lists, as identified by the MEI and 5VAR index ranks, were compared with the lists of ENSO events from Rasmusson and Carpenter (1982), Kiladis and van Loon (1988), Trenberth (1997), Larkin and Harrison (2001, 2002) and Reid (2003).

For the SH, the TC year was considered as the 12 month period from July to June inclusive. In general, the main historically-recognized ENSO events included in the lists of examined publications were identified by the MEI rank and the 5VAR index rank using the selected thresholds. As a result, the following TC years are considered as El Niño seasons: 1951/52, 1957/58, 1963/64, 1965/66, 1969/70, 1972/73, 1976/77, 1977/78, 1979/80, 1982/83, 1986/87, 1987/88, 1991/92, 1992/93, 1993/94, 1994/95, 1997/98, 2002/03, 2004/05, 2006/07 and 2009/10. The corresponding list of La Niña seasons is 1950/51, 1954/55, 1955/56, 1956/57, 1961/62, 1964/65, 1970/71, 1973/74, 1974/75, 1975/76, 1988/89, 1998/99, 1999/2000, 2007/2008 and 2010/11. The developed list of historical El Niño and La Niña events has been used for analysis of TC variability and change across the SH presented in the following sections.

4. Variability and change in tropical cyclone occurrences across the Southern Hemisphere

Analysis of data from the SHTC archive has confirmed that ENSO is a significant contributor to the year-to-year variability in TC activity in the SPO and SIO. This clearly raises important questions about the future direction of TC change under the enhanced greenhouse effect. For example, a more El Niño-like future may lead to substantially increased frequencies of TCs in some regions and reduced frequencies in others.

The number of times a TC crosses a particular longitude has been found to provide a useful statistics in TC track studies (Hall and Jewson, 2007; WMO 1993). The convention adopted here takes into account multiple crossings of the same longitude by the same TC. Average annual profiles of TC longitude crossings for El Niño and La Niña years are presented in Figure 1. Comparing the profiles, geographical shifts in TC occurrences and changes in intensity of the maxima are evident. In the Australian region, a higher TC impact on the coastal area (between 110°E and 155°E) in La Niña years than in El Niño years was observed. TC impacts on island nations in the SIO and SPO also vary. During El Niño years, TCs tend to occur further eastwards in the SPO. In the western SIO, the position of maximum TC activity shifts eastwards in La Niña years compared to El Niño years.

Spatial profiles for the three phases of ENSO (El Niño, La Niña and neutral) have been constructed, and expressed as departures from climatology (Figures 2A, 2B and 2C, respectively). The graph of anomalous average annual TC longitude crossings for El Niño years shows a general decrease in TC activity in the areas west of 60°E and between 80°E and 165°W, and an increase of TC activity in the SPO east of 165°W and in the SIO between 60°E and 80°E (Figure 2A). The departure from climatology for La Niña years indicates an almost directly opposite effect on TC activity: it is above climatology in the areas west of 55°E and between 70°E and 180°, and below climatology in the SPO east of 180° and in the SIO between 55°E and 70°E (Figure 2B). The graph for departure from climatology of TC longitude crossings for neutral years demonstrates that TC activity generally stays close to the climatology (Figure 2C). The mean number of average annual TC longitude crossings in the SH is 2.8 (climatology) decreasing to 2.7 in neutral years, which represents a less than 4% reduction in the mean. From this analysis it is evident that ENSO has a significant influence on changes in areas favourable for TC genesis and consequently on geographical variation in TC tracks in the SH.

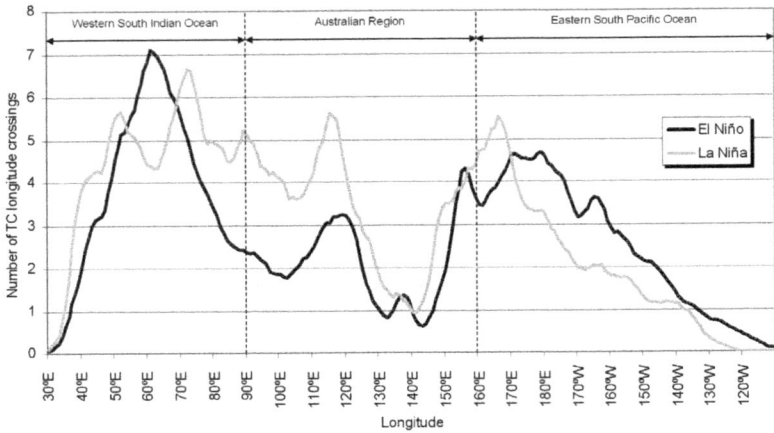

Fig. 1. Average annual profiles (5-point smoothed) for TC longitude crossings for El Niño and La Niña years.

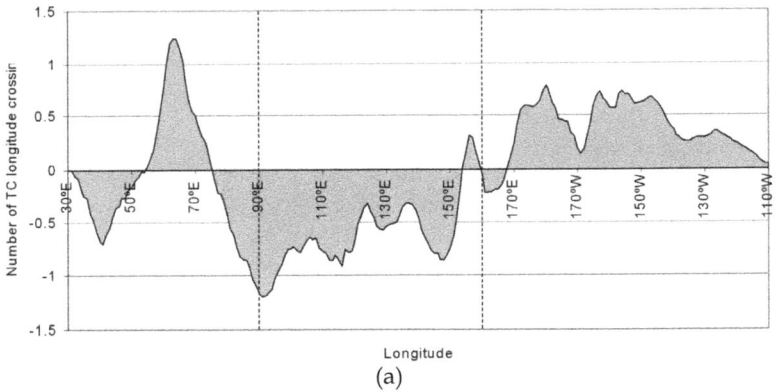

(a)

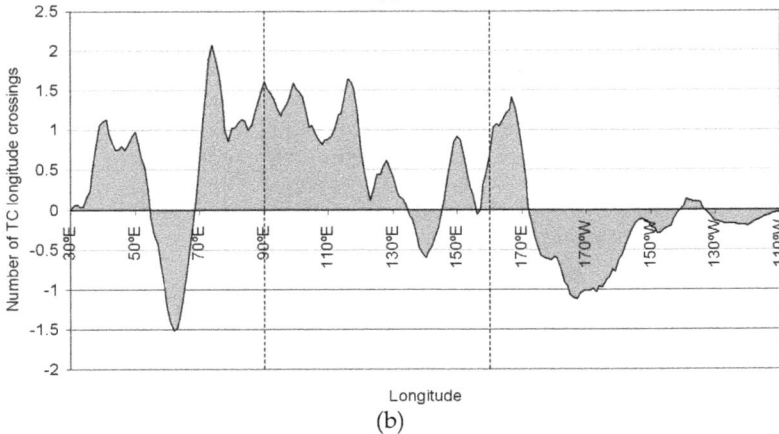

(b)

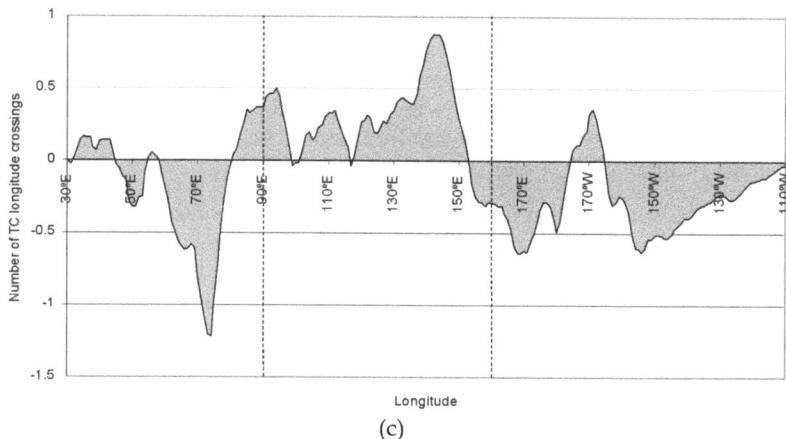

(c)

Fig. 2. Departure from climatology of average annual TC longitude crossings (5-point smoothed) for (A) El Niño, (B) La Niña and (C) neutral years. Vertical dashed lines indicate regional boundaries for the western South Indian Ocean, the Australian region and the eastern South Pacific Ocean as per Figure 1.

5. Large-scale environmental variables which influence tropical cyclone genesis and development

Influence of the ENSO phenomenon on TC activity in the various regions of the SH has been demonstrated in numerous studies (Nicholls, 1984; Hastings, 1990; Evans and Allen, 1992; Basher and Zheng, 1995; Nicholls et al., 1998; Camargo et al., 2007; Kuleshov et al., 2008). It is well known that the ENSO also plays an important role in TC interannual variability in ocean basins of the Northern Hemisphere (e.g. Chu, 2004), although the influence of large-scale environmental factors on TC genesis and development varies for different basins. Gray (1979) formulated six large-scale environmental parameters playing key roles in formation of TCs including the Coriolis parameter, low-level relative vorticity, relative humidity in the mid-troposphere, tropospheric vertical wind shear, ocean thermal energy, and difference in equivalent potential temperature between the surface and 500 hPa. Recently, Camargo et al. (2007) used a genesis potential index which is similar to Gray's seasonal genesis index (Gray, 1979) to diagnose ENSO effects on tropical cyclogenesis and concluded that in the basins of the SH vertical wind shear and mid-tropospheric relative humidity are especially important, vorticity anomalies contribute most significantly in the central Pacific, and thermodynamic variable of the genesis potential index described in terms of potential intensity (Bister and Emanuel, 1998) plays a secondary role. It is well established that vertical wind shear in the Northern Hemisphere, particularly in the North Atlantic, plays a primary role in TC variability related to the ENSO (e.g. Gray 1984). However, in the SH it is less obvious because of strong gradients in wind-shear over the regions of cyclogenesis. Also, changes in SSTs related to changes in ENSO phases are significant in the equatorial and near-equatorial regions of the Indian and Pacific Oceans, and it appears that the thermodynamic variable can be in addition one of the important environmental factors contributing to tropical cyclogenesis variability in this region. Thus, it is pertinent to further examine influence of changes in TC surrounding environment on TC genesis. Based on results of earlier studies, large scale environmental factors which influence TC genesis and development such as SST, relative humidity, vertical wind shear and relative vorticity were examined here through the National Center for Environmental Prediction (NCEP) composite re-analysis dataset (Kalnay et al., 1996).

The SIO is the least studied basin in terms of analysis of relationship between cyclones and their environment (Kuleshov *et al.* 2010b). Only a few studies attempted to investigate critical environmental factors which govern SIO cyclone activity. An attempt to establish relationship between ENSO and SIO cyclone numbers was made by Jury (1993); however, no statistically significant correlation was found between the number of cyclones that occur and an index of ENSO activity (the SOI). Examining variation in geographical distribution of cyclone variability, Ho *et al.* (2006) and Kuleshov *et al.* (2008) demonstrated that the tropical cyclogenesis area tends to shift west in El Niño when compared with La Niña years. A number of environmental factors critical for SIO cyclone development has been identified. For example, Kuleshov *et al.* (2009a) attribute shifts in cyclone tracks during ENSO phases to changes in vorticity and relative humidity and Camargo *et al.* (2007) showed that the track shifts are primarily due to changes in vertical wind shear.

Analysing geographical variation in TC activity in the SPO, Basher and Zheng (1995) and Kuleshov *et al.* (2009a) have demonstrated that cyclones tend to form in the southwest of the basin in La Niña years compared with El Niño years. A plausible explanation for this east-west shift is change in distribution of the sea surface temperatures over the Pacific during El Niño and La Niña. In El Niño years, positive anomalies of the sea surface temperatures occur in the central and eastern Pacific compared with La Niña years which explains the observed shift in tropical cyclogenesis (Kuleshov *et al.*, 2009a). Camargo *et al.* (2007) and Kuleshov *et al.* (2009a) have demonstrated that relative humidity and vertical wind shear are important factors of the environment in which TCs form and develop in the SH. Kuleshov *et al.* (2009a) also found that anomalously enhanced pre-existing cyclonic vorticity tends to occur in the southwest part of the basin in La Niña years, and this anomaly tends to exist further northeast in El Niño years.

In the following sub-sections, detailed analysis of TC activity and its relationship to ENSO is presented. For the SH, the TC year is considered as the 12 month period from July to June inclusive. Fields of the SST, relative humidity in mid-troposphere, vertical wind shear and lower tropospheric vorticity have been obtained from the NCEP re-analysis dataset (Kalnay et al., 1996). Geographical boundaries for the SIO basin are defined as south to the equator, west of 135°E, and for the SPO basin as south to the equator, east of 135°E.

5.1 Tropical cyclone genesis

Spatial distribution of TC genesis over the SIO basin during all TC seasons from 1969 to 2006, including El Niño, La Niña and neutral phases of ENSO is presented in Fig. 3. In this analysis, a cyclogenesis point is defined by locating a point along cyclone track where an estimated TC central pressure attained a threshold of 1000 hPa or lower. In the SIO, TC genesis is observed over the whole range of longitudes of the basin. The maximum occurrence is located in the area around [5-15°S, 60-130°E]. In the area between African coast and 60°E, there is a noticeable displacement of the points of TC genesis southwards to 10 -25°S.

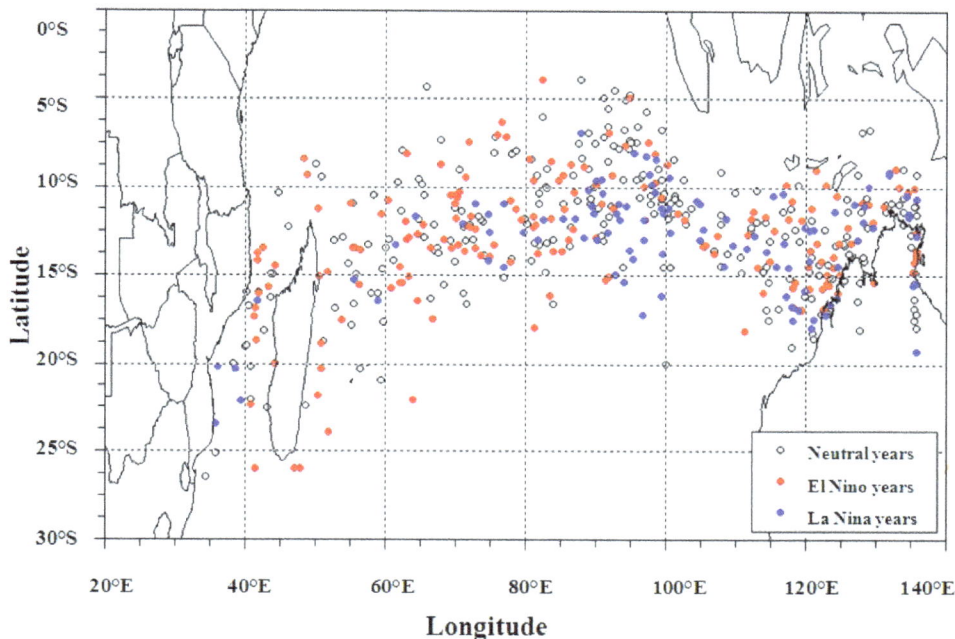

Fig. 3. Geographical distribution of tropical cyclogenesis events over the South Indian Ocean for the period 1969-2006.

Analysis of composite cyclogenesis anomalies demonstrates that during El Niño episodes (Fig. 4a) TC genesis is above climatology in the western part of the basin (west of around 85°E) with the maximum located around [15°S, 65°E] and it is below climatology in the eastern part of the basin (east of around 85°E) with the minimum in the area centred around [10°S, 95°E] (results in this area are significant at 90% level). To evaluate significance of the results, the Student t-test is employed in here, with the null hypothesis that the anomaly is not significantly different from zero. It should be noticed that the results from this significance test need to be treated cautiously because of the relatively small size of TC data sample. As mentioned in the section 2, the reliable TC records in the SH start with the beginning of the satellite era in late 1960s – early 1970s. As an effort to increase the sample size, a relatively large grid box, 6° by 6°, is selected for the analysis in this analysis. [Similar approach was applied by Chan (2000) for studying TC activity over the western North Pacific associated with El Niño and La Niña events.] A large area with test values greater than the specified significance threshold is likely to be true for individual seasons.

During La Niña episodes (Fig. 4b), increase of TC genesis is observed in the eastern part of the basin (east of about 85°E) with a primary cyclogenesis maximum in the area located off the northwest coast of Western Australia around [15°S, 120°E] and a secondary TC genesis maximum centred around [15°S, 95°E] (both passed the significance test at 90% level).

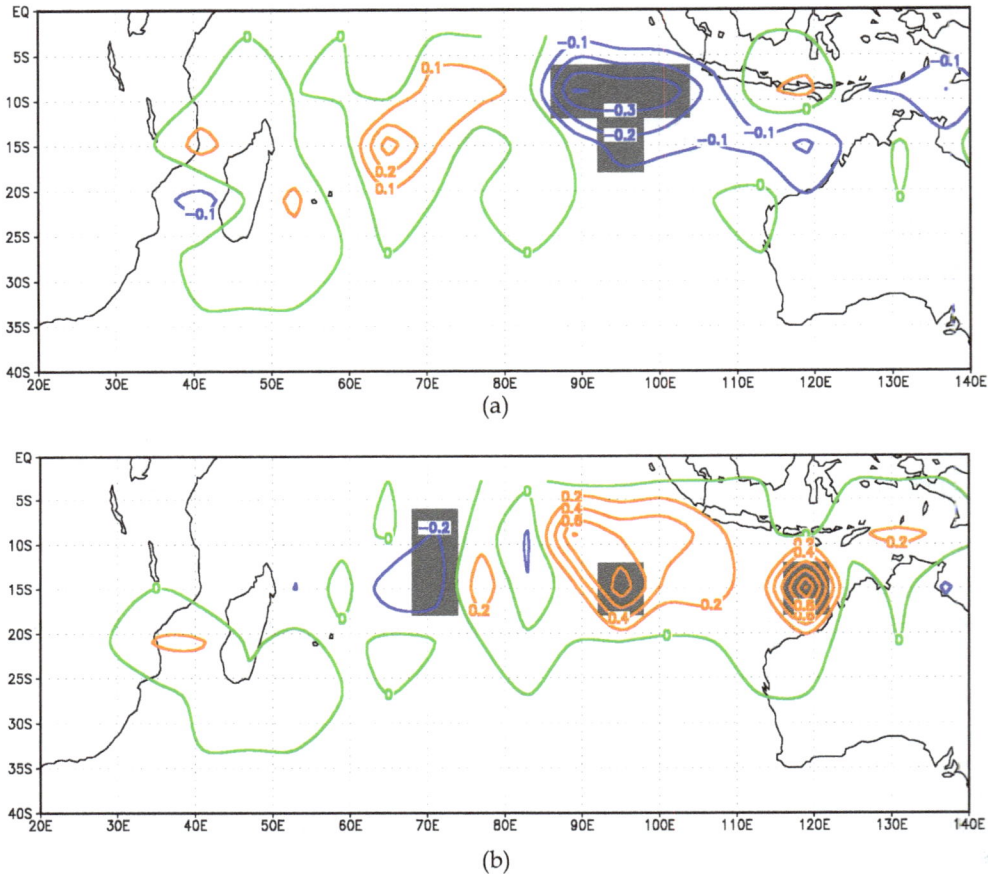

(a)

(b)

Fig. 4. Composite anomalies of the average annual number of tropical cyclogenesis in the South Indian Ocean during (a) El Niño and (b) La Niña in 6° x 6° boxes. The shades indicate areas where the statistical test is significant at the 90% level.

Spatial distribution of TC genesis over the SPO basin during all TC seasons from 1969 to 2006 is shown in Fig. 5. Points of cyclogenesis are observed over the range of longitudes between 135°E and 120°W. A comparison between cyclogenesis events during El Niño, La Niña and neutral seasons (Fig. 5) reveals that points of TC genesis are displaced towards southwest during La Niña seasons compared to El Niño seasons. This result is in agreement with previous studies on TCs in the Southwest Pacific Ocean (Gray, 1988; Basher and Zheng, 1995).

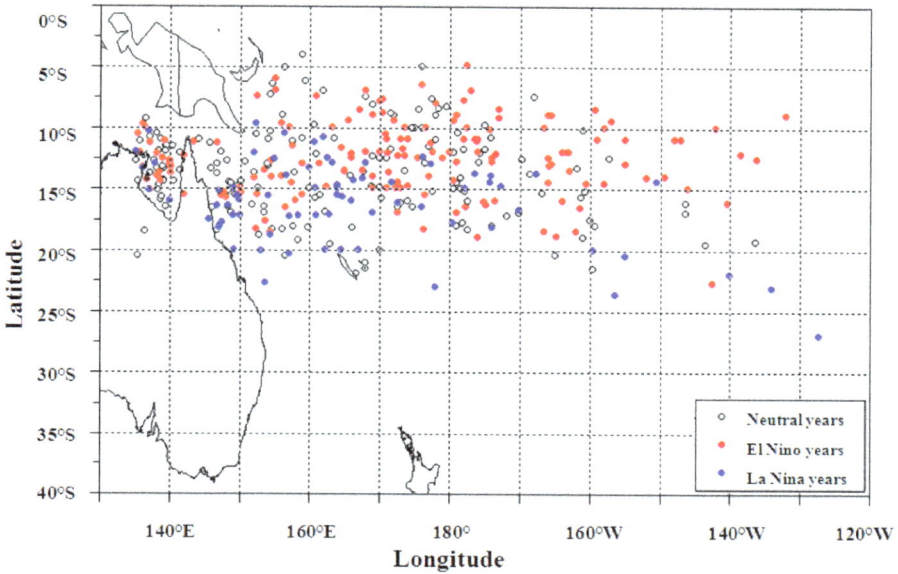

Fig. 5. Geographical distribution of tropical cyclogenesis events over the South Pacific Ocean for the period 1969-2006.

Analysis of spatial distribution of tropical cyclogenesis during the El Niño seasons (Fig. 6a) shows that a large area of above climatology TC genesis is located in the eastern part of the basin east of the line [10°S,165°E - 20°S,175°E]. During La Niña seasons (Fig. 6b), area of active TC genesis is displaced southwest to around [10 -20°S, 145-170°E] (statistically significant at 90% level) and TC activity in the central near-equatorial area of the basin around [5-15°S, 170°E - 170°W] is below the climatology.

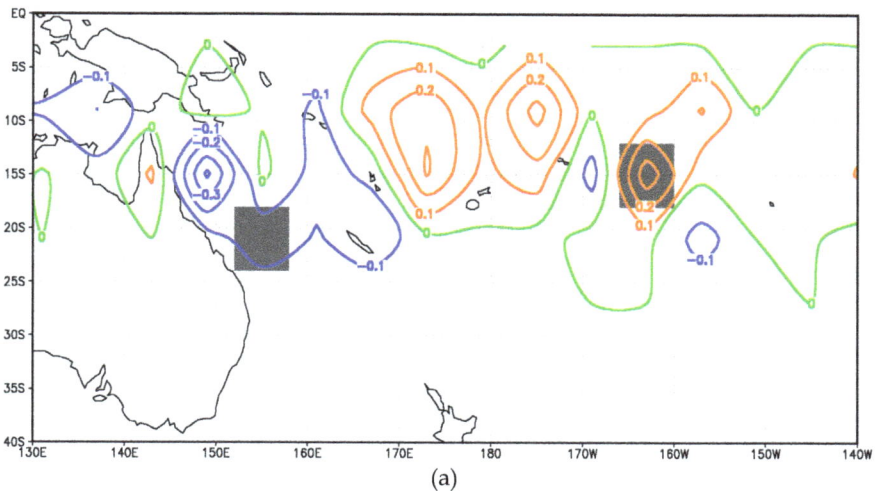

(a)

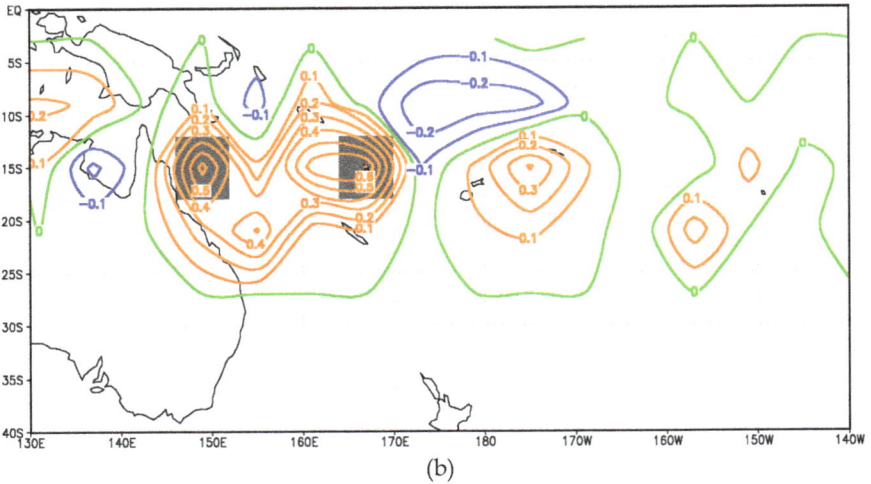

(b)

Fig. 6. Composite anomalies of the average annual number of tropical cyclogenesis in the South Pacific Ocean during (a) El Niño and (b) La Niña episodes in 6° x 6° boxes. The shades indicate areas where the statistical test is significant at the 90% level.

5.2 Sea surface temperature distribution

Ocean thermal energy is an important contributor to TC genesis and according to Gray (1979) ocean temperatures greater than 26°C to a depth of 60 meters is an essential climatological aspect of the seasonal frequency of TC formation at any location. To evaluate temperature's geographical distribution over the two oceanic basins, the SST climatology was firstly derived based on 1968 to 1996 period as defined by Physical Sciences Division, National Oceanic and Atmospheric Administration (NOAA). In the SIO, during both warm and cold ENSO phases oceanic conditions in terms of SSTs above 26°C (Gray 1979) are favourable over the near-equatorial belt between the equator and about 15°S~25°S (Figure not shown). Thus, it appears that changes in geographical distribution of SSTs over the basin related to warm and cold phases of ENSO rather than temperature values *per se* influence changes in TC genesis.

Analysing average SST anomaly fields (SSTAs) from November to April, it was found that positive SSTAs are present nearly over the whole range of longitudes between the equator and about 30-35°S during El Niño events except for a relatively small area around [35°S, 60°E] (Fig. 7a; shaded areas indicate where statistical tests are significant at 95% level). Strong positive anomalies are observed in the central parts of the basin with maximum values greater than 0.4°C, which can contribute to increased TC genesis in those areas in El Niño seasons. The area of enhanced cyclogenesis located around [15°S, 65°E] (Fig. 4a) matches the significant positive SSTAs in the same region quite well.

During La Niña seasons, positive SSTAs observed over the basin in El Niño seasons (Fig. 7a) are replaced by negative SSTAs which are found over the most areas in the SIO from the equator to about 30°S except for the central and southern areas off Western Australia coastline (Fig. 7b). This contrast in the SST distribution over the basin during La Niña seasons can partially explain displacement of area favourable for cyclogenesis to the eastern part of the basin closer to the coast of Western Australia (Fig. 4) where the warmer SSTs are present.

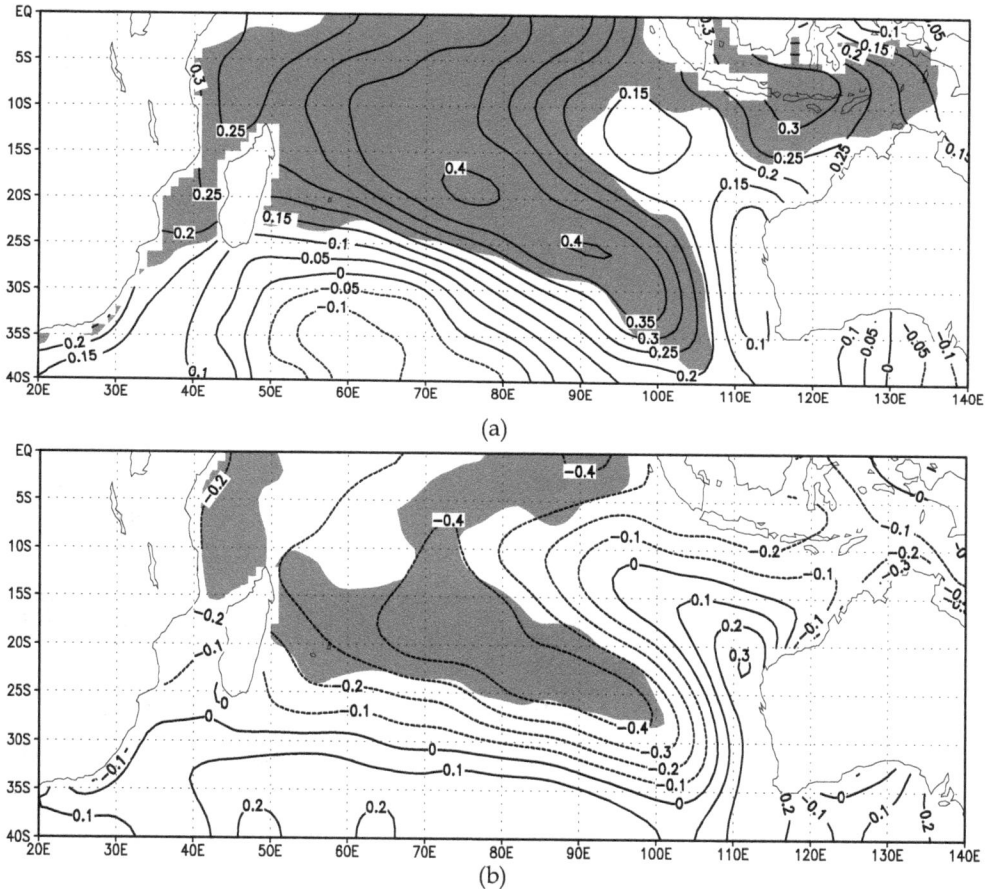

Fig. 7. SST anomalies (°C) from November to April in the South Indian Ocean during (a) El Niño and (b) La Niña episodes. The shades indicate areas where the statistical test is significant at the 95% level.

During El Niño seasons, the SSTAs are positive in the near-equatorial area and in the eastern part of the SPO basin, with maximum positive anomalies in the area around [0-15°S, 160°E-140°W] (Fig. 8a), while across the Coral Sea, there is a narrow zone of negative anomalies extending southeastward to the central and eastern parts of the basin. The positive SSTAs promote TC genesis and development further eastwards in the basin in El Niño seasons when compared with La Niña seasons (Fig. 6). During La Niña seasons, negative SSTAs replacing a large area between the equator and 15°S (Fig. 8b) inhibit TC genesis and contribute to displacing area preferable for TC genesis southwest compared to El Niño seasons (Fig. 6b). Warmer than climatological average waters found in the area from the

Coral Sea to the central and eastern parts of the basin during La Niña seasons favour increase of TC activity in the western part of the basin as well as in the central part.

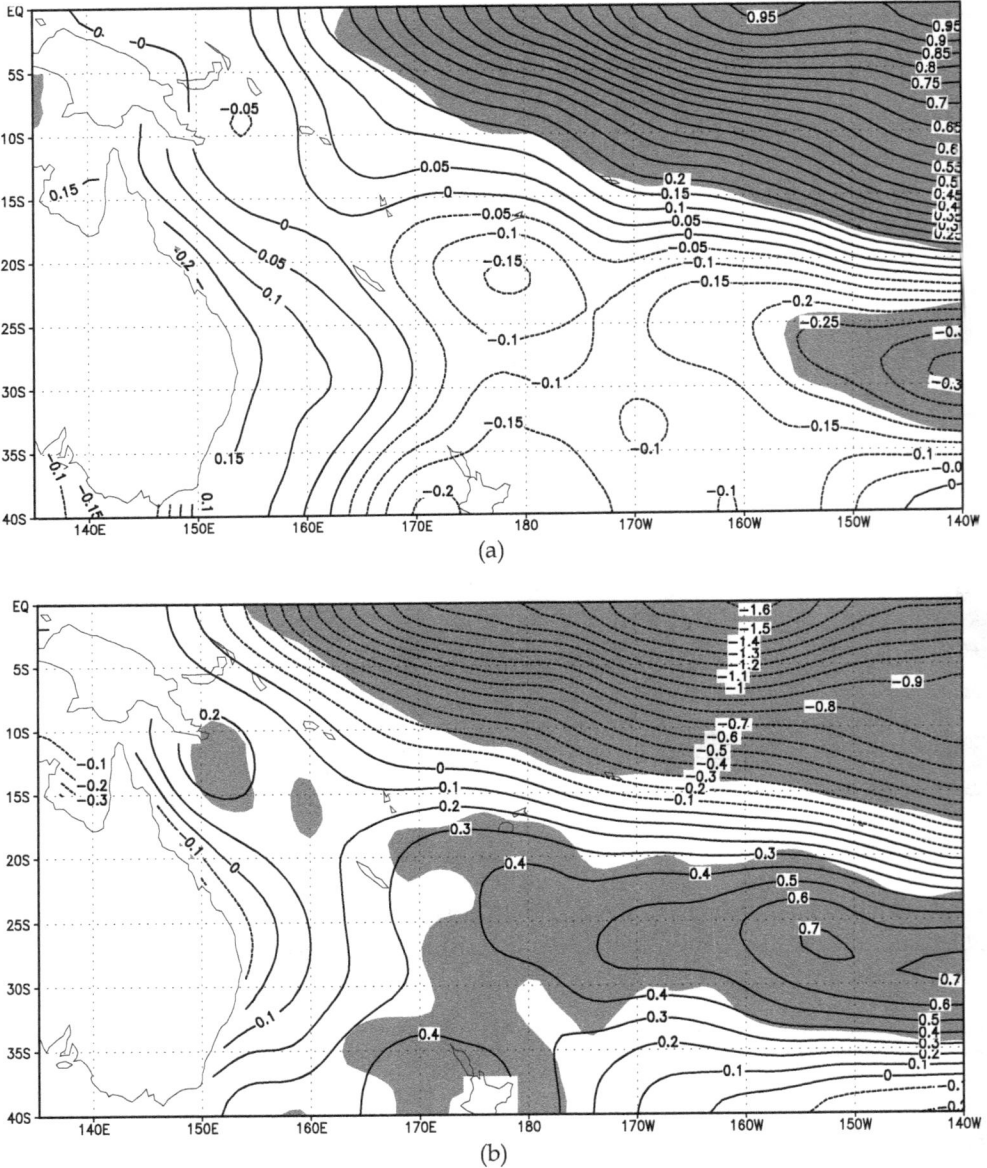

(a)

(b)

Fig. 8. SST anomalies (°C) from November to April in the South Pacific Ocean during (a) El Niño and (b) La Niña episodes. The shades indicate areas where the statistical test is significant at the 95% level.

5.3 Relative humidity

High relative humidity is one of the key factors influencing TC development as it provides sufficient moisture at the lower level and in the mid-troposphere (4 to 8 km) (Gray, 1979). Geographical distribution of mid-tropospheric relative humidity at 500hPa level during the period from November to April has been examined in this section. Results for El Niño events are shown in Fig. 9a. A large area with negative anomalies of relative humidity is observed between the equator and about 15°S (Fig. 9a), but it is noticed that most of the negative anomalies did not pass the significance test except for the some areas off Western Australia and over parts of Indonesia. The areas of negative anomalies of relative humidity are collocated with the areas of relatively weak anomalies of SSTs, which can additionally contribute to decreased TC activity in the eastern part of the basin (about 10°S, 95°E in Fig. 4a) during El Niño seasons.

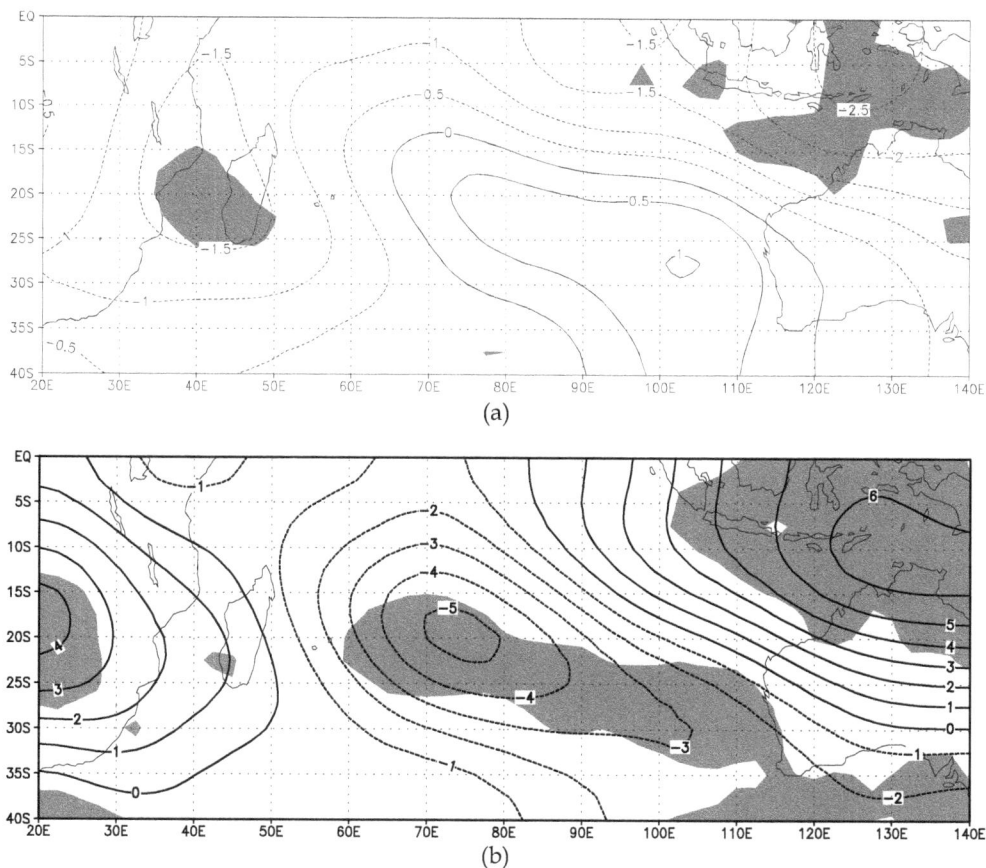

(a)

(b)

Fig. 9. Relative humidity anomalies (%) at 500hPa over the South Indian Ocean for (a) El Niño and (b) La Niña episodes. The shades indicate areas where the statistical test is significant at the 95% level.

Comparing results presented in Figs. 9a (El Niño) and 9b (La Niña), significant relative humidity decrease in the central part of the basin (about 70°E) was found during La Niña seasons, which is consistent with the decreased TC activity over this area evident from comparing Fig. 4a and Fig. 4b. During La Niña seasons, there is a large area with positive anomalies of relative humidity over the equatorial and tropical regions of the northwest of Australia and Indonesia from 85°E to 140°E, with another positive anomaly area to the west of Madagascar (Fig. 9b), and large area for negative anomalies over the central part (around 20°S, 75°E) of the basin. To the northwest of Australia, positive relative humidity anomalies and warmer SSTs both contributes towards increased TC activity during La Niña seasons.

Relative humidity anomalies at 500 hPa over the SPO during warm and cold phases of the ENSO are shown in Fig. 10a and 10b, respectively. During El Niño years, an area of positive relative humidity anomalies is located east of dateline between the equator and about 10°S, which is a classical pattern for El Niño events (Fig. 10a). A large drier zone of relative humidity is observed southwest to the positive anomalies, with a minimum centre in the central part of the basin. Over the Coral Sea, significant negative anomalies are also observed, where TC activity is weaker. During La Niña events, the geographical distribution of relative humidity anomalies is the opposite; they are positive over a large area extending from the Coral Sea to the central part of the basin, significant at 95% level (Fig. 10b), corresponding well with the increased TC activities over the western and central parts of the basin (Fig. 6b).

5.4 Vertical wind shear

The vertical wind shear plays an important role in tropical cyclogenesis and development: strong vertical wind shear inhibits tropical cyclogenesis, while weak vertical shear favours TC genesis and development (DeMaria 1996, Maloney and Hartmann 2000, Zehr 2003). It is well known that vertical wind-shear variations can have a significant impact on the interannual variability of TC activity in the Northern Hemisphere (*e.g.*, Gray 1984).

The vertical wind shear is defined here as the difference of magnitudes between the zonal wind fields at 200 hPa and 850 hPa. Results of analysis are shown for El Niño and La Niña seasons in Fig. 11a and 11b, respectively. The vertical wind shear over the whole basin is slightly weaker during El Niño seasons than that during La Niña seasons, in particular over the central and eastern parts of the basin. In general, wind shear is relatively weak in the near-equatorial area. Also, in the western part of the basin (west of 70°E) the wind shear is weaker than in the eastern part of the basin during both warm and cold ENSO episodes. Changes in this environmental factor are not easily interpreted in order to explain observed changes in TC activity.

During the ENSO warm phase (Fig. 12a), a zone of relatively weak vertical wind shear (<8 ms⁻¹) located over most of the equatorial and tropical areas of the SPO basin with the contour of 10 ms⁻¹ located at 13°S to 14°S in the western and central part of the basin. During the ENSO cold phase (Fig. 12b), the zone of relatively weak wind shear moves westwards with a minimum centre at about 160°E to 175°E. The contour of 10 m⁻¹ moves southwards to about 15°S to 20°S across the whole basin. This spatial distribution of vertical wind shear contributes to increased TC activity over the area west of the dateline and shifting area of enhanced cyclogenesis and TC development closer to Australia during La Niña seasons.

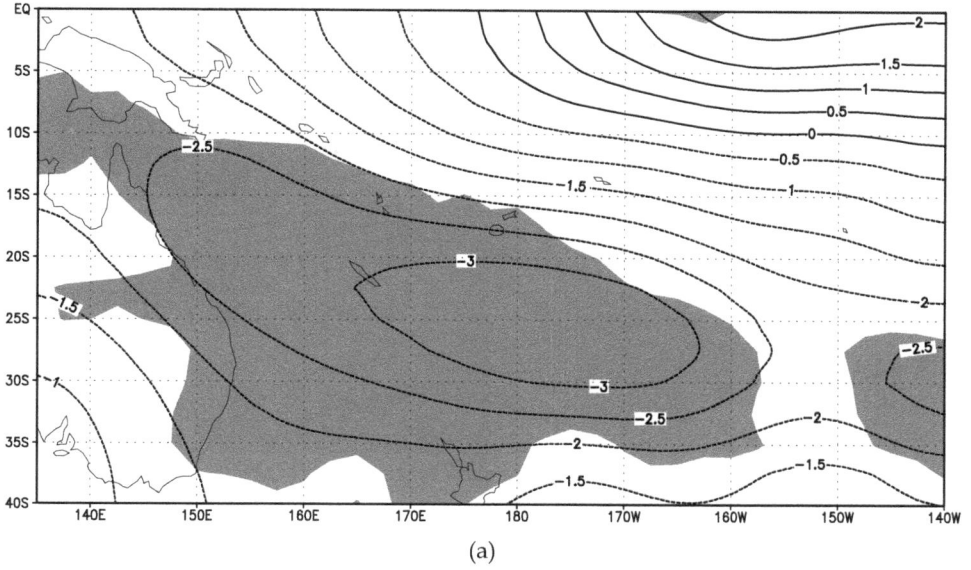

(a)

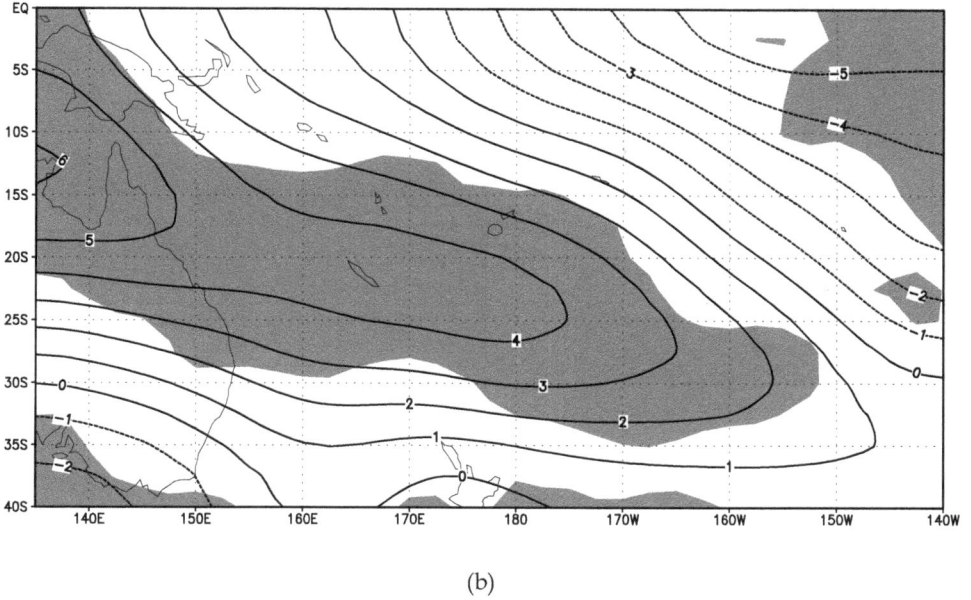

(b)

Fig. 10. Relative humidity anomalies (%) at 500hPa over the South Pacific Ocean for (a) El Niño and (b) La Niña episodes. The shades indicate areas where the statistical test is significant at the 95% level.

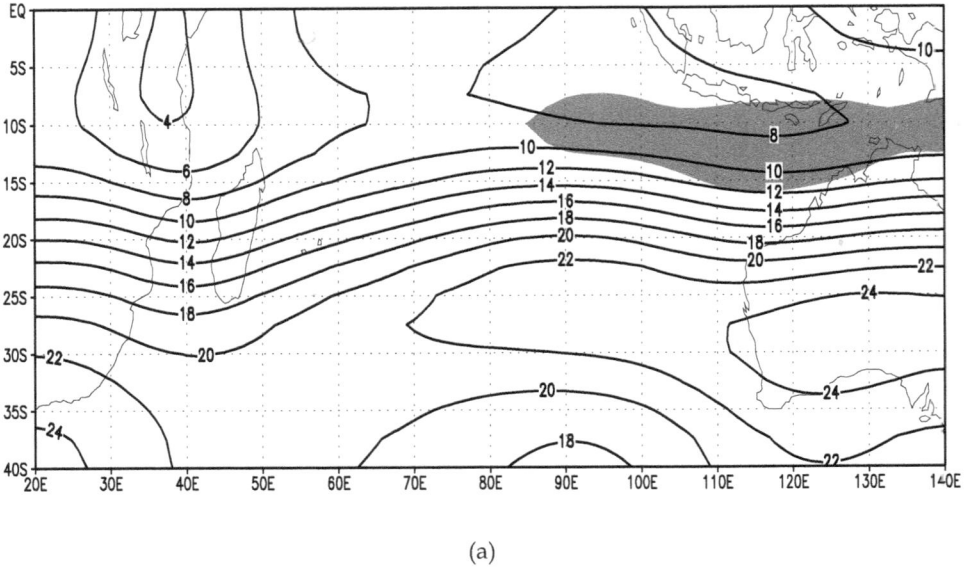

(a)

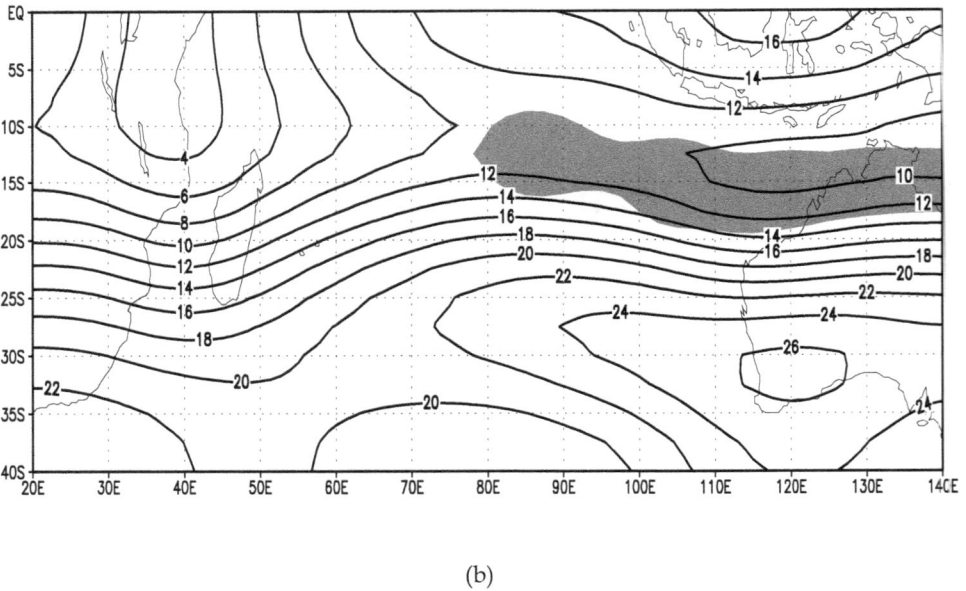

(b)

Fig. 11. Zonal wind shear (U_{200}-U_{850}) (m s^{-1}) over the South Indian Ocean for (a) El Niño and (b) La Niña episodes. The shades indicate areas where the statistical test is significant at the 95% level.

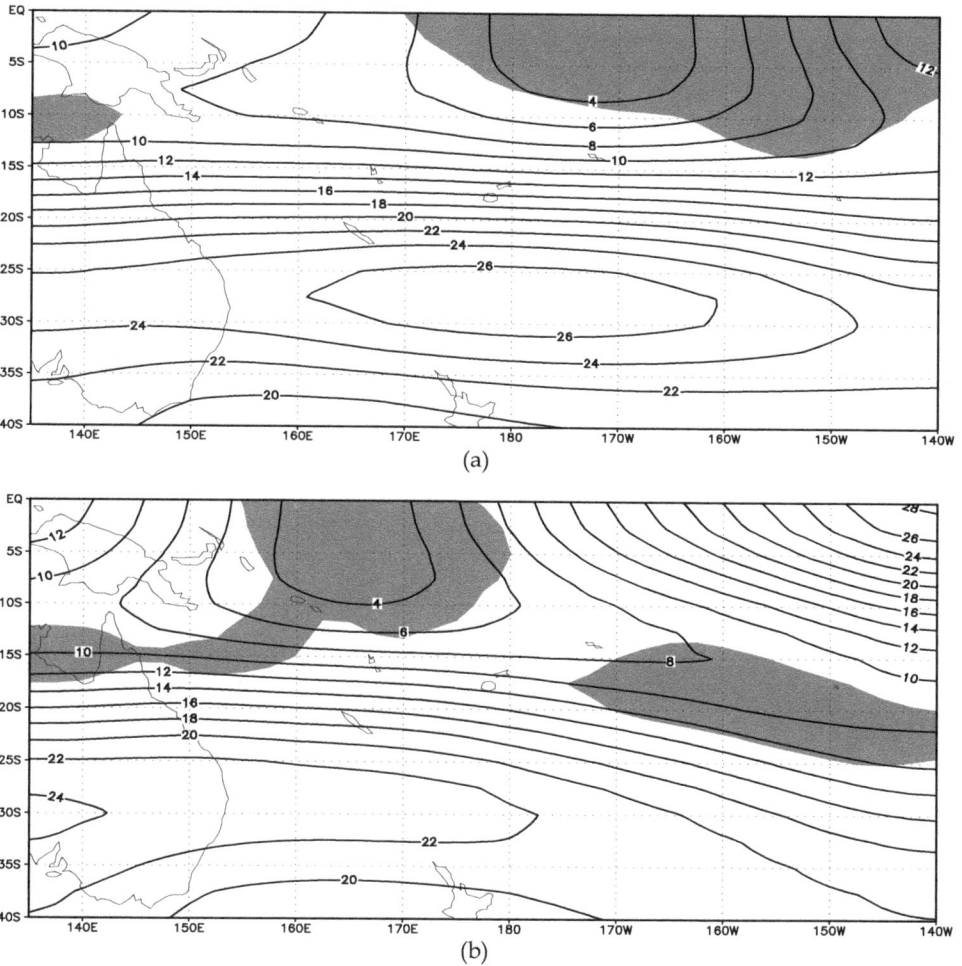

Fig. 12. Vertical wind shear (U_{200}-U_{850}) (m s^{-1}) over the South Pacific Ocean for (a) El Niño and (b) La Niña episodes. The shades indicate areas where the statistical test is significant at the 95% level.

5.5 Vorticity

The vorticity anomalies at 0.995 sigma level (surface level) from November to April during El Niño seasons over the SIO are shown in Fig. 13a. Large area of positive (anticyclonic) vorticity anomalies at [about 5-15°S, 70-135°E] which extends from the central part of the basin to the northwest coast of Australia is evident in Fig. 13a. However, centred around [25°S, 65°E], there is a large area of negative (cyclonic) vorticity anomalies which are more favourable to tropical cyclogenesis. During El Niño seasons, the areas to the north of 15°S and east of 70°E, are not favourable (in terms of vorticity impacts) to the TC genesis in comparison to the rest of the basin.

(a)

(b)

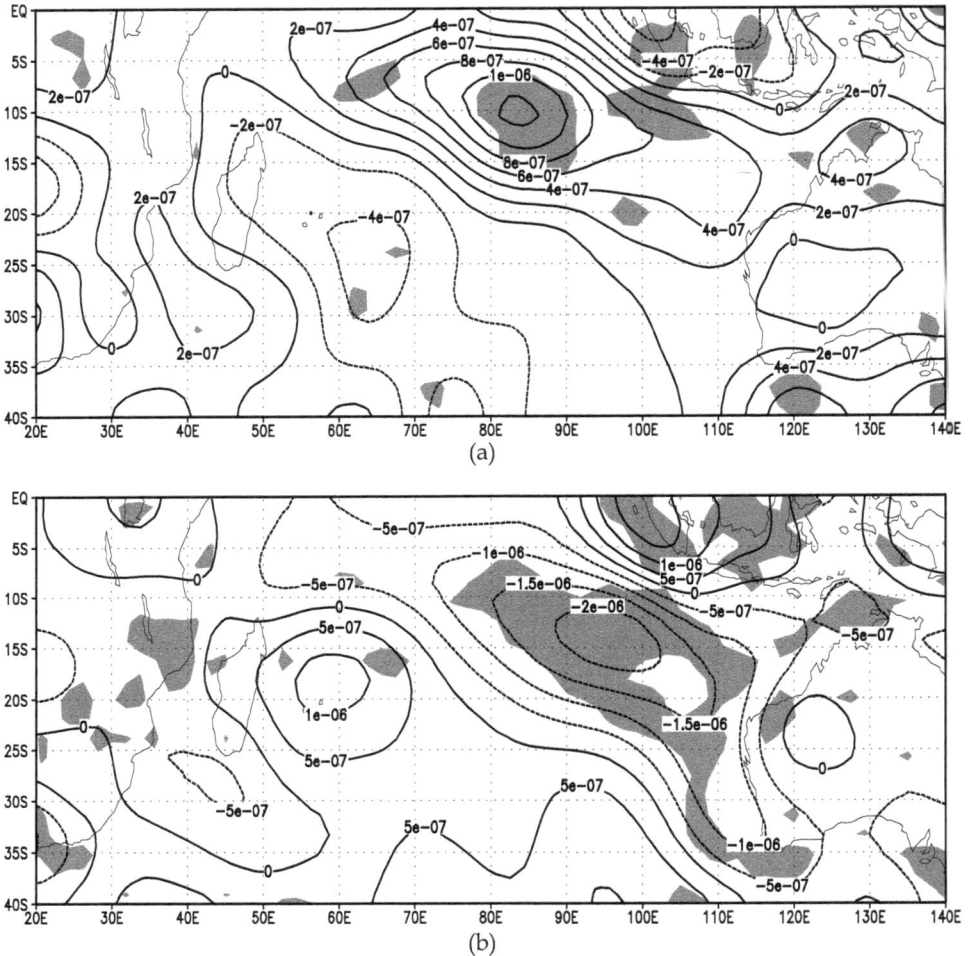

Fig. 13. Vorticity anomalies at 0.995 sigma level (s^{-1}) over the South Indian Ocean for (a) El Niño and (b) La Niña episodes. The shades indicate areas where the statistical test is significant at the 95% level.

During La Niña seasons, negative (cyclonic) vorticity anomalies over the tropical areas of the central and eastern parts of the basin (Fig. 13b) replace the anticyclonic vorticity anomalies observed in this area during El Niño seasons (Fig. 13a). It additionally contributes to increase in cyclone activity in the eastern part of the basin during La Niña seasons. In the western part, cyclonic vorticity anomalies are also observed between Madagascar and mainland Africa.

Vorticity anomalies at 0.995sigma level during warm and cold phases of the ENSO over the SPO are shown in Fig. 14a and 14b, respectively. For El Niño seasons, negative vorticity anomalies dominate over the near-equatorial areas, with a minimum centre located at about

10°S to 175°W, and positive anomalies of vorticity are observed to the south over latitudes from about 15°S to 35°S across the whole basin (Fig. 14a). Cyclonic (negative) vorticity anomalies in the area [5-15°S] favour cyclogenesis in the near-equatorial areas and contribute to increased TC activity in the eastern part of the basin in El Niño seasons. For La Niña seasons, spatial distribution of vorticity anomalies is the opposite: positive anomalies are located over the near-equatorial areas and negative anomalies are observed to the south (Fig. 14b). This geographical distribution of vorticity anomalies contributes to restraining TC genesis in areas close to the equator but promotes cyclogenesis in the Australian region in La Niña seasons.

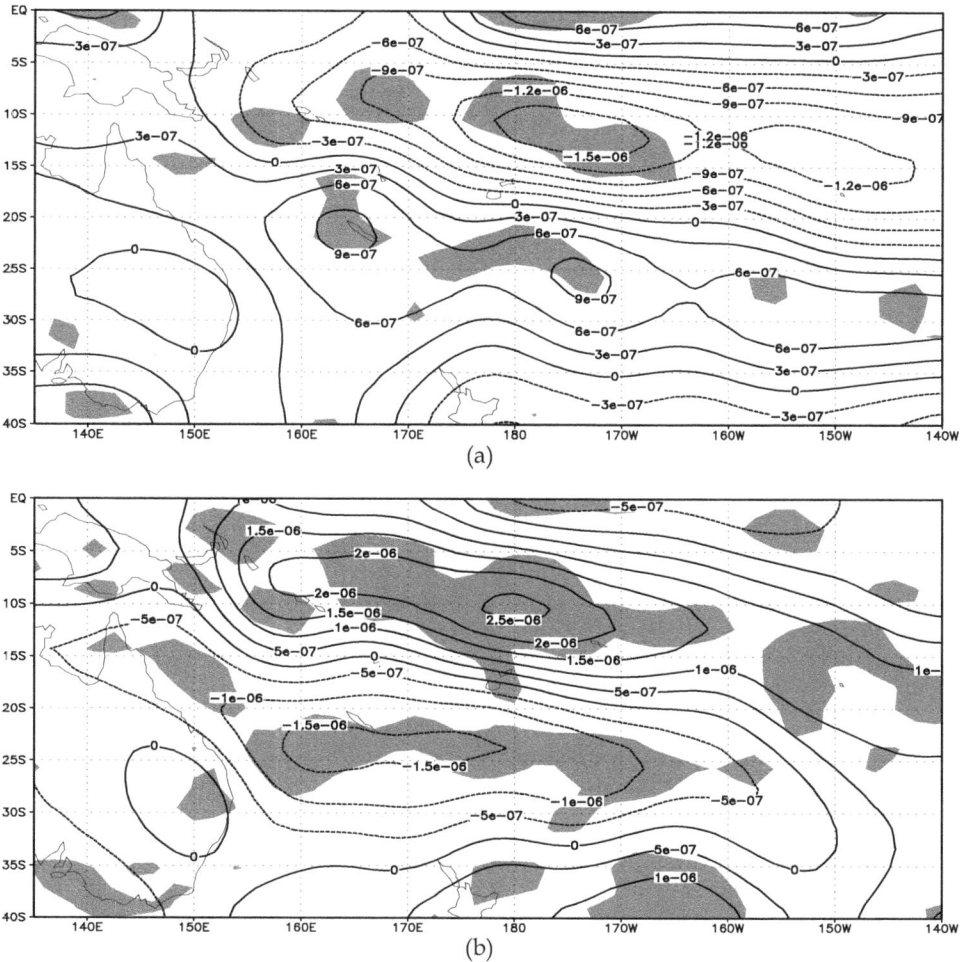

Fig. 14. Vorticity anomalies at 0.995 sigma level (s⁻¹) over the South Pacific Ocean for (a) El Niño and (b) La Niña episodes. The shades indicate areas where the statistical test is significant at the 95% level.

5.6 Discussion and summary

ENSO is a coupled ocean-atmosphere phenomenon and changes in oceanic conditions associated with El Niño and La Niña events are coupled with changes in atmospheric circulation. During the ENSO warm phase, an El Niño event, changes in large-scale environmental conditions such as weakening the Walker circulation (this manifests in weakening the easterly trade winds, rise in air pressure over the Indian Ocean, Indonesia and Australia and fall in air pressure over the central and eastern Pacific Ocean) and displacing warm water in the Pacific (i.e. cooling water in the western Pacific and warming water in the central and eastern equatorial Pacific) occur. During the ENSO cold phase, a La Niña event, the trade winds strengthen, warmer water accumulates in far western Pacific and cold pool in the eastern Pacific intensifies resulting in equatorial SSTs being cooler than climatological average.

In this analysis, changes in tropical cyclogenesis over the South Indian and South Pacific Oceans related to changes in large-scale environmental factors associated with the warm and cold phases of the ENSO phenomenon have been investigated. The physical mechanism by which changes in TC activity are related to the environment has been explored. Warm equatorial water provides thermal energy necessary for TC genesis (Gray 1979) and therefore variation in geographical distribution of SSTs related to changes of the ENSO phases was examined. Large-scale vertical motion which provides convection was examined through analysing relative humidity in mid-troposphere. Positive (*negative*) anomalies of mid-tropospheric relative humidity which characterize upward (*downward*) large-scale vertical motion enhance (*reduce*) TC genesis. The negative effect of vertical wind shear on TC genesis and intensification is well documented (Gray 1968, Merrill 1988). As ENSO affects the strength of easterly and westerly winds in near-equatorial area, these variations in zonal flow are associated with changes in low-level relative vorticity fields. Consequently, influence of the key environmental factors such as SSTs, relative humidity in mid-troposphere, vertical wind shear and lower tropospheric vorticity on TC genesis have been examined.

Over the SIOcean, TC genesis occurs over most of the basin with maximum cyclogenesis in the area [5-15°S, 60-130°E]. In the central part of the basin, only a small number of cyclogenesis events are found south of 15°S. However, in the eastern part of the basin (east of 105°E) maximum TC genesis is located in the area between 10°S and 20°S, and in the western part of the basin (between African coast and 60°E) area favourable to TC genesis occupies range of latitudes from 10°S to 25°S.

During El Niño episodes, TC genesis is above climatology over the western part of the basin and below climatology over the eastern part, while during La Niña episodes it is the opposite. TC genesis also tends to occur further away from the equator during La Niña episodes compared to El Niño episodes.

In the eastern part of the basin near the coast of Western Australia, enhanced TC activity is observed in La Niña seasons compared to El Niño episodes. A number of large-scale environmental factors contribute to increase in cyclogenesis. During El Niño episodes, positive (anticyclonic) vorticity anomalies, negative relative humidity anomalies, and relatively weak positive anomalies of SSTs are observed over this area; all these factors contribute to suppressing TC genesis. On the contrary, negative (cyclonic) vorticity anomalies, positive relative humidity anomalies, and positive anomalies of SSTs observed during La Niña episodes enhance cyclogenesis in this area.

Positive vorticity anomalies contribute to inhibiting TC cyclogenesis in the western part of the basin during La Niña episodes. Negative SSTAs in the central part of the basin (around 70°E) during La Niña episodes additionally reduce TC genesis in this part the SIO in La Niña seasons compared to El Niño seasons.

In the SPO, tropical cyclogenesis is observed over the range of longitudes between 135°E and 120°W. During El Niño episodes, area of above climatology TC genesis is located in the eastern part of the basin east of the line [10°S,165°E - 20°S,175°E]. During La Niña episodes, TC genesis occurs further away from the equator compared to El Niño episodes, and TC activity is displaced to the western part of the basin closer to Australia. In general, southwest (*northeast*) shift of points of cyclogenesis during La Niña (*El Niño*) episodes is observed.

During El Niño (*La Niña*) episodes, positive (*negative*) SSTAs in the eastern part of the basin and in the near-equatorial region promote (*inhibit*) TC genesis and development in the eastern part of the basin and contribute to displacing area preferable for TC genesis closer to (*away from*) the equator. During La Niña episodes, warmer waters in the southwestern part of the basin favour increase of TC activity in the western part of the basin, closer to Australia. The variations of the relatively humidity fields at 500hPa which consistent with variations in SSTs contribute significantly to changes in cyclogenesis over the basin.

Cyclonic vorticity anomalies are another factor which favours increased TC activity in the central and eastern part of the basin in El Niño episodes. In La Niña episodes, anticyclonic vorticity anomalies located over the near-equatorial region contribute to decrease in TC genesis in the area close to the equator while cyclonic vorticity anomalies located between latitudes around 15-25°S contribute to increase in TC activity in the Australian region.

Relatively low vertical wind shear observed in the central equatorial and tropical areas of the basin also enhances TC activity in the central and eastern part of the basin in El Niño episodes. During La Niña episodes, low vertical wind shear west of 180° enhances cyclogenesis in areas closer to Australia.

This analysis demonstrates that over the whole SIO basin, the changes in geographical distribution of relative humidity and vorticity are primary contributors to the variations in TC genesis related to the ENSO phenomenon. Positive anomalies of SSTs observed during La Niña seasons in the eastern part of the basin also contribute to enhanced cyclogenesis near the Western Australia.

In the Pacific, large-scale environmental changes associated with the ENSO (*i.e.* changes in near-equatorial SSTs and the Walker Circulation) significantly affect TC genesis. Changes in geographical distribution of relative humidity and vorticity are primary factors influencing tropical cyclogenesis over the entire South Pacific, with changes in SSTs and vertical wind shear additionally favouring enhanced TC activity in the eastern (*western*) part of the basin during El Niño (*La Niña*) years.

The general conclusions of Camargo et al. (2007) and Kuleshov et al. (2009a) that mid-tropospheric relative humidity and vertical wind shear are important contributors to variation in tropical cyclogenesis in the basins of the SH; however the key role of vorticity and additional contribution of changes in SSTs in the South Indian and the South Pacific Oceans has been also demonstrated.

During El Niño events TC activity in the South Pacific is displaced away from the Australian coast further eastward (Gray 1988, Basher and Zheng 1995) and it reports that relative humidity and vorticity are the key factors of the South Pacific TC environment.

In summary, using the data from recently developed SHTC archive, changes in tropical cyclogenesis over the South Indian and South Pacific Oceans related to changes in the ENSO phases have been investigated and significant variations in geographical distribution of areas of enhanced TC genesis have been identified. To explain these changes, large-scale environmental factors which influence TC genesis and development such as sea surface temperature, relative humidity in mid-troposphere, vertical wind shear and lower tropospheric vorticity have been examined. In the SIO, reduction of TC genesis in the western part of the basin and its increase in the eastern part as well as displacement of the area favourable for TC genesis further away from the equator during La Niña events compared to El Niño events can be explained by changes in geographical distribution of relative humidity and vorticity across the basin as primary contributors; positive anomalies of SSTs observed during La Niña seasons in the eastern part of the basin also contribute to enhanced cyclogenesis near the Western Australia. In the SPO, changes in geographical distribution of relative humidity and vorticity appear to be the key large-scale environmental factors responsible for enhanced TC genesis in the eastern (*western*) part of the basin as well as for the northeast (*southwest*) shift of points of cyclogenesis during El Niño (*La Niña*) events, with vertical wind shear and SSTs as additional contributing environmental factors.

6. Spatial variability of tropical cyclones

Using the best track data from the SH TC archive, various aspects of TC climatology for the SIO and SPO were examined in detail. In particular, Figure 15 shows the climatology of TC days for the entire period of available data (1969/1970 to 2009/2010), and Figure 16 shows the climatology of TC genesis for the period of available central pressure data (1981/1982 to 2009/2010) (Dowdy and Kuleshov, 2011). The region of TC activity stretches from the east coast of Africa throughout the SIO and into the eastern regions of the SPO. Numerous spatial features are apparent in the climatological fields of TC days and TC genesis, broadly consistent with the results of other previous studies (e.g. Kuleshov *et al.* 2008, 2009a, 2010b).

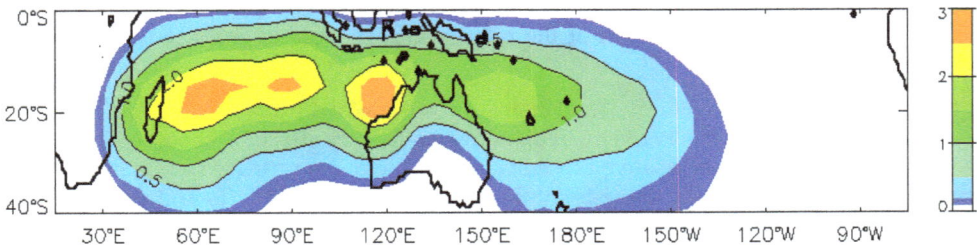

Fig. 15. Average annual number of TC days for the Southern Hemisphere.

As it was described in previous sections, ENSO phenomenon has significant impact on TC spatial variability. Based on the above considerations, the SHTC data were stratified between warm and cold ENSO phases to construct maps of average annual number of TCs in El Niño (Fig. 17) and La Niña (Fig. 18) years (Kuleshov et al., 2008). A comparison of the maps demonstrates substantial shifts in the positions of maximum TC occurrences as well as changes in the intensity of the occurrence maxima. On average, around 25 and 29 TCs annually occur in the SH during El Niño and La Niña years, respectively. In general, TC activity is higher in the SIO than in the SPO. This difference is especially pronounced in La Niña years, with an average annual number of around 18 TCs (*11 TCs*) in the SIO (*SPO*).

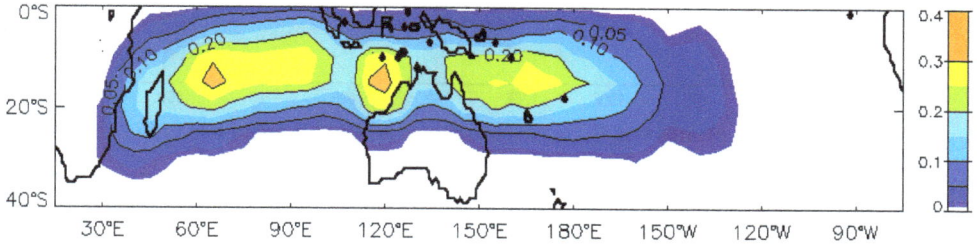

Fig. 16. Average annual number of TC genesis events for the Southern Hemisphere.

Tropical cyclogenesis (defined here as the point at which central pressure attains a threshold of 1000 hPa or lower) in El Niño (Fig. 19) and La Niña (Fig. 20) years has also been examined (Kuleshov et al., 2008). In the SIO, one preferred area of cyclogenesis stays centred around 120°E during El Niño and La Niña years. However, another area of cyclogenesis which is located between 60°E and 85°E in El Niño years shifts eastwards to between 85°E and 105°E in La Niña years. In the SPO, the focus for cyclogenesis shifts eastwards in El Niño years in comparison to La Niña years, which is consistent with Basher and Zheng (1995).

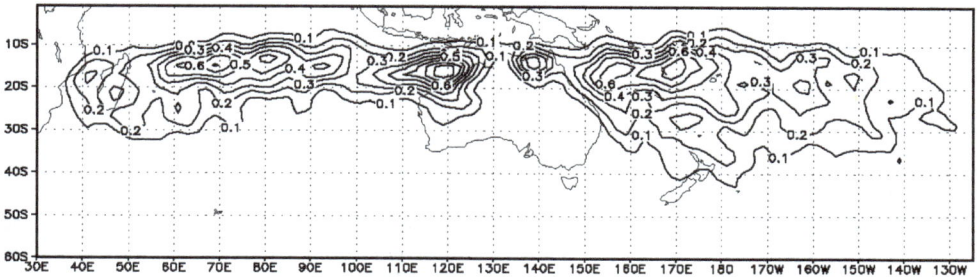

Fig. 17. Average annual number of TCs in the Southern Hemisphere in El Niño years.

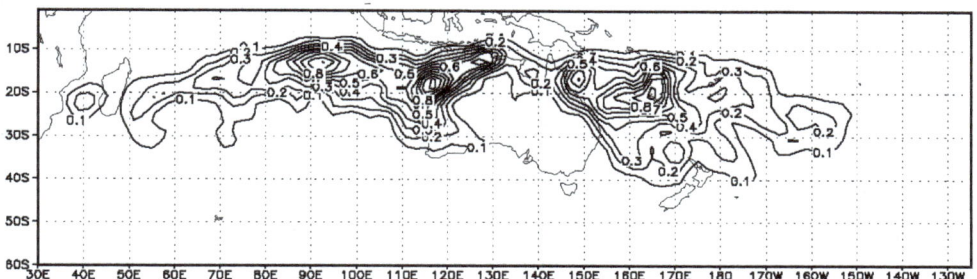

Fig. 18. Average annual number of TCs in the Southern Hemisphere in La Niña years.

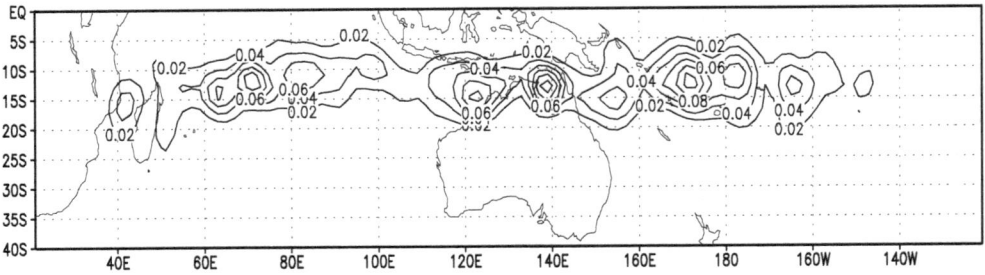

Fig. 19. Average annual number of TC genesis in the Southern Hemisphere in El Niño years.

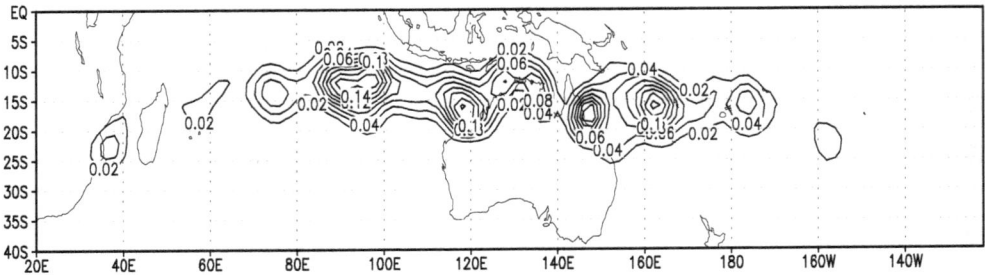

Fig. 20. Average annual number of TC genesis in the Southern Hemisphere in La Niña years.

7. Temporal variability of tropical cyclones

This section builds on previous work through the examination of the TC temporal variability, in particular focusing on intraseasonal variability in multiple sub-regions of each of the two TC basins of the SH. Figure 21 shows monthly distributions of TC days for the four different sub-regions: the west SIO (WSIO) for the longitude range 30-80°E, the east SIO (ESIO) for the longitude range 80-135°E, the west SPO (WSPO) for the longitude range 135-170°E and the central SPO (CSPO) for the longitude range 170°E-120°W (Dowdy and Kuleshov, 2011). This is shown separately for the different ENSO phases. The location of these regions is based on the results of prior studies (e.g. Kuleshov, 2003; Kuleshov et al., 2008) showing regions of strong commonality in their variation to factors such as ENSO as well as being regions of commonality in relation to the underlying climatology based on all available data. Data have only been used equatorward of 25°S, following the results of studies such as by Sinclair (2002) who indicated that, poleward of this latitude, the average TC becomes more characteristic of a baroclinic midlatitude storm than a tropical cyclone.

The shapes of the monthly distributions are consistent with the TC season in the SH - generally defined as from November to April, while noting that TCs can occur outside this range with a relatively lower frequency of occurrence. However, numerous features apparent in the monthly distributions are examined in detail below.

In the WSIO, there is very little difference in the monthly distributions of TC days between the different ENSO phases. Given the inherent small-scale variability expected as a result of the finite period of available data, the similarity between the different phases is quite remarkable. The vast majority of the TC days occur between November and April, with the highest and second highest monthly values of TC days being January and February for each of the three different phases.

In the ESIO there is an indication of a double peak (December and March) for La Niña conditions being the predominant feature. The number of TC days during these two peaks is about 50-100% higher than during the other phases of ENSO. A consistent feature between the three ENSO phases is that they all have peak TC days occurring during March.

In the WSPO, there is reasonably little difference in the monthly distributions of TC days between the different ENSO phases, with the exception of February and March for neutral conditions for which the number of TC days is higher than for the other ENSO phases. The skewness of the distribution for neutral conditions is also indicated by the peak month for TC days being March, in contrast to January and February for El Niño and La Niña conditions, respectively.

In the CSPO, the number of TC days is higher for virtually all months of the year during seasons characterized by El Niño conditions, than for the other two ENSO phases, with peak activity occurring in February. The peak activity occurs slightly earlier for La Niño conditions, in January, and slightly later for neutral conditions, in March.

Figure 22 shows monthly distributions of TC genesis, from 1981/1982 to 2009/2010. As for Figure 21, this is shown separately for the different ENSO phases and data have only been used equatorward of 25°S. Many of the features apparent in the distributions of TC days can also be seen in the genesis distribution, although some differences are also notable.

In the WSIO, the different phases of ENSO are once again broadly similar to each other, as was the case in the monthly distributions of TC days (Figure 21). There is a slightly higher degree of variability between the different ENSO phases for TC genesis than was the case for TC days, potentially relating to the shorter time period used to produce the genesis distributions compared to the TC days distributions.

In the ESIO, a double peak in TC activity for La Niña seasons is apparent, with maxima occur in December and March. As this is consistent with the case for TC days, which is based on twice the number of La Niña seasons, this gives more weight to the likelihood that this is a real feature. The other ENSO phases are broadly similar to the case for TC days, although the peak in TC genesis for El Niño years in January is notable in that it is more comparable in magnitude to the peaks for La Niña conditions in December and March than was the case for TC days.

In the WSPO, there appears to be virtually no genesis during March for La Niña conditions, suggesting that the TC days in this region are due to TCs which have formed elsewhere and then been transported to this region. However, in the following month, more TCs form during La Niña conditions than for the other phases of ENSO. This apparently high degree of variability could potentially be related to the fact that the period of available data has relatively few La Niña seasons (eight) compared with 17 El Niño seasons and 18 neutral seasons. In addition to this, only four La Niña seasons occur during the period of available central pressure data required by the genesis definition used in this paper. It is therefore not clear whether or not small-scale features such as the variability discussed above are representative of the underlying climatology, and so they should be interpreted with a degree of caution. This highlights the need for increased availability of data, as is expected to be the case as the available period of satellite-era data grows with time.

In the CSPO, TC genesis is clearly higher during El Niño conditions than during the other phases of ENSO. This is the same as was also the case for the monthly distribution of TC days. In general throughout the SH, the broad similarities between the monthly distributions of TC days and the TC genesis suggests that although there may be some differences in the lifetime and transport of TCs, the overall climatologies of TC characteristics have a considerable degree of inter-relationship.

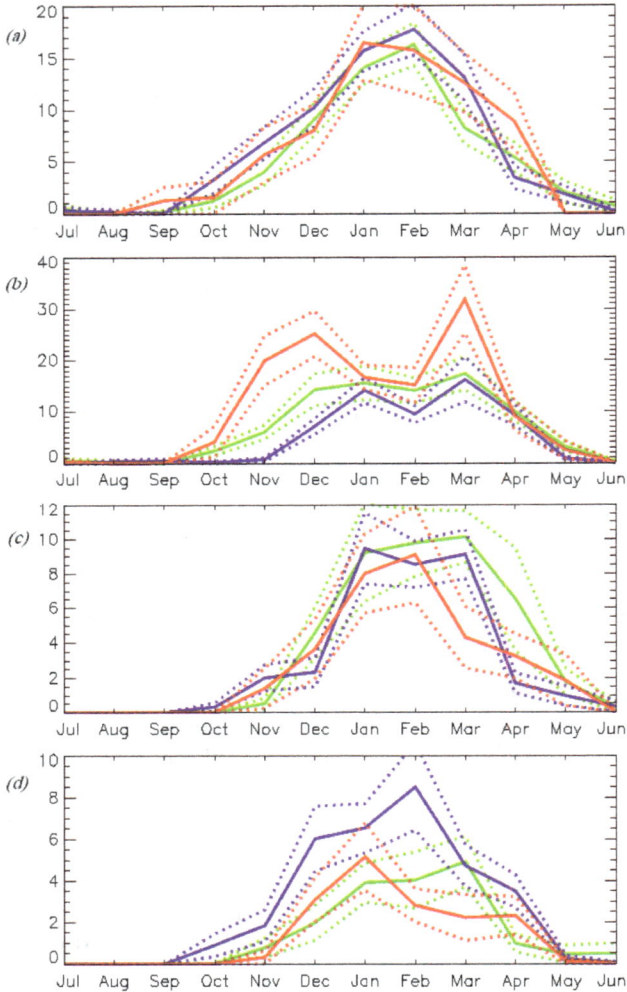

Fig. 21. Monthly distribution of TC days in the Southern Hemisphere, for neutral (green lines), El Niño (blue lines) and La Niña (red lines) phases of ENSO. This is shown for different longitudinal ranges: western SIO 30°E to 80°E (a), eastern SIO 80°E to 135°E ESIO (b), western SPO 135°E to 170°E (c) and central SPO 170°E to 120°W (d). Dotted lines represent the standard error of the mean.

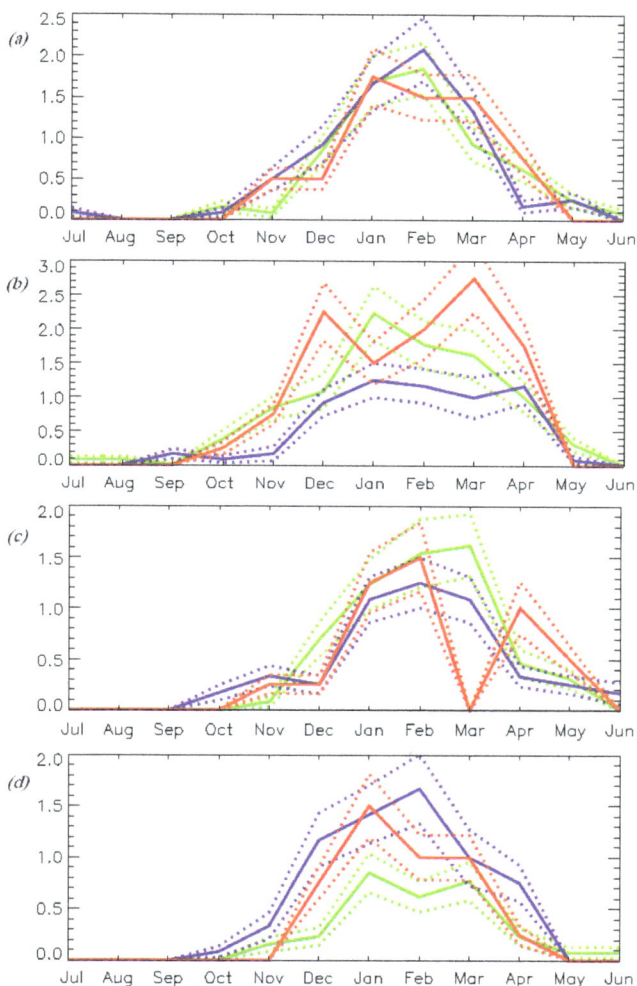

Fig. 22. As for figure 4, but for TC genesis.

8. Trends in tropical cyclone occurrences in the Southern Hemisphere

Trends in TC occurrences and intensity, and possible physical mechanisms for change, have been discussed widely in recent years. Webster et al. (2005) reported that the global number of very intense TCs (Saffir-Simpson categories 4 and 5) had almost doubled over recent decades. Using the TC potential dissipation index as a measure of TC activity, Emanuel (2005) arrived at similar conclusions for two of the major TC basins: the North Atlantic and the North West Pacific. Other authors have rejected these findings, mainly based on the argument that changes have been so great in observation technologies and analysis techniques that the reported changes are artificial, and not due to any actual trends (Landsea et al., 2006; Chan, 2006; Kossin et al., 2007). For the Australian region (AR) and the SH, trends have been reported in the number of TCs and in the proportion of very intense TCs by various authors, including Nicholls et al. (1998) and Harper et al. (2008).

It is well documented that the sea surface temperatures over the major TC basins have increased over recent decades (Knutson et al., 2006; IPCC, 2007). Theoretical and modelling studies would indicate that in response there should be a concomitant increase in the TC limiting upper intensity referred to in the literature as the maximum potential intensity (Emanuel, 1987; Holland, 1997). There is substantial discussion, however, on the magnitude of the response and whether it should be distinguishable in current data where the signal to noise ratio may be small. The small signal to noise ratio in the data is due to a number of effects including: (i) the large interannual variability of TC activity in each basin, (ii) the fact that only a small fraction of TCs actually approach their maximum potential intensity and (iii) limitations in the underlying data.

Recently, trends in activity of TCs in the SH (the region south of the equator, 30°E to 120°W) were estimated by Kuleshov et al. (2010a) analysing the data from SHTC archive. In that study, a TC is considered a tropical system which attains minimum central pressure of 995 hPa or lower. The statistical significance of trends was examined based on non-parametric Monte Carlo methods and based on the test of whether a constant model, a linear model or a simple break-point model represents a best fit to the data. The purpose of this examination was to determine whether there are trends in the SH TC occurrence and intensity time series beyond what can be attributed to inter-annual variability and changes in observing procedure. The data set used is the SHTC archive which was compiled from the best track data sets of the NMHSs with WMO responsibility for TC forecasts and warnings across the SH, in consultation with these offices. A documentation of trends in this data set thus provides baseline information for detection and attribution studies towards projections of expected changes in TC activity under global warming.

Changes in TC occurrences in the SH, the SIO and the SPO were analysed over the 26 seasons 1981/82 to 2006/07. Over this period, there are no significant trends in the annual numbers of TCs (SPO, SIO, SH) attaining an lifetime mean central pressure (LMCP) of 995 hPa or lower (Figure 23).

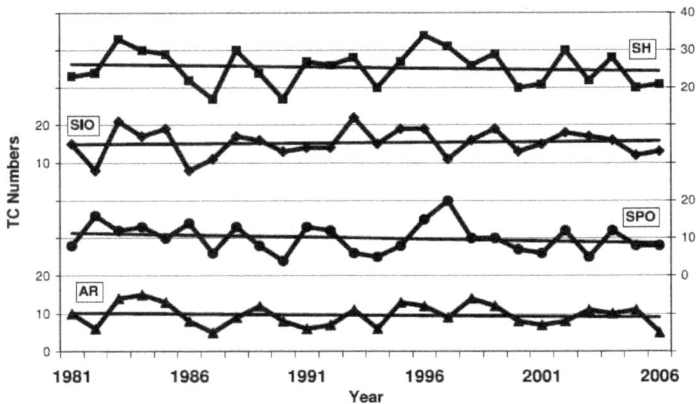

Fig. 23. Annual numbers of TCs with LMCP of 995 hPa or lower for the SH (squares, right axis), SIO (diamonds, left axis), SPO (circles, right axis) and AR (triangles, left axis), 1981/82 to 2006/07 seasons, with linear trends.

For (970 hPa or lower) severe TCs (Figure 24), there are no significant trends in the SIO and the SH, although the declining trend (−0.096 TCs/yr) in the SPO is border-line significant (11% two-tailed).

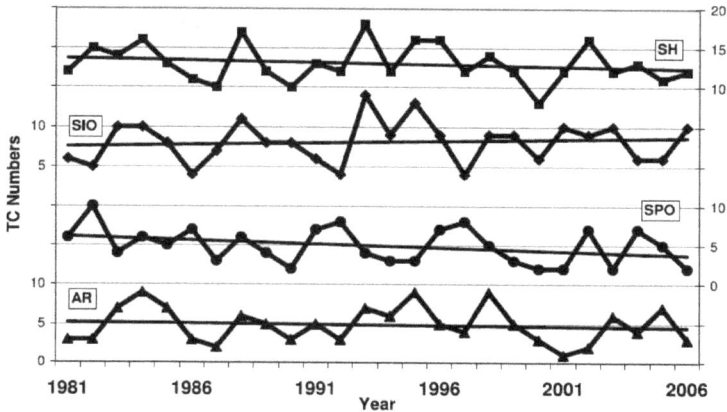

Fig. 24. Annual numbers of TCs with LMCP of 970 hPa or lower for the SH (squares, right axis), SIO (diamonds, left axis), SPO (circles, right axis) and AR (triangles, left axis), 1981/82 to 2006/07 seasons, with linear trends.

For the most intense (950 hPa or lower) TCs (Figure 25), there is no significant trend in the SPO, but the trends are significant in the SIO (+0.15 TCs/yr; 1% two-tailed) and in the SH as a whole (+0.14 TCs/yr; 3% two-tailed). Similar results were obtained in the annual 945 hPa counts, but not in the 955 hPa counts.

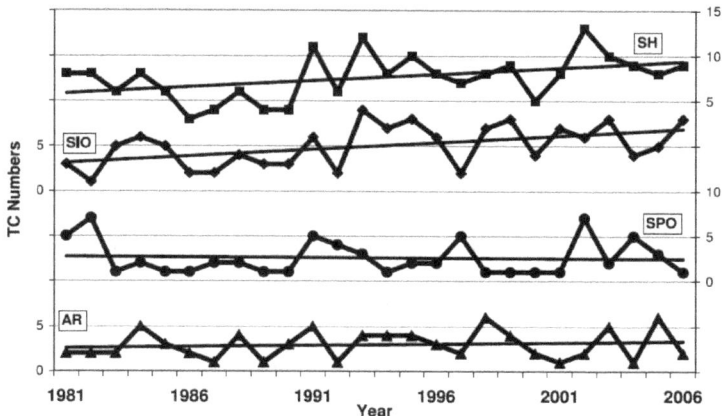

Fig. 25. Annual numbers of TCs with LMCP of 950 hPa or lower for the SH (squares, right axis), SIO (diamonds, left axis), SPO (circles, right axis) and AR (triangles, left axis), 1981/82 to 2006/07 seasons, with linear trends.

Analysing changes in TCs in the AR over the 26-year period (1981/82 to 2006/07), there were no significant trends in the numbers of TCs.

Quality of historical TC data is highly important when analysing trends. Worldwide the underlying technique for determining TC intensity is the Dvorak analogue procedure based on patterns of infra-red brightness temperature (Dvorak, 1984; Velden et al., 2006). The original version of the technique, applied to visible satellite imagery, was published in 1975, with its present form, based on digital infra-red imagery, published in 1984. It was only during the mid-1980s that high-resolution multi-channel imagery became available at the AR forecast offices. In addition, supplementary data sources have increased during the past 26 years, including deployment of automatic weather stations along the Australian coastline and on small islands, and the advent of satellite-based scatterometer surface wind estimates. Various authors have discussed the potential impact of these changes on our ability to accurately determine the intensity of the more intense TCs, including Trewin (2008) and Harper et al. (2008). On a global basis, the impact of data quality on our ability to determine trends has been discussed by Landsea et al. (2006). A further compounding issue is that during forecast operations, the Dvorak technique output and the classification of TCs into intensity classes are both based on estimated sustained wind speeds. Intensity in terms of central pressure is then obtained through a wind-pressure relationship. Different wind-pressure relationships are in use in different forecast offices and warning centres across the SH (Knaff and Zehr, 2007; Harper et al., 2008), and there have been changes in the wind-pressure relationships used through the period of study. The importance of this for determination of trends is that the SH track archive (in its current state) contains only the derived central pressure information and does not contain the wind-speed intensity estimates used operationally.

For the Western SIO where the tracks are maintained by Météo-France (La Réunion), operational meteorologists consider the intensity data insufficiently reliable for trend estimation prior to the establishment of the RSMC La Réunion in 1993 [P. Caroff, pers. comm.]. A further improvement in data quality occurred with the beginning of geostationary satellite coverage in 1998, as a result of the launch of the MeteoSat-7 satellite (Kossin et al., 2007).

In identifying TC positions and estimating intensities, operational forecasters from Fiji and New Zealand (responsible for the eastern SPO tracks) faced similar challenges, through gradually improved observational data and understanding of TC development [S. Ready, pers. comm.]. Within the satellite era, low-resolution geostationary satellite imagery for the South Pacific (west of about 155°W) became available to those forecasters from 1980. However, only from the early 1990s, with high-resolution imagery from the GOES-West satellites now available to the RSMC Nadi (established in 1993), did the Dvorak technique become increasingly used for estimating storm intensity. Throughout the 1980s and in the first half of the 1990s, there still was a reluctance to assign intensities beyond 80 knots (central pressures below about 955 hPa). From the mid-1990s, as forecasters in Fiji became more proficient at the Dvorak technique, there was a greater range of intensities assigned in TC warnings. It appears that for the SPO east of 160°E the most reliable estimates of TC intensity start in the early 1990s, with reliable estimates of LMCP of the most intense TCs from the mid-1990s. It is likely that prior to this time the number of TCs with reported intensities below 950 or 945 hPa in the eastern SPO is underestimated.

While some of the trends in the TC data appear to be artificial to a degree, due to changes in TC observation practices and analysis techniques as discussed above, it is possible that TC occurrences are subject to variability associated with low-frequency climate modes, such as ENSO (e.g., Kuleshov et al., 2008; Kuleshov et al., 2009a), the Indian Ocean Dipole (e.g.,

Chan and Liu, 2009) and the Pacific Decadal Oscillation (e.g., Goh and Chan, 2009), to the extent of having a noticeable impact on the trends. The inclusion of these low-frequency climate modes in the trend analysis is beyond the scope of this study, but will be a topic of further investigation.

9. Southern Hemisphere tropical cyclone historical data portal

To provide a user-friendly means for accessing detailed information and data on historical tropical cyclones for the Southern Hemisphere, the first version of a specialised website for disseminating results and data "Tropical Cyclones in the Southern Hemisphere" was developed in 2007. Recently, a new design for the website has been developed (Figure 26; http://www.bom.gov.au/cyclone/history/tracks/). Main features of this design are briefly presented in this section.

The new tropical cyclone website has been developed using OpenLayers platform. This allows dynamic map navigation, presenting detailed information for a selected region in the Southern Hemisphere and the display of changes in tropical cyclone intensity over the lifetime of a cyclone. The features of the new web site are presented below using tropical cyclone *Heta* as an example.

Tropical cyclone *Heta* formed on December 25, 2003 and reached a maximum intensity of 260km/h and an estimated pressure of 915 hPa on January 6, 2004. Cyclone *Heta* caused catastrophic damage to the islands of Tonga, Niue, and American Samoa estimated at $170 million dollars (2009 USD). Best track of tropical cyclone *Heta* is presented in Figure 27. In Figure 28, track of tropical cyclone *Heta* is displayed over the "Elevation and bathymetry" layer. Orange dots represent best track 6-hourly positions of the cyclone when its central pressure was estimated above 970 hPa and red dots represent the cyclone's positions when pressure was estimated as equal to or below 970 hPa (an approximate pressure threshold for *severe* tropical cyclones in the Southern Hemisphere).

The dynamic map navigation feature of the new web site allows one to examine the cyclone track over a selected region in detail. Dashed line connects points of the tropical cyclone best track when the system was at a stage of Tropical Depression (i.e. estimated pressure is above 995 hPa) and solid line connects the points when the storm attained Tropical Cyclone intensity (a threshold of 995 hPa or below is selected for display purposes). Users can obtain information about characteristics of the tropical cyclone at particular location by clicking on a dot point. As a result, the following information is displayed: tropical cyclone name, geographical coordinates of the selected position, time and intensity in terms of an estimated value of pressure in hectopascals.

Other features of the website include displaying multiple tropical cyclone tracks to satisfy users' requirements to present cyclone tracks over the whole tropical cyclone season, or a number of seasons, and enabling a down-load facility. Specifically, "Tropical Cyclone Track Details" feature allows users to display tropical cyclone data for a selected cyclone (date and time, latitude and longitude, and intensity in terms of central pressure) in a separate window. The "Report on Specific Location" feature allows users to display tracks for cyclones crossing within specified distance (e.g. within a radius of 50, 100, 200 and 400 km) from a specified position. The specified position can be defined as a point with specific geographical coordinates (e.g. 12.42°S 130.89°E) or as a location selected from a list (e.g. Darwin).

Fig. 26. Front page of the Southern Hemisphere tropical cyclone specialised website.

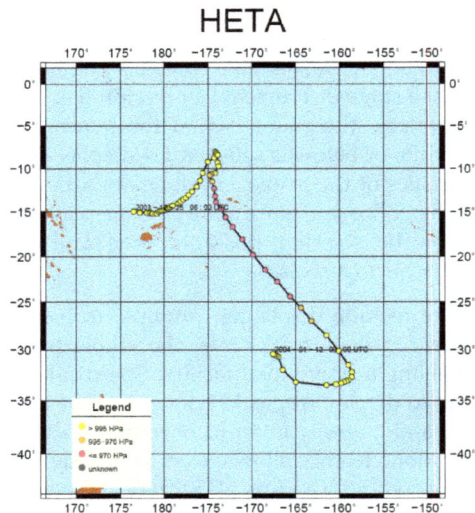

HETA

Fig. 27. Best track of tropical cyclone *Heta*.

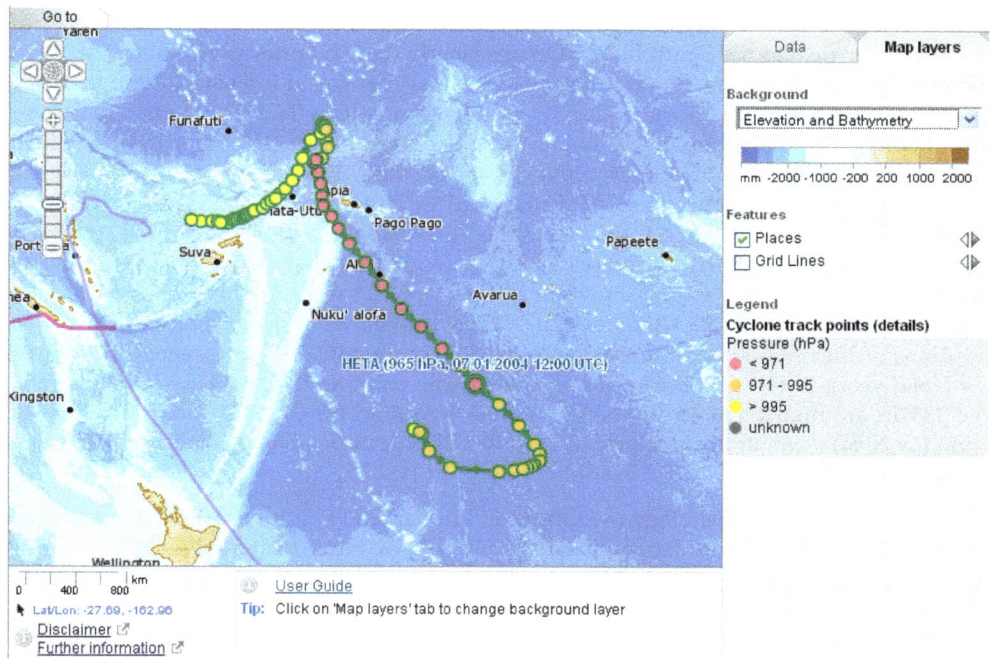

Fig. 28. Detailed presentation of tropical cyclone *Heta* best track over the selected area where the cyclone reached 915 hPa intensity on January 6, 2004.

10. Summary

Tropical cyclones (TCs) in the South Indian Ocean and the South Pacific Ocean are frequent and intense, and they dramatically affect maritime navigation and the lives of communities in coastal areas. To understand TC variability and changes in this region, "Climate change and Southern Hemisphere tropical cyclones" international initiative has been established in 1999. Over more than a decade, significant progress has been made through dedicated international efforts. A new high-quality TC data set has been created and subsequently used to examine variability in TC activity in the Southern Hemisphere (SH). Data from a number of NMHSs were combined in a high-quality tropical cyclone dataset, available through a specialised website "Tropical Cyclones in the Southern Hemisphere". Previously, many of these data have been only available to a small number of individuals, and tended to be fragmented and inconsistent across national borders. Analyses have been undertaken to understand the variability and change in TCs across the SH, with a view to providing a better scientific basis for understanding the current risks posed by TCs, and possible changes as a result of climate change.

A particularly important finding is that much of the variability of TCs can be understood (and even predicted in advance) using broad-scale indices which describe the El Niño-Southern Oscillation (ENSO) phenomenon. Climatological analysis of TC occurrences in the South Pacific and the South Indian Oceans clearly demonstrates geographical shifts in the positions of maximum TC occurrences as well as changes in intensity of the maxima. This suggests that a key to understanding future changes in TCs lies in understanding how ENSO will change under the enhanced greenhouse effect. Different indices which describe ENSO (the SOI, the SSTA NINO3.4 index, the MEI and the 5VAR index) have been examined, and a collective list of historical El Niño and La Niña events has been developed. Based on this list, TC data from the TC archive for the SH have been stratified, and significant changes in TC occurrences (in terms of geographical distribution and intensity of maxima) depending on warm and cold phases of ENSO have been identified. The TC climatology for El Niño and La Niña years shows a geographical shift in the positions of maxima of TC occurrences in the SH. In the SIO, TC occurrences in El Niño (*La Niña*) years in the area between the east African coast and 75°E (*between around 75°E and 135°E*) are higher than in La Niña (*El Niño*) years. In the SPO, TC occurrences in El Niño (*La Niña*) years in the area east of around 170°E (*between around 140°E and 170°E*) are higher than in La Niña (*El Niño*) years.

Tropical cyclogenesis climatology over the South Indian and South Pacific Oceans has been developed and changes in geographical distribution of areas favourable for TC genesis related to changes in the ENSO phases have been investigated. To explain these changes, large-scale environmental variables which influence TC genesis and development such as sea surface temperatures (SSTs), relative humidity in mid-troposphere, vertical wind shear and lower tropospheric vorticity have been examined. In the South Indian Ocean, reduction of TC genesis in the western part of the basin and its increase in the eastern part as well as displacement of the area favourable for TC genesis further away from the equator during La Niña events compared to El Niño events can be explained by changes in geographical distribution of relative humidity and vorticity across the basin as primary contributors; positive anomalies of SSTs observed during La Niña seasons in the eastern part of the basin additionally contribute to enhanced cyclogenesis near the Western Australia. In the South Pacific Ocean, changes in geographical distribution of relative humidity and vorticity appear to be the key large-scale environmental factors responsible for enhanced TC genesis in the eastern (*western*) part of the basin as well as for the northeast (*southwest*) shift of points of cyclogenesis during El Niño (*La Niña*) events, with vertical wind shear and SSTs as additional contributing large-scale environmental variables.

The statistical significance of trends in TCs in the South Indian Ocean and the South Pacific Ocean has been examined. Calculation of significance was based on non-parametric Monte Carlo methods, and additionally it was explored whether a constant model, a linear model or a simple break-point model represents a best fit to the data. For the 1981/82 to 2006/07 TC seasons, there are no apparent trends in the total numbers of TCs (those tropical systems attaining a minimum central pressure of 995 hPa or lower), nor in numbers of 970 hPa TCs (such TCs being called *severe* in the Southern Hemisphere). Positive trends in the numbers of 945 hPa and 950 hPa TCs in the South Indian Ocean are significant, but appear to be influenced to some extent by changes in data quality. In the

Australian region, no significant trends in the total numbers of TCs, or in the proportion of the most intense TCs, have been found.

These findings are important as the data set used constitutes the official best track data archive for the Southern Hemisphere, even though there are uncertainties in TC intensity estimates (mainly prior to the 1990s). Despite all this, the archive represents the current best estimate of recent SH TC climatology. Attempts have been made to prepare consolidated global datasets and there are plans to continue these efforts. However, consolidation of historical data from various regions is currently limited by the inhomogeneity of TC observation and analysis practice, and there is evidently a considerable need for re-analysis of the historical TC data in order to obtain globally homogeneous records. These homogeneity issues place limit our ability at the present time to answer the important question of how TC activity is changing and its possible relationship to global climate change more generally.

11. Acknowledgements

Since establishing "Climate Change and Southern Hemisphere Tropical Cyclones" international initiative in 1999, its various aspects were supported by the Australian Greenhouse Office and the Laboratory of the Atmosphere and Cyclones at the University of La Réunion and Météo-France. The Meteorological Services of Australia, Fiji, France and New Zealand provided the TC best track data for the areas of responsibilities of RSMC La Réunion, RSMC Nadi and TCWC Wellington, respectively. Author thanks Philippe Caroff, Jim Davidson and Steve Ready for discussions on quality of regional TC data sets. The research discussed in this chapter was conducted with the partial support of the Pacific Climate Change Science Program (PCCSP) and Pacific-Australia Climate Change Science and Adaptation planning Program (PACCSAP), which are supported by the Australian Agency for International Development (AusAID), in collaboration with the Department of Climate Change and Energy Efficiency (DCCEE), and delivered by the Bureau of Meteorology and the Commonwealth Scientific and Industrial Research Organisation (CSIRO).

12. References

Australian Government (1977), *Report on cyclone Tracy December 1974.* Australian Government Publishing Service, Canberra, 82 pp.

Basher, R.E. and Zheng, X. (1995), Tropical cyclones in the Southwest Pacific: Spatial patterns and relationships to Southern Oscillation and sea surface temperature. *Journal of Climate,* 8, pp. 1249-1260.

Bister, M. and Emanuel, K.A. (1998), Dissipative heating and hurricane intensity, Meteor. Atmos. Phys., 52, 233-240.

Broomhall, M., Grant, I., Majewski, L., Willmott, M., Jones, D. and Kuleshov, Y., (2010), Improving the Australian tropical cyclone database: Extension of GMS satellite image archive. In *Indian Ocean Tropical Cyclones and Climate Change*, Y. Charabi (Ed.), pp. 199- 206 (NY, Springer) doi: 10.1007/978-90-481-3109-9_24.

Camargo, S. J., Emanuel, K.A. and Sobel, A.H., (2007), Use of a genesis potential index to diagnose ENSO effects of tropical cyclone genesis. *Journal of Climate*, 20, pp. 4819-4834.

Chan, J.C.L., (2000) Tropical cyclone activity over the western North Pacific associated with El Niño and La Niña events. *J. Climate*, 13, 2960-2972.

Chan, J. C. L. and K. S. Liu (2009), Interannual variations of tropical cyclone activity in the Southern Hemisphere, Proceeding of the 9th International Conference on Southern Hemisphere Meteorology and Oceanography (9ICSHMO), American Meteorological Society and the Australian Meteorological and Oceanographic Society, Melbourne, Australia (http://www.bom.gov.au/events/9icshmo).

Chan, J.C.L., J.E. Shi and K.S. Liu, (2001) Improvements in the seasonal forecasting of tropical cyclone activity over the western North Pacific. *Wea. Forecasting*, 16, 491-498.

Chan, J. C. L. (2006), Comment on "Changes in Tropical Cyclone Number, Duration, and Intensity in a Warming Environment", *Science*, 311, 1713.

Chu, P.-S. (2004), ENSO and tropical cyclone activity. Hurricanes and Typhoons : Past, Present and Potential, R.J. Murnane and K.B. Liu ed., Columbia University Press, New-York, 297-332.

DeMaria M., (1996) The effect of vertical shear on tropical cyclone intensity change. J. Atmos. Sci., 53, 2076–2087.

Dowdy, A. and Y. Kuleshov (2011), An analysis of tropical cyclone occurrence in the Southern Hemisphere derived from a new satellite-era dataset, *Int. J. of Remote Sensing* (in press).

Dvorak, V. F. (1984), Tropical cyclone intensity analysis using satellite data. NOAA Tech. Report NESDIS 11. 47pp.

Emanuel, K. A. (1987), The dependence of hurricane intensity on climate, *Nature*, 326, 483–485.

Emanuel, K. A. (2005), Increasing destructiveness of tropical cyclones over the past 30 years, *Nature*, 436, 686-688.

Evans, J.L. and Allan, R.J., (1992), El Niño/Southern Oscillation modification to the structure of the monsoon and tropical cyclone activity in the Australian region. *Int. J. Climatol.*, 12, pp. 611–623.

Goh, A. Z.-C. and J. C. L. Chan (2009), Interannual and interdecadal variations of tropical cyclone activity in the South China Sea. *Int. J. Climatol.* doi: 10.1002/joc.1943.

Gray, W.M. (1968), Global view of the origin of tropical disturbances and storms, Mon Wea. Rev., 96, 669-700.

Gray, W.M., (1979), Hurricanes: Their formation, structure and likely role in the tropical circulation. *Meteorology over the Tropical Oceans*, D.B. Shaw, Ed., Royal Meteorological Society, 155-218.

Gray, W.M., (1984), Atlantic seasonal hurricane frequency. Part I: El Niño and 30 mb quasi-biennial oscillation influences. *Mon. Wea. Rev.*, 112, 1649-1668.

Gray, W.M., (1988), Environmental influences on tropical cyclones. *Aust. Meteor. Mag.*, 36, 127-139.

Hall, T.M. and S. Jewson, (2007), Statistical modelling of North Atlantic tropical cyclone tracks. *Tellus*, 59A, 486-498.

Harper, B. A., S. A. Stroud, M. McCormack, and S. West (2008), A review of historical tropical cyclone intensity in northwestern Australia and implications for climate change trend analysis, *Aust. Met. Mag., 57*, 121-141.

Hastings, P.A. (1990), Southern Oscillation influences on tropical cyclone activity in the Australian/ South-west Pacific region, Int. J. Climatol., 10, 291-298.

Ho, C.H., Kim, J.H., Jeong, J.H. and Kim, H.S., (2006), Variation of tropical cyclone activity in the South Indian Ocean: El Niño – Southern Oscillation and Madden-Julian Oscillation effects. *Journal of Geophysical Research*, 111, D22191, doi:10.1029/2006JD007289.

Holland, G. J. (1997), The maximum potential intensity of tropical cyclones, *J. Atmos. Sci., 54*, 2519-2541.

Holland, G.J., (1981), On the quality of the Australian tropical cyclone data base. *Aust. Meteor. Mag.*, 29, 169-181.

Holland, G.J., (1984), On the climatology and structure of tropical cyclones in the Australian / southwest Pacific region: I. Data and tropical storms. *Australian Meteorological Magazine*, 32, pp. 1-15.

IPCC (2007), IPCC WG1 AR4 Report: ipcc-wg1.ucar.edu/wg1/wg1-report.html

Jury, M. R., (1993), A preliminary study of climatological associations and characteristics of tropical cyclones in the SW Indian Ocean. *Meteor. Atmos. Phys*, 51, 101–115.

Kalnay, E., Kanamitsu, M., Kistler, R., Collins, W., Deaven, D., Gandin, L., Iredell, M., Saha, S., White, G., Woollen, J., Zhu, Y., Leetmaa, A., Reynolds, B., Chelliah, M., Ebisuzaki, W., Higgins, W., Janowiak, J., Mo, K.C., Ropelewski, C., Wang, J., Jenne, R. and Joseph, D. (1996), The NCEP/NCAR 40-Year Reanalysis Project, Bull. Amer. Meteor. Soc.: 77, 437-472.

Kerr, I.S., (1976), Tropical storms and hurricanes in the southwest Pacific, November 1939 to April 1969. *NZMS Misc. Publication 148*, New Zealand Meteorological Service, pp. 114.

Kiladis, G.N., and H. van Loon, (1988), The Southern Oscillation. Part VII: Meteorological anomalies over the Indian and Pacific sectors associated with extremes of the oscillation. *Monthly Weather Review*, 116, 120-136.

Knaff, J.A., and R. M. Zehr (2007). Reexamination of tropical cyclone wind-pressure relationships, *Wea. Forecasting, 22*, 71-88.

Knutson, T. R., T. L. Delworth, K. W. Dixon, I. M. Held, J. Lu, V. Ramaswamy, M. D. Schwarzkopf, G. Stenchikov and R. J. Stouffer (2006), Assessment of twentieth-century regional surface temperature trends using the GFDL CM2 coupled model, *J. Clim., 19*, 1624-1651.

Kossin, J. P., K. R. Knapp, D. J. Vimont, R. J. Murnane, and B. A. Harper (2007), A globally consistent reanalysis of hurricane variability and trends, *Geophys. Res. Lett., 34*, L04815, doi:10.1029/2006GL028836.

Kuleshov, Y. (2003), Tropical cyclone climatology for the Southern Hemisphere. Part I. Spatial and temporal profiles of tropical cyclones in the Southern Hemisphere. *Commonwealth of Australia, Bureau of Meteorology*, 20 pp.

Kuleshov, Y., and G. de Hoedt, (2003), Tropical cyclone activity in the Southern Hemisphere. *Bull. Austral. Met. Ocean. Soc.*, 16, 135-137.

Kuleshov, Y. Qi, L., Fawcett, R., Jones, D., (2008), On tropical cyclone activity in the southern hemisphere: Trends and the ENSO connection. *Geophysical Research Letters*, 35, L14S08, doi:10.1029/2007GL032983.

Kuleshov, Y., Ming, F.C., Qi, L., Chouaibou, I., Hoareau, C. and Roux, F., (2009a), Tropical cyclone genesis in the Southern Hemisphere and its relationship with the ENSO. *Ann. Geophys.*, 27, pp. 2523-2538.

Kuleshov, Y. Qi, L., Fawcett, R., Jones, D., (2009b), Improving preparedness to natural hazards: Tropical cyclone prediction for the Southern Hemisphere. In *Advances in Geosciences*, 12 *Ocean Science*, J. Gan (Ed.), pp. 127-143 (Singapore: World Scientific Publishing).

Kuleshov, Y., Fawcett, R., Qi, L., Trewin, B., Jones, D., McBride, J. and Ramsay, H., (2010a), Trends in tropical cyclones in the South Indian Ocean and the South Pacific Ocean. *Journal of Geophysical Research*, 115, doi:10.1029/2009JD012372.

Kuleshov, Y. Qi, L., Jones, D., Fawcett, R., Chane-Ming, F., McBride, J. and Ramsay, H. (2010b), On developing a tropical cyclone archive and climatology for the South Indian and South Pacific Oceans, In *Indian Ocean Tropical Cyclones and Climate Change*, Y. Charabi (Ed.), pp. 189-197 (NY: Springer), doi: 10.1007/978-90-481-3109-9_23.

Landsea, C.W., (2000a), Climate Variability of Tropical Cyclones: Past, Present and Future Storms, 2000 Ed. R.A. Pielke, Sr. and R.A. Pielke, Jr, Routledge, New York, 220-241.

Landsea, C.W., (2000b), El Niño/Southern Oscillation and the Seasonal Predictability of Tropical Cyclones. In *El Niño and the Southern Oscillation: Multiscale Variability and Global and Regional Impacts*, H.F. Diaz and V. Markgraf, Ed., Cambridge University Press, 149-181.

Landsea, C. W., B. A. Harper, K. Hoarau, and J. A. Knaff (2006), Can we detect trends in extreme tropical cyclones, *Science*, 313, 452-454.

Larkin, N.K., and D.E. Harrison, (2001), Tropical Pacific ENSO cold events, 1946-95, SLP, and surface wind composite anomalies, *J. Climate*, 14, 3904-3930.

Larkin, N.K., and D.E. Harrison, (2002), ENSO warm (El Niño) and cold (La Niña) event life cycles: ocean surface anomaly patters, their symmetries, asymmetries, and implications, *J. Climate*, 15, 1118-1140.

Maloney E. D. and Hartmann D. L. (2000), Modulation of Eastern North Pacific Hurricanes by the Madden–Julian Oscillation. *J. Climate*, 13, 1451-1460.

Merrill, R. T. (1988), Environmental influences on hurricane intensification. *J. Atmos. Sci.*, 45, 1678-1687.

Nicholls, N., (1984), The Southern Oscillation, sea-surface temperature, and inter-annual cyclone activity in the Australian region. *J. Climatol.*, 4, 661-670.

Nicholls, N., (1985) Predictability of interannual variations of Australian seasonal tropical cyclone activity. *Mon. Wea. Rev.*, 113, 1144-1149.

Nicholls (1992), Recent performance of a method for forecasting Australian seasonal tropical cyclone activity. *Aust. Meteor. Mag.*, 40, 105-110.

Nicholls, N., Landsea, C.W. and Gill, J. (1998), Recent trends in Australian region tropical cyclone activity, Meteor. Atmos. Phys., 65, 197-205.

Rasmusson, E.M. and T.H. Carpenter, (1982), Variations in tropical sea surface temperature and surface wind fields associated with the Southern Oscillation/El Niño, *Monthly Weather Review*, 110, 354-384.

Reid, P.A. (2003), An operational definition of El Niño. *Proc. 7th Int. Conf. on Southern Hemisphere Meteorology and Oceanography*, Wellington NZ, 24-28 March 2003, 128-130.

Revell, C.G., (1981), Tropical cyclones in the Southwest Pacific, November 1969 to April 1979. *NZMS Misc. Publication 170*, New Zealand Meteorological Service, pp. 53.

Revell, C.G. and Goulter, S.W., (1986), South Pacific tropical cyclones and the Southern Oscillation, *Mon. Wea. Rev.*, 114, pp. 1138-1145.

Saunders, M.A., R.E. Chandler, C.J. Merchant, and F.P. Roberts, (2000), Atlantic hurricanes and NW Pacific typhoons: ENSO spatial impacts on occurrence and landfall. *Geophys. Res. Lett.*, 27, 1147-1150.

Shapiro, L.J., (1987), Month-to-month variability of the Atlantic tropical circulation and its relationship to tropical storm formation. *Mon. Wea. Rev.*, 115, 2598-2614.

Sinclair M. R., (2002), Extratropical transition of Southwest Pacific Tropical Cyclones. Part 1: Climatology and Mean Structure Changes. *Monthly Weather Review*, 130, pp. 590-609.

Trenberth, K.E., (1997), The definition of El Niño. *Bull. Amer. Meteor. Soc.*, 78, 2771-2777.

Trenberth, K.E., (2007), Warmer oceans, stronger hurricanes. *Scientific American*, July 2007, 45-51.

Trewin, B. (2008), An enhanced tropical cyclone data set for the Australian region, *20th Conference on Climate Variability and Change*, Abstract JP3.1.

Troup, A.J., (1965), The Southern Oscillation. *Q. J. Royal Met. Soc.*, 91, 490-506.

Velden, C., B. Harper, F. Wells, J. L. Beven II, R. Zehr, T. Olander, M. Mayfield, C. (Chip) Guard, M. Lander, R. Edson, L. Avila, A. Burton, M. Turk, A. Kikuchi, A. Christian, P. Caroff and P. McCrone (2006), The Dvorak tropical cyclone intensity estimation technique: A satellite-based method that has endured for over 30 years, *Bull. Am. Meteorol. Soc.*, 87, 1195–1210.

Webster, P. J., G. J. Holland, J. A. Curry and H.-R. Chang (2005), Changes in tropical cyclone number, duration, and intensity in a warming environment, *Science, 309*, 1844-1846.

WMO (1993), WMO/TD No. 560. Global Guide to Tropical Cyclone Forecasting. G.J. Holland, Ed. Report No. TCP-31

WMO (2002a), WMO/TD No. 292. Tropical cyclone operational plan for the South Pacific and South-east Indian Ocean. Report No. TCP-24.

WMO (2002b), WMO/TD No. 577: Tropical cyclone operational plan for the South-west Indian Ocean. Report No. TCP-12.

Wolter, K., (1987), The Southern Oscillation in surface circulation and climate over the tropical Atlantic, Eastern Pacific, and Indian Oceans as captured by cluster analysis. *J. Climate Appl. Meteor.*, 26, 540-558.

Zehr, R.M., (2003), Environmental vertical wind shear with Hurricane Bertha (1996), Wea. Forecasting, 18, 345-356.

3

Characteristics of the Quasi-16 Day Wave in the Mesosphere and Lower Thermosphere (MLT): A Review Over an Equatorial Station Thumba (8.5°N, 76.5°E)

Siddarth Shankar Das

Space Physics Laboratory, Vikram Sarabhai Space Centre ISRO-PO, Trivandrum
India

1. Introduction

To understand the climate variability and weather prediction in the Earth's atmosphere, measurements of winds, temperature and wave activities are very crucial. The Earth's atmosphere is believed to act as a source and sink for the waves of a broader spectrum with periods from few seconds to years. Generally, the Earth's atmosphere is stably stratified except the planetary boundary layer and thus makes a reasonable assumption for the presence of atmospheric waves. A barotropic atmosphere in a resting basic state is able to support these spectra of waves. These waves move diagonally upward or downward and horizontally. The mean zonal circulation is mainly driven by these atmospheric waves, which are believed to be generated in the troposphere and propagates horizontally and vertically in to the middle and upper atmosphere. These waves transport energy and momentum from one region to another without the transport of material medium thereby impinging the signature of the source region on to the sink region. The waves propagating in Earth's atmosphere are expected to be both anisotropic and dispersive. The anisotropic characteristics of these waves mean that the properties of the waves are not uniform in all the directions. The propagating waves can be characterized by the amplitude and phase, which depends on time and space. When the wave frequency (ω) is depends on the wavelength ($2\pi/k$) then the wave is dispersive. For such waves the group velocity is different from the phase velocity. A better understanding of the vertical coupling by these wave activities will provide a deeper insight into the processes that control the dynamics and energetics of the whole atmosphere.

The atmospheric waves can be classified on the basis of (i) extra tropical modes against equatorially trapped modes, (ii) free mode and forced modes, (iii) external mode and internal modes and (iv) modes that interact with the mean flow through wave dissipation. In middle atmosphere, internal force mode is important due to its generation in the mid-troposphere and propagates vertically into the middle atmosphere. These waves carry forward the information over thousands of kilometer horizontally and tens of kilometers vertically. The three major oscillations in existing in the middle atmosphere (~10-100 km) are gravity waves, atmospheric tides and planetary waves, distinguished mainly by wave periodicity (*Holton*, 1975; *Andrew et al.*,1987). Gravity waves are oscillation with a period between minutes to hours, whose restoring force is buoyancy; tidal waves are daily oscillations with larger horizontal structures and are excited through direct absorption of sunlight by water vapour in the troposphere and ozone in the stratosphere; planetary waves are disturbances having zonal wavelengths of global scale, often maintained by internal dynamics.

Among these waves and oscillations, the planetary waves play an important role in the dynamics of the mesosphere and lower thermosphere (MLT) region. The quasi 16 day wave is a planetary wave which has period of 16 ± 4 days with westward propagating phase and zonal wave number = 1. This is a global-scale phenomenon of tropospheric origin, and is frequently observed in the MLT region. These oscillations become quite important in the equatorial latitudes as they exhibit unique propagation characteristics (*Salby*, 1984). The characteristics of these planetary scale waves are significantly modified by the energetics and dynamical state prevailing in the MLT region through which they propagate.

The quasi-16 day wave was first reported by *Kingsley et al.* (1978) using meteor wind radar and by now, there are ample of observational evidences of quasi 16 day wave in winds at MLT region using space-borne, ground-based, and in-situ measurements (*Forbes et al.*, 1995; *Espy and Witt*, 1996; *Mitchell et al.*, 1999; *Espy et al.*, 1997; *Luo et al.*, 2002; *Manson et al.*, 2003; *Pant et al.*, 2004; *Jiang et al.*, 2005; *Lima et al.*, 2006; *Jacobi et al.*, 2007; *Vineeth et al.*, 2007a and 2011; *Das et al.*, 2010). There are two proposed mechanisms for the existence of the quasi-16 day wave in the MLT region: (1) a 16 day wave is generated in the winter hemisphere, ducted in the vicinity of equator and then propagates vertically toward the summer pole by following the westerly mean winds, and (2) a 16 day wave is generated by the oscillatory deposition of energy and momentum into the summer mesosphere due to gravity wave breaking, which propagates upward after modulation by a 16 day wave in the summer troposphere and lower stratosphere (*Lima et al.*, 2006).

Investigations of quasi 16 day wave are also been carried out on the seasonal characteristics (*Mitchell et al.*, 1999; *Pancheva et al.*, 2004; *Lima et al.*, 2006; *Das et al.*, 2010). *Jacobi et al.* (2007) have established the phase relation between wind and temperature of quasi 16 day wave in MLT region over mid-latitude. Simultaneous observation of these planetary waves at two different locations is also reported by *Luo et al.* (2002) and the authors have established a phase relation between them using medium frequency (MF) radar. *Manson et al.* (2003) have shown the modulation of gravity waves by planetary waves of 2 and 16 day periodicity using MF radar network over North American Pacific region. *Luo et al.* (2001) have suggested that the solar activity can also modulate the existing planetary waves or even trigger wave like perturbations through changing radiation fluxes in the MLT region. Earlier studies have shown that these planetary scale waves can have global scale structure in the mesopause region (e.g. *Miyoshi*, 1999). Over equatorial station Thumba (8.5°N, 76.5°E), *Reddi and Ramkumar* (1995) and *Das* et al. (2010) have shown the presence of quasi 16 day wave in mesospheric winds and temperature. *Pant et al.* (2004) have shown the evidence of direct solar control of these planetary waves in mesospheric temperature using multi-wavelength day glow photometer and MF radar. *Vineeth et al.* (2007a and 2011) have shown that quasi 16 day wave in mesopause temperature and equatorial electrojet. However, most of the previous studies have investigated the seasonal variations of 16 day planetary wave and their characteristics in winds over mid-latitude MLT region and very few studies are carried out over lower latitudes, especially using mesospheric temperature and wind simultaneously. Moreover, the propagation characteristics of these planetary waves above 80 km are not well understood.

The present study discusses on overall characteristics of the quasi-16 day wave in the MLT region over an Equatorial station Thumba (8.5°N, 76.5°E). Figure 1 shows the geographical location of Thumba. It is envisaged that the present results will contribute to our understanding of the dynamics of the MLT region over low latitudes.

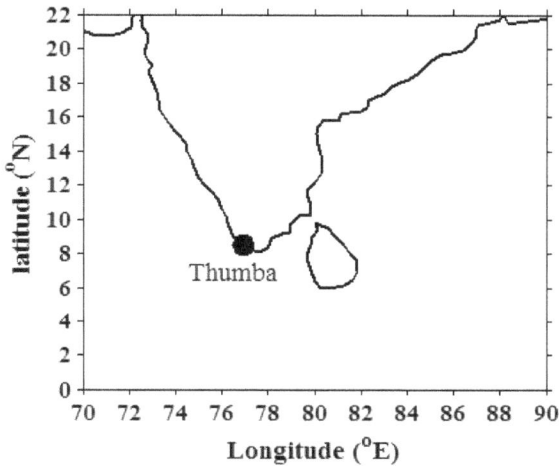

Fig. 1. Map showing the geographical location of the Thumba (8.5°N, 76.5°E), where SKiYMET meteor wind radar is installed.

2. Methodology

The All sky Interferometric meteor wind radar installed at Thumba, commercially named as SKiYMET, operates at 35.25 MHz with a peak power of 40 kW and duty cycle upto 15% to derive the three-dimensional winds and temperature at MLT region (82-98 km), which provides a unique opportunity to study the dynamics of the middle atmosphere. The system has a multi-channel coherent receiver utilizing a meteor detection algorithm to acquire, detect, analyze and display the entrance of meteors. The inbuilt software is used to perform the various calculations in real-time on the detected meteors (Hocking et al., 2001). The winds are derived by observing how the meteor trails drift with time, and the decay time of meteor trails are used to determine the absolute temperature in the height region of 86-90 km (Das et al., 2011). Details of wind derivation can be found in Hocking et al. (2001) and the temperature derivation can be found in Hocking (1999) and Das et al. (2011). The Thumba meteor radar has been operational since June 2004 and still in operational. The data collected from the meteor radar is for ~ 7 years. Figure 2 shows the panoramic view of transmitter/receiver antenna system and in-house transmitter/ receiver system of the radar at Thumba. The specification of Thumba SKiYMET meteor wind radar is given in Table 1.

In principle, hourly mean zonal and meridional winds data can be obtained but for the present study the data used are daily mean at 6-height levels, viz., 82, 85, 88, 91, 94, and 98 km. The wind data derived from SKiYMET radar were well comparable with the MF radar at Tirunelveli (8.7°N, 77.8°E) (Kumar et al. 2007a). Daily mean temperature in the height

Parameters	Specification
Frequency	35.25 MHz
Peak power	40 kw (solid state)
Duty cycle	Upto 15 %
Pulse width	13.3 μs
Pulse repetition frequency	2144 Hz
Band width	~ 1.5 MHz
Dynamics range	62-122 dB
Transmitter	4-circular polarized 3-elements Yagi (Cross-elements)
Receiver	5-circular polarized 2-elements Yagi (Cross-elements)

Table 1. Experimental mode of Thumba SKiYMET meteor wind radar.

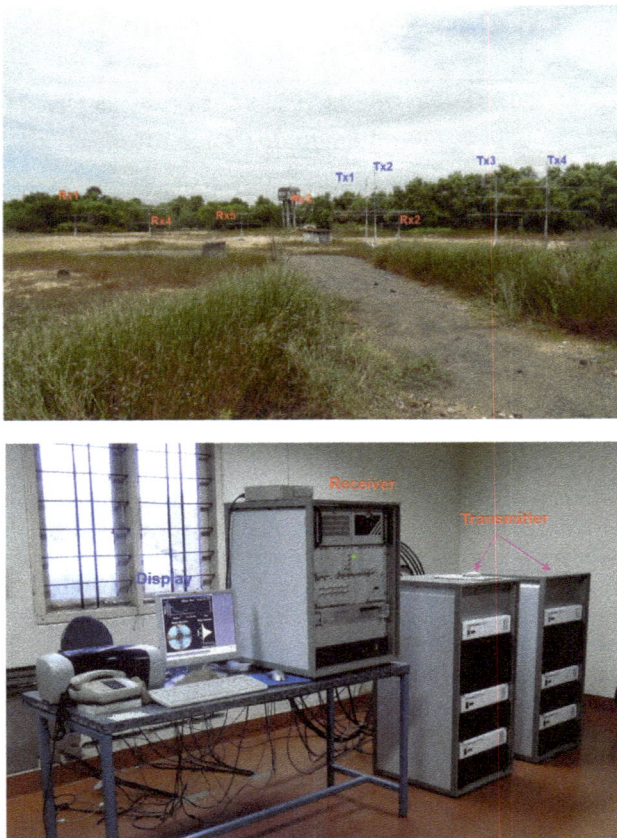

Fig. 2. Panoramic view of transmitter/receiver antenna system (top panel) and in-house transmitter/receiver system of SKiYMET meteor wind radar at Thumba.

region of 86-90 km, derived from Thumba SKiYMET (*Hocking* 1999; *Das et al.*, 2011) radar is fairly comparable with the space borne measurement by Sounding the Atmosphere using Broadband Emission Radiometry (SABER) instrument on board the TIMED satellite and day-glow multi-wavelength photometer in the height range of 86-90 km (*Vineeth et al.*, 2007b; *Kumar* 2007b, *Das et al.*, 2011 and reference therein).In the present study, data gaps in both winds and temperature were interpolated with a spline interpolation method. The observed data gaps are very few and they are not affecting the time-series.

3. Results and discussion

- *Seasonal property*

Mesosphere and Lower Thermosphere (MLT) winds were measured over the equatorial station Thumba using SKiYMET meteor wind radar during the period of June 2004 to September 2011, while SKiYMET meteor radar is still in operation. Figure 3 shows height-time intensity plot of monthly mean zonal and meridional winds at MLT region from June 2004 to September 2011. Most of the time eastward propagating winds were observed in the month of November to January and May to July. However, a weak eastward propagating wind is observed during November 2007 to January 2008. Similarly, northward propagating winds were observed during November to January. It is interesting to note that the eastward propagating wind shows semi-annual cycle (SAC), whereas the northward propagating wind shows only annual cycle (AC). Figure 4 shows the time series plot of mesospheric temperature at 88±2 km derived from SKiYMET meteor wind radar. The variability of mesospheric temperature is between 185 and 225 K. Both AC and SAC are observed in the time series of temperature (please referee Fig.8 of *Das et al.*, (2011) for amplitude spectrum of temperature).

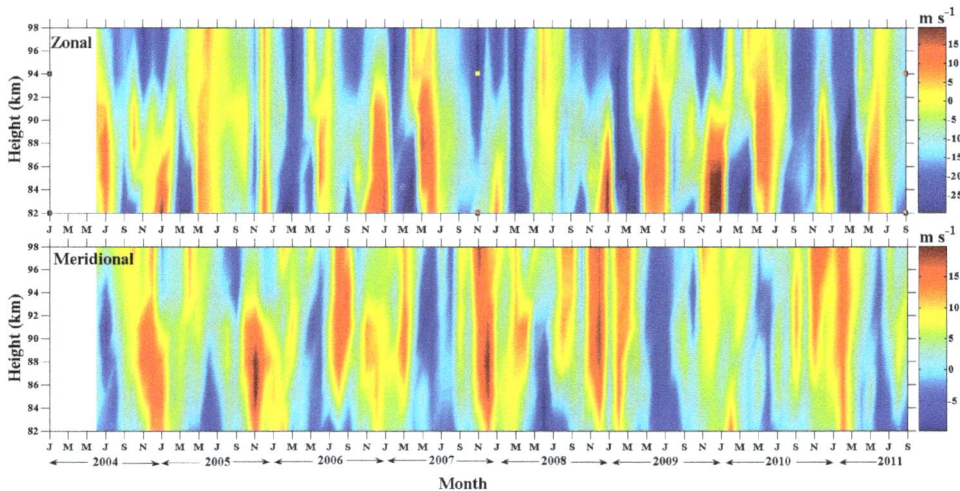

Fig. 3. Height-time intensity plot of monthly mean zonal (top panel) and meridional (bottom panel) winds at Mesosphere and Lower Thermosphere (MLT) region from Jun.-2004 to Sep.-2011.

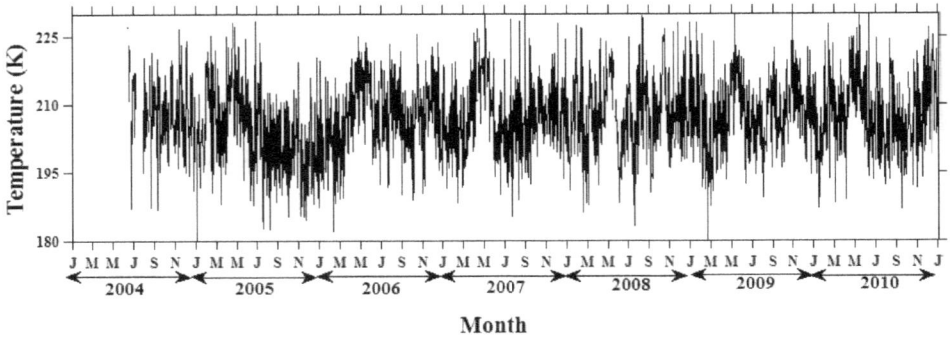

Fig. 4. Daily mean mesospheric temperature between 88±2 km from Jun.2004 to Dec. 2010.

The time-series of both the daily mean winds and temperature, discussed above, were further analyzed to examine for the existence of planetary wave with a period of quasi 16 day. In this context, the perturbation of winds and temperature were obtained by removing annual mean from the time series of each year. The perturbation time series of winds and temperature is then subjected to Fast Fourier Transform (FFT) to explore the prominent periodicities. Figure 5 and 6 show the amplitude spectra of zonal and meridional winds at 88 km for the year 2005-2010. It is interesting to note that all the 6 years have dominant period of quasi 16 day (12-20 day) in both the wind components. However, the amplitude of this oscillation varies from year

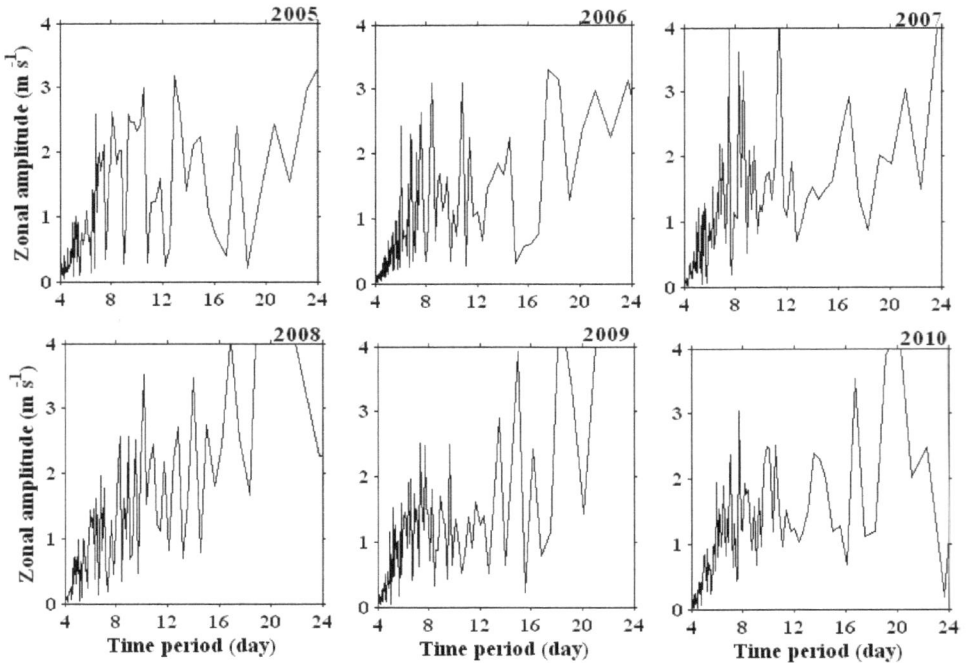

Fig. 5. Amplitude spectra of daily mean zonal wind at 88 km for the year 2005-2010.

to year with small variation in the dominant periodicities. The amplitude of both the zonal and meridional winds is found to be 3-4 m s $^{-1}$. Maximum amplitude of zonal wind is found to be 4 m s $^{-1}$ for 2007 and 2008, whereas for meridional wind it is found be in 4 m s $^{-1}$ for 2007 only. Figure 7 shows the amplitude spectra of mesospheric temperature at 88±2 km for the year 2005-2010.The amplitude of temperature is found to be ~1-1.5 K. Maximum amplitude of temperature is found to be ~2 K for the year 2007. Thus, the above observations clearly reveal the presence of quasi 16 day wave in the MLT region over the equatorial station Thumba. In the year 2007, both the winds and temperature shows maximum amplitude of quasi 16 day, indicating the presence of strong planetary wave activity.

In order to study the seasonal characteristics of quasi 16 day wave, we have extracted the amplitude of zonal and meridional wind averaged between 12 and 20 day at different heights using wavelet analysis. Figure 8 shows the height-time amplitude plot of quasi 16 day wave in zonal and meridional winds from 2004-2010. These plots do not show any seasonal dependence of the planetary wave of the order quasi 16 day in both the wind components. However, maximum amplitudes were observed in the month of January to February. It is also noted that the zonal component is larger in amplitude than the meridional component. Earlier observations have shown that these planetary scale wave exhibits year-to-year variability but do not show any seasonal behavior (e.g. *Namboothri et al.* 2002; *Lima et al.*, 2006). However, these authors have shown that stronger wave activities occur during January-February and this is attributed to vertical propagation from lower atmosphere. The present analysis also shows the enhanced wave activity during January-February.

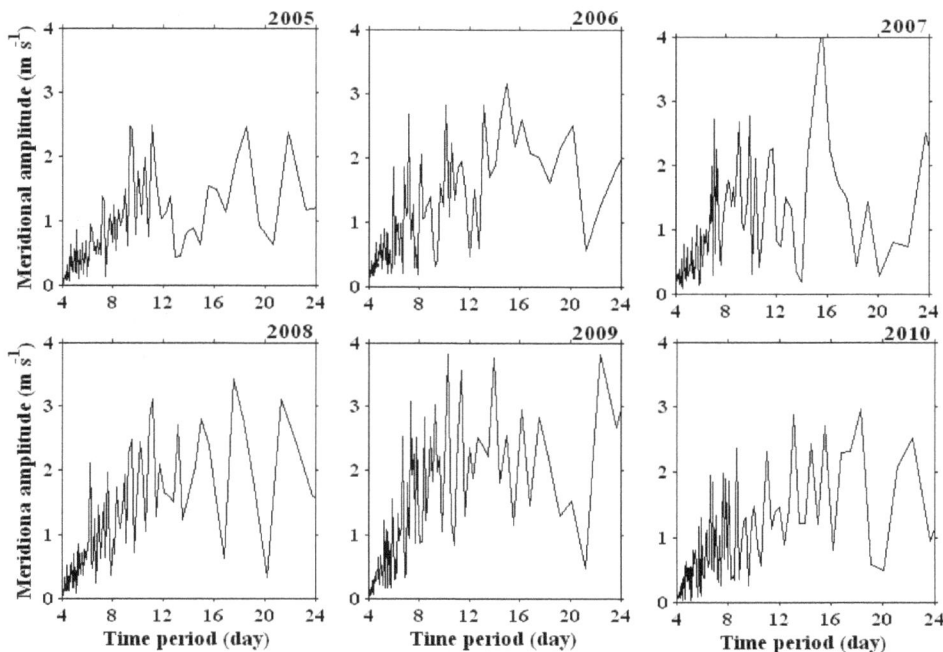

Fig. 6. Same as Fig.3, but for meridional wind at 88 km.

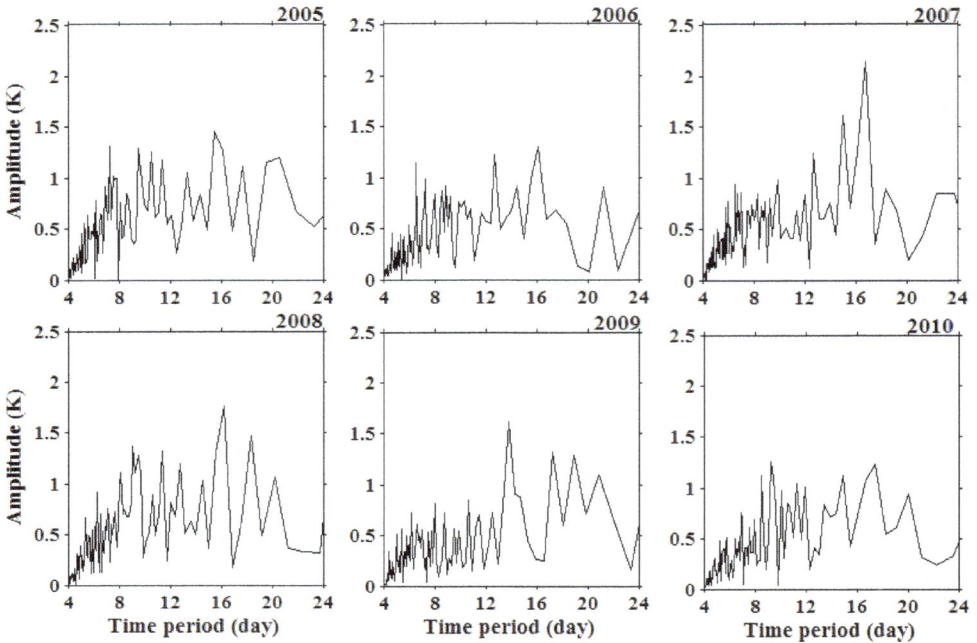

Fig. 7. Same as Fig.3 but for mesospheric temperature at 88±2 km for the year 2005-2010.

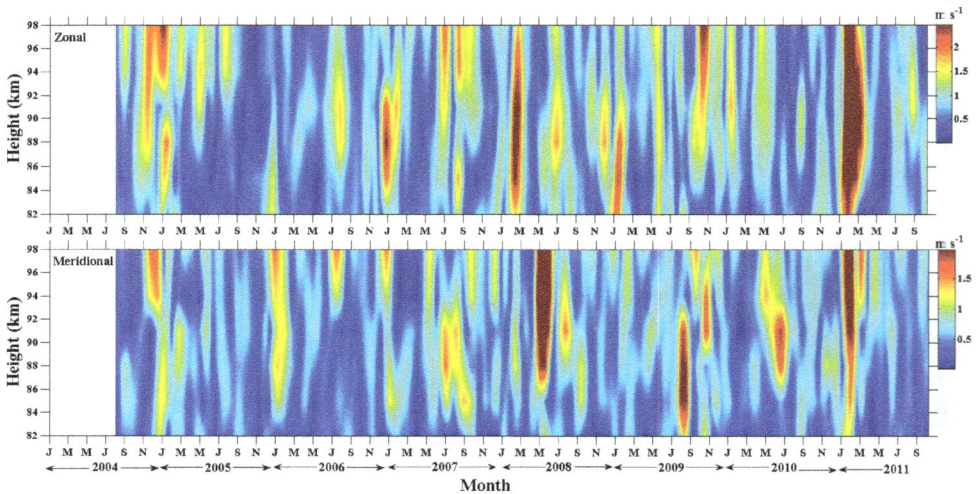

Fig. 8. Amplitude of quasi 16 day oscillation in zonal (top panel) and meridional (bottom panel) winds from 2004-2010.

In addition to winds, amplitude is also been extracted for temperature measurements. In the present analysis, we have used a band-pass Butterworth filter of 12-20 day to extract the temperature amplitude. Figure 9 shows the filtered (12-20 day) time series plot of temperature at 88±2 km over Thumba. This plot clearly shows the seasonal dependence of quasi 16 day wave amplitude. It is remarkable to note that the quasi 16 day wave in temperature is modulated by SAO. The maximum peak of SAO in 16 day wave activity is observed between January-March and August-September with amplitudes around ±5 K. The presence of summer wave activity is due to the inter-hemispheric ducting of quasi 16 day wave or by the in-situ generation through gravity wave, which is dissipated in summer hemisphere. Since quasi 16 day wave is westward propagating (e.g. *Das et al.*, 2010 and reference therein), eastward phase of zonal wind is favorable for its propagation in the MLT region. It is also clear that the intensity of quasi 16 day amplitude in temperature is stronger when the background wind exhibit eastward phase. As discussed above that the eastward phase of background wind is favourable for westward propagating waves (in the present case 16 day oscillation). However, for some years, phase delay of few months is also observed. The phase delay is generally found to be 1-2 months between temperature and background winds. In earlier studies, it is found that these planetary scale waves are sensitive to background winds and also to quasi-biennial oscillation (QBO) (e.g. *Luo et al.*, 2000 and *Espy et al.*, 1997). This can further effectively moves the zero wind line closer to the summer hemisphere during the eastward phase of QBO. Thus, the widening or narrowing equatorial duct due to QBO can modulate the normal mode structure and possible forcing source linking to the phase delay of 1-2 months between temperature and background wind in our present observations.

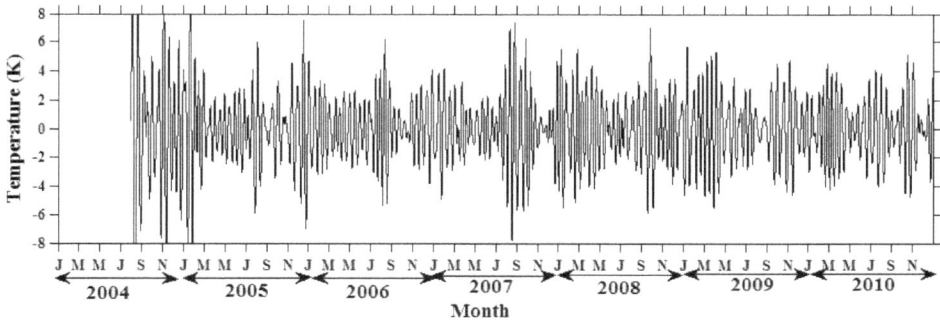

Fig. 9. Filtered (12-20 day) mesospheric temperature from 2004-2010.

Vineeth et al. (2007a) have shown the presence of quasi 16 day wave in mesospheric temperature measured from day glow multi-wavelength photometer and the electrojet-induced surface magnetic field over the same equatorial region Thumba. The day glow multi-wavelength photometer measurements are weighted average in 80-98 km height region. The important findings of these authors were (i) amplification of the quasi 16 day wave in the equatorial mesopause temperature and the equatorial electrojet (EEJ) induced magnetic field, and (ii) occurrence of counter electrojet (CEJ) with periodicity of quasi 16 day, which further decreases its strength with decrease of amplitude of these wave oscillation. It is explained that the occurrence of consecutive CEJ events of quasi 16 day is due to the interaction of intensified planetary wave with the prevailing tidal winds. *Vineeth et al.* (2011) have studied the role of quasi 16 day wave in controlling day-to-day variability of EEJ in association with the planetary wave-tidal interaction in the MLT region and found it is more during the winter month. The present observation also shows that the enhancement of quasi 16 day wave in both winds and temperature during winter months.

- *Interannual property*

Espy et al. (1997) have shown the inter-annual variation of quasi 16 day wave oscillation in the polar summer mesospheric temperature and found the amplitude as high as ~5 K. The quasi 16 day wave observed over the equatorial station Thumba is in good agreement with the observations at other latitudes. The modulation of 16 day wave by SAO was first reported by *Das et al.* (2010) over this latitude (Thumba). The authors have also extracted the phase and amplitude for both the wind components during the intense burst of quasi 16 day. The maximum amplitude is found at 88-92 km and the vertical wavelength is found to be 30-50 km. However, most of the phase profiles shows constant phase with height which also indicates the presence of longer vertical wavelength. *Das et al.* (2010) have also established for the first time the phase relationship between winds and temperature of quasi 16 day wave during the intense events over the equatorial station Thumba. The intense events mean the period when the amplitude of quasi 16 day is maximum. The intese event corresponding to quasi 16 day is between January-February. It is found that the zonal and meridional winds are in phase, whereas temperature leads zonal wind by 5 ±1 days. The present observation over this latitude brought out the seasonal characteristics and the effect of background wind on quasi 16 day wave in the MLT region. However, the exact mechanism for the enhanced quasi 16 day wave activity over Thumba is yet to be explored, which require the observational data from the other locations simultaneously and will be our focus in the near future.

4. Summary

Simultaneous observational evidence of planetary scale wave with a periodicity of quasi 16 day wave in winds and temperature at Mesosphere and Lower Thermosphere (MLT) region are reported in the equatorial station Thumba (8.5°N, 76.5°E) using SKiYMET meteor wind radar. The mesospheric temperature shows a seasonal characteristic and follows semi-annual oscillation (SAO) with peak during January-March and August-September. Unlike mesospheric temperature, both the wind components, i.e. zonal and meridional winds do not exhibit any clear seasonal characteristics. During the eastward phase of background wind in MLT region, the intensity of quasi 16 day wave in mesospheric temperature is stronger. It is also noticed that a phase delay of 1-2 months is observed between the maximum events of such planetary wave activities and peak eastward phase of background wind. The maximum amplitude is found to be 88-92 km and most of the time with constant phase with height. The vertical wavelength is estimated to be 30-50 km. It is also found zonal and meridional winds are in phase, whereas the temperature leads zonal wind by 5±1 days.

As discussed in the introductory section, 16 day wave generated by heating due to moist convection in the troposphere and thus many of the planetary waves appearing in the MLT region are not in-situ excited but comes from the lower atmospheric source. These planetary scale waves propagate upwards from lower stratosphere to mesosphere under certain condition and while propagating, these waves interact and couple between various layers of the atmosphere through their generation, propagation and dissipation. Further, these waves affect the global redistribution of momentum and energy and finally change the thermodynamics and even electrodynamics characteristics in these regions. Studies have also showed that the 16 day wave significantly contributes to the polar summer mesopause temperature.

5. Acknowledgement

This study was performed at Space Physics Laboratory (SPL), Vikram Sarabhai Space Centre (VSSC), Trivandrum, India. Dr.K.Kishore Kumar is highly acknowledged for his many useful discussions during this study. Thanks are also due to Prof. R. Sridharan, Dr. K. Krishnamoorthy, Dr. Geetha Ramkumar and Prof. W. K. Hocking for establishing the facility of SKiYMET meteor wind radar at Thumba (Trivandrum). All the scientific and technical staffs of SPL are highly acknowledged for their support in operating and maintaining the radar.

6. References

Andrews, D.G., Holton, J.R., Leovy, C.B., (1987), Middle Atmosphere Dynamics, *Academic Press*, New York.

Das, S. S., K. K. Kumar, S. B. Veena, and G. Ramkumar (2011), Simultaneous observation of quasi 16 day wave in the mesospheric winds and temperature over the low latitudes with the SKiYMET radar, *Radio Sci.*, 45, RS6014, doi : 1029/2009RS004300.

Das, S. S., K. K. Kumar, S. K. Das, C.Vineeth, T. K. Pant and G. Ramkumar (2011), Variability of mesopause temperature derived from two independent meteor radar methods and its comparison with SABER and EOS-MLS and a co-located multi-wavelength day-glow photometer, over the equatorial station Thumba (8.5°N, 76.5°E), *Int. J. Remo. Sen.*, (in-press)

Espy, P. J., and G. Witt (1996), Observation of a quasi 16-day oscillation in the polar summer mesospheric temperature, *Geophys. Res. Lett.*, 23, 1071-1074.

Espy, P. J., J. Stegman, and G. Witt (1997), Interannual variations of the quasi-16-day oscillation in the polar summer mesospheric temperature, *J. Geophys. Res.*, 102, 1983-1990.

Forbes, J. M., M. E. Hagan, S. Miyahara, F. Vial, A. H. Mason, C. E. Meek, and Y. I. Portnyagin (1995), Quasi 16-day oscillation in the mesosphere and lower thermosphere, *J. Geophys. Res.*, 100, 9149–9163.Hocking, W.K., T. Thayaparan, and J. Jones (1997), Meteor decay times and their use in determining a diagnostic mesospheric temperature-pressure parameter: methodology and one year of data, *Geophy. Res. Lett.*, 24 (23), 2977-2980.

Hocking, W.K. "Temperatures using radar-meteor decay times". Geophys. Res. Lett. 26: 3297-3300,1999.

Hocking, W. K., B. Fuller, and B. Vandapeer (2001), Real-time determination of meteor-related parameters utilizing modern digital technology, *J. Atmos. Sol. Terr. Phys.*, 63, 155–169.

Holton, J. R. (1975), The dynamic meteorology of the stratosphere and mesosphere, *Meteorol. Mongr.Ser.*, 37, 216, Amer. Meteorol.Soc., Boston, Mas.

Jacobi, Ch., K. FrÖhlich, C. Viehweg, G. Stober, and D.Kürschner (2007), Midlatitude mesosphere/lower thermosphere meridional winds and temperatures measured with meteor radar, *Adv. Space Res.*, 39, 1278-1283.

Jiang, G., J.G. Xiong, W. X. Wan, B.Q. Ning, L. B. Liu, R. A. Vincent, and I. Reid (2005), The 16-day waves in the mesosphere and lower thermosphere over Wuhan (30.6oN, 114.5oE), *Adv. Space Res.*, 35, 2005-2010.

Kumar, K.K., G.Ramkumar, and S.T.Shelbi (2007a), Initial results from SKiYMET meteor radar at Thumba (8.5oN, 77oE): 1. Comparison of wind measurements with MF spaced antenna radar system, *Radio Sci.*, 42, RS6008, doi: 10.1029/2007GL030704.

Kumar, K. K. (2007b), Temperature profiles in the MLT region using radar-meteor trail decay times: Comparison with TIMED/SABER observations, *Geophys. Res. Lett.*, 34, L16811, doi:10.1029/2007GL030704.

Kingsley, S. P., H. G. Muller, L Nelson, A. Scholefield (1978), Meteor winds over Sheffield (53oN, 2oW), *J. Atmos. Terr. Phys.*, 40, 917-922.

Lima, L. M., P. P. Batista, B. R. Clemesha, H. Takahashi (2006), 16-day wave observed in the meteor winds at low latitudes in the Sothern hemisphere, *Adv. Space Res.*

Luo, Y., A. H. Manson, C. E. Meek, C. K. Meyer, and J. M.Forbes (2000), The quasi 16-day oscillations in the mesosphere and lower thermosphere at Saskatoon (52°N, 107°W),1980–1996, J. Geophys. Res., 105, 2125–2138, doi:10.1029/1999JD900979.

Luo, Y., A. H. Manson, C.E.Meek, T.Thayaparan, J. MacDougall, and W. K. Hocking (2002), The 16-day wave in the mesosphere and lower thermosphere: simultaneous observations at Saskatoon (52oN, 107oW), Canada, *J. Atmos. Sol. Terr. Phys.*, 64, 1287-1307.

Manson, A. H., C.E.Meek, Y. Luo, W. K. Hocking, J. MacDougall, D. Riggin, D.C. Fritts, R.A. Vincent (2003), Modulation of gravity waves by planetary waves (2 and 16 d): observations with the North American-Pacific MLT-MFR radar network, *J. Atmos. Sol. Terr. Phys.*, 65, 85-103.

Mitchell, N. J., II. R. Middleton, A. G. Beard, P.J.S. Williams, II. G. Muller (1999), The 16-day planetary wave in the mesosphere and lower thermosphere, *Ann. Geophys.*, 17, 1447-1456.

Miyoshi, Y. (1999), Numerical simulation of the 5-day and 16-day waves in the mesopause region, *Earth Plan. Spac.* 51, 763-772.

Namboothiri, S. P., P. Kishore, and K. Igarashi (2002), Climatological studies of quasi 16-day oscillation in the mesosphere and lower thermosphere at Yamagawa (31.2oN, 130.6oE), Japan, *Ann. Geophys.*, 20 (8),1239-1246.

Pancheva, D., N. J. Mitchell, and P. T. Younger (2004), Meteor radar observations of atmospheric waves in the equatorial mesosphere/lower thermosphere over Ascension Island, *Ann. Geophys.*, 22, 387-404.

Pant, T. K., D. Tiwari, S. Sridharan, R. Sridharan, S. Gurubaran, K. S. V. Subbarao, and R.Sekar (2004), Evidence for direct solar control of the mesopause dynamics through dayglow and radar measurements, *Ann. Geophys.*, 22, 3299-3303.

Reddi, C. R., and G. Ramkumar (1995), Long period wind oscillations in the meteor region over Trivandrum (8oN, 77oE), *J. Atmos. Terr. Phys.*, 57, 1-12.

Salby, M. L. (1984), Survey of planetary-scale traveling waves: the state of theory and observations, *Rev. Geophys. Space Phys.*, 22, 209-236.

Vineeth, C., T. K. Pant, C. V. Devasia, and R. Sridharan (2007a), Atmosphere-Ionosphere coupling observed over the dip equatorial MLTI region through the quasi 16-day wave, *Geophys. Res. Lett.*, 34, L12102, doi: 10.1029/2007GL030010.

Vineeth,C., T. K. Pant, C. V. Devasia, and R. Sridharan (2007b), Highly localized cooling in daytime mesopause temperature over the dip equator during counter electrojet events: First results, Geophys. Res. Lett., 34, L14101, doi:10.1029/2007GL030298.

Vineeth, C., T. K. Pant, S. G. Sumod, K. K. Kumar, S. Gurubaran, and R. Sridharan (2011), Planetary wave-tidal interactions over the equatorial mesosphere-lower thermosphere region and their possible implications for the equatorial electrojet, *J. Geophys. Res.*, 116, A01314, doi : 10.1029/2010JA015895.

4

Thunderstorm and Lightning Climatology of Australia

Yuriy Kuleshov

National Climate Centre, Bureau of Meteorology, Melbourne
School of Mathematical and Geospatial Sciences, Royal Melbourne
Institute of Technology (RMIT) University, Melbourne
Australia

1. Introduction

Thunderstorms are spectacular but hazardous weather phenomena and the associated lightning and wind gusts can be very hazardous to people, buildings, and industry and utility assets. The thunderstorm hazards in Australia are in most respects similar to those in other countries, but some are worthy of special comment. In some instances, the effects of lightning-initiated bushfires are so extreme that they are classified as natural disasters. The wildfires started by lightning are known locally in Australia as bushfires and grass fires, and these can cause extensive damage and loss of life. The ignition is caused by the lightning current in ground flashes, and firing is associated with low moisture contents of (potential) fine fuels such as duff in trees and dense grass, and the occurrence of multiple stroke currents and continuing current. During dry periods, lightning initiated grass fires are a major problem in inland areas, as are lightning initiated bushfires in remote forest areas. Once ignited, grass fires spread quickly, whereas in bushfires, ignition often starts as a small localised fire in or near the crown of the tree, and it may take a few hours for the fire to spread to other trees and to become an uncontrollable wildfire. Given the potential hazards associated with lightning, knowledge about spatial and temporal distributions of thunderstorm and lightning activity is of great importance for developing comprehensive protective measures.

Traditionally, thunderstorm activity is recorded at meteorological sites as a number of thunder-days per year. Thunder-day records are commonly used as proxy information for describing lightning activity. To obtain instrumental records of lightning incidence, many lightning flash counters (LFC) and lightning location systems (LLS) have been installed at meteorological and other sites worldwide. Recently, remotely-sensed lightning data gathered by the National Aeronautics and Space Administration (NASA) satellite-based instruments became available.

In this chapter, a review of thunderstorm and lightning climatology of Australia, with emphasis on spatial distribution and frequency of thunder-days and lightning flash density over the Australian continent, is presented. Geographical distribution of thunderstorm and lightning activity for the Australian continent has been analysed in a number of studies based on long-term records from LFCs and thunder-day records, as well as recently acquired NASA satellite-based lightning data.

2. Spatial distribution and frequency of thunderstorms and lightning in Australia

Thunderstorm occurrence at a particular location can be expressed in terms of thunder-days, T_d, defined as 'an observational day (any 24-hour period selected as the basis for climatological or hydrological observations) during which thunder is heard at the station. Precipitation need not occur' (Huschke, 1959). The 24-hour period selected at the Australian Bureau of Meteorology (ABM) for the phenomenon registration is midnight to midnight (local standard time). The requirement that thunder should actually be heard limits the area covered by each observing point, under favourable listening conditions, to a circular area with a radius of some 20 km (WMO, 1953).

Using the Australian thunder-day records, the thunderstorm distribution across the country was analyzed in detail and an updated average annual thunder-day map of Australia (Fig. 1) was prepared at the National Climate Centre, the ABM (Kuleshov et al., 2002). Clearly thunderstorms are most frequent over the northern half of the country, and generally decrease southward, with lowest frequencies in southeast Tasmania. A secondary maximum is also apparent in southeast Queensland and over central and eastern New South Wales, extending into the north-eastern Victorian high country.

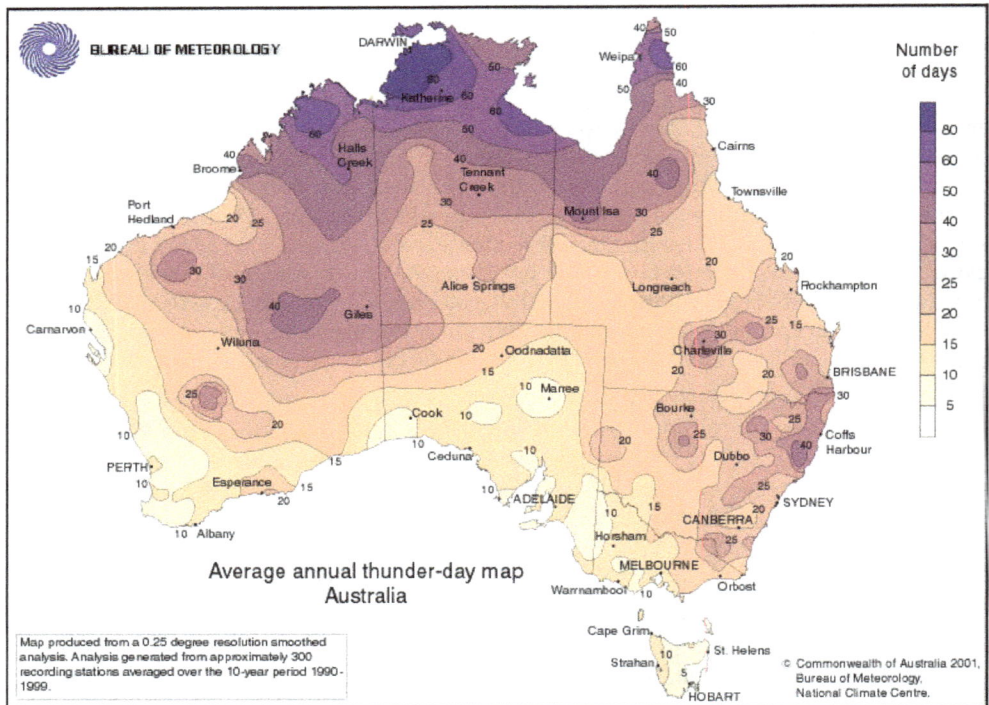

Fig. 1. Map of average annual thunder-days for Australia (1990-1999).

To understand this distribution, it is important to note that thunderstorm development generally requires three factors:

1. An unstable convective atmosphere, with generally high surface temperatures and a strong vertical lapse rate (i.e., temperature falls rapidly with height), which provides a favourable environment for the strong vertical atmospheric motions that produce thunderstorms.
2. A trigger for this vertical motion, such as low level convergence of airstreams, a frontal system, local differences in heating or orography.
3. High atmospheric moisture levels.

High moisture levels, especially near the coast, affect the tropical north of Australia in the summer half of the year. Lower pressure lies across northern and central Australia, and gives rise to the vertical motion and low level convergence that favour thunderstorm development. The northern half of Australia is thus very favourable for thunderstorms in the warmer months October through March, and especially in the far north, thunderstorms are frequent and often associated with heavy rainfall and intense lightning. In the remainder of the year dry, stable outflow from the subtropical high pressure belt, which normally lies over the continent, inhibits convective showers and storms.

Thunderstorm frequencies generally decrease in the southern parts of the tropics and the adjacent desert areas of central Australia. This is because the air, though often very hot, is generally drier. The exception appears to be over inland western Australia where a wide area experiences over 30 thunder-days a year. However, many of these would be so-called "dry" thunderstorms, with little or no rain, because low-level relative humidity tends to be low and acts to evaporate any falling precipitation.

3. Estimating lightning flash density from thunder-day records

The lightning flash density, defined as the number of flashes of a specific type occurring on or over unit area in unit time, is commonly used to describe lightning activity. The lightning flash densities are denoted as N_g, N_c and N_t for cloud-to-ground (CG, or simply ground), intra-cloud (IC, or simply cloud) and total (TF = CG + IC) flashes, respectively, and expressed as a number of flashes per square kilometre per year ($km^{-2}\ yr^{-1}$).

In the absence of information about lightning activity obtained directly from lightning detectors, the lightning ground flash density, N_g, may be estimated from thunder-days, T_d, using an equation of the form $N_g = aT_d^b$, in which a and b are empirically derived constants that depend on the meteorological conditions at a given location. The earliest estimates of the equation for Australia were by Mackerras (1978), who derived values of a = 0.01 and b = 1.4. This study was based on results from 26 sites for the period 1965 to 1977 and the ABM thunder-day map based on data from 1954 to 1963. Anderson and Eriksson (1980), based on 120 observations over two years in South Africa, derived values of a = 0.023 and b = 1.3. The corresponding equation has since become known as 'Eriksson's Formula'. A subsequent study using 62 stations over a longer period of five years from 1976 to 1980 yielded the values a = 0.04 and b = 1.25 (Anderson et al., 1984). The equation using these two values of the constants is generally known as the 'CIGRE Formula'. Both the CIGRE and Eriksson's equations are used in the literature, although they were both derived at the same location. But there are very few systematic studies enabling one to compare their results with those from other parts of the world. Recently, using long-term (up to 22 years) LFC registrations and thunder-day observations for 17 Australian localities, Kuleshov and Jayaratne (2004) updated Mackerras' empirical formula and compared the results with Eriksson's and CIGRE formulas. It was concluded that the empirical formula, $N_g = 0.012\ T_d^{1.4}$, gives the best estimate of N_g in Australia. However, lightning flash densities derived from such empirical relationships to thunder-day records are less accurate than those derived from direct measurements of N_g.

4. Instruments of lightning registration

In this section, an overview of instrumental methods used for recording lightning activity in Australia is presented. The first instrumental records of lightning activity using LFCs were made in the 1950s (Pierce, 1956); later, several types of LFCs known as CIGRE-500 Hz, CIGRE-10kHz (Barham and Mackerras, 1972; Anderson et al., 1979) and CGR3 (Mackerras, 1985) were developed and have been used widely around the world. LLSs are also used in most developed countries, beginning in America in the 1980s and then worldwide, including two commercial systems in Australia, GPATS (GPATS website, www.gpats.com.au) and Kattron (Kattron website, www.lightning.net.au). The GPATS network provides coverage of the Australian continent while spatial coverage of the Kattron network is limited to selected areas in South-East Australia.

The role of a LFC differs from that of a LLS. The LFC records the occurrence of local lightning flashes with effective ranges between about 10 and 30 km, whereas the LLS produces information about lightning strokes occurring within the bounds of its network of sensors. In such a network, each sensor is optimally separated by no more than 300 to 400 km from neighbouring sensors. The LFC registrations are more accurate than LLSs data for ranges below 30 km (Kuleshov et al., 2010).

The CIGRE-500 counter is the instrument used by the ABM for its network of LFCs (the LFC site locations are shown in Fig. 2). Another type of counter, primarily developed for research purposes, is the CGR3 (Cloud-to-Ground Ratio #3) instrument (Mackerras, 1985). CGR3 counters were installed in 11 countries and were used for the derivation of total, intra-cloud and ground flash density estimates around the world (Mackerras and Darveniza, 1994; Mackerras et al., 1998). In Australia, these counters have been installed in Brisbane (27.5° S, 153.0° E) and Darwin (12.3° S, 131.0° E), and were used for estimating ground, cloud, and total flash densities at those localities. A later generation CGR4 counter (Mackerras et al., 2009) use the same (or similar) type of aerial for detecting electric field changes caused by lightning as the CGR3 LFCs; the main difference between the two instruments is that the CGR4 LFC uses a microprocessor to implement the flash discrimination procedures, the counting of flashes and the display of registrations.

The GPATS LLS operates on the principle of accurately measuring the time-of-arrival of the electric field component of lightning impulses at each sensor, by using GPS-synchronised local clocks to timestamp observations to sub-100 ns accuracy (GPATS website). These measurements are transferred to a central processing system at GPATS, where they are correlated and combined to produce estimates for the location (latitude and longitude), time, peak current (kA), type (cloud-ground or cloud-cloud) and polarity of each stroke. The sensors have the ability to discriminate individual strokes in a CG flash comprising multiple strokes.

Australian Government
Bureau of Meteorology

Darwin Gove

Centre Island

Kununurra

Mareeba South Jonstone

Tennant Creek Townsville

Mt. Isa

Port Hedland .Hughenden

Emerald .Blackwater

Callide Dam

Meekatharra

Brisbane

Geraldton Lismore. Mt.Gravatt

Three Springs Kalgoorlie Grafton

Moora Ceduna Port Augusta Cobar Bowraville. Coffs Harbour

Perth. Taree

Waikerie Nowra

Albany Mulwala

Ballarat .Cooma North

Melbourne Whitlands

Mt. Burnett

Analysis generated from NASA Optical Transient Detector and Lightning Imaging Sensor data (0.5 degrees grid resolution) averaged over the 8-year period 1995-2002. The satellite data were calibrated against the ground-based Lightning Flash Counter data and adjusted accordingly.

© Commonwealth of Australia 2005,
Bureau of Meteorology,
National Climate Centre

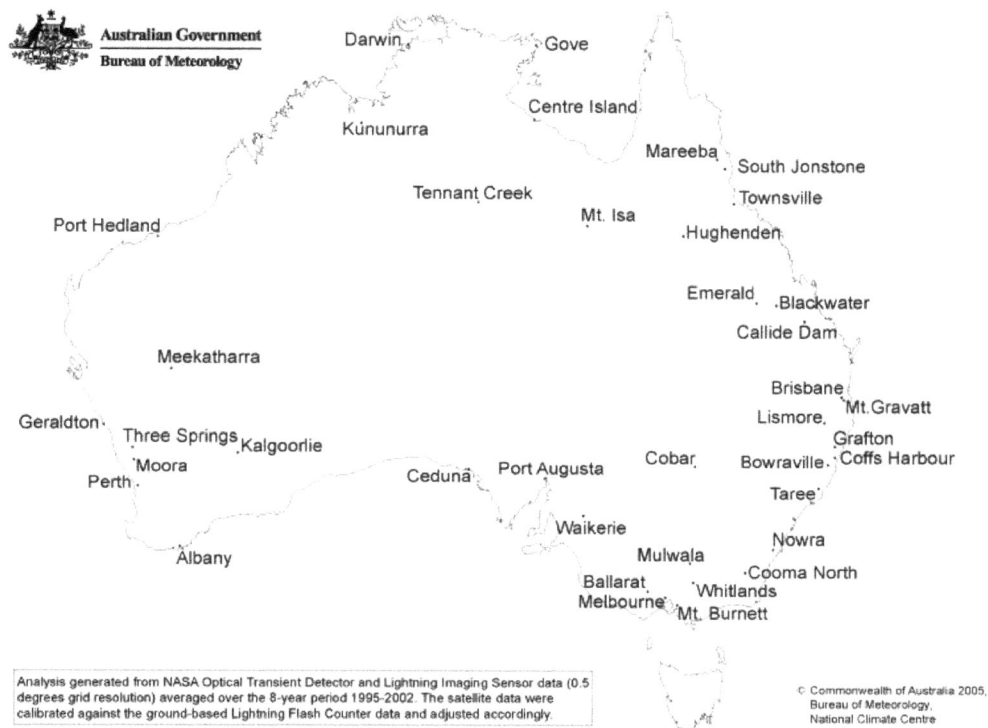

Fig. 2. Geographical distribution of the CIGRE-500 counters over the Australian continent.

LFCs have proved to be useful and reliable sources of lightning data. However, their sparse spatial distribution makes it difficult to prepare a map of lightning distribution without supplementary information. Data from two NASA satellite-based instruments, the MicroLab-1 Optical Transient Detector (OTD) (Christian et al., 1996) and the Tropical Rainfall Measuring Mission (TRMM) Lightning Imaging Sensor (LIS) (Christian et al., 1999), became recently available for estimating global lightning activity.

Satellite observations provide better spatial coverage of lightning which is, by its nature, widely distributed geographically. The remotely-sensed data are a valuable source of information for the areas with little or no local observation data (i.e. LFC or thunder-day observations). Using five years of the satellite data, a world map of average annual distribution of total lightning activity was constructed (Christian et al., 2003) and it was demonstrated that the spatial distribution of flash density is in broad general agreement with the world thunder-day map (WMO, 1953).

4.1 Comparison of lightning detection results obtained by lightning flash counters and lightning location network

Accurate long-term data are vital for developing reliable lightning climatology. To compare lightning detection results obtained over Australia by various instruments, in particular by the CGR3 and CGR4 LFC and GPATS LLS, a case study for Brisbane was examined by Kuleshov et al. (2011). In Australia, long term records obtained by the CGR3 LFCs are available for Brisbane and Darwin, and short term records from the CGR4 LFCs are available for Brisbane. Based on registrations obtained by the CGR4 LFC, a data set for Brisbane (2005-08) has been prepared. This data set consisted of daily total registrations of negative ground flashes (NGF), positive ground flashes (PGF), cloud flashes (CF) and total flashes (TF = GF + CF). Flash density for specific types of flashes was derived from the counter registrations: the lightning flash densities are denoted as N_{ng}, N_{pg}, N_g, N_c and N_t for NGF, PGF, GF, CF and TF, respectively, and expressed as a number of flashes per square kilometre per year (km^{-2} yr^{-1}).

Records of lightning activity measured in Brisbane by a CGR4 LFC in 2005-08 are presented in Tables 1 and 2. For these four years, significant variation in lightning activity was recorded, with flash densities of TF, N_t, varying from around 2 to 6 km^{-2} yr^{-1}, and percentage of GF varying from about 43 to 77% (Table 1). For the specific types of lightning flashes, flash densities for NGF, N_{ng}, varied from around 1.3 to 3.7 km^{-2} yr^{-1} and for PGF, N_{pg}, from 0.06 to 0.12 km^{-2} yr^{-1} (Table 2). However, little variation was found in the percentages of NGF and PGF: for NGF it was between 95 and 98% and for PGF it was between 2 and 5% (Table 2).

Long-term measurements using CGR3 instruments provided additional information concerning lightning occurrence that was used by Kuleshov et al. (2011) as a check on the validity of results derived from short-term CGR4 data. In Brisbane, for the 9 years July 1995 to June 2004 (for reporting purposes the years are from July to June, covering complete thunderstorm seasons), the mean ratio of positive to negative ground flashes was 0.04 (Kuleshov et al., 2006).

Year	N_g	N_c	N_t	GF, %	CF, %
2005	2.55	3.44	5.99	42.6	57.4
2006	1.85	1.79	3.64	50.8	49.2
2007	1.37	0.8	2.16	63.2	36.8
2008	3.76	1.17	4.89	77	23

Table 1. A comparison of flash densities (km^{-2} yr^{-1}) of GF, CF, TF and the percentage of GF and CF recorded.

Year	N_{ng}	N_{pg}	N_g	NGF, %	PGF, %
2005	2.4263	0.1234	2.55	95.16	4.84
2006	1.7848	0.0633	1.85	96.57	3.43
2007	1.3038	0.0623	1.37	95.44	4.56
2008	3.6769	0.0872	3.76	97.68	2.32

Table 2. A comparison of flash densities (km^{-2} yr^{-1}) of NGF, PGF, GF and a percentage of NGF and PGF recorded by a CGR4 LFC in Brisbane in 2005 – 2008.

For Darwin, using CGR3 instruments, it was reported that the mean ratio of positive to negative ground flashes was 0.02, for a 3-year period between 1987 and 1991 (Mackerras and

Darveniza, 1994). For the period August 2003 to June 2004, a CGR4 instrument has been in use at Darwin Airport and the proportion of positive ground flashes was 0.04.

On examining GPATS LLS data for 2005 – 2008 (Table 3), Kuleshov et al. (2011) found that GPATS' detection of cloud flashes was somewhat lower than expected – cloud flashes constituted a small proportion of total flashes (0.4, 3.8 and 1.6% for 2005, 2007 and 2008, respectively) while the CGR4 LFC data indicated 57.4, 36.8 and 23% for those years, respectively (Table 1). This resulted in significantly lower values of N_c as derived from GPATS LLS data (0.0028, 0.0594 and 0.0156 km^{-2} yr^{-1}, respectively, Table 3) compared with N_c values as derived from the CGR4 data (3.44, 0.8 and 1.17 km^{-2} yr^{-1}, respectively, Table 1). For 2006, GPATS-derived N_c = 0.31 km^{-2} yr^{-1} (21% of total flashes) was still lower compared with N_c = 1.79 km^{-2} yr^{-1} (49% of total flashes) obtained from the CGR4 data.

Year	N_{ng}	N_{pg}	N_g	N_c	NGF,%	PGF, %
2005	0.4584	0.2631	0.7215	0.0028	63.54	36.47
2006	0.5857	0.5687	1.1544	0.3056	50.74	49.26
2007	1.4784	0.0212	1.4996	0.0594	98.59	1.41
2008	0.9521	0.0141	0.9662	0.0156	98.54	1.46

Table 3. A comparison of flash densities (km^{-2} yr^{-1}) of NGF, PGF, GF, CF and a percentage of NGF and PGF derived from GPATS stroke data for Brisbane in 2005 – 2008.

Comparing detection of ground flashes, Kuleshov et al. (2011) found from the CGR4 LFC data that GF constituted 43 to 77% of TF (Table 1). As a result of under-detecting CF by the GPATS LLS, derived proportion of GF in the GPATS data was higher ranging from 79 to 99% of TF for those four years. The N_g values as derived from GPATS LLS data (0.72, 1.15 and 0.96 km^{-2} yr^{-1} for 2005, 2006 and 2008, respectively) were lower than N_g values as derived from GGR4 LFC data (2.55, 1.85 and 3.76 km^{-2} yr^{-1}, respectively) with exception of 2007 when the N_g values were comparable (1.37 km^{-2} yr^{-1} for the CGR4 LFC and 1.5 km^{-2} yr^{-1} for the GPATS LLS).

The ratio of NGF to GF estimated from GPATS LLS data was significantly lower than that for the CGR4 for 2005 and 2006 at 63.5% and 50.7% respectively, compared to 95.2% and 96.6% for the CGR4. On the other hand, the GPATS-derived ratio of PGF to GF was 36.5% and 49.3% for 2005 and 2006, respectively, compared to 4.8% and 3.4% for the CGR4.

The results for 2007 and 2008 indicate that a change in processing of raw GPATS sensor data which was implemented in January 2007 moved the above ratios significantly closer to that observed by the CGR4. The ratio of NGF to GF as derived from GPATS data for 2007 and 2008 was 98.6% and 98.5% respectively, compared to 95.4% and 97.7% for the CGR4. The GPATS-derived ratio of PGF to GF was 1.4% and 1.5% for 2007 and 2008 respectively, compared to 4.6% and 2.3% for the CGR4.

Comparing CGR4 LFC and GPATS LLS data prior to 2007 on a daily basis, Kuleshov et al. (2011) found that for individual storms GPATS LLS typically underestimated a number of NGF and overestimated a number of PGF (example of data for Brisbane for December 2005 is given in Table 4). Also note the low detection efficiency of GPATS LLS to CF – only one intra-cloud flash was recorded for December 2005 compared to 483 records by a CGR4 LFC (Table 4).

Day	NGF		PGF		CF		TF	
	CGR4	GPATS	CGR4	GPATS	CGR4	GPATS	CGR4	GPATS
1	0	11	0	7	0	0	0	18
8	172	21	2	4	109	1	283	26
13	0	3	0	0	46	0	46	3
16	1	1	0	4	47	0	48	5
17	118	27	2	70	92	0	212	97
25	110	35	3	29	189	0	302	64

Table 4. A comparison of a number of NGF, PGF, CF and TF (total flashes, NGF + PGF +CF) recorded by a CGR4 LFC and derived from GPATS stroke data for Brisbane in December 2005.

4.2 Proportion of negative and positive ground flashes

The significant difference in the proportion of NGF and PGF as recorded by the CGR4 LFC and GPATS LLS prior to 2007, and possible reasons for these discrepancies, were discussed by Kuleshov et al. (2011). Typically, most GF are negative, and in tropical and subtropical areas, the proportion of PGF is small (*e.g.*, Uman 1987). However, PGF are reported to occur more frequently in winter thunderstorms in parts of Japan, Sweden and Norway (Takeuti et al., 1978; Brook et al., 1982). Nevertheless, results from LFCs demonstrate that PGF typically constitute 2% to 10% of GF. Proportion of PGF at 14 sites located in 11 countries covering latitudes from 60°N to 27°S recorded by CGR3 instruments has been reported by Mackerras and Darveniza (1994). At 13 sites, percentage of PGF was in a range from 2% to 15%. A higher percentage of PGF (28%) was recorded in Kathmandu, Nepal (Mackerras and Darveniza, 1994). These findings were discussed in detail in (Baral and Mackerras, 1993); possible reasons for the relatively high rate of occurrences of PGF were given as the site altitude, thundercloud charge heights, vertical wind shear and the mountainous nature of the terrain.

Recent results obtained by LLSs in different countries are in agreement with early results obtained by LFCs. Kuleshov et al. (2011) compared long-term lightning statistical data obtained by LLSs in Austria (10 years) (Schulz et al., 2005), Brazil (6 years) (Pinto et al., 2006), Italy (7 years) (Bernardi et al., 2002), Spain (10 years) (Soriano et al., 2005), and the USA (5 and 10 years) (Zajac and Rutledge, 2000; Orville and Huffines, 2001). This comparison is presented in Table 5.

Based on results obtained by the LLSs, it was confirmed that PGF usually constitute a small proportion of all cloud-to-ground discharges, in the range of about 5% to 10% in all countries where studies have been conducted apart from Austria where a higher value (17%) was observed (Schulz et al., 2005). Comparing the monthly distribution of cloud-to-ground lightning flashes as observed by LLSs in different countries, it was suggested in (Pinto et al., 2006) that the large difference of the values for Austria with respect to the other countries can be explained by the very short base line of the network in Austria compared to the others.

The detection efficiency of the LLS was reported as 80-90% for GF with peak currents above 5 kA (Cummins et al., 1998). However, during the EULINOX experiment, signals detected by the LLS in Germany were compared with a VHF interferometer, and it was found that 32% of LLS NGF were in fact CF and 61% of LLS PGF were found to be CF (Thery, 2001).

In Australia, during the Down Under Doppler and Electricity Experiment (DUNDEE) which was conducted near Darwin during the wet seasons of November 1988 through February

1989, and November 1989 through February 1990, about 90% of GF were reported as NGF (Petersen and Rutledge, 1992). However, results obtained by the GPATS LLS in Australia are different from results in other countries. Conducting the BIBLE experiment near Darwin, the Northern Territory, in December 2000 it was noted that about 80% of GF were reported by GPATS LLS as positive (Koike et al., 2007).

Country / Data source	Peak current of PGF, kA	Peak current of NGF, kA	Percentage of PGF, %
Austria (Schulz et al., 2005)	10 (for 2001)	10 (for 2001)	17.0
Brazil (Pinto et al., 2006)	29.5	26.5	8.5
Italy (Bernardi et al., 2002)	35.8	22.7	5.0
Spain (Soriano et al., 2005)	35.3	23.5	9.0
USA (Zajac and Rutledge, 2000)			10% [1] (>20% [2])
USA (Orville and Huffines, 2001)	20	22	3 - 9 [3]

Table 5. A comparison of median peak current values of NGF and PGF and a percentage of PGF recorded by lightning location networks in different countries.

The high proportion of PGF (up to 50%) recorded by the GPATS LLS in Australia prior to 2007 was attributed by Kuleshov et al. (2011) to one or more possible factors, including (i) changes in the processing of raw GPATS sensor data, implemented in January 2007, (ii) upgrades to sensor firmware, and (iii) the installation of more GPATS sensors. CF could be misclassified as PGF (Mackerras and Darveniza, 1992), and sensors might detect the reflected sky wave component rather than the ground wave. The latter situation may come about due to large baseline distances between sensors in the GPATS network resulting in a highly attenuated ground wave signal going undetected by majority of sensors contributing to a solution. By comparison, the corresponding (one-hop) sky wave impulse would be relatively unattenuated and therefore detectable, but inverted in polarity due to its reflection from the ionosphere. As a result of this reversal, the impulse signature of what was originally an NGF may be interpreted as a PGF. A careful analysis of the three possible factors indicates that reduction in the excessive proportion of PGF is mostly explained by changes in processing GPATS sensor data.

A comparison with CGR4 LFC data done by Kuleshov et al. (2011) for a location in Brisbane over the period 2005–2008 has shown that prior to January 2007 the GPATS LLS was reporting an excessive proportion of total GF as being PGF, with a correspondingly low proportion being reported as NGF. Over the same period the CGR4 LFC reported PGF and NGF ratios consistent with those obtained by other researchers both in Australia and elsewhere using a range of instrumentation, including the CGR4's predecessor, the CGR3.

---

[1] Over most of the contiguous United States.

[2] Over the north-central United States and along and near the Pacific coast.

[3] The annual percentage of positive lightning has increased from 3% in 1989 to 9% in 1998. Orville and Huffines (2001) attributed the increase to improved sensor detection capability.

After January 2007, the PGF and NGF ratios reported by the GPATS changed significantly to become more aligned with the CGR4 observations. Indeed, the ratio of PGF to total GF for 2007 and 2008 is somewhat lower than that derived from the CGR4 data and lightning cloud flash density values, N_c, are lower by an order of magnitude than those estimated from the CGR4 measurements indicating that further investigation of later data may be required. This transition coincided with an upgrade to the GPATS' stroke processing software. It is therefore concluded that the polarity of GPATS stroke data prior to January 2007 not in agreement with records obtained by lightning location networks in different countries. Consequently, only GIGRE-500, CGR3 and CGR4 data records which are consistent with lightning data obtained internationally were used for analysis of lightning climatology over the Australian continent presented in following sections.

5. Variation of lightning distribution

Geographical distribution and seasonal variability of lightning activity for the Australian continent have been analysed in detail based on long-term records from LFCs and NASA satellite-based lightning data (Kuleshov et al., 2006). For the first time, a total lightning flash density map (Fig. 3) and a lightning ground flash density map (Fig. 4) for Australia have been prepared. The maps are contoured in units of flash density (km^{-2} yr^{-1}). The N_g map is now the standard reference on lightning ground flash density in Australia (Lightning Protection, 2007).

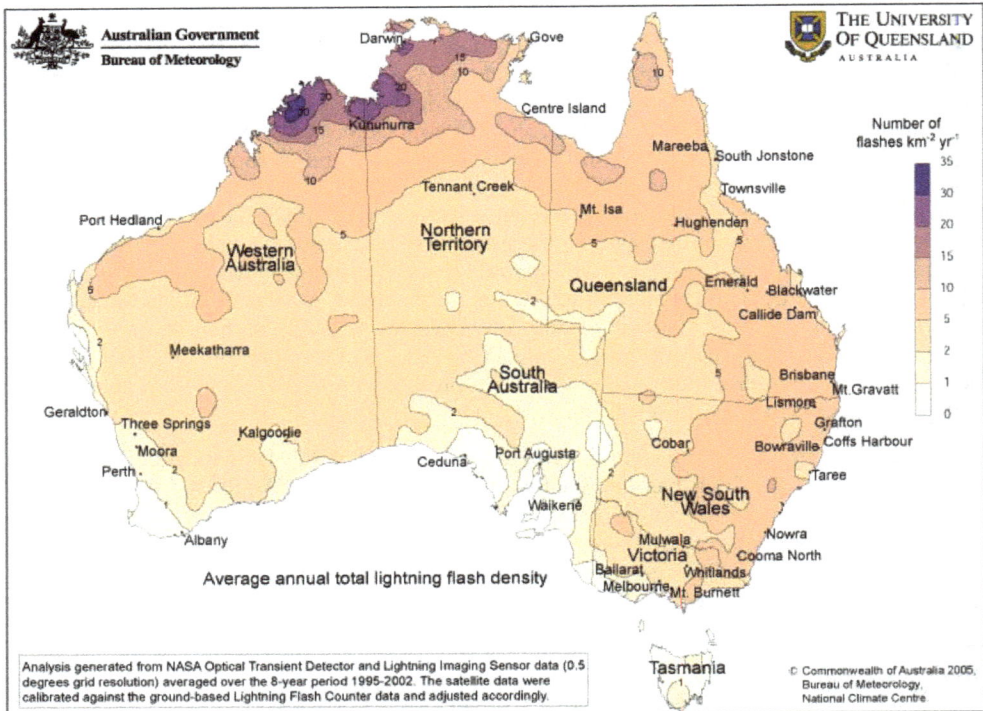

Fig. 3. Map of average annual total lightning flash density for Australia.

5.1 Geographic distribution

The spatial distribution of average annual total flash density, N_t, (Fig. 3) is in general qualitative agreement with the thunderstorm climatology (Fig. 1) as represented by the average annual thunder-day map of Australia (Kuleshov et al., 2002). Both maps demonstrate high annual thunderstorm ($T_d>40$ day yr^{-1}) and lightning activity ($N_t>10$ km^{-2} yr^{-1}) in the northern parts of Australia, and decreases in annual thunder-days and total flash density southwards.

The peak lightning occurrence is in the north-western part of the Australian continent with N_t values above 15 km^{-2} yr^{-1} and it is in the region of the peak in thunderstorm occurrence ($T_d>50$ day yr^{-1}) as it appears in the thunder-day map. The two peaks do not exactly coincide, however this finding is not unexpected. Discussing the world distribution of lightning as observed by the OTD, Christian et al. (2003) reported values of total flash density of about 23 km^{-2} yr^{-1} at Entebbe (T_d=206) and Kampala (T_d=242), Uganda, while the planet's "lightning hot spot", Kamembe, Rwanda, had 83 km^{-2} yr^{-1} and 221 thunder-days recorded.

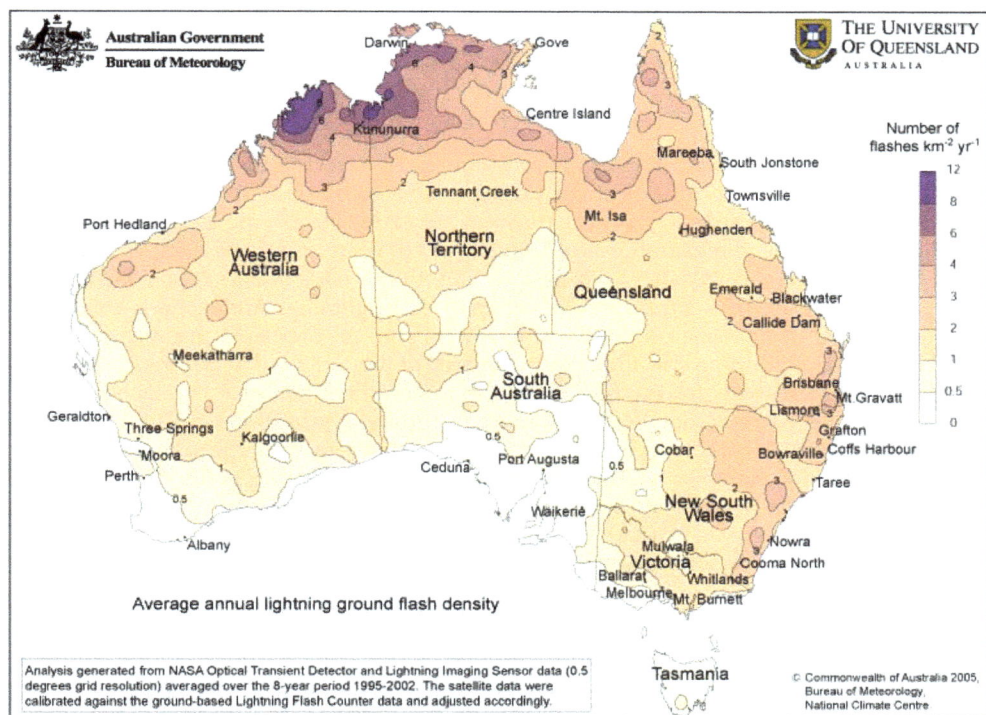

Fig. 4. Map of average annual lightning ground flash density for Australia.

The maximum average annual number of $T_d > 80$ is in the vicinity of Darwin, where storms triggered over the Arnhem Land ranges and over Bathurst Island are very common (Fig. 1). High frequency of thunderstorms (40-60 thunder-days per year) is observed at the King Leopold and Durack Ranges in the north of Western Australia. This approximately corresponds to the area of the maximum average annual total lightning flash density with $N_t > 30$ km^{-2} yr^{-1} centred around 16° S 126° E (Fig. 3). The highest value of N_t as derived from the satellite data is about 35 km^{-2} yr^{-1}.

The secondary maximum of thunderstorm and lightning occurrence is over the north-eastern part of the Northern Territory and northern part of Queensland, with values of T_d between 40 and 60 days per year, and with N_t values between 5 and 15 km^{-2} yr^{-1}. Another secondary maximum is evident over eastern part of Australia, with local maxima over the Great Dividing Range. Average annual number of thunder days is between 20 and 40 in this area, and lightning activity is also high with N_t values up to 10 km^{-2} yr^{-1}. Lightning occurrence generally decreases southward, and number of total lightning flashes is small in Victoria, South Australia, the southern parts of Western Australia and New South Wales, and Tasmania, with the values of $N_t < 5$ km^{-2} yr^{-1}.

5.2 Seasonal variations

Analysis of annual distributions of thunderstorms and lightning occurrences, as registered by number of thunder-days and lightning days, demonstrates dominance of these phenomena in Australia in summer months (typical distributions are presented in Figs. 5A and 5B for Darwin and Melbourne, respectively).

Data obtained by the CIGRE 500 LFCs demonstrate that on average, 90% of annual lightning registrations have been recorded between October and April, with 55% recorded in three months December to February. Comparison of seasonal distributions of standardized monthly mean lightning registrations for Kununurra (10 years of data for 1987-92 and 1999-2002), Brisbane (15 years of data for 1980-94) and Perth (21 years of data from 1982 to 2002) in Fig. 6 demonstrates this summer peak occurrence of lightning. Atmospheric conditions (high boundary-layer moisture levels and lower surface pressure) are favorable for thunderstorm and lightning development in this part of the continent in wet season months. The seasonal distributions for the sites located in the northern half of Australia show clear seasonality of the phenomenon: 99.8% of annual lightning registrations in Kununurra and 97.5% in Brisbane were recorded in warmer months October to April. Seasonal distribution of lightning activity in higher latitudes is more uniform. The distribution for Perth (Fig. 6) demonstrates that in higher latitudes, lightning is still more frequent in warmer months (65% of annual lightning registrations in Perth), but it also occurs in cooler months (May - September) in association with active frontal systems.

In general, over the Australian continent the lightning intensity and the number of thunder-days are highest in the tropics and decrease steadily away from the equator. The Inter Tropical Convergence Zone approaches north of Australia in November, giving rise to high moisture levels and precipitation over the next two or three months. During the Southern Hemisphere winter, dry stable outflow from the subtropical high over the Australian continent inhibits convective storm activity in the north while frontal systems sweeping in over the southern parts of the continent continue to provide moisture right through the year (Jayaratne and Kuleshov, 2006a).

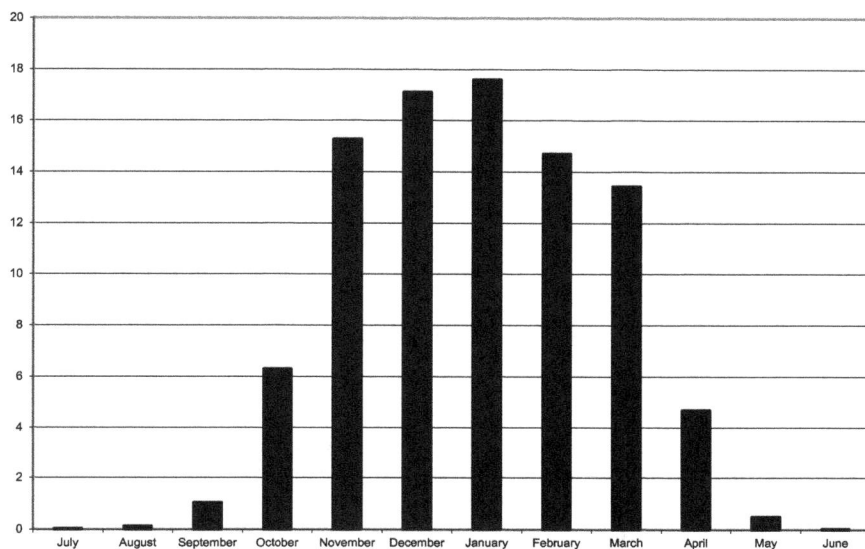

a)

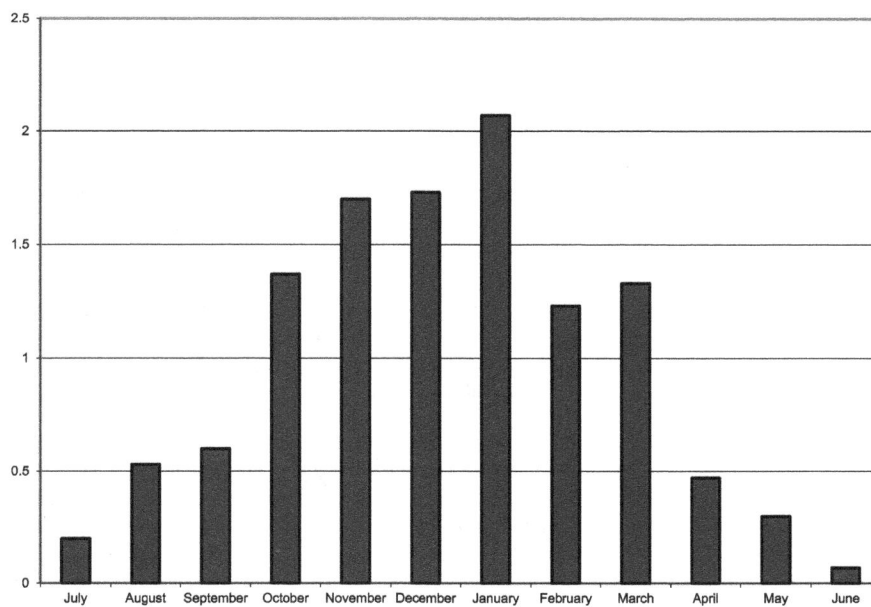

b)

Fig. 5. Seasonal distributions of monthly mean thunder-days in Darwin (A) and Melbourne (B).

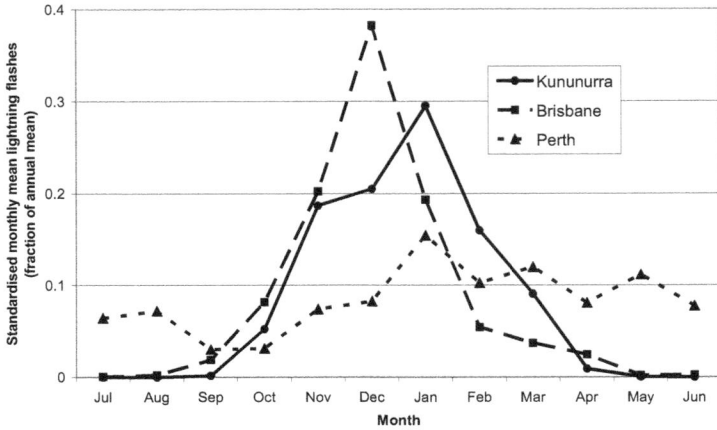

Fig. 6. Seasonal distributions of monthly mean lightning ground flashes in Kununurra, Brisbane and Perth, expressed as a fraction of the annual mean.

5.3 Variation of total flash density with latitude

Latitudinal dependence of lightning characteristics is a subject of long-term investigations. Orville and Spencer (1979) reported satellite-based optical observations of total lightning flash density at local dusk and midnight over a latitude range 60° N to 60° S and demonstrated a reduction in flash frequency by a factor of 10 for a 40° increase in latitude in the Southern Hemisphere. Results of another satellite-based observations using radio-frequency detection of lightning indicated a similar rate of change of total flash density – a reduction in lightning flash rate by a factor of about 10 for a change in latitude from 20° S to 60° S (Kotaki et al., 1981). Mackerras and Darveniza (1994) analyzed latitudinal variation in lightning activity using data from ground-based lightning flash counters recorded at 14 sites in 11 countries covering latitudes from 60° N to 27° S, and concluded that total flash density falls by a factor of 10 for every 30° increase in latitude. Using the data gathered by the satellite detectors over the landmass of Australia, (Kuleshov et al., 2006) described the variation of N_t for the range of latitudes from 10° S to 40° S by the empirical relationship $N_t = \exp(3.85 - 0.088\lambda)$ km^{-2} yr^{-1}, where λ is the magnitude of the latitude in degrees. Once again, a reduction in lightning total flash density, N_t, by a factor of about 10 for a change in latitude from 10° S to 40° S was found, which was in agreement with the earlier studies.

One of the explanations of variation in lightning rate is traditionally sought in convective available potential energy (CAPE) which is a driving force for a thunderstorm development (Williams, 1995). CAPE determines the updraft velocity in deep convective systems and therefore determines electrical charge separation rates and rate of lightning occurrence. Based on results of the Down-Under Doppler and Electricity Experiment (DUNDEE) conducted during the periods of November 1988 - February 1989 and November 1989 - February 1990 near Darwin, Rutledge et al. (1992) and Williams et al. (1992) demonstrated a strong increase in lightning activity with CAPE as well as nearly linear relationship between CAPE and wet bulb potential temperature. In view of the observations by Rutledge et al. (1992), Williams (1992) and Williams et al. (1992), Mackerras and Darveniza (1994) suggested that the physical basis for the annual variation in monthly total lightning flash density in Darwin is to be sought mainly in the annual variation in wet bulb temperature, T_w, and CAPE. Investigating further the relationship between lightning activity and surface wet bulb temperature and its variation with latitude in Australia, Jayaratne and Kuleshov (2006a) examined data from ten LFCs gathered over a sufficiently long period (ranging from 15 to 21 years of records) and widely distributed across continental Australia. It was demonstrated that at each of the stations the monthly total of lightning ground flashes increased with the monthly mean daily maximum wet bulb temperature. The dependence was most pronounced in the tropics (in Darwin, a modest 3-4°C increase in wet bulb temperature increased the lightning activity by over two orders of magnitude) and decreased in temperate latitudes (in Melbourne, an increase of about half an order of magnitude in the monthly total of ground flashes within a 10°C range of wet bulb temperature was observed). Further examining relationship between thunderstorm occurrences and T_w, strong increase in monthly mean number of thunder-days, T_d, with increase in monthly mean daily maximum wet bulb temperature, T_w, for tropical localities in Australia was found (example for Mt Isa is given in Fig. 7.A). However, only weak increase or even decrease was observed for non-tropical localities (e.g. Perth, Fig. 7.B). These results are in line with earlier findings of Mackerras and Darveniza (1994) and Jayaratne and Kuleshov (2006a) about variation in lightning rate in Australia.

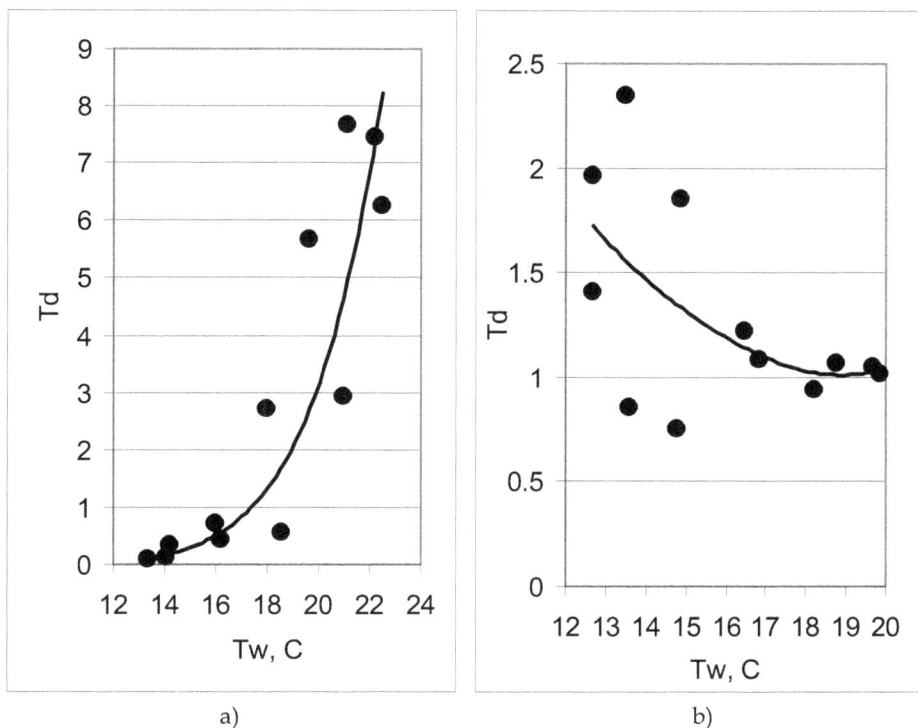

a)   b)

Fig. 7. Monthly mean number of thunderdays, T_d, plotted against monthly mean daily maximum wet bulb temperature, T_w, for Mt Isa (A) and Perth (B).

It appears that the main explanation for the annual variation in monthly flash density in tropical parts of the Australian continent is found in the variation in wet bulb temperature and CAPE. However, other factors such as variation in freezing level (Williams et al., 2005; Mushtak et al., 2005), in aerosol loading (Williams and Stanfill, 2002), as well as variation in topography (Bocippio et al., 2001), may also influence the geographical and seasonal variations of lightning parameters in other parts of the continent.

5.4 Yearly variations in thunder-days

Climatic hazards such as thunderstorms and lightning can accompany destructive weather, particularly in the tropical regions. It is important to examine how the frequencies of thunderstorms have changed over the periods for which thunder-day records are available in Australia and, perhaps more importantly, how they are likely to change in the future.

Based on the results of earlier studies which demonstrated that sensitivity of thunderstorm and lighting activity to environmental changes is high in the tropics (section 5.3), time series of T_d for several Australian tropical and subtropical observational sites were examined - 50 year records for Darwin (Fig. 8A), Townsville (Fig. 8B), Mt Isa (Fig. 8C), and the 118 year record for Brisbane (Fig 9). While the correlation coefficients (R^2) are rather low for all the data sets (less than 0.1), it is possible to perceive small trends in the long-term T_d data. For about the 50 year period 1940 to 1990, it would appear that there has been a general upward trend in T_d for all sites, possibly followed by a subsequent downward trend. However, the longer-term T_d data for Brisbane appear to indicate an upward trend to about 1930, then a downward trend to about 1950, an upward trend to about 1995, followed by a downward trend, with an overall yearly mean of 32.8 and standard deviation of 10.2 thunder-days.

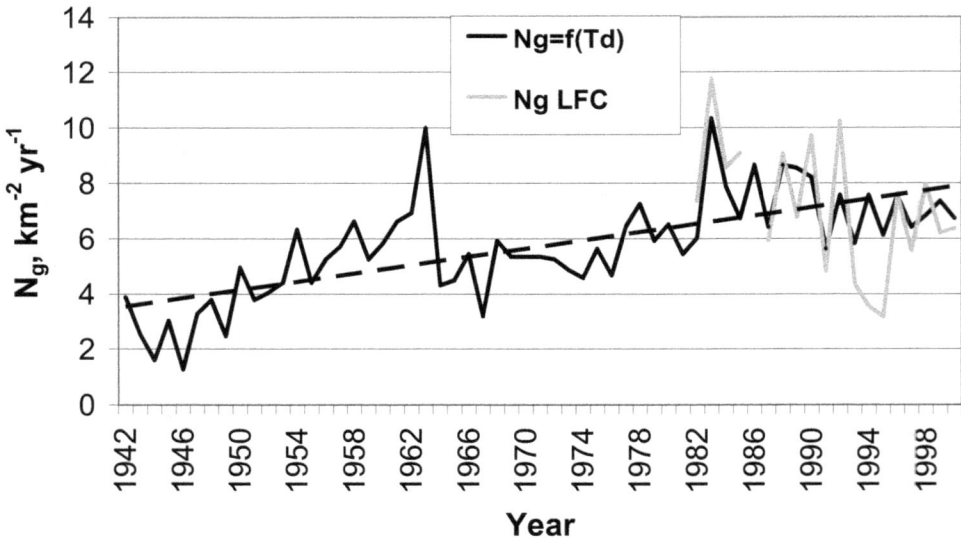

a)

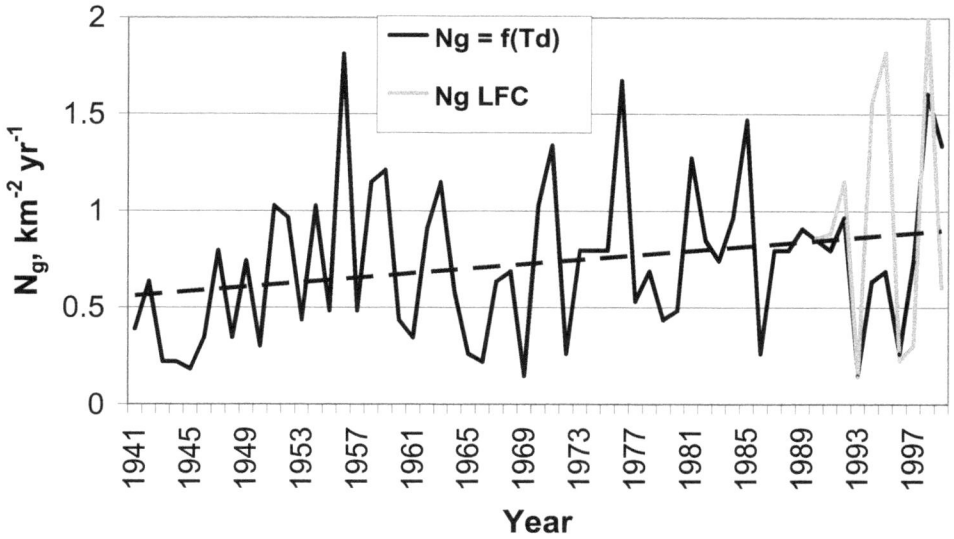

b)

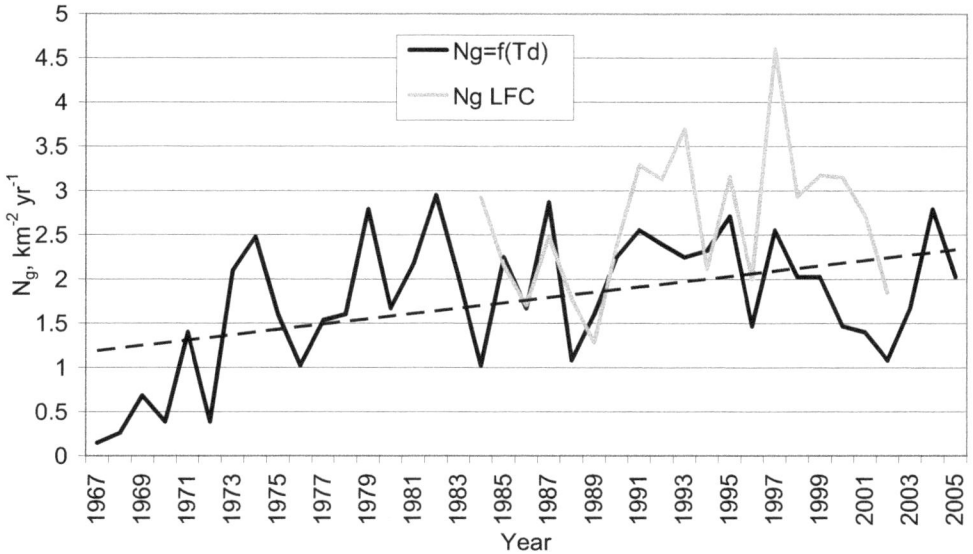

c)

Fig. 8. Time series of lightning ground flash density N_g (as derived from T_d records and measured by the LFCs) for Darwin (A), Townsville (B) and Mt Isa (C).

$$y = 2E\text{-}09x^6 - 2E\text{-}05x^5 + 0.0877x^4 - 8.54x^3 + 334920x^2 - 3E\text{+}08x + 9E\text{+}10 \quad R^2 = 0.3091$$

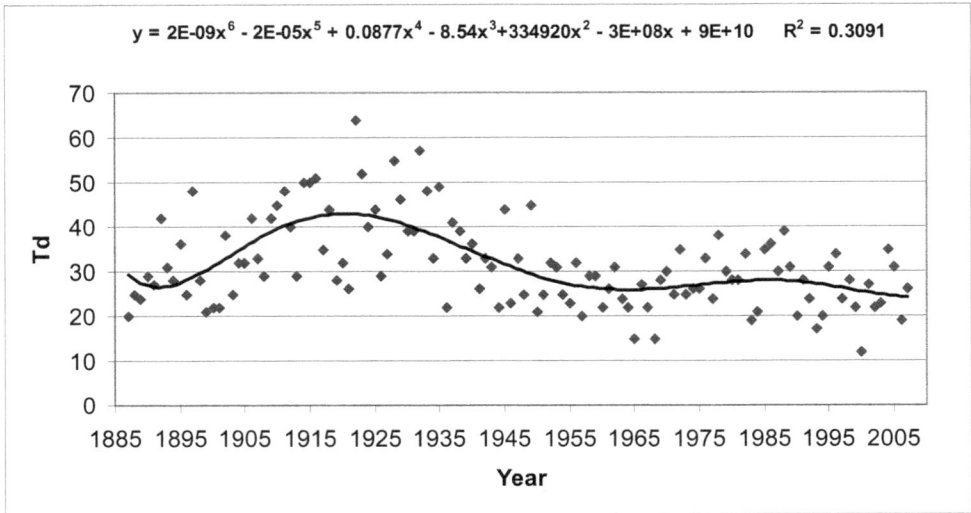

Fig. 9. Time series of yearly thunder-days for Brisbane 1887 to 2007 with sixth-order polynomial trend line.

It would be of interest to attempt to relate these T_d data to *Climate Change 2007*, the Fourth Assessment Report of the United Nations Intergovernmental Panel on Climate Change (IPCC). It states that "Warming of the climate system is unequivocal" and "Most of the observed increase in globally averaged temperatures since the mid-20th century is very likely due to the observed increase in anthropogenic greenhouse gas concentrations" (IPCC 2007). Clearly, the effects of climate change on thunderstorms in Australia need to be studied in much greater detail, because thunderstorms and lightning are temperature-sensitive weather phenomena.

6. Relationship between lightning activity and surface wet bulb temperature

Convective Available Potential Energy (CAPE) is the driving force for thunderstorm development. CAPE is closely controlled by wet bulb temperature. The relationship between lightning activity and surface wet bulb temperature and its variation with latitude has been studied by Jayaratne and Kuleshov (2006a) utilizing data obtained by network of CIGRE-500 lightning flash counters widely distributed across continental Australia.

The Down-Under Doppler and Electricity Experiment (DUNDEE), conducted during the wet seasons of 1988-1990 near Darwin, Australia, showed that, while lightning activity was very intense in isolated continental convective thunderstorms, it was at least an order of magnitude lower in the ocean-based convective storms that produced heavy rain during the monsoonal periods (Rutledge et al, 1992). Williams et al (1992) attributed these dramatic differences in lightning activity between the two types of storms, both of which produced heavy rainfall, to differences in conditional instability – the mechanism by which thunderstorms are formed. The energy that drives conditional instability is CAPE and can be represented on a thermodynamic diagram by the area bounded by the environmental lapse rate and the saturated adiabatic lapse rate (SALR). Field experiments have demonstrated a strong increase in lightning activity with CAPE (Williams et al., 1992). It has also been shown that CAPE increases linearly with potential wet bulb temperature, T_w (Williams et al, 1992; Williams and Renno, 1993; Peterson et al, 1996). This is not surprising in view of the fact that the SALR is determined by T_w. Williams et al (1992) showed that a 1°C change in T_w resulted in a change in CAPE of about 1 kJ kg^{-1}. The mean daily maximum surface T_w in Darwin during the 1988-89 wet season dropped by about 2°C from the break periods to the monsoon periods. While the high lightning incidence continental storms occurred during the break periods, little lightning was observed in the monsoonal storms. The calculated CAPE showed an increase from about 800 J kg^{-1} to about 2000 J kg^{-1} from the monsoonal period to the break period. The corresponding lightning flash density over an area of 40,000 km^2 increased from about 100 to over 1000 per day. Analysis of the two-season data set indicated that the monthly mean lightning count in Darwin increased by more than two orders of magnitude as T_w increased by 2°C. Jayaratne (1993) showed a similar behaviour in Botswana in Africa where the lightning activity doubled for every 1°C increase in T_w. Peterson et al (1996) studying data collected during TOGA COARE over the western Pacific Ocean, concluded that a 0.5°C increase in T_w led to a 5 to 8 factor increase in the cloud to ground lightning activity. A detailed explanation for the increase of lightning activity with CAPE and T_w may be found in Williams (1995).

In Australia, the ABM maintains a network of about 40 lightning flash counters scattered widely around Australia (Fig 2). The availability of a reasonably large lightning flash data set obtained at several sites over a range of geographical latitudes on the Australian continent allowed to investigate the reported sensitive relationship between lightning activity and T_w. (Jayaratne and Kuleshov, 2006a). Of the 40 stations, ten sites were selected for their reliability and availability of data over sufficiently long periods of observation and to represent a wide geographical distribution across the continent.

The monthly total of lightning ground flashes, N, were related to the monthly mean daily maximum wet bulb temperature, $T_{w,max}$. (Jayaratne and Kuleshov, 2006a). Each of the ten stations showed a roughly exponentially increase in number of ground flashes with $T_{w,max}$. Fig. 10 shows a typical example for Coffs Harbour where the $T_{w,max}$ values have been classified into 1°C bins in order to plot a distribution histogram. The exponential trend line is very clear, with a correlation coefficient of 0.99. An important point to note was that the rate of increase of N with $T_{w,max}$ was clearly dependent on the latitude. Stations closer to the equator (Darwin, Kununurra and Centre Island) showed very sharp increases, while the rates of increase for higher-latitude stations were lower. To evaluate this latitudinal dependence, log(N) was plotted as a function of $T_{w,max}$ for each of the ten stations. Fig 11 shows the data for the three widely spaced stations (a) Darwin, (b) Coffs Harbour and (c) Melbourne. Each point represents a calendar month during the period of observation. In Darwin (latitude 12°S), the overall increase in monthly total of lightning ground flashes was over three orders of magnitude over the 7°C range of $T_{w,max}$ for which lightning activity was recorded. However, a sharp increase of over two orders of magnitude was observed within a modest 3-4°C increase in wet bulb temperature in the range between 23°C and 27°C. Similar increases were found in Kununurra and Centre Island. The dependence with wet bulb temperature decreased as we moved away from the equator. An increase of about one and a half order of magnitude was observed for Coffs Harbour (latitude 30°S) over a 10°C range of the monthly mean daily maximum wet bulb temperature (Fig 11b). In Melbourne (latitude 38°S), the effect was less obvious, but still gave an increase of about half an order of magnitude in the range of $T_{w, max}$ for which lighting flashes were registered (Fig 11c).

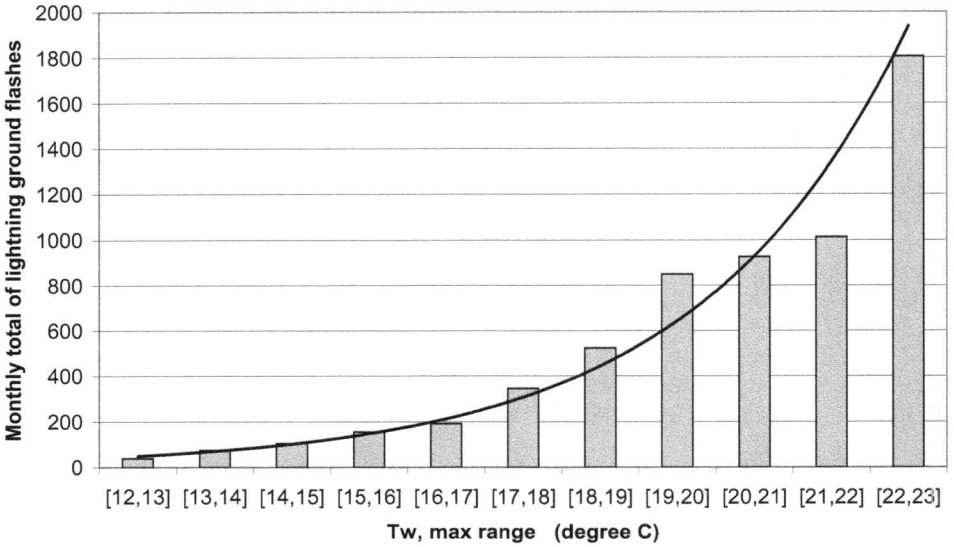

Fig. 10. Monthly total of lightning ground flashes N as a function of monthly mean daily maximum wet bulb temperature $T_{w,max}$ for Coffs Harbour.

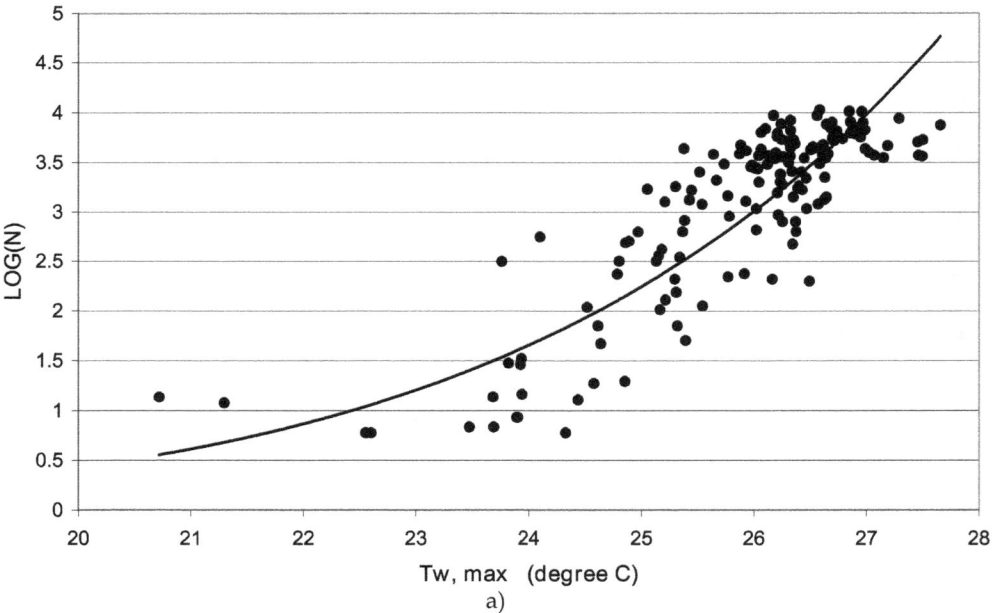

a)

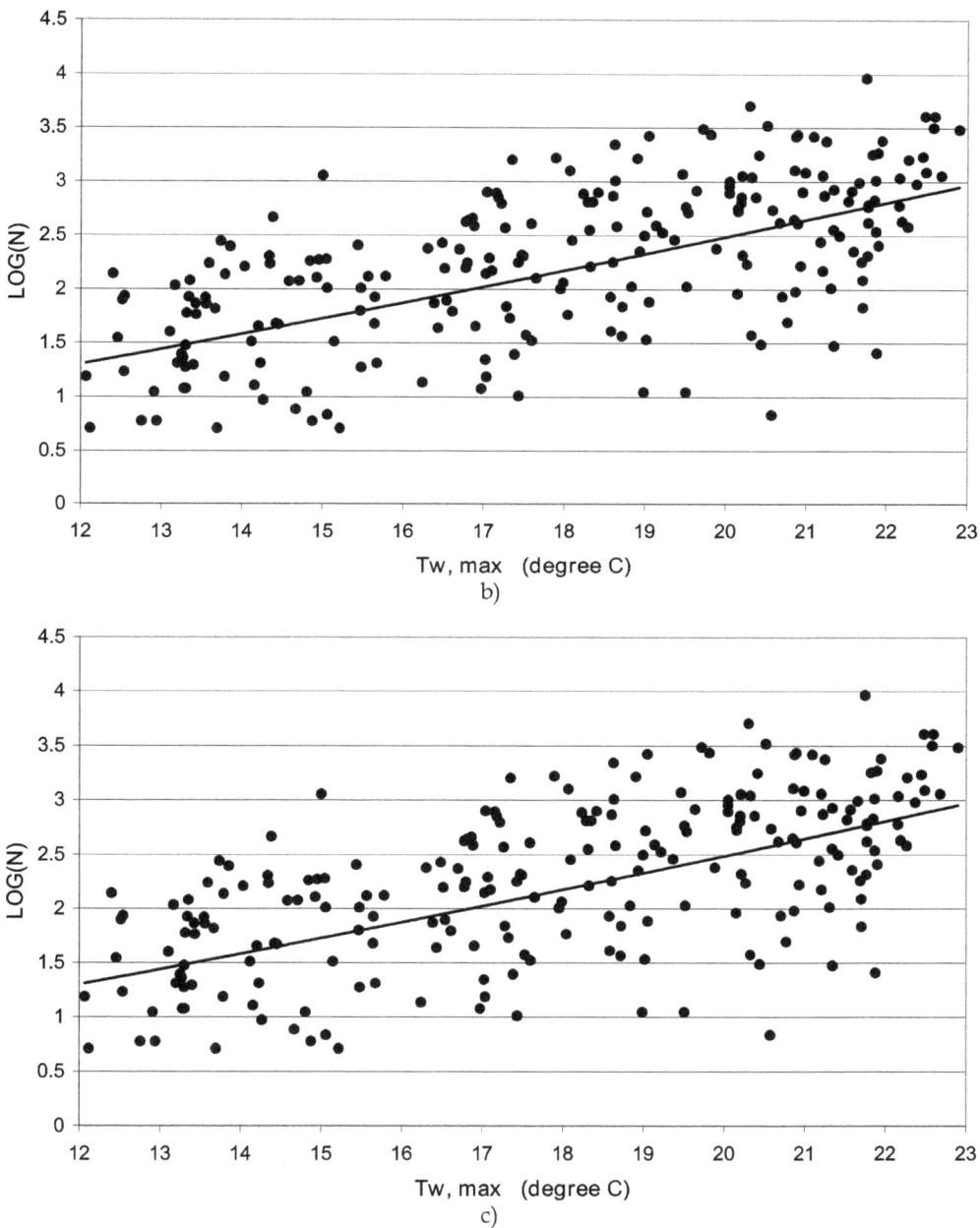

Fig. 11. Monthly total of lightning ground flashes presented as log(N) versus the monthly mean daily maximum wet bulb temperature $T_{w,max}$ at (a) Darwin, (b) Coffs Harbour and (c) Melbourne.

In order to compare the increase in lightning activity with increase in $T_{w, max}$ within a range of Australian latitudes, Jayaratne and Kuleshov (2006a) derived best fit approximations for the data from each of the ten studied stations using a power law approximation $\log(N) = a\ T_{w,max}^P$ after Williams and Renno (1991). The value of P for Darwin using data covering the 20-year period 1982-2001 was 7.45. The corresponding value of P found by Williams and Renno (1991) for Darwin, 1988 only, was 7.5. The value of P falls with latitude and at the highest latitude investigated (Melbourne, 38°S), P had fallen to 0.21. In Fig 12 P-values as a function of the latitude of the stations are presented. P showed a sharp decrease with increasing latitude in a range approximately between 10°S and 20°S. The decrease was more gradual at higher latitudes. Overall, the best trend was exponential with a correlation coefficient of 0.95.

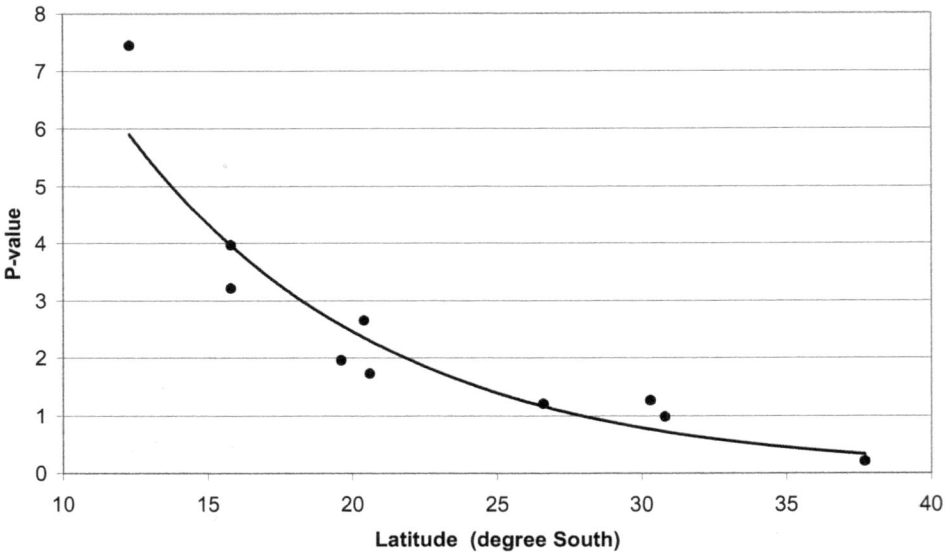

Fig. 12. Variation of P-value with latitude of the stations.

It is well known that CAPE and lightning activity are strongest in the tropics. Using satellite observations, Orville and Spencer (1979) showed that total lightning activity decreased exponentially with latitude. Based on the analysis of Australian soundings during the Australian Monsoon Experiment (AMEX), Williams and Renno (1993) showed that the rate of change of CAPE with wet bulb was greatest in the tropics and decreased southwards away from the equator. The values ranged from about 1300 J kg^{-1} °C^{-1} at Darwin (12°S) to about 750 J kg^{-1} °C^{-1} at 20°S. Considering monthly mean values of total lightning counts and maximum wet bulb temperatures for specific years at several locations around the world, Williams (1994) concluded that the local sensitivity of lightning to temperature was highest near the equator and diminished at higher latitudes. In the near-equatorial region, the observed sensitivities, expressed as a percentage increase of total flash count, were 400%-800% per 1°C, falling to 200%-400% at the edge of the tropics. The sensitivity at mid latitude was only 100%-200%. The value for Darwin (12°S), using total lightning count data gathered during the year 1988, was close to 500%. From Fig 11(a), the value using ground flash data over a period of 20 years, calculated over the temperature range 22°C to 27°C was 400%, in good agreement with the above. It should be noted that the increase within the narrower temperature range of 24°C to 27°C, corresponding to the intense summer thunderstorms, gave a much higher factor of 1000%. The values for Coffs Harbour (30°S) and Melbourne (38°S) were significantly lower at 160% and 130% respectively. Williams (1994) did not present results from Southern Hemisphere stations at these higher latitudes but our values compare well with their results for Northern Hemisphere stations at the same distance from the equator.

The reason for the decrease of P-values as the location moved away from the equator (Fig 12) is not very obvious. Williams and Renno (1991) suggested that, although CAPE still plays a major role in the temperate latitudes, CAPE is strongly influenced by the presence of cold dry air aloft in the environment and less influenced by the wet bulb temperature of boundary layer air. Williams (1994) stresses that a more fundamental understanding of the strongly nonlinear relationship between lightning and CAPE must await a more complete understanding of the microphysics of charge separation within thunderstorms.

7. Relationship between lightning ground flash density and rainfall

The close relationship between rain and lightning is well known. Battan (1965) and Piepgrass et al. (1982) related number counts of cloud to ground (CG) lightning from nearby thunderstorms to rain gauge readings and found them to be well correlated. Many studies have found the intensity of lightning to be positively correlated to rainfall estimated from radar measurements (Kinzer, 1974; Reap and MacGorman, 1989; Williams et al.., 1992; Cheze and Sauvageot, 1997). Intense falls of rain associated with nearby CG lightning have been documented by Shackford (1960), Moore et al. (1962), Piepgrass et al. (1982) and Jayaratne et al. (1995).

However, there are numerous documented observations of heavy rain accompanied by little or no lightning activity (Williams et al, 1992; Jayaratne, 1993). Kuleshov et al (2002) studied thunderstorm distribution and frequency in Australia using data from 300 sites around the country. In particular, they showed that the section of eastern coast in Queensland between Cooktown (15°S) and Innisfail (18°S), reputed to be the wettest part of Australia, had fewer than 25 thunder-days per year, while other tropical regions of the country recorded over 40 thunder-days annually. Thorough analysis and comparison of time series of thunder-day frequency with rainfall variations over the period 1970-1999 allowed them to conclude that thunderstorm frequency in Australia does not, in general, appear to vary in any consistent way with rainfall. To study geographical and seasonal characteristics of the relationship between lightning ground flash density and rainfall in detail, Jayaratne and Kuleshov (2006b) analysed observations of annual rainfall and lightning incidence collected at 23 stations around the continent of Australia.

7.1 Rain yields

Although there is a high positive correlation between rainfall and lightning, the ratio of rain mass to CG lightning flash count over a common area, with units of kilograms of rain per flash, quantitatively defined as the "rain yield", varies considerable with location. In general, heavy rain associated with monsoon or oceanic convection show rain yields of the order of $10^9 - 10^{10}$ kg fl[-1], while continental convective thunderstorms show much smaller values of $10^7 - 10^8$ kg fl[-1] (Williams et al., 1992; Petersen and Rutledge, 1998). Zipser (1994) found that the number of thunder days associated with heavy rain in tropical monsoon and oceanic storm regions was significantly lower than that in continental rainfall regimes. Observing the results of several studies at various geographical locations, Petersen and Rutledge (1998) concluded that the rain yield varied by a factor of 10 or more at any given location and by a factor of up to 10^3 between different locations and rainfall regimes. At the lower end, values of around 5×10^7 kg fl[-1] were found in the arid south-western United States. A wide section of the mid-continental United States showed remarkably stable values clustered near 10^8 kg fl[-1], as did a landlocked station in Botswana within the African subcontinent. In tropical locations, the rain yields increased systematically from a continental value of 4×10^8 kg fl[-1] to a maritime value of 10^{10} kg fl[-1] in the western Pacific Ocean. Williams et al. (1992) identified two distinct rainfall regimes in Darwin in continental northern Australia. Rain yields for tropical continental break period thunderstorms and tropical oceanic thunderstorms differed by almost an order of magnitude, being 3×10^8 kg fl[-1] and 2×10^9 kg fl[-1] respectively. Similarly, rain yields for break and monsoon period convection that occurred offshore over Melville and Bathurst Islands near Darwin showed values of 8×10^8 kg fl[-1] and 8×10^9 kg fl[-1] respectively.

Williams et al. (1992) attributed the contrasting lightning activity in the two types of rainfall regimes to differences in convective available potential energy (CAPE). Many studies have demonstrated a strong increase in lightning activity with CAPE (Williams et al., 1992; Petersen et al., 1996). This is not surprising as CAPE bears a strong relationship to the potential wet bulb temperature, T_w, - a parameter that increases with temperature and humidity – both of which lead to an increase in lightning activity (Williams and Renno, 1993). Williams et al. (1992) showed that a 1°C change in T_w resulted in a change in CAPE of about 1 kJ kg[-1]. The mean daily maximum surface T_w in Darwin during the 1988-89 wet season dropped by about 2°C from the break periods to the monsoon periods. Highly active lightning storms occurred during the break periods while relatively little lightning was observed in the monsoonal storms. The mean values of CAPE during the two periods were 2000 J kg[-1] and 800 J kg[-1] respectively. The corresponding lightning flash rates observed over an area of 40,000 km[2] were about 1000 and 100 per day respectively.

Continental land surface is systematically hotter than the sea. This gives rise to greater CAPE, atmospheric instability and stronger air motions that are vital for deep convection and thunderstorm formation. Although the total rainfall is about the same, lightning activity over land is an order of magnitude greater than over the oceans (Orville and Henderson, 1986). Thus, maritime stations, in general, have a higher rain yield than continental stations.

An alternative hypothesis for the land-ocean contrast in lightning is based on differences in boundary layer aerosol concentrations (Rosenfeld and Lensky, 1998). Continental air is more polluted than ocean air and contains more cloud condensation nuclei. Typical concentrations range from 100-200 cm[-3] over the oceans to values greater than 1000 cm[-3] over land. The resultant larger numbers of smaller cloud droplets at continental locations give rise to a dominance of diffusional droplet growth and suppressed coalescence. This leads to a reduction in rainfall and allows liquid water to ascend to the higher mixed phase region of thunderclouds where strong electrification takes place. The net result is increased lightning activity, reduced rainfall and reduced rain yields at continental stations when compared to maritime stations.

A further possible explanation for the contrast in lightning and rainfall characteristics between land and ocean thunderstorms is based on cloud base height (Williams and Stanfill, 2002). They argue that higher cloud base heights provide larger updraught widths and reduced dilution by mixing – two factors that promote lightning activity. It has been shown that lightning flash rate increases with cloud base height (Williams et al, 2004). Typically, cloud base heights over the maritime and continental locations are about 500 m and 3000 m respectively and the associated lightning flash rates between these two locations differed by an order of magnitude.

7.2 Rainfall climatology of Australia

The continent of Australia contains a diverse range of climatic zones. The tropical northern and eastern coastal rim is generally humid and experiences heavy rainfall in the summer. The continental interior is largely arid and the southern regions are mostly temperate. The average rainfall in Australia is 450 mm. Around 80% of the landmass has a median rainfall less than 600 mm per year with 50% less than 300 mm. Large areal pockets within South and West Australia have less than 150 mm. The vast interior of the continent has a median annual rainfall of less than 200 mm. This region is not normally exposed to moist air masses for extended periods and rainfall is irregular. However, in favourable synoptic situations, which occur infrequently over extensive parts of the region, up to 400 mm of rain may fall within a few days and cause widespread flooding. The region with the highest annual rainfall is the east coast of Queensland near Cairns, with some stations recording over 3000 mm per year.

Owing to its low relief, compared to other continents, Australia causes little obstruction to the atmospheric systems that control the climate. However, as outlined earlier, the rainfall pattern is strongly seasonal in character, with a winter rainfall regime in the south and a summer regime in the north. During the Southern Hemisphere winter (May-October), huge anticyclonic high pressure systems transit from west to east across the continent and may remain almost stationary over the interior for several days. Northern Australia is thus influenced by mild, dry south-east winds, while southern Australia experiences cool, moist westerly winds. During the winter, frontal systems passing from the west to the east over the Southern Ocean have a controlling influence on the climate of southern Australia, causing rainy periods. In the summer months (November-April), the anticyclones move in a more southerly track along the coast, directing easterly winds over the continent and providing fine, hot weather in southern Australia. During this season, northern Australia is heavily influenced by the intertropical convergence zone. The associated intrusion of warm moist air gives rise to hot and humid conditions. Heavy rain may be prevalent for two to three weeks at a time due to tropical depressions caused by monsoonal low-pressure troughs. Thus, in contrast to the wet summer/ dry winter typical of Darwin and Brisbane, Adelaide and Perth show the wet winter/dry summer pattern whereas Sydney, Melbourne, Canberra and Hobart show a relatively uniform pattern of rainfall throughout the year.

Rainfall frequency can be described in terms of rain-days. A rain-day is defined as a 24-hour period, usually from 9 am to 9 am the next day, when more than 0.2 mm of rain is recorded. The frequency of rain-days does not necessarily correlate well with the annual rainfall. For example, the frequency exceeds 150 rain-days per year in parts of the north Queensland coast where the annual rainfall is over 2000 mm, as well as in much of southern Victoria and in the extreme south-west of Western Australia where it is not more than about 600 mm. Over most of the continent the frequency is less than 50 rain-days per year. In the high rainfall areas of northern Australia, the number of rain-days is about 80 per year, but much heavier falls occur in this region than in southern regions.

7.3 Geographical and seasonal distribution of rainfall yield in Australia and its relationship with lightning ground flash density

To investigate geographical and seasonal distribution of rainfall yield in Australia and its relationship with lightning ground flash density, 23 Australian sites were selected by Jayaratne and Kuleshov (2006b) for their reliability and availability of lightning and rainfall data over sufficiently long periods of observation and to represent a wide geographical distribution across the continent. These sites are shown on the map in Fig 13.

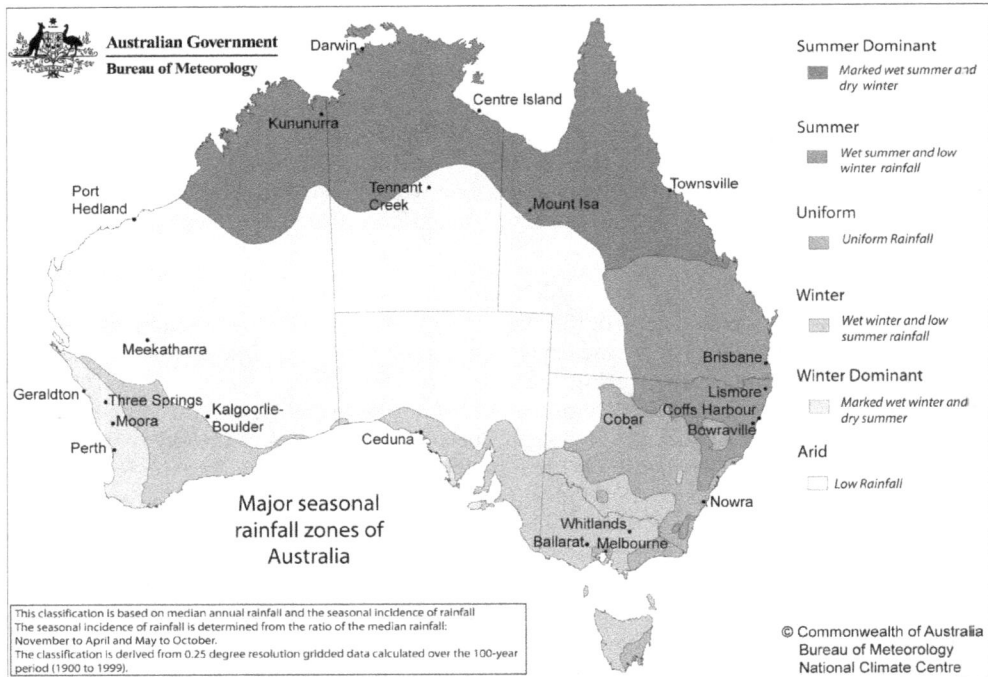

Fig. 13. Major seasonal rainfall zones of Australia.

The mean annual rainfall of the stations against the mean annual ground flash density is presented in Fig 14. The two straight lines show the constant rain yields of 10^8 kg fl^{-1} (lower line) and 10^9 kg fl^{-1} (upper line). Williams et al. (1992) showed that the rain yield from

Darwin break period storms was below the lower line, while for the monsoonal periods at the same location the value was above the upper line. Petersen and Rutledge (1998) showed that most of the arid mid-continental stations lay close to or below the lower line while the maritime and tropical oceanic stations showed values above the upper line. These results covered a wide range of locations around the world. In agreement, results for Australia indicate a similar trend.

In Fig 14, most of the coastal stations lie close to the upper rain yield line, while almost all of the arid continental stations lie closer to the lower line. For example, the six stations with the highest rain yields (above the upper line) are all situated on the coast or in proximity to the coast (1, 4, 5, 15, 20 and 21). At the other end of the scale, all five dry inland stations (shown as triangles) are found clustered together near the lower line (2, 3, 10, 13 and 17). Two coastal stations with relatively low rain yields are Kununurra (8) and Port Hedland (6). Both these stations are situated along the dry north-west coast. Although coastal fringes are comparatively moist, this is less evident along the north west coast where continental effects are marked. Kununurra, in fact, shows the lowest rain yield of all the stations investigated. The station with the highest rain yield is Nowra (20), which is a temperate station on the east coast.

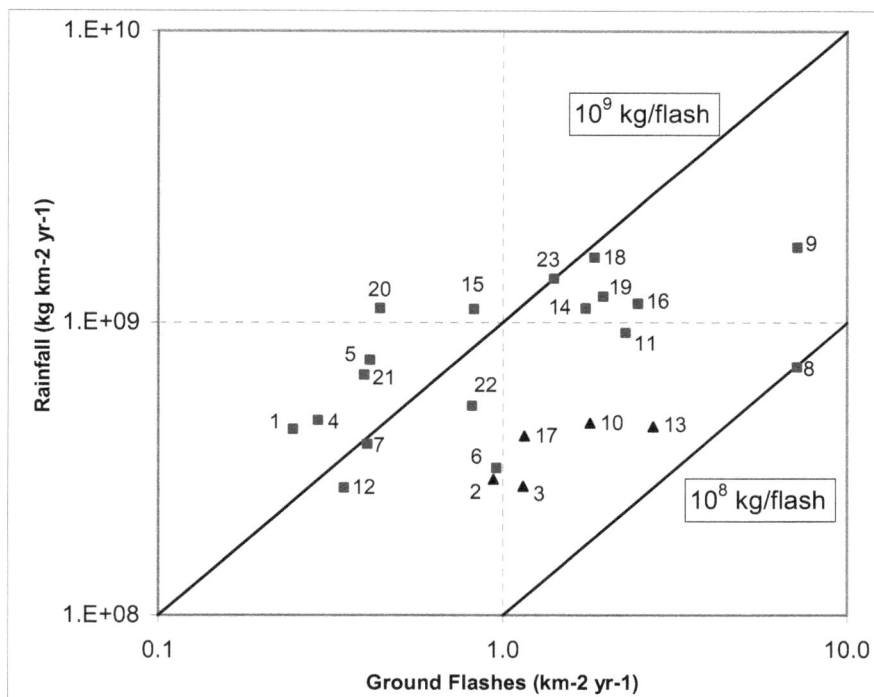

Fig. 14. The mean annual rainfall versus the mean annual ground flash density.

Another interesting feature that emerges from the figure is the relatively high rain yields in the low lightning coastal locations in Western Australia (1, 4, 5 and 7), South Australia (12) and Victoria (21, 22 and 23). These are the locations that receive winter rain. Despite the high winter rainfall, lightning activity during this time is low owing to the low surface temperatures and CAPE. The coastal stations in New South Wales and Queensland (14, 15, 16, 17, 18, 19 and 20) receive most of its annual rainfall during the summer months when high surface temperatures give rise to higher CAPE and lightning activity.

In order to investigate a seasonal difference, the rainfall and lightning data for each station were separated into two seasons – arbitrarily defined as winter half of the year (Apr-Sep) and summer half of the year (Oct-Mar) and calculated the corresponding rain yields. The winter-half rain yield versus the summer-half rain yield is presented in Fig 15. Each point represents a station. The straight line shows equality. As expected, every one of the 23 stations showed a winter-half rain yield greater than the summer-half yield. At many stations, especially in the tropics, the winter-half values were over an order of magnitude greater than the summer-half values. The mean winter-half and summer-half values were 2.46 x 10^9 kg fl-1 and 4.53 x 10^8 kg fl-1 respectively. A Students t-test analysis showed that the winter-half mean value was significantly greater than the summer-half mean at a confidence level of 99%. This difference was also obvious in the monthly variation.

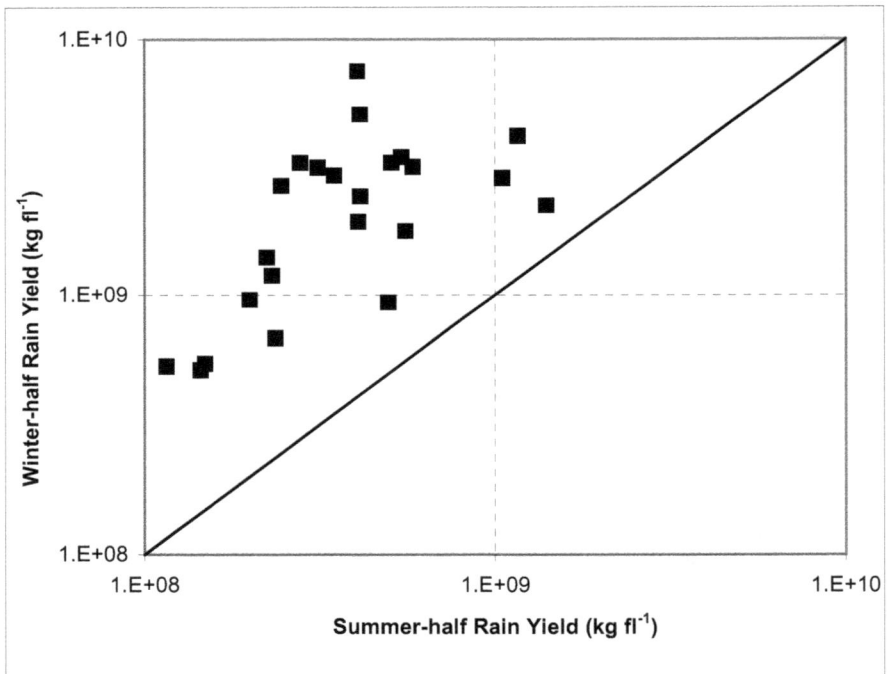

Fig. 15. The winter-half rain yield versus the summer-half rain yield.

Using a purely geographical classification, the stations were grouped into two zones – coastal (near or on the coastline) and inland (mid-continental; at least 500 km from the nearest coast). In good agreement with other studies (Williams et al., 1992; Petersen and

Rutledge, 1998), the mean rain yield in Australia (Table 6) for the mid-continental stations (2.64 x 10^8 kg fl⁻¹) was lower than the corresponding overall mean value of the coastal stations (9.91 x 10^8 kg fl⁻¹) (Jayaratne and Kuleshov, 2006b).

In attempting to look for trends in relation to climatic conditions, it was necessary to identify the various climatic zones in Australia. Jayaratne and Kuleshov (2006b) used the classification proposed by the ABM based on Gaffney (1971) and grouped the stations into six seasonal rainfall zones as shown in the map in Fig 13. The zones are listed in Table 7, together with the stations in each zone and the corresponding zonal station mean annual rainfalls, ground flash densities and rain yields. The annual rainfall in the two summer rainfall zones, S and SD, were significantly higher than in the two winter rainfall zones, W and WD. However, the converse was true for the rain yield, with the W and WD values being significantly higher than those for S and SD. The mean rain yields for groups S and SD was 5.44 x 10^8 kg fl⁻¹ and for groups W and WD it was 1.28 x 10^9 kg fl⁻¹ – a difference of over a factor of two. Statistical analyses showed that the winter zone rain yields were significantly higher than both the summer zone and arid zone values at a confidence level of 95%. The rainfall and rain yield in the arid zone, A, was significantly lower than in all other zones. As expected, the ground flash densities were highest in the two seasonal rainfall zones S and SD.

Zone	Notation	Mean Rain Yield (x10^8 kg fl⁻¹)
Inland	I	2.64
Northern Coastal	NC	2.52
Eastern Coastal	EC	10.9
Southern Coastal	SC	10.3
Western Coastal	WC	12.9
All Coastal		9.91

Table 6. The stations classified according to geographical location. Also shown in the last row is the overall mean for all coastal stations.

Zone	Notation	Mean Rainfall (x10^8 kg km⁻² yr⁻¹)	Mean Ground Flash Density, N_g (fl km⁻² yr⁻¹)	Mean Rain Yield (x10^8 kg fl⁻¹)
Summer	S	12.9	2.00	6.59
Summer Dominant	SD	9.96	4.02	4.52
Uniform	U	7.64	0.80	14.5
Winter	W	7.16	0.74	10.3
Winter Dominant	WD	5.06	0.34	15.3
Arid	A	3.35	1.21	2.84

Table 7. The stations classified according to the six major seasonal rainfall zones.

In agreement with previous studies, the present results in Australia show increased rain yields at coastal stations over inland stations and during the winter over the summer seasons. These observations are broadly explicable in terms of the differences in surface heating between the two seasons and between the geographical locations. Greater surface heating is related to higher cloud base height and greater CAPE, both of which have been linked with higher lightning activity (Williams et al, 2005; Williams and Renno, 1993). Moreover, the air over inland stations contain higher aerosol and cloud condensation nuclei concentrations than coastal stations and, by the explanation given in the introduction, may exhibit higher lightning activity, lower rainfall and lower rain yields.

8. Summary and conclusions

In this chapter, a review of thunderstorm and lightning climatology of Australia, with emphasis on spatial distribution and frequency of thunder-days and lightning flash density over the Australian continent, was presented. Thunderstorms are most frequent over the northern half of the country, and generally decrease southward, with lowest frequencies in southeast Tasmania. A secondary maximum is also apparent in southeast Queensland and over central and eastern New South Wales, extending into the north-eastern Victorian highlands.

Long-term lightning data obtained by Australian ground-based lightning flash counters as well as lightning data gathered by the NASA satellite-based instruments (OTD and LIS), have been used in order to analyze spatial distribution and frequency of lightning activity in Australia. The lightning data were used to produce lightning flash density maps, the first for Australia. Long-term records of thunder-days registered at the Australian stations were also used to produce updated thunder-day map.

In general, the geographical distribution of lightning incidence expressed as total lightning flash density, N_t, (i.e. cloud-to-ground and intracloud flashes) demonstrates high level of lightning activity ($N_t > 10$ km^{-2} yr^{-1}) in the northern parts of Australia and a decrease in total flashes southwards ($N_t < 5$ km^{-2} yr^{-1} in Victoria, the southern parts of Western and Southern Australia, and Tasmania). The peak lightning occurrence is in the north-western part of the Australian continent with N_t values above 30 km^{-2} yr^{-1} centred around 16° S 126° E. Secondary maxima are evident over the north-eastern part of the Northern Territory and northern part of Queensland, and over the eastern parts of Queensland and New South Wales, with N_t values up to 10 km^{-2} yr^{-1}. Spatial distribution of lightning occurrence is in general agreement with the spatial distribution of thunderstorms as it is presented in the map of average annual thunder-days for Australia.

Seasonal distributions of lightning occurrence demonstrate predominance of lightning in Australia in warmer months, with peak occurrence of lightning in January. In the northern half of the country, development of thunderstorms and lightning is enhanced by high boundary-layer moisture levels and lower surface pressure in the wet season months (October – April). Seasonal distribution of lightning activity in higher latitudes is more uniform; lightning is still more frequent in warmer months, but it also occurs in cooler months (May - September) in association with active frontal systems.

A reduction in lightning total flash density, N_t, by a factor of about 10 for a change in latitude from 10° S to 40° S was found, and this is in agreement with the earlier studies. The variation of N_t for the range of latitudes over the Australian continent is approximately expressed by the empirical relationship $N_t = \exp(3.85 - 0.088\lambda)$ km^{-2} yr^{-1}, where λ is the magnitude of the latitude in degrees. It appears that the main explanation for the annual variation in monthly flash density and latitudinal variation in annual flash density across the Australian continent is found in the variation in wet bulb temperature and convective available potential energy (CAPE).

Time series of T_d for several Australian tropical and subtropical observational sites have been examined. For about the 50 year period, it would appear that there has been a small upward trend in T_d for all sites, possibly followed by a subsequent downward trend. However, the longer-term T_d data (118 years) for Brisbane appear to indicate both upward and downward trends. The effects of climate change on thunderstorms in Australia need to be studied in much greater detail, because thunderstorms and lightning are temperature-sensitive weather phenomena.

CAPE is the driving force for thunderstorm development and it is closely controlled by wet bulb temperature. The lightning activity over continental Australia was studied as a function of wet bulb temperature and it was found that the monthly total of lightning ground flashes, N, increased sharply with the increase of the monthly mean daily maximum wet bulb temperature, $T_{w,max}$. The dependence was strongest in the tropics and became less pronounced at temperate latitudes. In Darwin, located at latitude 12°S, a modest 3-4°C increase in wet bulb temperature increased the lightning activity by over two orders of magnitude. Similar increases were found at other tropical stations. The corresponding increases for Coffs Harbour (latitude 30°S) and for Melbourne (latitude 38°S) were about one and a half orders of magnitude and about half an order of magnitude respectively, each over a 10°C range of $T_{w,max}$. Power law approximations were derived for each of the ten stations and showed that the logarithm of N was directly proportional to the power, P, of $T_{w,max}$. The value of P showed a sharp exponential decrease with increasing latitude away from the equator. A possible reason attributed for this behaviour is the sensitive dependence of wet bulb temperature on CAPE – the driving force for thunderstorm development.

Ground-based observations of annual rainfall and lightning incidence around the continent of Australia were used to compute values of 'rain yield', defined as the mass of rain produced per lightning ground flash (units: kg fl^{-1}) over a given area of ground. The rain yield was found to vary considerably with geographical location, season and climatic conditions. Mid-continental stations showed a mean rain yield of 2.64 x 10^8 kg fl^{-1} in contrast to the coastal and near-coastal stations that showed a corresponding mean value of 9.91 x 10^8 kg fl^{-1}. When the stations were classified according to seasonal climate zones, the winter and winter-dominant rainfall stations showed a rain yield of 1.28 x 10^9 kg fl^{-1} while the summer and summer-dominant rainfall stations showed a significantly lower value of 5.44 x 10^8 kg fl^{-1}. Mean winter rain yields were significantly higher than the summer values. These differences are attributed to surface heating which controls such parameters as cloud base height and CAPE in the atmosphere. In terms of the behaviour of the rain yield with geographical, seasonal and climatic conditions, the Australian observations are in good agreement with studies in other parts of the world.

9. Acknowledgement

Author expresses his gratitude to long term collaborators on Australian lightning studies Professors Mat Darveniza, David Mackerras and Rohan Jayaratne.

10. References

Anderson, R. B., and A. J. Eriksson (1980), Lightning parameters for engineering application, *Electra, 69,* 65-102.

Anderson, R. B., A. J. Eriksson, H. Kroninger, D. V. Meal, and M. A. Smith (1984), Lightning and thunderstorm parameters, *Paper presented at IEEE Int. Conf. Lightning and Power Systems,* London, UK.

Anderson, R. B., H. R. van Niekerk, S. A. Prentice, and D. Mackerras (1979), Improved Lightning Flash Counter, *Electra, 66,* 85-98.

Baral, K. N., and D. Mackerras (1993), Positive cloud-to-ground lightning discharges in Kathmandu thunderstorms, *J. Geophys. Res.,* 98, 10,331-10,340.

Barham, R. A., and D. Mackerras (1972), Vertical-aerial CIGRE-type lightning flash counter, *Electron. Lett., 8,* 480-482.

Battan, L.J. (1965). Some factors governing precipitation and lightning from convective clouds. *J. Atmos. Sci.,* vol. 22, pp. 79-84.

Bernardi, M., A. Pigini, G. Diendorfer, and W. Schulz (2002), Long term experience on lightning acquisition in Italy and Austria and data application to the improvement of lightning performance, *CIGRE Rep. 33-205,* Cons. Int. des Grands Reseaux Electr., Paris.

Boccippio, D. J., K. L. Cummins, H. J. Christian, and S. J. Goodman (2001), Combined satellite- and surface-based estimation of the intracloud-cloud-to-ground lightning ratio over the continental United States, *Mon. Weather Rev., 129,* 108-122.

Brook, M., M. Nakano, and P. Krehbiel (1982), The electrical structure of the Hokuriku winter thunderstorms. *J. Geophys. Res.* 87, C2, 1207-1215.

Cheze, J.L. and Sauvageot, H. (1997). Area average rainfall and lightning activity. *J. Geophys. Res.,* vol. 102, 1707-1715.

Christian, H. J., K. T. Driscoll, S. J. Goodman, R. J. Blakeslee, D. A. Mach, and D. E. Buechler (1996), The Optical Transient Detector (OTD), paper presented at 10th Int. Conf. on Atmospheric Electricity; Osaka, Japan; 10-14 June.

Christian, H. J., R. J. Blakeslee, S. J. Goodman, D. A. Mach, M. F. Stewart, D. E. Buechler, W. J. Koshak, J. M. Hall, W. L. Boeck, K. T. Driscoll, and D. J. Bocippio (1999), The Lightning Imaging Sensor, paper presented at 11th Int. Conf. on Atmospheric Electricity, Guntersville, Alabama, 7-11 June.

Christian, H. J., R. J. Blakeslee, D. J. Boccippio, W. L. Boeck, D. E. Buechler, K. T. Driscoll, S. J. Goodman, J. M. Hall, W. J. Koshak, D. M. Mach, and M. F. Stewart (2003), Global frequency and distribution of lightning as observed from space by the Optical Transient Detector, *J. Geophys. Res., 108*(D1), 4005, doi:10.1029/ 2002JD002347.

Cummins, K.L., M.J. Murphy, E.A. Bardo, W.L. Hiscox, R.B. Pyle, and A.E. Pifer (1998), A combined TOA/MDF technology upgrade of the U.S. National Lightning Detection Network, *J. Geophys. Res.*, 103, 9035-9044.

Gaffney, D.O. (1971). Seasonal rainfall zones in Australia. Bureau of Meteorology. Working papers. No. 141, 9 p.

GPATS website, www.gpats.com.au

Huschke, R. E. (Ed.) (1959), *Glossary of Meteorology*, 638 pp., Am. Meteorol. Soc., Boston, Massachusetts.

Intergovernmental Panel on Climate Change (IPCC) (2007), *Climate Change 2007*, the Fourth Assessment Report, WMO and UNEP, Geneva, Switzerland.

Jayaratne ER (1993) Conditional instability and lightning incidence in Gaborone, Botswana. Meteorol Atmos Phys 52: 169-175

Jayaratne, E. R. and Y. Kuleshov (2006a), The relationship between lightning activity and surface wet bulb temperature, *Meteorol. Atmos. Phys.*, 91, 17-24, doi:10.1007/s00703-004-0100-0.

Jayaratne, E. R. and Kuleshov (2006b), Geographical and seasonal characteristics of the relationship between lightning ground flash density and rainfall within the continent of Australia, *Atmospheric Research*, vol 79, no 1, pp. 1-14.

Jayaratne, E.R., Ramachandran, V., Devan, K.R.S. (1995). Observations of lightning flash rates and rain gushes in Gaborone, Botswana. J. Atmos. Terr. Phys. 57 (4), 325-331.

Kattron website, www.lightning.net.au

Kinzer, G. (1974). Cloud-to-ground lightning versus radar reflectivity in Oklahoma thunderstorms. J. Atmos. Sci. 31, 787-799.

Koike, M.; Y. Kondo, K. Kita, N. Takegawa, N. Nishi, T. Kashihara, S. Kawakami, S. Kudoh, D. Blake, T. Shirai, B. Liley, M. Ko, Y. Miyazaki, Z. Kawasaki, and T. Ogawa (2007), Measurements of reactive nitrogen produced by tropical thunderstorms during BIBLE-C, *J. Geophys. Res.*, 112, D18304, doi:10.1029/2006JD008193.

Kuleshov, Y., G. de Hoedt, W. Wright, and A. Brewster (2002), Thunderstorm distribution and frequency in Australia, *Austral. Meteorol. Mag. 51*, 145-154.

Kuleshov, Y., and E. R. Jayaratne (2004), Estimates of lightning ground flash density in Australia and its relationship to thunder-days, *Austral. Meteorol. Mag. 53*, 189-196.

Kuleshov, Y., D. Mackerras, and M. Darveniza (2006), Spatial distribution and frequency of lightning activity and lightning flash density maps for Australia. *J. Geophys. Res.*, 111, D19105, doi:10.1029/2005JD006982.

Kuleshov, Y., D. Mackerras, and M. Darveniza (2010), Spatial Distribution and Frequency of Thunderstorms and Lightning in Australia. Chapter 8, in *Lightning: Principles, Instruments and Applications*. H.D. Betz et al. (eds), Springer Science + Business Media B.V., 189-209, doi 10.1007/978-1-4020-9079-0_8.

Kuleshov, Y., P. Hettrick, D. Mackerras, M. Darveniza and E. R. Jayaratne (2011), Occurrence of positive and negative polarity cloud-to-ground lighting flashes: Case study of CGR4 and GPATS data for Brisbane, Australia. *Australian Meteorological and Oceanographic Journal*, 61,107-112.

Lightning Protection (2007), Australian Standard/New Zealand Standard 1768-2007, 188 pp., Standards Australia, Sydney, Australia and Standards Association of New Zealand, Wellington, New Zealand.

Mackerras, D. (1978), Prediction of lightning incidence and effects in electrical systems, *Electr. Eng. Trans. Int. Eng. Austral.* 73-77.

Mackerras, D. (1985), Automatic short-range measurement of the cloud flash to ground flash ratio in thunderstorms, *J. Geophys. Res. 90*, 6,195-6,201.

Mackerras, D., and M. Darveniza (1992), Design and performance of CGR3 instruments for measuring the cloud flash-to-ground flash ratio in thunderstorms, *Internal Report EE92/2*, 99 pp., Univ. of Queensland, St. Lucia, Australia.

Mackerras, D. and M. Darveniza (1994), Latitudinal variation of lightning occurrence characteristics, *J. Geophys. Res. 99*, 10,813-10,821.

Mackerras, D., M. Darveniza, R. E. Orville, E. R. Williams, and S. J. Goodman (1998), Global lightning: Total, cloud and ground flash estimates, *J. Geophys. Res. 103*, 19,791-19,809.

Mackerras, D., M. Darveniza and P. Hettrick (2009), The CGR4 lightning sensor, *Austral. Meteorol. and Oceanograph. J.*, 58, 263-273.

Moore, C.B., Vonnegut, B., Machado, J.A., Survilas, H.J. (1962). Radar observations of rain gushes following overhead lightning strokes. J. Geophys. Res. 67, 207-220.

Mushtak, V., E. Williams, and D. Boccippio (2005), Latitudinal variations of cloud base height and lightning parameters in the tropics , *Atm. Res.* 76, 222-230.

Orville, R. E., and G. R. Huffines (2001), Cloud-to-ground lightning in the United States: NLDN results in the first decade, 1989-98, *Mon. Weather Rev.*, 129, 1179-1193.

Orville, R.E., and W. Spencer (1979), Global lightning flash frequency, *Mon. Weather Rev.*, 107, 934-943.

Orville, R.E., Henderson, R.W. (1986). Global distributions of midnight lightning: September 1977 to August 1978. Mon. Weather Rev. 114, 2640-2653.

Piepgrass, M.V., Krider, E.P., Moore, C.B. (1982). Lightning and surface rainfall during Florida thunderstorms. J. Geophys. Res. 87, 11193-11201.

Pierce, E. T. (1956), The influence of individual variations in the field changes due to lightning discharges upon the design and performance of lightning flash counters, *Archiv für Meteorologie, Geophysik und Bioclimatologie, Serie A: Meteorologie Und Geophysik, Band 9*, 78-86.

Petersen, W.A., and S.A. Rutledge (1992), Some characteristics of cloud-to-ground lightning in tropical northern Australia. *J. Geophys. Res.*, 97, 11553-11560.

Peterson WA, Rutledge SA (1998), On the relationship between cloud to ground lightning and convective rainfall. J Geophys Res 103: 14025-14040

Peterson WA, Rutledge SA, Orville RE (1996), Cloud to ground lightning observations from TOGA COARE: selected results and lightning location algorithms. Mon Wea Rev 124: 602-620.

Pinto, O., Jr., K. P. Naccarato, I. R. C. A. Pinto, W. A. Fernandes, and O. Pinto Neto (2006), Monthly distribution of cloud-to-ground lightning flashes as observed by lightning location systems. *Geophys. Res. Lett.*, 33, L09811, doi:10.1029/2006GL026081.

Reap, R.M., MacGorman, D.R. (1989), Cloud-to-ground lightning: climatological characteristics and relationships to model fields, radar observations, and severe local storms. Mon. Wea. Rev. 117, 518-535.

Rosenfeld, D., Lensky, I.M. (1998), Satellite-based insights into precipitation formation processes in continental and maritime convective clouds. Bull. Amer. Meteorol. Soc. 79, 2457-2476.

Rutledge S. A., E. R. Williams, and T. D. Keenan (1992), The Down Under Doppler and Electricity Experiment (DUNDEE): Overview and Preliminary results, *Bull. Am. Meteorol. Soc.*, 73(1), 3-15.

Schulz, W., K. Cummins, G. Diendorfer, and M. Dorninger (2005), Cloud-to-ground lightning in Austria: a 10-year study using data from a lightning location system. *J. Geophys. Res.* 110, D09101, doi:10.1029/2004JD005332.

Shackford, C.R. 1960. Radar indications of a precipitation-lightning relationship in New England thunderstorms. J. Meteor. 17, 15-19.

Soriano, L. R., F. de Pablo, and C. Tomas (2005), Ten-year study of cloud-to-ground lightning activity in the Iberian Peninsula, *J. Atmos. Terr. Phys.*, 67(16), 1632-1639.

Takeuti, T., M. Nakano, M. Brook, D. J. Raymond, and P. Krehbiel (1978), The anomalous winter thunderstorms of the Hokuriku Coast. *J. Geophys. Res.* 83, C5, 2385-2394.

Thery, C. (2001), Evaluation of LPATS data using VHF interferometric observations of lightning flashes during the Eulinox experiment, *Atmos. Res.*, 56, 397-409.

Williams, E. R. (1992), The Schumann resonance: A global tropical thermometer, *Science*, 256, 1184-1187.

Williams, E.R. (1994), Global circuit response to seasonal variations in global surface air temperature. *Mon Wea Rev* 122: 1917-1929.

Williams, E. R. (1995), Meteorological aspects of thunderstorms, in *Handbook of Atmospheric Electrodynamics*, edited by H. Volland, pp. 27-60, CRC Press, London, UK.

Williams, E. R., S. A. Rutledge, S. G. Geotis, N. Renno, E. Rasmussen, and T. Rickenbach (1992), A radar and electrical study of tropical "hot towers", *J. Atmos. Sci.* 49: 1,386-1,395.

Williams, E., and S. Stanfill (2002), The physical origin of the land-ocean contrast in lightning activity, *C. R.-Acad. Sci.*, Phys. 3, 1277-1292.

Williams, E., V. Mushtak, D. Rosenfeld, S. Goodman, and D. Boccippio (2005), Thermodynamic conditions favorable to superlative thunderstorm updraft, mixed phase microphysics and lightning flash rate, *Atm. Res.* 76, 288-306.

Williams, E.R., and Renno, N. (1991), Conditional instability, tropical lightning, ionospheric potential and global change. 19th *Conf on Hurricanes and Tropical Meteorology*, Miami, Florida

Williams, E.R., and Renno, N. (1993), An analysis of the conditional instability of the tropical atmosphere. *Mon. Wea. Rev.* 121, 21-36.

Williams, E.R., Mushtak, V., Rosenfeld, D., Goodman, S., Boccippio, D. (2005). Thermodynamic conditions favourable to superlative thunderstorm updraft, mixed phase microphysics and lightning flash rate. *Atmos. Res.* 76, 288-306.

World Meteorological Organization (WMO) (1953), World distribution of thunderstorm days. Part 1: Tables. *WMO Publ. 21, TP 6*, WMO, Geneva, Switzerland.

Uman, M.A. (1987), The lightning discharge, Dover Publication Inc

Zajac, B., and S. A. Rutledge (2000), Cloud-to-ground lightning activity in the contiguous United States from 1995 to 1999, *Mon. Weather Rev.*, 129, 999-1019.

Zipser, E.J. (1994), Deep cumulonimbus cloud systems in the tropics with and without lightning. *Mon. Wea. Rev.* 122, 1837-1851.

Part 2

Regional Climatology

5

The South American Monsoon System: Climatology and Variability

Viviane B. S. Silva and Vernon E. Kousky
NOAA/National Weather Service
Climate Services Division
Climate Prediction Center
USA

1. Introduction

A typical Monsoon System is characterized by a reversal in the low-level wind direction between summer and winter seasons, and distinct wet (summer) and dry (winter) periods. The changes in low-level atmospheric circulation are related to changes in the thermal contrast between oceans and continents. During summer, the air over continents is warmer and more convectively unstable than air over adjacent oceanic regions. Consequently, lower pressure occurs over land and higher pressure occurs over nearby oceanic areas. This pressure pattern causes low-level moist air to converge onto the land, resulting in precipitation, especially during the late afternoon and evening hours. During winter the temperature contrasts and low-level atmospheric circulation are reversed, resulting in dry conditions over continents.

The regions on the globe that show distinct monsoon characteristics include 1) western Sub-Saharan Africa, 2) Asia (India, southern China, Korea and parts of Japan), 3) Northern Australia, 4) South America (Brazil, Bolivia, Paraguay) and 4) North America (Southern US and Mexico). The focus of this chapter is on the South American Monsoon System (SAMS).

South America has several important geographical features that contribute to the climate of the region (Fig. 1). The entire continent is surrounded by water, with the high Andes Mountains stretching along the entire west coast. South America also contains the world's largest rainforest (the Amazon) and driest Desert (Atacama in Chile). The core of the SAMS includes the Brazilian "Planalto" (BP), which contains the headwaters of major rivers flowing into the Amazon, La Plata and São Francisco basins. Those basins contain major agricultural areas and provide most of Brazil's hydroelectric energy production.

The SAMS displays considerable variability on time scales ranging from diurnal to inter-annual. Prolonged periods of wetter-than-average or drier-than-average conditions can have significant impacts on agriculture, energy production, and society in general. Due to the accentuated topography near Brazil's east coast, heavy rainfall can result in disastrous flooding, with loss of life, property and infrastructure. In many cases, the poorest inhabitants suffer the greatest impacts from heavy rainfall events, since they often reside in the most vulnerable areas, such as along streams and on steep unstable slopes. In addition, persistent rainfall deficits (droughts) can have negative impacts on agriculture and also on Brazil's energy production, leading to restrictions on energy usage affecting large sections of Brazil (e.g., Silva et al., 2005).

Fig. 1. South American key topographic features and major river basins [Amazon basin (light yellow), Sao Francisco basin (light green) and La Plata basin (light blue)]. The Brazilian "Planalto" (BP) is indicated by the red oval.

The objective of this chapter is to provide an overview of 1) the characteristic features of SAMS, including the evolution of precipitation and atmospheric circulation during the wet season, 2) the variability of SAMS on time scales ranging from diurnal to inter-annual, and 3) extreme rainfall events and their impacts.

2. Data sets

Precipitation data used to show the characteristic features of SAMS are derived from gridded daily precipitation analyses available from the NOAA/Climate Prediction Center (Silva et al., 2007; Chen et al., 2008). Prior to selecting a data set as the basis for an analysis of mean circulation features, an inter-comparison among six re-analyses was made for South America during the period 1979-2000. The selected re-analyses for comparison are: the NCEP/Climate Forecast System Reanalysis (CFSR) (Saha et al., 2010), the NCEP/NCAR CDAS-Reanalysis (R1) (Kalnay et al., 1996), the NCEP/Department of Energy-DOE Reanalysis (R2) (Kanamitsu et al., 2002), the European Centre for Medium-Range Forecasts (ECMWF) Reanalysis (ERA-40) (Uppala et al., 2005), the NASA /Global Modeling and Assimilation Office (GMAO) Reanalysis (MERRA) (Rienecker et al., 2011), and the Japanese Meteorological Agency Reanalysis (JRA-25) (Onogi, et al., 2005, 2007).

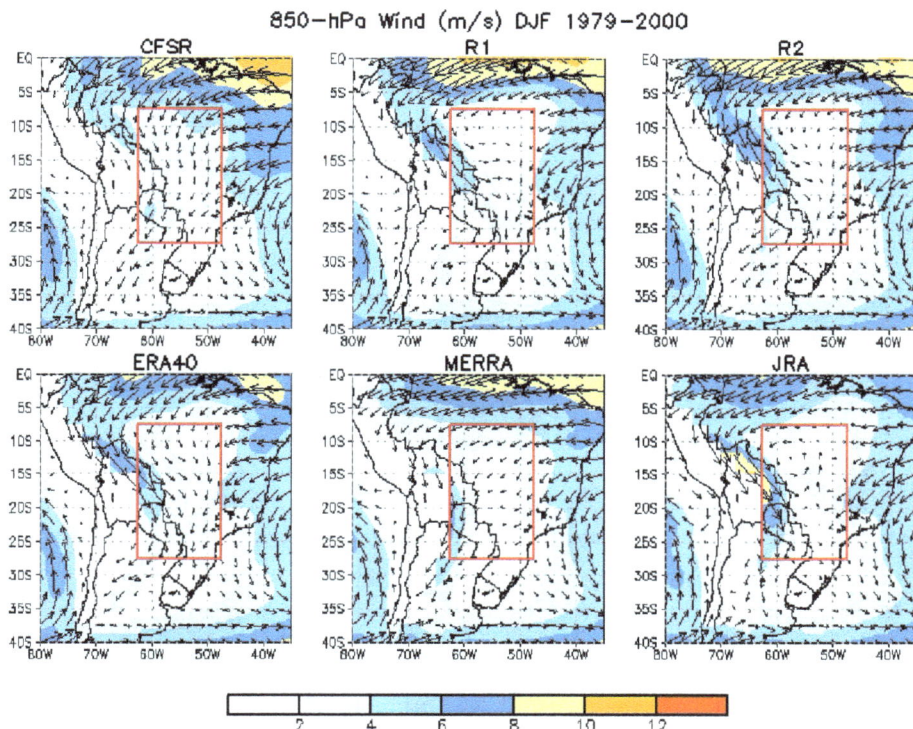

Fig. 2. Mean 850-hPa wind direction (vectors) and magnitude (shading, ms⁻¹) taken from six reanalysis data sets for December-February 1979-2000.

The upper-tropospheric circulation (200-hPa wind, not shown) is in good agreement among the various re-analyses over the entire region. Therefore, any of the reanalysis data sets could be used to qualitatively describe upper-tropospheric circulation features. In contrast, the December-February (DJF) mean (1979-2000) lower-tropospheric circulation (850-hPa wind) shows considerable variability among the re-analyses in the orientation and strength of the low-level flow (low-level jet) east of the Andes within the area 7.5°-20°S, 45°-65°W (red boxes in Fig. 2). Consequently, there is considerable uncertainty in the analyzed low-level flow characteristics in this region and in derived quantities, such as moisture flux, convergence, and vertical motion within the core region of the SAMS.

Comparing the DJF 1979-2000 mean precipitation patterns in the re-analyses to the analyzed station-based precipitation (Fig. 3) it is evident that the CFSR pattern is more similar to the observation-based pattern [lower right panel - OI (T62)] than any of the other reanalysis patterns. CFSR improvements include the proper location of a maximum in precipitation over the southern Amazon basin, and an absence of the spurious maximum over northeastern Brazil that is evident in the other reanalysis patterns. Silva et al., 2011 found that the pattern correlation between the DJF CFSR mean precipitation pattern and the observed precipitation pattern is much higher for CFSR than for either R1 or R2.

Furthermore, the DJF CFSR 500-hPa mean vertical motion pattern is much better correlated with the observed precipitation pattern than are the 500-hPa vertical motion patterns in R1 and R2 (Silva et al., 2011). Consequently, in the remainder of this chapter we will use the CFSR data to describe the mean circulation features related to SAMS. R1 analyses will be used in a qualitative manner to describe features related to extreme events.

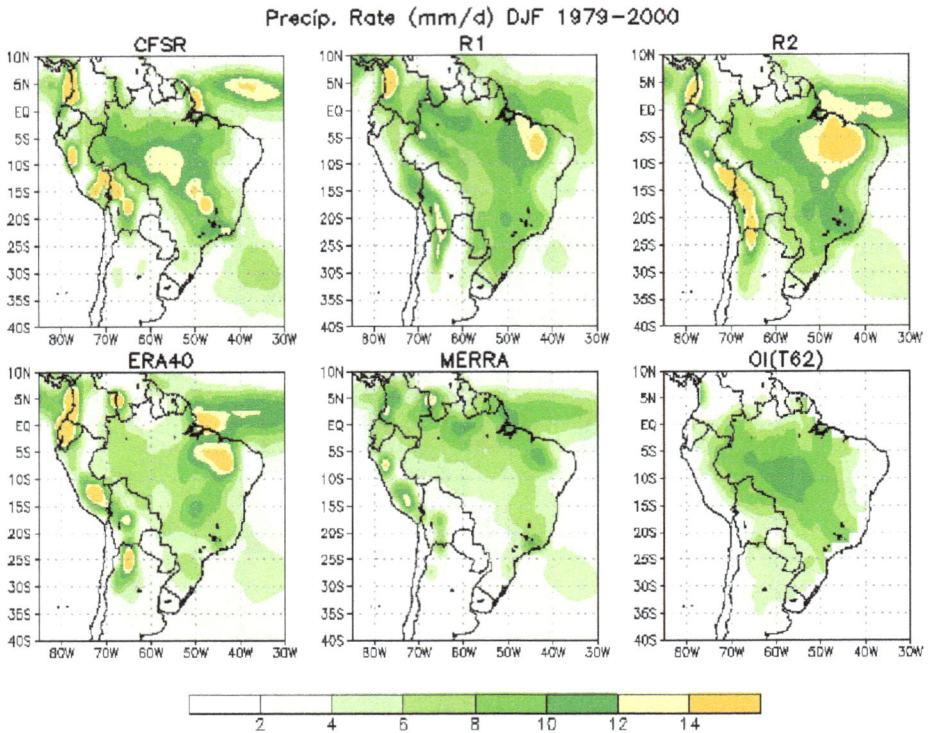

Fig. 3. Mean precipitation (mm d⁻¹) from five reanalysis data sets, and a station-based analysis (OI-T62, lower right panel) for December-February 1979-2000.

3. Characteristic features of SAMS

3.1 Major large-scale elements affecting the South American Monsoon System

The South American Monsoon System (SAMS) has been a major focus of the CLIVAR/VAMOS (Variability of the American Monsoon System) program. [CLIVAR is the World Climate Research Programme (WCRP) project that addresses Climate Variability and Predictability, with a particular focus on the role of ocean-atmosphere interactions in climate]. Several studies and reviews on the SAMS have described major features and phenomena that affect the behavior of SAMS on various time scales (e.g., Gan et al., 2004; Grimm et al., 2005; Vera at al., 2006; Gan et al., 2009; Liebmann & Mechoso, 2010; Marengo et al., 2010). The CLIVAR/VAMOS Panel developed a schematic diagram showing the major large-scale features related to the South American Monsoon System (Fig. 4). The

Andes mountains and Amazon basin play important roles in the South America monsoon. The Andes act as a barrier to the low-level easterly flow, which is deflected to the south over western Brazil, Bolivia and Paraguay during the austral summer. Intense summertime convection and latent heating over the continent contribute to the formation of an upper-tropospheric anticyclone, often referred to as the "Bolivian High". The rising air motion over the continent is compensated by sinking motion over the adjacent Pacific and Atlantic Oceans. These oceanic regions feature an absence of deep convection and the presence of upper-tropospheric cyclonic circulation (troughs).

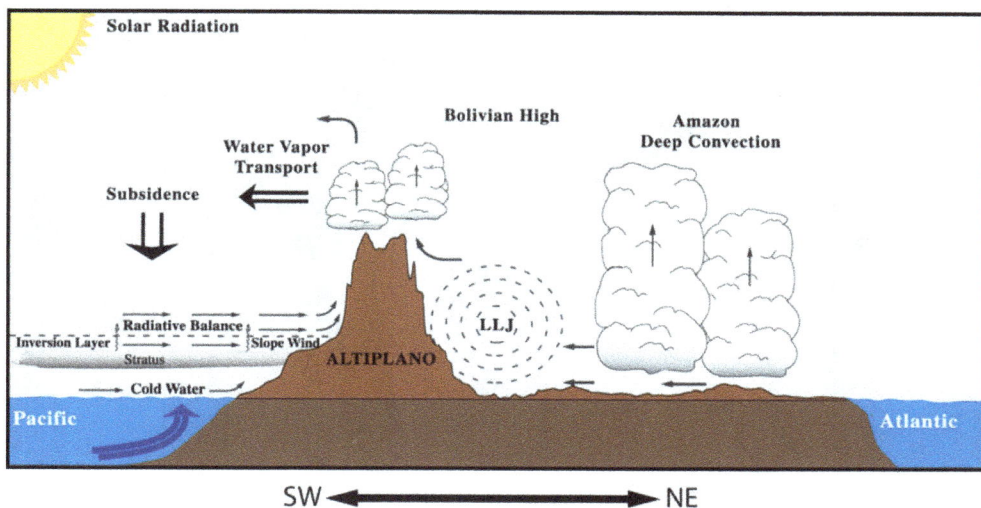

Fig. 4. Section across South America displaying schematically the major large-scale elements related to the South American Monsoon System. Source: Climate Variability & Predictability Program (CLIVAR)
(http://www.clivar.com/publications/other_pubs/clivar_transp/pdf_files/av_g3_0106.pdf)

3.2 Life cycle of the SAMS wet season

The annual cycle of precipitation, over tropical South America, features distinct wet and dry seasons between the equator and 25°S. Portions of central and eastern Brazil and the Andes Mountains between 12°S and 32°S receive more than 50% of their total observed annual precipitation during the austral summer (December-February: DJF) (Fig. 5). These same regions receive less than 5% of their total annual precipitation during the austral winter (June-August: JJA). The area from the mouth of the Amazon River to northern Northeast Brazil experiences a maximum in precipitation during austral fall (March-May: MAM).

During the wet season an upper-tropospheric anticyclone dominates the circulation over tropical and subtropical South America, while cyclonic circulation dominates the upper-tropospheric circulation over low latitudes of the eastern South Pacific and central South Atlantic (Fig. 6, top panel). The position of the upper-level anticyclone (southwest of the region of most intense precipitation and latent heating) is consistent with the atmospheric circulation response to tropical forcing (heating) (e.g., Webster 1972; Gill 1980). Prominent low-level features (Fig. 6, bottom panel) include: 1) surface high pressure systems and anticyclonic circulation over the subtropical oceans (Pacific and Atlantic), 2) a surface low-pressure system (Chaco Low) centered over northern Argentina, and 3) a low-level northwesterly flow (low-level jet) extending from the southwestern Amazon to Paraguay and northern Argentina. Throughout the region one notes a reversal of circulation features between the lower troposphere and the upper troposphere (Fig. 6, compare bottom and top panels), which is typical of the global Tropics.

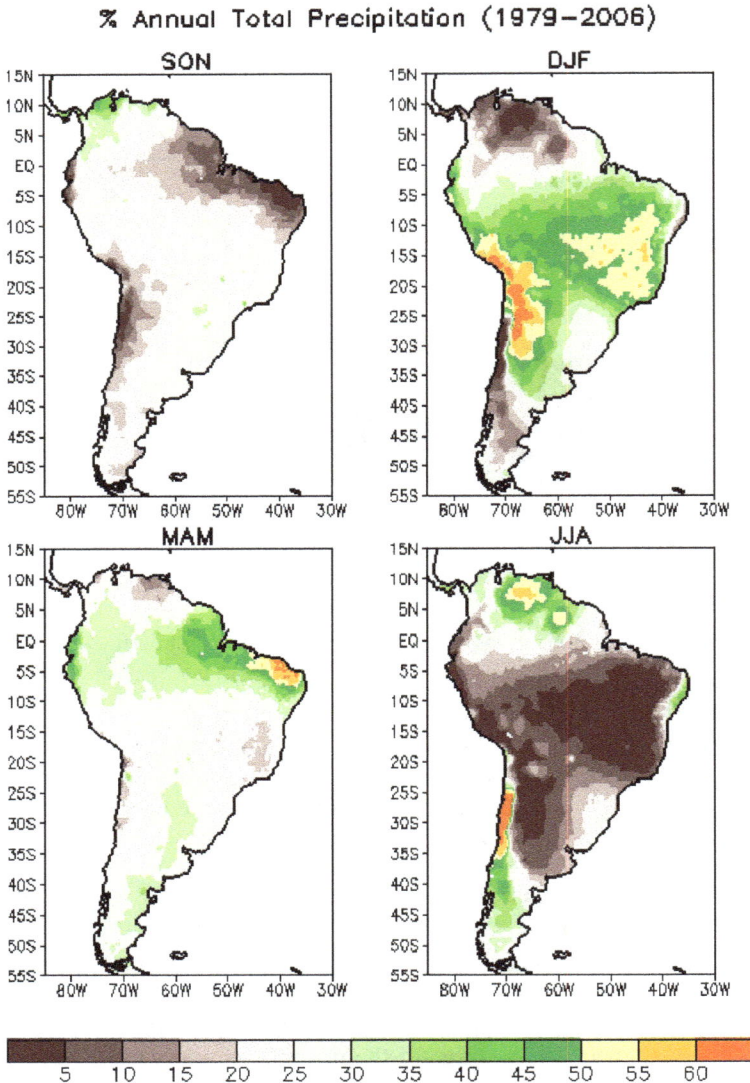

Fig. 5. Percent of observed mean (1979-2006) annual precipitation for each season.

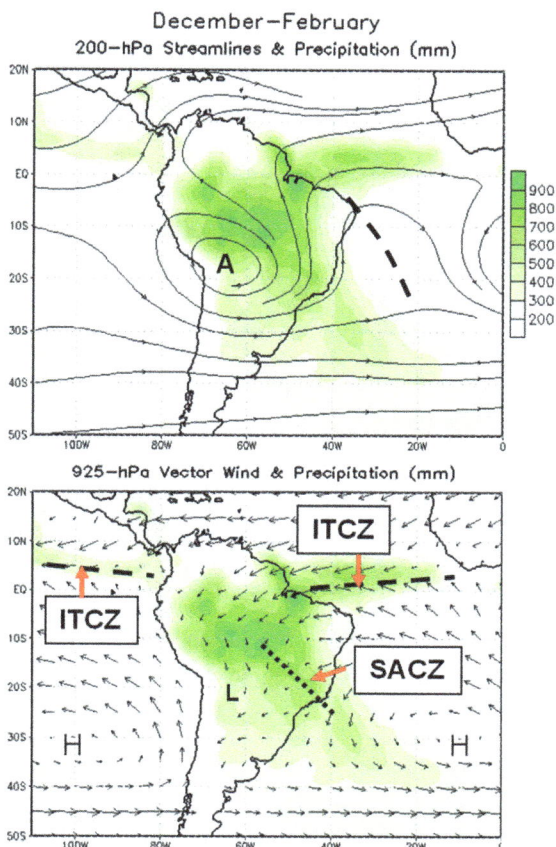

Fig. 6. Mean 200-hPa vector wind/ streamlines and estimated precipitation (mm) (top) and 925-hPa vector wind/ streamlines and estimated precipitation (bottom). The climatology period for the circulation fields is 1979-2010 and for precipitation is 1979-1995 (CAMS OPI).

The annual cycle of upper-tropospheric circulation features over South America is intimately linked to the seasonally varying horizontal temperature gradients, which arise from differential heating due to the difference in the thermal capacity between land and water. During summer, temperatures over the continent become warmer than the neighboring oceanic regions. This results in a direct thermal circulation with low-level (upper-level) convergence (divergence), mid-tropospheric rising motion and precipitation over the continent, and low-level (upper-level) divergence (convergence), mid-tropospheric sinking motion and dry conditions over the neighboring oceanic areas (Fig. 7, top left panel). These features are typical of summertime monsoons. During winter, temperatures over the continent and nearby oceanic regions are more uniform in the zonal (east-west) direction, which gives rise to a more zonally symmetric upper-tropospheric circulation pattern over the region (Fig. 7, lower right panel) and little or no evidence of any east-west direct thermal circulation (Fig. 7, top right panel).

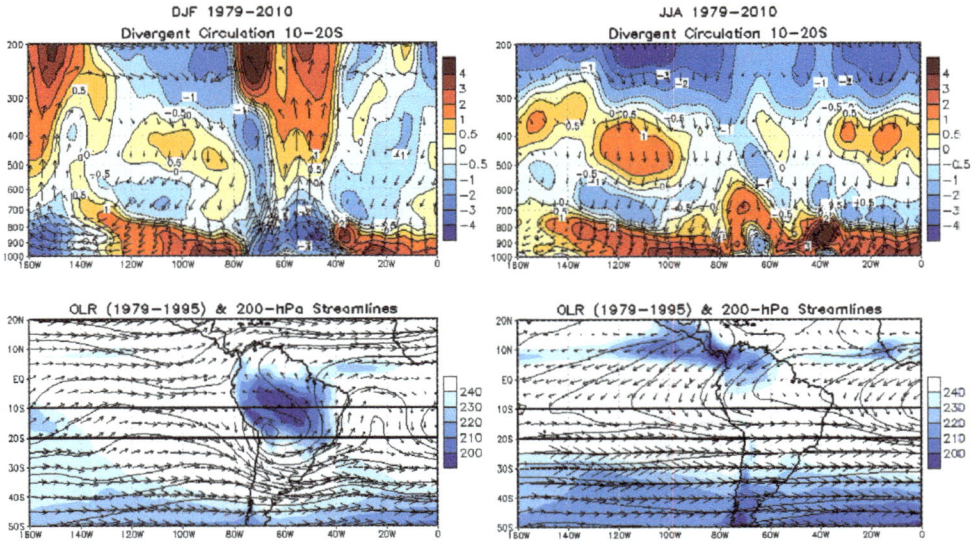

Fig. 7. Height-longitude cross-sections of the mean (1979-2010) divergent circulation (vectors) for the latitude band 10°-20°S (top panels) and mean (1979-2010) 200-hPa vector wind, streamlines and OLR (bottom panels) for December-February (left panels) and June-August (right panels). Units are 10^{-6} s^{-1} for divergence (contours and shading in top panels) and W m^{-2} for OLR (shading in bottom panels).

3.3 Onset, mature and demise phases

The development of the South American warm season Monsoon System during the austral spring is characterized by a rapid southward shift of the region of intense convection from northwestern South America to the southern Amazon Basin and Brazilian highlands (Altiplano) (Kousky, 1988; Horel et al., 1989; Marengo et al., 2001; Liebmann & Marengo, 2001; Nogues-Paegle et al., 2002) (Fig. 8). Deep convection increases over the western Amazon Basin in September and subsequently expands southward and southeastward, reaching central Brazil in October and Southeast Brazil in November. Lower-tropospheric (850-hPa) temperatures reach their annual maximum over the southern Amazon and BP region in early September, just prior to the onset of the rains (Fig. 9).

Transient synoptic systems at higher latitudes play an important role in modulating the southward shift in convection. Cold fronts that enter northern Argentina and southern Brazil are frequently accompanied by enhanced deep convection over the western and southern Amazon and an increase in the southward flux of moisture from lower latitudes (e.g., Garreaud & Wallace, 1998). These cold fronts are also important in the formation of the South Atlantic Convergence Zone (SACZ) (e.g., Garreaud & Wallace, 1998), which becomes established in austral spring over Southeast Brazil and the neighboring western Atlantic (see Fig. 8, middle column). During spring an upper-tropospheric anticyclone (Bolivian High) becomes established near 15°S, 65°W (Fig. 8), as the monsoon system develops mature-phase characteristics. Upper-level troughs and dry conditions are found over oceanic areas to the east and west of the Bolivian High. The deep convection over central Brazil and the Bolivian High reach their peak intensities during December-March. These features shift northward and weaken during April and May, as the summer monsoon weakens and a transition to drier conditions occurs over subtropical South America.

OLR Climatology

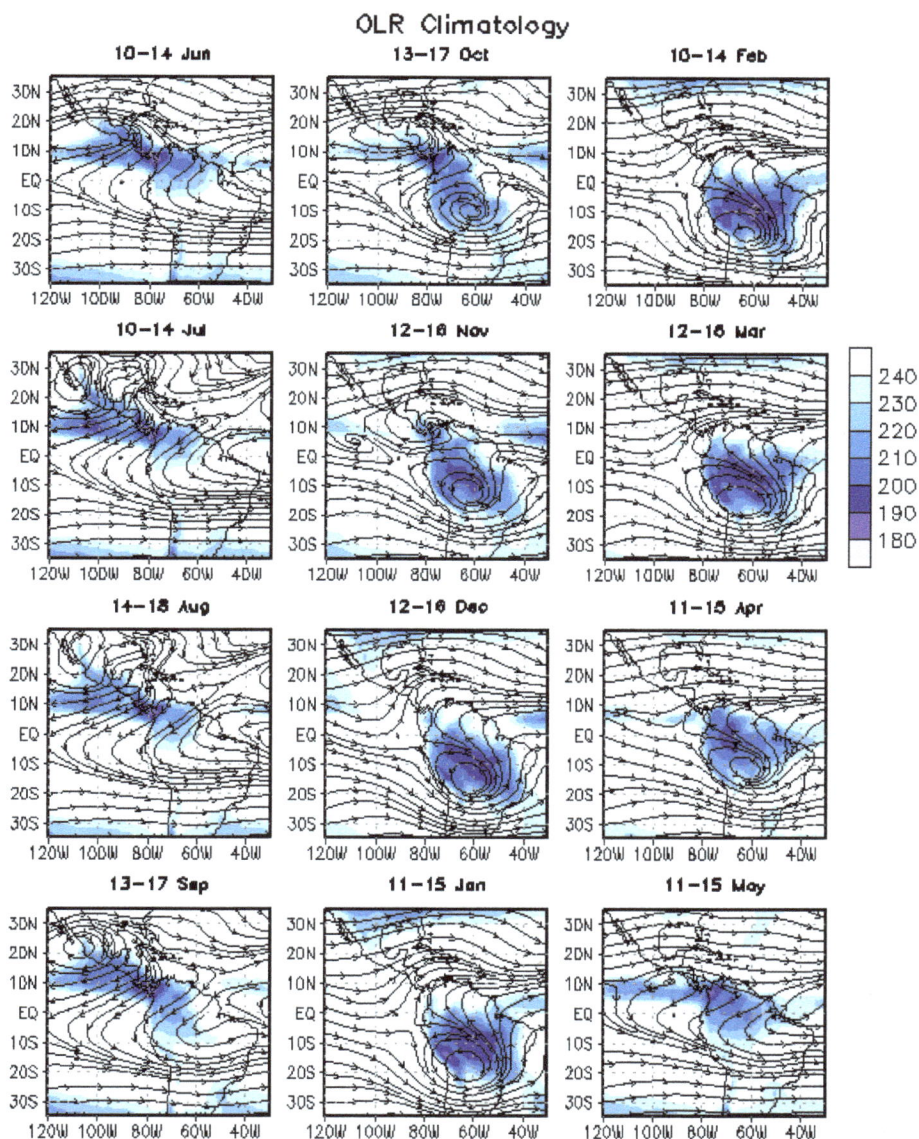

Fig. 8. Mean (1979-1995) seasonal cycle of OLR and 200-hPa streamlines. Units for OLR are W m-2. Low values of OLR indicate cold cloud tops (deep convection) in the Tropics.

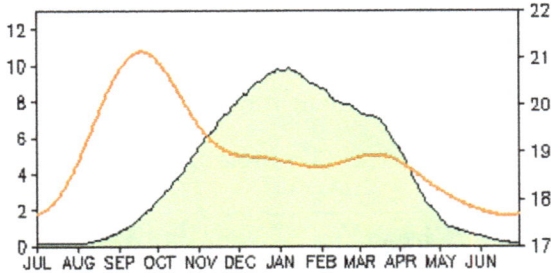

Fig. 9. Mean (1979-1995) daily precipitation (mm) for central Brazil (12.5°-17.5°S, 47.5°-52.5°W (shaded in green), and 850-hPa temperature (degrees C, red curve). The 850-hPa temperature data are taken from the R1 data archive.

There are many indexes in the literature that define the SAMS onset, each one with its unique characteristics (eg., Kousky, 1988; Marengo et al., 2001; Gan et al., 2006; Gonzales et al., 2007; Silva & Carvalho, 2007; Raia & Cavalcanti, 2008; Garcia & Kayano, 2009; Nieto-Ferreira & Rickenbach, 2010). The onset/end dates of the SAMS wet season, based on outgoing longwave radiation (OLR, a proxy for deep convection in the Tropics) (Kousky, 1988), are shown in Fig. 10. The wet season onset occurs in mid-September over the western Amazon basin, in mid-October over central Brazil (including the BP region), and in mid-November in Southeast Brazil. The end of the wet season occurs in early April over central Brazil, and in mid- to late May over the southern Amazon basin.

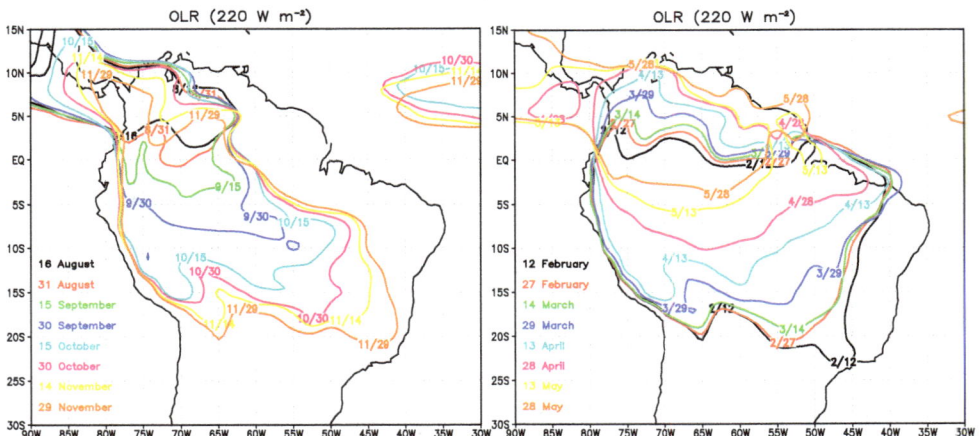

Fig. 10. Time onset and end dates for the wet season in the monsoon core region (Central Brazil) based on OLR less than 220 W m-2.

4. Variability of SAMS

4.1 Interannual variability

The phases of the EL Nino Southern Oscillation (ENSO) cycle (moderate to strong El Nino and La Nina episodes) have significant impacts on SAMS and the rainfall pattern over tropical South America (e.g. Hastenrath & Heller, 1977; Ropelewski & Halpert, 1987; Aceituno, 1988; Kousky & Kayano, 1994; Silva et al., 2007). Mossman (1924) was one of the first to notice the relationship between the Southern Oscillation and rainfall over central South America. He showed that the Paraná River level increases during the negative (warm) phase of the Southern Oscillation (El Niño). Subsequent studies (Streten, 1983; Kousky et al., 1984; Ropelewski & Halpert, 1987, 1989; Grimm et al., 1998) demonstrated that wetter-than-average conditions occur over southeast South America during El Niño, consistent with the results of Mossman (1924).

Since the extreme phases of the ENSO cycle tend to peak during the austral summer, Silva et al. (2007) elected to use water-year (July–June) rainfall departures to show ENSO-related interannual variability over Brazil. The pattern of anomalous precipitation during El Niño episodes (Fig. 11) shows considerable event-to-event variability, especially in the magnitude of the departures. The strongest El Niño episodes (1982/83, 1991/92, and 1997/98) feature

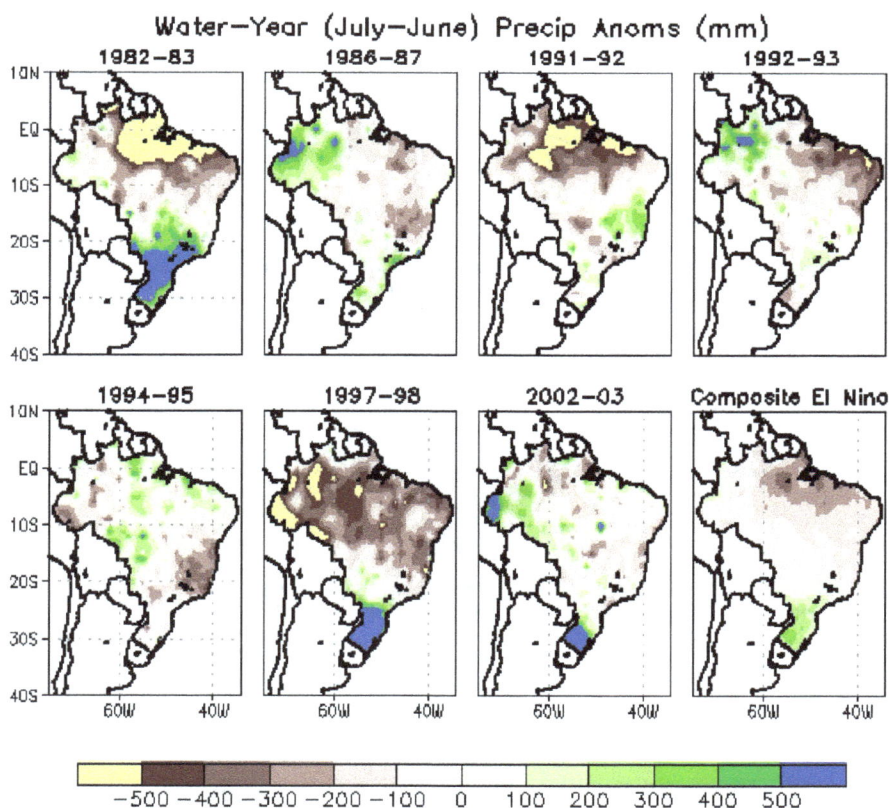

Fig. 11. Precipitation anomalies for water years (July–June) during El Niño episodes. The composite for the seven episodes is shown in the lower right-hand panel. Anomalies (mm) are computed with respect to the July 1977–June 2004 base period means. (Figure taken from Silva et al., 2007)

large precipitation deficits over the Amazon basin. The weaker events tend to have weaker precipitation anomalies. Most of the events also feature excess precipitation in southern Brazil, a region that sometimes experiences disastrous flooding related to strong El Niño episodes such as 1982/83 (Kousky et al., 1984). The composite for the seven El Niño episodes shows precipitation deficits in the central and eastern Amazon, and over northeast Brazil, and precipitation surpluses in southern Brazil, consistent with previous studies on ENSO cycle impacts (e.g., Ropelewski & Halpert, 1987, 1989; Grimm et al., 1998).

The precipitation anomaly patterns during La Niña episodes (Fig. 12) show more event-to-event consistency compared to those for El Niño. Above-average precipitation is evident over the northern part of Brazil in all six La Niña episodes. There is also a tendency for wetter-than-average conditions (four out of six cases) to occur over northeast Brazil. The composite pattern for the water-year precipitation anomalies during La Niña episodes does not reflect substantial dryness in southern Brazil, which is a feature associated with La Niña at certain times of the year (e.g., Ropelewski & Halpert, 1989; Grimm et al., 1998).

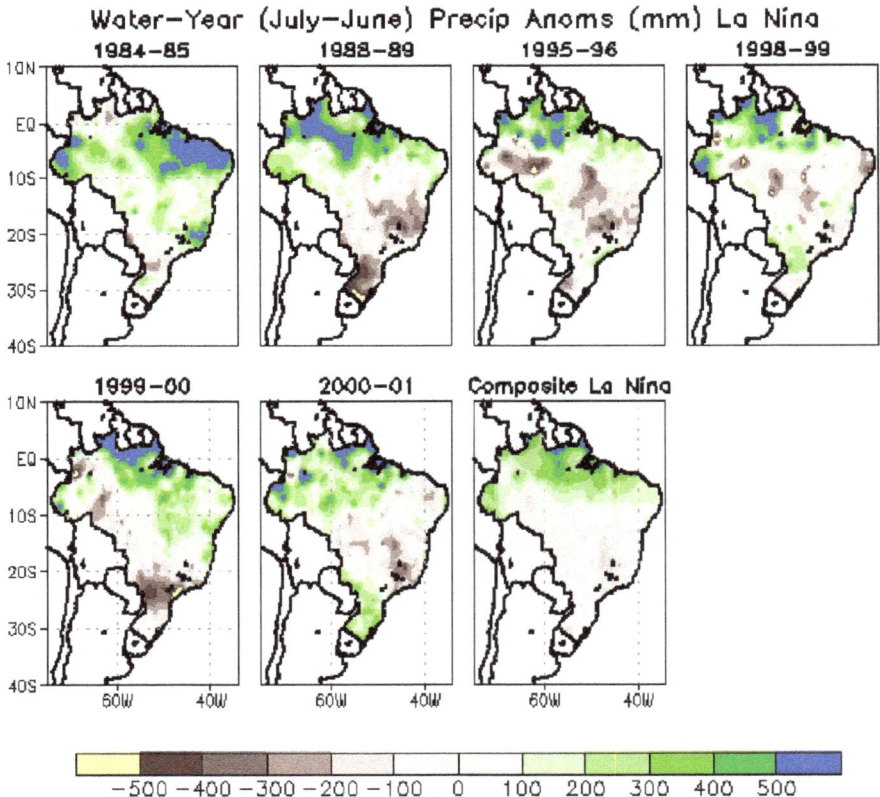

Fig. 12. Precipitation anomalies for water years (July–June) during La Niña episodes. The composite for the six episodes is shown in the lower right-hand panel. Anomalies (mm) are computed with respect to the July 1977–June 2004 base period means. (Figure taken from Silva et al., 2007)

The Atlantic SST anomaly dipole pattern (e.g., Hastenrath & Heller, 1977; Markham & McLain, 1977; Moura & Shukla, 1981; Servain, 1991; Nobre & Shukla, 1996) has a profound influence on rainfall over northeastern Brazil (the eastern flank of the SAMS). The dipole pattern usually consists of SST anomalies of one sign north of the equator and SST anomalies of the other sign south of the equator, which results in an anomalous displacement of the equatorial trough and Intertropical Convergence Zone (ITCZ). Rainfall deficits and drought over northern Northeast Brazil accompany positive SST anomalies north of the equator, negative SST anomalies south of the equator, and an anomalously northward displaced ITCZ. In contrast, above-average rainfall in northern Northeast Brazil accompanies negative SST anomalies north of the equator, positive SST anomalies south of the equator, and an anomalously southward displaced ITCZ (Fig. 13).

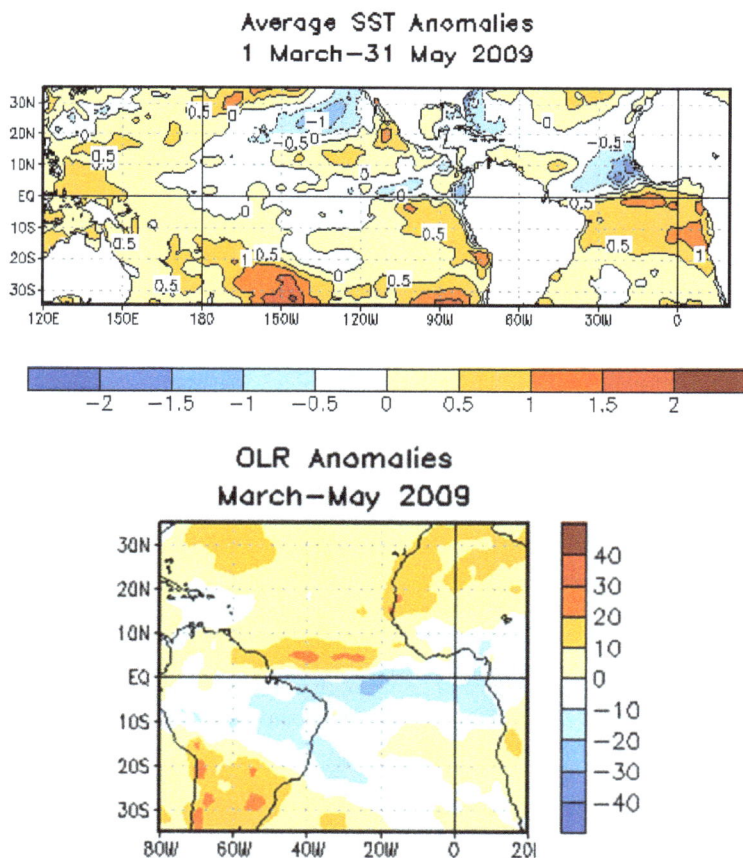

Fig. 13. Sea surface temperature anomalies (°C) (left panel) and outgoing longwave radiation anomalies (W m-2) for March-May 2009 (right panel). SST anomalies are departures from the 1971-2000 base period means and OLR anomalies are departures from the 1979-1995 base period means. Negative OLR anomalies in the Tropics indicate enhanced convection and above-average rainfall.

4.2 Intraseasonal variability

4.2.1 SAMS and the Madden-Julian Oscillation (MJO)

Several studies have shown that the SAMS can be influenced by the Madden-Julian Oscillation (MJO). The MJO is a naturally occurring intraseasonal fluctuation in the global tropics, with a typical period of 30-60 days (Madden & Julian, 1971, 1972; Madden & Julian, 1994; Zhang, 2005). The MJO is a significant cause of weather variability in the Tropics and Subtropics that affects several important atmospheric and oceanic parameters, including lower- and upper-level wind speed and direction, cloudiness, rainfall, sea surface temperature (SST), and ocean surface evaporation. The enhanced rainfall phase of the MJO can affect both the timing of a monsoon onset and monsoon intensity. Moreover, the suppressed phase of the MJO can prematurely end a monsoon and initiate breaks during monsoon wet seasons.

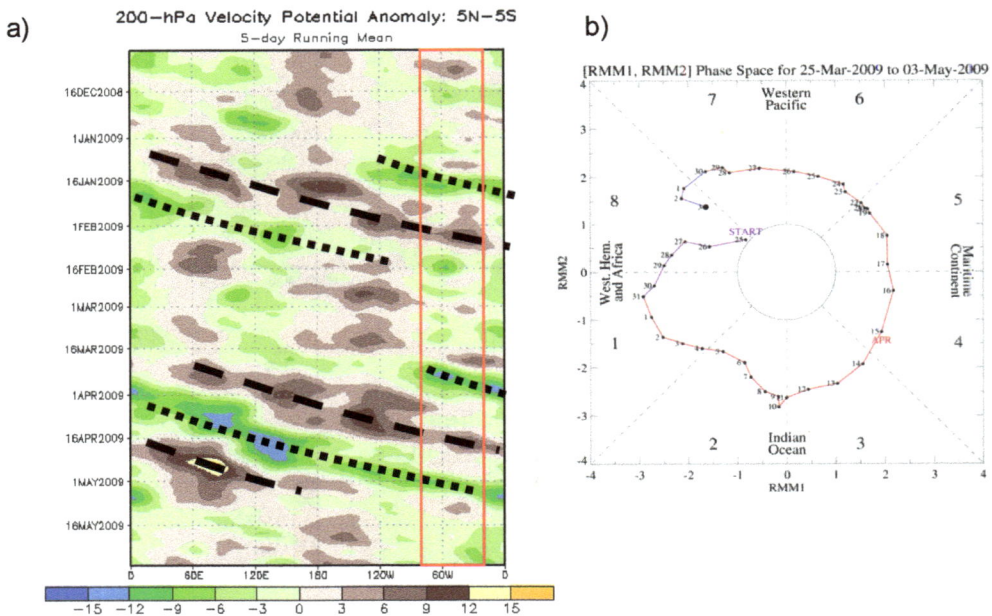

Fig. 14. a) Time longitude section of 200-hPa velocity potential anomalies in the latitude band 5°N-5°S, and b) the NOAA/Climate Prediction Center version of the Wheeler and Hendon (2004) daily MJO index.

The MJO modulates summer rainfall over northern and northeastern South America. It is very important for Northeast Brazil (semi-arid region), which experiences a short (3-4 months) wet season. An example of MJO-related heavy rainfall events occurred during March-May 2009. From mid-March to early May 2009, eastward propagating velocity potential anomalies (Fig. 14a) indicate moderate-to-strong MJO activity. The active phase of the MJO was in the South American sector (Fig. 14b, phase 8) during the end of March and again in the beginning of May, contributing to excessive rainfall and flooding over portions of northeastern Brazil, especially in early May (Fig. 15). Another factor contributing to the excessive seasonal rainfall was the presence of the Atlantic SST dipole (discussed in the previous section), which favored an anomalously southward displaced ITCZ and enhanced rainfall over Northeast Brazil.

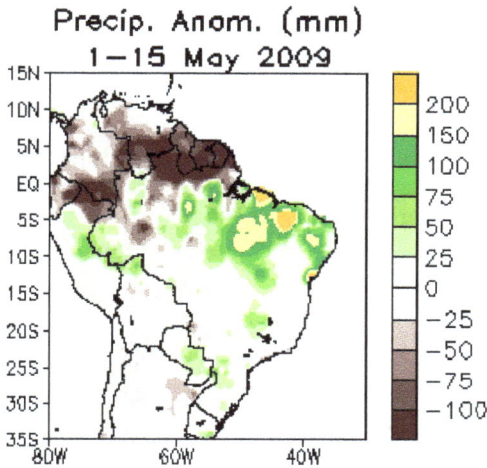

Fig. 15. Precipitation anomalies (mm) for 1-15 May 2009. Anomalies are departures from the 1979-2010 base period.

4.2.2 SAMS and the South Atlantic Convergence Zone

A characteristic feature of anomalous precipitation over South America is the tendency for a dipole pattern to occur, with anomalies of one sign located in the region of Southeast Brazil, the climatological position of the South Atlantic Convergence Zone (SACZ), and anomalies of the other sign situated over southeastern South America (southern Brazil, Uruguay, Paraguay and northeastern Argentina (e.g., Casarin & Kousky, 1986; Kousky & Cavalcanti, 1988; Kayano & Kousky, 1996; Nogues-Paegle & Mo, 1997; Herdies et al., 2002; Nogues-Paegle et al., 2002; Diaz & Aceituno, 2003; Silva & Berbery, 2005; Marengo et al., 2010). This dipole pattern has been shown to be partly related to the phasing of synoptic waves with the phase of the MJO (e.g., Casarin and Kousky, 1986; Nogues-Paegle et al., 2002; Liebmann et al., 2004; Carvalho et al., 2004; Cunningham & Cavalcanti, 2006).

Mid-latitude frontal systems (e.g., Kousky 1979, 1985; Garreaud & Wallace 1998) have an important effect on the intensity and distribution of deep convection over tropical and subtropical South America, and the location of the SACZ. As cold fronts move northward over southern Brazil, they organize a band of intense convection stretching along the front, often extending from the slopes of the Andes eastward to the western Atlantic. This band of intense convection shifts northward accompanying the advance of the front, and may eventually reach as far north as the Amazon basin and Northeast Brazil.

Subtropical upper-level cyclonic vortices (Kousky & Gan, 1981) also affect the distribution and intensity of rainfall, particularly over eastern Brazil. These systems typically form within the Atlantic mid-oceanic trough, near the coast of Northeast Brazil. Once formed they tend to drift slowly westward with time, often moving over Northeast Brazil. These vortices are cold core systems characterized by a central region of relatively dry sinking air, while on the western, northern, and sometimes eastern flanks of these systems convection is often enhanced, resulting in heavy rainfall.

4.3 SAMS diurnal variability

Major features of the summertime diurnal cycle for the South American region, as depicted in the NOAA/Climate Prediction Center Morphing technique (CMORPH) precipitation analyses (Joyce et al. 2004), include an afternoon (18-21 UTC) maximum in precipitation over the Andes and the high terrain in central and eastern Brazil, a nocturnal (06-09 UTC) maximum in precipitation over areas just east of the Andes (over western Argentina, central Bolivia, and western Paraguay), and an early morning (12-15 UTC) maximum over the Atlantic Ocean in the vicinity of the South Atlantic Convergence Zone (Fig. 16, and as described in

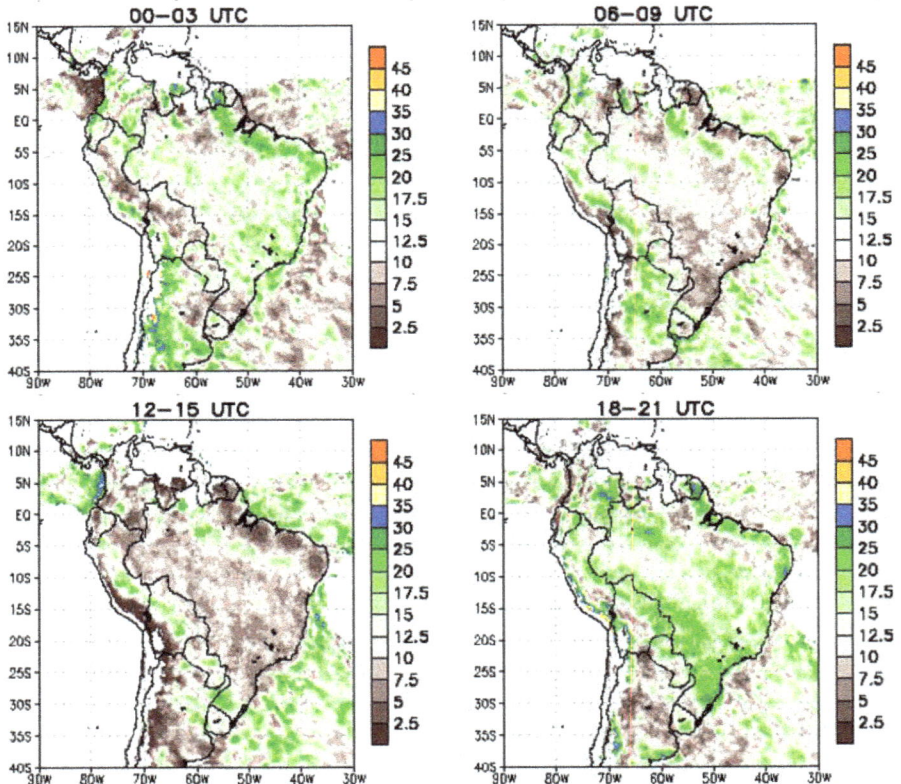

Fig. 16. Mean percentage of daily total precipitation for 00-03 UTC (20-23 LST), 06-09 UTC (02–05 LST), 12-15 UTC (08–11 LST), and 18-21 UTC (14-17 LST). Local standard time (LST) is for the center longitude of the domain. The mean is computed for the combined December–February periods for 2002–2003 and 2003–2004. Note: if rainfall were distributed equally throughout the 24-hour period, then 12.5% would be the expected percentage of the daily total for each 3-hour interval. Percentages have been masked out in regions where rainfall average is less than 1 mm day^{-1}.

Janowiak et al. 2005). A nocturnal or early morning (12-15 UTC) precipitation maximum also occurs along the immediate coast and offshore in the vicinity of the Atlantic ITCZ and over the Pacific near the coast of Colombia, consistent with the hypothesis presented by Silva Dias et al. (1987) that an out-of-phase relationship exists in the diurnal cycles of continental and nearby oceanic regions.

A remarkable diurnal cycle in precipitation occurs in coastal areas of northern and northeastern South America. With daytime heating, precipitation rapidly develops along and just inland from the coast (Fig. 16, lower right panel), probably related to the sea breeze (Kousky, 1980; Garstang et al., 1994; Negri et al,. 2002). This precipitation advances westward and southward with time, producing a nocturnal maximum in areas approximately 500 km inland from the coast (Fig. 16, upper right panel). The average diurnal cycle for equatorial South America (equator to 5°N) for March–May 2003–2004 (Fig. 17) indicates that sea-breeze-induced precipitation systems propagate westward (dashed lines), reaching the western Amazon Basin in about two days. As these systems propagate inland, they contribute to a nocturnal precipitation maximum in some areas and a diurnal precipitation maximum in other areas. The nocturnal maximum in precipitation over the central Amazon basin and the inland propagation of sea-breeze-induced rainfall systems are most often observed during January-May (Fig. 18), when the diurnal cycle in the central Amazon basin displays two maxima (one nocturnal and the other diurnal). Propagating features can also be found east of the Andes Mountains over northern Argentina. Daytime heating initiates convection along the east slopes of the Andes, which subsequently propagates eastward over the low lands of northern Argentina and Paraguay resulting in a nocturnal maximum in those regions. Similar propagating features have been observed over the central Plains east of the Rocky mountains of the United States (e.g., Carbone et al. 2002; Janowiak et al. 2005).

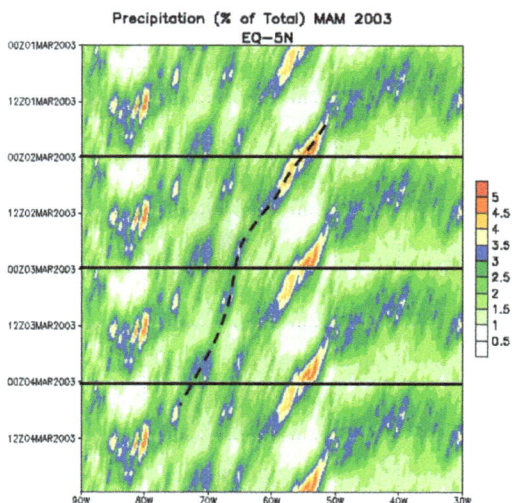

Fig. 17. Time-longitude section of the mean (March-May 2003) percentage of daily precipitation for the latitude band 0°-5°N. The mean diurnal cycle is repeated 4 times. The dashed line indicates the westward propagation with time associated with sea-breeze-induced convection along the northeast coast of South America (vertical dashed line near 50°W).

Precipitation (% of total) 0-13S, 52-70W

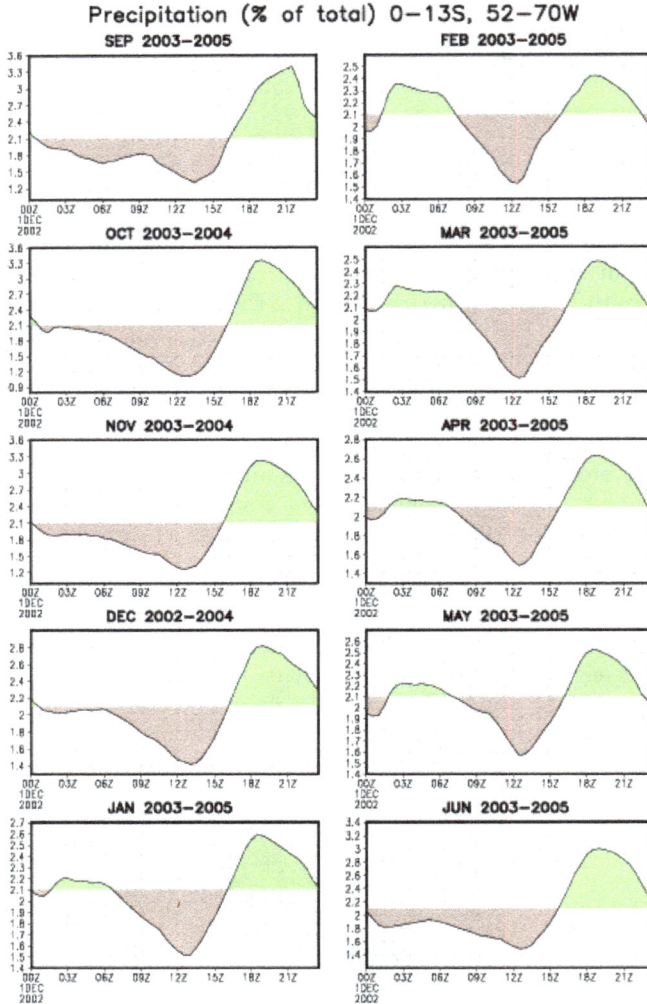

Fig. 18. The annual cycle of the mean diurnal cycle for the central Amazon basin (0°-13°S, 52°-70°W. Local time is approximately 4 hours less than UTC (Z).

4.4 Extreme precipitation events

As mentioned in the introduction, heavy rainfall events near Brazil's heavily populated east coast can result in disastrous flooding, with loss of life, property and infrastructure. A climatology of intense rainfall events is presented for three regions in eastern Brazil shown in Fig. 19. The daily average precipitation in each of the 5x5 degree boxes was computed during the 32-year period 1979-2010. Next the daily data were stratified by month and the number of cases for precipitation equal to or exceeding selected thresholds

(10, 15, 20, 25, 30, 35, 40, 45, and 50 mm) was computed. The results are shown in Tables 1-3. The greatest number of cases among the three regions and for all thresholds occurs in region-3 (Southeast Brazil, Table 1) during the height of the SAMS wet season (December-January). A similar peak, but with fewer cases is observed in region-2 (Table 2), which also has a secondary peak in March. Region-1 (Table 3) has fewer cases than in the other two regions for all thresholds, and, as in region-2, features two peaks (January and March) for thresholds below 20 mm.

The daily values for each month were ranked (highest to lowest) and the top 20 cases for each region (Table 4) were selected for further analysis. To determine the independent events in each region, cases where the dates are close together (within 5 days) are considered as a single event. Thus, the number of events in the top 20 cases (Table 4) is 14 for region-3, 12 for region-2 and 7 for region-1. This indicates a tendency for extreme events to persist for longer periods of time at lower latitudes over eastern South America.

The sea level pressure (SLP) and precipitable water (PW) analyses for the top 12 events in region-3 (indicated by the red asterisks in Table 1) are shown in Fig. 20. All of the events show high values of PW and a SLP trough in the vicinity of Southeast Brazil. In most cases, high PW values and a pressure trough extend in a band eastward/southeastward over the Atlantic Ocean. Since PW depends primarily on moisture available in the lowest layers of the atmosphere, bands of high PW are usually co-located with regions of low-level convergence, which accompany surface cold fronts or remnant pressure troughs. Of the 12 events shown in Fig. 20, only two events (17 January 1980 and 9 January 2004) do not show any apparent relationship with fronts.

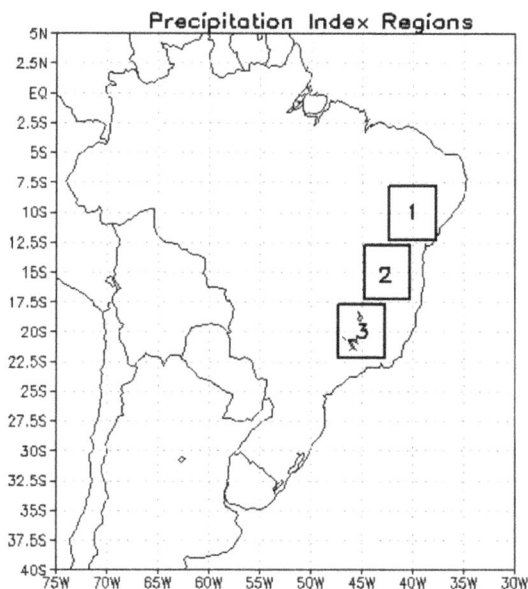

Fig. 19. Precipitation Index for selected regions.

Region-3 (17.75-22.25S, 42.75, 47.25W)								
	≥10mm	≥15mm	≥20mm	≥25mm	≥30mm	≥35mm	≥40mm	≥45mm	≥50mm
JUL	8	3	0	0	0	0	0	0	0
AUG	6	1	1	1	0	0	0	0	0
SEP	35	17	5	1	0	0	0	0	0
OCT	87	35	12	3	1	0	0	0	0
NOV	229	99	37	10	3	3	3	0	0
DEC	349	182	77	35	16	4	2	1	0
JAN	366	190	79	40	14	7	4	3	1
FEB	207	100	36	14	4	3	2	0	0
MAR	178	71	31	6	1	0	0	0	0
APR	43	19	5	3	1	0	0	0	0
MAY	21	11	7	3	2	0	0	0	0
JUN	6	4	3	1	0	0	0	0	0

Table 1. Number of events P ≥ given thresholds for Region-3.

Region-2 (12.75-17.25S, 40.25, 44.75W)								
	≥10mm	≥15mm	≥20mm	≥25mm	≥30mm	≥35mm	≥40mm	≥45mm	≥50mm
JUL	0	0	0	0	0	0	0	0	0
AUG	0	0	0	0	0	0	0	0	0
SEP	5	1	1	0	0	0	0	0	0
OCT	40	18	7	4	0	0	0	0	0
NOV	153	62	24	11	5	2	0	0	0
DEC	204	97	41	13	4	2	1	0	0
JAN	162	84	36	12	3	2	0	0	0
FEB	81	42	21	6	3	2	0	0	0
MAR	113	50	16	7	0	0	0	0	0
APR	23	6	0	0	0	0	0	0	0
MAY	0	0	0	0	0	0	0	0	0
JUN	1	1	0	0	0	0	0	0	0

Table 2. Number of events P ≥ given thresholds for Region-2.

Region-1 (7.75-12.25S, 37.75, 42.25W)								
	≥10mm	≥15mm	≥20mm	≥25mm	≥30mm	≥35mm	≥40mm	≥45mm	≥50mm
JUL	0	0	0	0	0	0	0	0	0
AUG	0	0	0	0	0	0	0	0	0
SEP	2	0	0	0	0	0	0	0	0
OCT	11	1	1	0	0	0	0	0	0
NOV	30	11	5	1	0	0	0	0	0
DEC	58	22	7	2	1	0	0	0	0
JAN	67	34	15	7	2	2	0	0	0
FEB	60	25	7	4	1	0	0	0	0
MAR	87	26	11	2	0	0	0	0	0
APR	40	12	4	1	1	0	0	0	0
MAY	6	3	1	0	0	0	0	0	0
JUN	2	0	0	0	0	0	0	0	0

Table 3. Number of events P ≥ given thresholds for Region-1.

	Region-3	Precipitation (mm)	Region-2	Precipitation (mm)	Region-1	Precipitation (mm)
1	3-Jan-97*	52.3	30-Jan-92	38.8	14-Jan-04	38.8
2	3-Jan-00*	47.3	17-Jan-04	35.1	28-Jan-92	36.2
3	4-Jan-97	45.7	16-Jan-02	31.9	20-Jan-04	28.2
4	23-Jan-92*	41.4	13-Jan-92	28.9	17-Jan-04	28.0
5	24-Jan-92	39.2	9-Jan-79	28.6	3-Jan-02	26.8
6	2-Jan-00	37.2	10-Jan-79	28.2	30-Jan-92	26.1
7	9-Jan-85*	35.8	5-Jan-09	26.7	29-Jan-92	25.5
8	3-Jan-82*	34.5	15-Jan-80	26.6	18-Jan-04	22.9
9	6-Jan-83*	34.0	28-Jan-92	26.3	24-Jan-04	22.6
10	25-Jan-85*	33.3	15-Jan-02	26.3	11-Jan-99	22.6
11	26-Jan-85	32.3	17-Jan-79	25.4	18-Jan-02	22.5
12	6-Jan-97	31.8	14-Jan-80	25.1	15-Jan-79	21.9
13	12-Jan-81*	30.2	2-Jan-02	24.7	19-Jan-04	20.7
14	30-Jan-08*	30.1	11-Jan-85	24.2	17-Jan-79	20.6
15	17-Jan-80*	29.9	1-Jan-83	24.1	21-Jan-00	20.4
16	9-Jan-04*	29.4	8-Jan-79	24.1	5-Jan-02	19.9
17	24-Jan-82*	29.3	1-Jan-81	23.7	7-Jan-02	19.7
18	12-Jan-91	29.1	27-Jan-92	23.7	26-Jan-92	19.3
19	16-Jan-91	29.0	25-Jan-92	23.6	21-Jan-04	19.2
20	28-Jan-91	28.8	6-Jan-09	23.4	8-Jan-02	19.0

Table 4. Top 20 precipitation cases during January (1979-2010) for the three regions in Fig. 19. The red asterisks indicate the 12 events used in Fig. 20.

The austral summer 1999-2000 featured considerable intraseasonal variability in the intensity and location of convection over eastern Brazil (Silva & Kousky, 2001). The lowest values of OLR (strongest deep convection) during the period were observed over Southeast Brazil during 1-5 January 2000 (3 January 2000 is ranked number two for the extreme events in region-3, Table 4). During this period, precipitation exceeded 150 mm (Fig. 21, left panel) over a large portion of Southeast Brazil, resulting in mudslides, flooding and loss of life. The average OLR for this period (Fig. 21, right panel) shows a well-defined band of low OLR (intense convection) extending from the Amazon basin southeastward over Southeast Brazil and the neighboring western Atlantic. The corresponding average vertical motion for this period (Fig. 22, left panel) shows that rising motion (negative omega) accompanied the band of low OLR. The average upper-tropospheric (200-hPa) wind (Fig. 22, right panel) shows a trough over costal sections of southern and southeastern Brazil, and a well-defined subtropical jet stream located near 25°S. This jet stream has a maximum over the western Atlantic, and its left rear entrance region is located near the bands of heavy rainfall, rising motion and low OLR over Southeast Brazil. The rising motion, associated with the low OLR is accompanied by sinking motion farther southwest over southern Brazil and Paraguay (Fig. 22, left panel). These features are consistent with the typical dipole pattern in precipitation, discussed in section 4.2.2, and an intense SACZ near its climatological position over Southeast Brazil.

SLP & PW

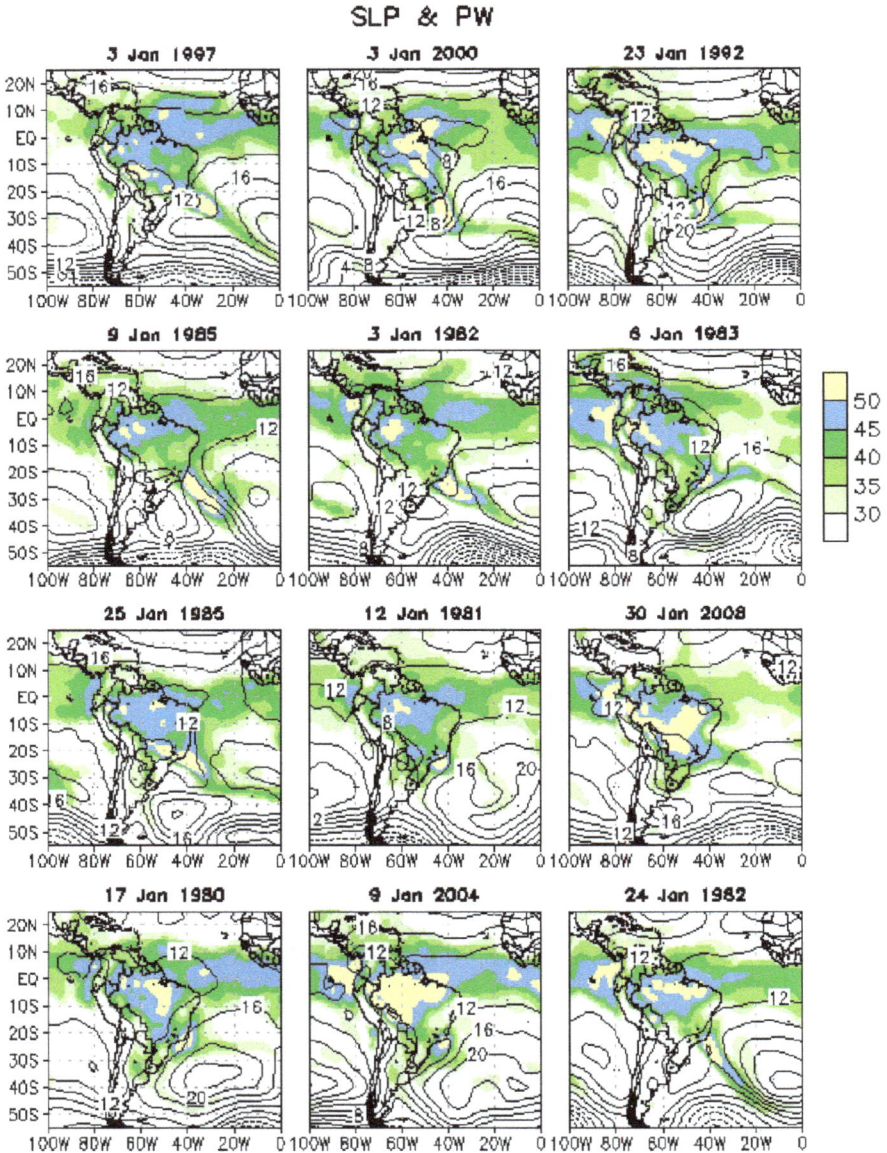

Fig. 20. Sea level pressure (contours, hPa-1000) and precipitable water (shading, mm) for 12 extreme events in region-3 (see Table 4).

The circulation features for active convection over southeastern Brazil indicate a strong coupling between the bands of enhanced convection/rising motion and the left rear entrance region of an upper-tropospheric jet stream (Silva & Kousky, 2001; Carvalho et al., 2002). Once established, these patterns tend to persist for several days. The possible evolution leading to persistence is as follows: 1) the synoptic-scale pattern provides a mechanism (surface front and upper-level trough) to enhance convection over the high terrain regions of eastern Brazil; 2) this convection is strongly modulated by the diurnal cycle and the topography of the region; 3) local thermal contrasts, due to the distribution of clouds and precipitation, tend to favor a persistence of convection within the region; 4)

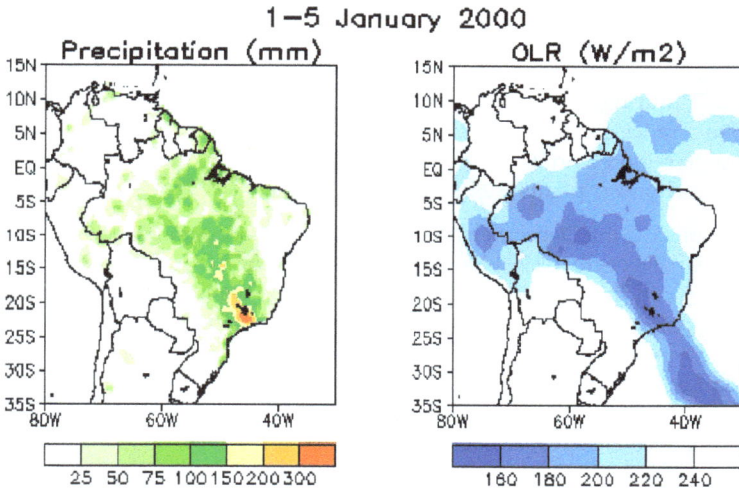

Fig. 21. Total precipitation (mm) (left panel) and average outgoing longwave radiation (OLR, W m^{-2}) (right panel) for the period 1-5 January 2000.

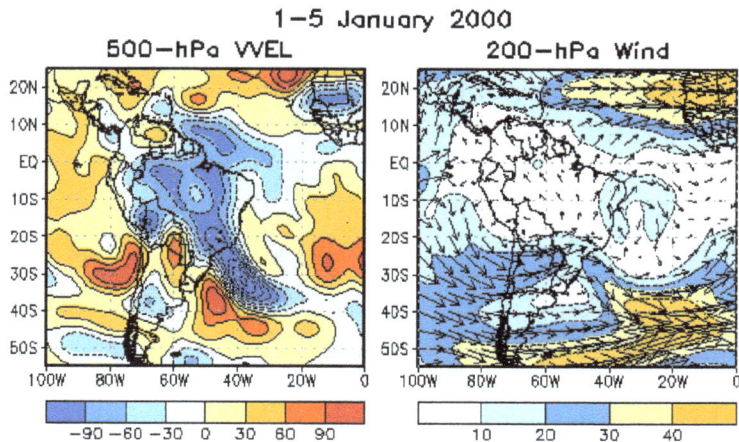

Fig. 22. Average vertical motion (omega, hPa d^{-1}) (left panel) and 200-hPa vector wind (m s^{-1}) (right panel) for the period 1-5 January 2000.

thus, anomalous latent heating in the middle and upper troposphere continues in the same area; and 5) this tends to maintain the upper-level jet stream and related low-level circulation features, such as a surface pressure trough and baroclinic zone, in approximately the same position for several days.

During December 2000-February 2001 (the peak of the wet season), large rainfall deficits (up to 400 mm, Fig. 23, right panel) were observed over Southeast Brazil (region-3 in Fig. 19) and the BP region (see Fig. 1), which are important regions for water storage and hydroelectric energy generation. As a result, Brazil experienced a major energy crisis in 2001 that led to the implementation of restrictions on energy usage throughout the country in order to avoid large-scale blackouts. Three major factors contributed to the energy crisis: 1) large rainfall deficits (Fig. 23), during the peak of the SAMS wet season, in the upper portions of the Tocantins, São Francisco and La Plata/Paraná river basins (BP region in Fig. 1, and Southeast Brazil region-3 in Fig. 19), 2) increasing energy demands, and 3) delays in implementing new power plants (Kelman et al., 2001).

The dry conditions during December 2000-February 2001 are remarkable when compared to the previous year (December 1999-February 2000, Fig. 23, left panel), which featured near- or above-average conditions over Southeast Brazil. The mean daily rainfall rate during DJF 2000-2001 was only about half the rate observed during DJF 1999-2000 (Fig. 24). The differences between the two wet seasons cannot be attributed to the ENSO cycle, since both years featured La Niña conditions in the tropical Pacific. Further investigation is necessary to identify the causes for the exceptionally dry conditions over the BP region and Southeast Brazil during the 2000-2001 wet season.

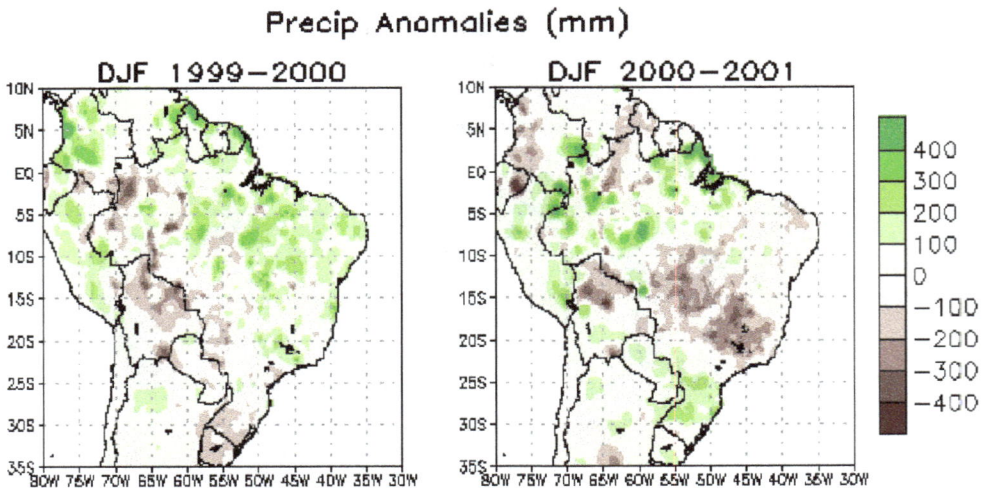

Fig. 23. Anomalous precipitation for December 1999-February 2000 (left panel) and December 2000-February 2001 (right panel). Data are derived from the daily gridded analyses of precipitation produced by the NOAA/ Climate Prediction Center. Anomalies are departures from the 1979-2010 base period mean.

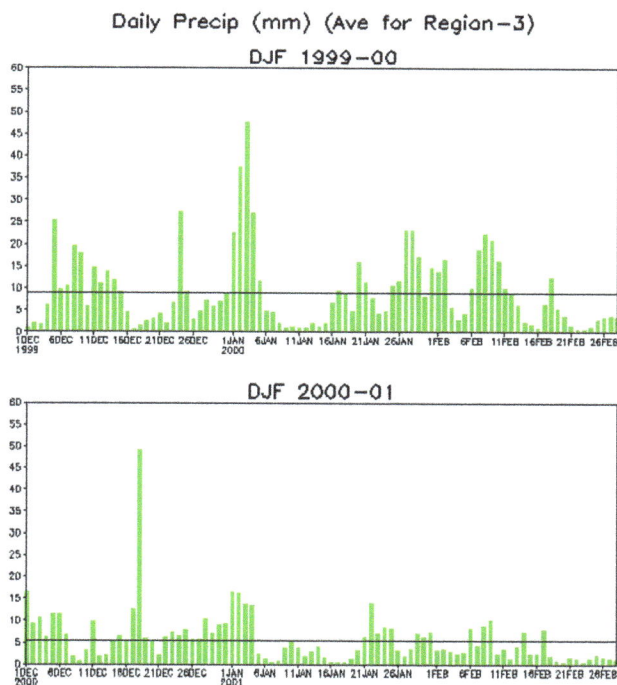

Fig. 24. Time series of daily rainfall (mm), averaged over region-3 (Fig. 19), for December 1999-February 2000 (left panel) and December 2000-February 2001 (right panel).

5. Conclusion

This chapter highlights some of the results presented in previous reviews and studies on the South American Monsoon System (SAMS), regarding circulation features, the evolution of the wet season and variability on time scales ranging from diurnal to inter-annual. In addition, a brief inter-comparison among several re-analyses during December-February indicates uncertainties in the low-level circulation features and related derived quantities, such as moisture flux, convergence and vertical motion within the core region of SAMS. These uncertainties are undoubtedly related to differences in the December-February precipitation patterns between the re-analyses and observations, which display large biases over many areas in South America. It is extremely important that future re-analyses emphasize bias reduction, in order to reduce these uncertainties.

Examples of extreme events on a variety of time scales illustrate the large range of variability associated with the SAMS wet season. Although much progress has been made in understanding the phenomena responsible for those events, further research is necessary to document their dynamical and thermodynamical causes, frequency of occurrence and predictability. Benchmark studies of this type are extremely important for decision makers as they develop plans to mitigate present and future impacts of weather and climate variability on society.

6. References

Aceituno, P. (1988). On the functioning of the Southern Oscillation in the South American sector. Part I: Surface climate. *Mon. Wea. Rev.*, Vol. 116, pp. 505–524.

Carbone, R. E.; Tuttle, J. D.; Ahijevych, D. A. & Trier, S. B. 2002. Inferences of predictability associated with warm season precipitation episodes. *J. Atmos. Sci.*, Vol. 59, pp. 2033-2056.

Carvalho, L. M. V.; Jones, C. & Liebmann, B. (2002). Extreme precipitation events in Southeastern South America and large-scale convective patterns in the South Atlantic Convergence Zone. *J. Climate*, Vol. 15, pp. 2377-2394.

Casarin, D. P. & Kousky, V. E. (1986). Precipitation anomalies in Southern Brazil and related changes in the atmospheric circulation. *Revista Brasileira de Meteorologia*, Vol. 1, pp. 83-90.

Chen, M.; Shi, W., Xie, P., Silva, V. B. S., Kousky, V E., Higgins, R. W. & Janowiak, J. E. (2008). Assessing objective techniques for gauge-based analyses of global daily precipitation, *J. Geophys. Res.*, Vol. 113, D04110, doi:10.1029/2007JD009132.

Cunningham, C. C. & Cavalcanti, I. F. A. (2006). Intraseasonal modes of variability affecting the South Atlantic Convergence Zone. *Int. J. Climatol.*, Vol. 26, pp. 1165-1180.

Diaz, A. & Aceituno, P. 2003: Atmospheric Circulation Anomalies during Episodes of Enhanced and Reduced Convective Cloudiness over Uruguay. *J. Climate*, Vol. 16, 3171-3185.

Gan M. A.; Rodrigues L. R., Rao V. B. (2009). Monção na America do Sul. Chapter 19 in *Tiempo y Clima no Brasil*, Cavalcanti I, Ferreira N.J., Justi M. A. G. , Silva Dias M. A. F., (eds) Editora. Oficina de Textos, S˜ao Paulo: Brazil, pp. 297–312.

Gan, M. A.; Rao, V. B. & Moscati, M. C. L. (2006). South American monsoon indices. *Atmospheric Science Letters*, Vol. 6, pp. 219–223.

Gan, M. A.; Kousky, V. E. & Ropelewski, C. F. (2004). The South American Monsoon circulation and its relationship to rainfall over west-central Brazil. *J. Climate*, Vol. 17, pp. 47-66.

Garreaud, R. D. & Wallace, J. M. (1998). Summertime incursions of midlatitude air into subtropical and tropical South America. *Mon. Wea. Rev.*, Vol. 126, pp. 2713-2733.

Garstang, M.; Massey Jr., H. L.; Halverson, J.; Greco, S. & Scala, J. (1994). Amazon coastal squall lines. Part I: Structure and kinematics. *Mon. Wea. Rev.*, Vol. 122, pp. 608-622.

Gill, A. E. 1980. Some simple solutions for heat induced tropical circulation. *Quart. J. Roy. Meteor. Soc.*, Vol. 106, pp. 447-462.

Gonzalez, M.; Vera, C., Liebmann B., Marengo, J., Kousky, V. E. & Allured, D. (2007). The nature of the rainfall onset over central South America. *Atmosfera*, Vol. 20, pp. 379–396.

Grimm, A. M.; Ferraz, S. E. T. & Gomes J. (1998). Precipitation Anomalies in Southern Brazil Associated with El Niño and La Niña Events. *J. Climate*, Vol. 11, pp. 2863–2880.

Grimm A. M.; Vera C., Mechoso R. (2005). The South American Monsoon System, Chang C-P, Wang B, Lau NC-G, (eds) The Global Monsoon System: Research and Forecast, WMO/TD 1266 – *TMRP*: pp. 542. Available at http://www.wmo.int/pages/prog/arep/tmrp/documents/globalmonsoonsyste mIWM3.pdf.

Hastenrath, S. & Heller, L. (1977). Dynamics of climatic hazards in northeastern Brazil. *Quart. J. Roy. Meteor. Soc.*, Vol. 110, pp. 77-92.

Herdies, D. L.; da Silva A., Silva Dias, M. A. F. & Nieto Ferreira, R. (2002). Moisture budget of the bimodal pattern of the summer circulation over South America. *J. Geophys. Res.*, Vol. 107 (D17), 10.1029/2001JD000997.

Horel, J. D.; Hahmann, A. & Geisler, J. (1989). An investigation of the annual cycle of convective activity over the tropical Americas. *J. Climate*, Vol. 2, pp. 1388-1403.

Janowiak, J. E.; Kousky, V. E. & Joyce, R. J. (2005). Diurnal cycle of precipitation determined from the CMORPH high spatial and temporal global precipitation analyses. J. Geophys. Res., Vol. 110, D23105, doi:10.1029/2005JD006156.

Joyce, R. J.; Janowiak, J. E.; Arkin, P. A. & Xie, P. (2004). CMORPH: A method that produces global precipitation estimates from passive microwave and infrared data at high spatial and temporal resolution. *J. Hydrometeorol.*, Vol. 5, pp. 487-503.

Kalnay, E., and Coauthors (1996) The NCEP/NCAR 40-year reanalysis project. *Bull. Amer. Met Soc.*, Vol. 77, pp. 437-471.

Kanamitsu, M.; Ebisuzaki, W., Woolen, J., Yang, S. K., Hnilo, J. J., Fiorino, M. & Potter, G. L. (2002). NCEP-DOE AMIP-II reanalysis (R-2). *Bull. Amer. Met. Soc.*, Vol. 83, pp. 1631-1643.

Kayano, M. T. & Kousky, V. E. (1996). Tropical circulation variability with emphasis on interannual and intraseasonal time scales. *Rev. Bras. Meteor.*, Vol. 11, pp. 6-17.

Kelman, J.; Venture, A., Bajay, S. V., Penna, J. C., and Haddad, C. L. S. (2001). Relatório da Comissão de Análise do Sistema Hidrotérmico de Energia Elétrica: O desequilíbrio entre oferta e demanda de energia elétrica.: *Agência Nacional de Águas/ANA*. Brasília 21 July 2001.

Kousky, V. E. (1979). Frontal Influences on Northeast Brazil. *Mon. Wea. Rev.*, Vol. 107, pp. 1140-1153.

Kousky, V. E. (1980). Diurnal rainfall variation in Northeast Brazil. *Mon. Wea. Rev.*, Vol. 108, pp. 488-498.

Kousky, V. E. & Gan, M. A. (1981). Upper Tropospheric Cyclonic Vortices in the Tropical South Atlantic. *Tellus*, Vol.33, pp. 538-551.

Kousky, V. E.; Kayano, M.T., & Cavalcanti, I. F. A. (1984). A review of the Southern Oscillation: oceanic-atmospheric circulation changes and related rainfall anomalies. *Tellus*, Vol. 36A, pp. 490-504.

Kousky, V. E. (1985). Atmospheric circulation changes associated with rainfall anomalies over tropical Brazil. Mon. *Wea. Rev., Vol.* 113, pp. 1951-1957.

Kousky, V. E. (1988). Pentad outgoing longwave radiation climatology for the South America sector. *Revista Brasilera de Meteorologia*, Vol. 3, pp. 217–231.

Kousky, V. E. & Cavalcanti, I. F. A. (1988). Precipitation and atmospheric circulation anomaly patterns in the South American sector. *Revista Brasileira de Meteorologia*, Vol. 3, pp. 199-206.

Liebmann B. & Mechoso C. R. (2010). *The South American Monsoon System, Chapter 8 in The Global Monsoon System: Research and Forecast*, 2nd Edition. C. P Chang *et al.*, (eds.) World Scientific Publishing: Singapore. 550 p.

Liebmann, B. & Marengo, J. A. (2001). The seasonality and interannual variability of rainfall in the Brazilian Amazon basin. *J. Climate*, Vol. 14, pp. 4308-4318.

Liebmann, B.; Kiladis, G.N., Vera, C.S., Saulo, A.C. & Carvalho, L.M.V. (2004). Subseasonal variations of rainfall in South America in the vicinity of the low-level jet east of the Andes and comparison to those in the South Atlantic convergence zone. *J. Climate*, Vol. 17, pp. 3829-3842.

Madden, R. & Julian, P. (1972). Description of global-scale circulation cells in the tropics with a 40-50 day period. *J. Atmos. Sci.*, Vol. 29, pp. 1109-1123.

Madden, R. & Julian P. (1994). Observations of the 40-50 day tropical oscillation: A review. *Mon. Wea.Rev.*, Vol. 112, pp.814-837.

Madden, R. A. & P. R. (1994). Observations of the 40-50-day tropical oscillation - A review. *Mon. Wea. Rev.*, Vol. 122, pp. 814-837.

Marengo, J. A.; Liebmann, B., Kousky, V., Filizola, N. & Wainer I. (2001). On the onset and end of the rainy season in the Brazilian Amazon basin. *J. Climate*, Vol. 14, pp833-852.

Marengo, J. A.; Liebmann, B.; Grimm, A. M.; Misra, V.; Silva Dias, P. L.; Cavalcanti, I. F. A.; Carvalho, L. M. V.; Barbery, E. H.; Ambrizzi, T.; Vera, C. S.; Saulo, A. C.; Nogues-Paegle, J.; Zipser, E.; Seth, A. & Alves, L. M. (2010). Review: Recent developments on the South American monsoon system. *Int. J. Climatol.*, DOI: 10.1002/joc.2254

Markham, C. G. & McLain, D. R. (1977). Sea Surface temperature related to rain in Ceara, northeastern Brazil. *Nature*, Vol.265, pp. 320-325.

Mossman, R. C. (1924). Indian Monsoon rainfall in relation to South America weather. *Mem. Ind. Meteor. Dept*, Vol. 23, pp. 157-242.

Moura, A. D. & Shukla, J. (1981). On the dynamics of droughts in northeast brazil: observations, theory and numerical experiments with a general circulation model. *J. Atmos. Sci.*, Vol. 38, pp. 2653–2675. doi: 10.1175/1520-0469(1981)038<2653:OTDODI>2.0.CO;2

Negri, A. J.; Xu, L. & Adler, R. F. (2002). A TRMM-calibrated infrared algorithm applied over Brazil. J. Geophys. Res., 107(D20), 8048, doi:10.1029/2000JD000265.

Nieto-Ferreira, R. & Rickenbach, T. M. (2010). Regionality of monsoon onset in South America: A three-stage conceptual model. *Int. J. Climatol.*, DOI:10.1002/joc.2161.

Nobre, P. & Srukla, J. (1996).Variations of Sea Surface Temperature, Wind Stress, and Rainfall over the Tropical Atlantic and South America. *J. Climate*, Vol. 9, pp. 2464–2479. doi: 10.1175/1520-0442(1996)009<2464:VOSSTW>2.0.CO;2

Nogues-Paegle, J. & Mo, K. C. (1997). Alternating wet and dry conditions over South America during summer. *Mon Wea. Rev.*, Vol. 125, pp. 279-291.

Nogues-Paegle, J. & Coauthors, (2002). Progress in Pan American CLIVAR Research: Understanding the South American Monsoon. *Meteorologica*, Vol. 27, pp. 3-32.

Onogi, K. & Coauthors, (2005). Japanese 25-year reanalysis project - Progress and status. *Quart. J. Roy. Meteor. Soc.*, Vol. 131, pp. 3259-3268.

Onogi, K. & Coauthors (2007). The JRA-25 reanalysis. *J. Meteor. Soc. Japan*, Vol. 85, pp. 369-432.

Raia, A. & Cavalcanti, I. F. A. (2008). The Life Cycle of the South American Monsoon System. *J. Climate*, Vol. 21, pp. 6227–6246. doi: 10.1175/2008JCLI2249.1

Rienecker, M.M.; Suarez, M.J., Gelaro, R., Todling, R., Bacmeister,J., Liu, E., Bosilovich, M.G., Schubert, S.D., Takacs, L., Kim, G.-K., Bloom, S., Chen, J., Collins, D., Conaty, A., da Silva, A., et al., 2011. MERRA - NASA's Modern-Era Retrospective Analysis for Research and Applications. *J. Climate*, Vol. 24, 3624-3648, doi: 10.1175/JCLI-D-11-00015.1.

Ropelewski, C. F. & Halpert, M. S. (1987). Global and regional scale precipitation patterns associated with the El Niño-Southern Oscilation. *Mon. Wea. Rev.*, Vol. 115, pp. 1606-1626.

Ropelewski, C. F. & Halpert, M. S. (1989). Precipitation Patterns Associated with the High Index Phase of the Southern Oscillation. *J. Climate*, Vol. 2, pp. 268–284.

Saha, S., & Coauthors (2010). The NCEP Climate Forecast System Reanalysis. *Bull. Amer. Meteor. Soc.*, Vol. 91, pp. 1015–1057.

Servain, J. (1991). Simple Climatic Indexes for the Tropical Atlantic Ocean and some applications. *Journal of Geophysical Research-Oceans*, Vol. 96(C8), pp. 15137-15146.

Silva, A. E. & Carvalho, L. M. V. (2007). Large-scale index for South America Monsoon (LISAM). *Atmospheric Science Letters* 8: pp. 51–57.

Silva Dias, P. L., Bonatti, J. P. & Kousky, V. E. (1987). Diurnally Forced Tropical Tropospheric Circulation over South America. *Mon. Wea. Rev.*, Vol. 115, pp. 1465–1478.

Silva, V. B. S. & Kousky, V. E. (2001). Intraseasonal precipitation variability over eastern Brazil during the summer of 1991–2000 (in Portuguese). *Rev. Bras.Meteorol.* Vol. 16, pp. 187–199.

Silva, V. B. S.; Kousky, V. E. & Busalacchi, A. J. (2005). CLIVAR Science: Application to energy - The 2001 Energy Crisis in Brazil, *CLIVAR Exchanges*, Vol. 10, No.2.

Silva, V. B. S. & Berbery, E. H. (2006). Intense Rainfall Events Affecting the La Plata Basin. *J. Hydrometeor.*, Vol. 7, pp. 769–787. doi: 10.1175/JHM520.1

Silva, V. B. S.; Kousky, V. E., Shi, W. & Higgins, R. W. (2007. An Improved Gridded Historical Daily Precipitation Analysis for Brazil. J. Hydrometeor., Vol. 8, pp. 847-861.

Silva, V. B. S.; Kousky, V. E. & Higgins, R. W. 2011. Daily Precipitation Statistics for South America: An Intercomparison between NCEP Ranalyses and Observations. *J. Hydrometeor.*, Vol. 12, pp. 101-117.

Streten, N. A. (1983). Southern Hemisphere circulation contrasts in the winters of 1972 and 1973: *Preprints First International Conference on Southern hemisphere Meteorology.*, Sao Jose dos Campos, Brazil., 108-111.

Uppala, S. M. & Coauthors (2005). The ERA-40 re-analysis. *Quart. J. Rot. Meteor. Soc.*, Vol. 131, pp.2961-3012.

Vera C. S.; Baez J., Douglas M., Emanuel C., Marengo J. A., Meitin J., Nicolini M., Nogues-Paegle J., Paegle J., Penalba O., Salio P., Saulo C., Silva Dias M. A. F., Silva Dias P. L. , Zipser E. (2006a). The South American Low-Level Jet Experiment. *Bulletin of the American Meteorological Society* 87: 63–77.

Webster, P. J. 1972. Response of the tropical atmosphere to local, steady forcing. *Mon. Wea. Rev.*, Vol. 100, pp. 518-541.

Wheeler, M. & Hendon, H. (2004). An All-Season Real-Time Multivariate MJO Index: Development of an Index for Monitoring and Prediction. *Mon. Wea. Rev.*, Vol. 132, pp. 1917-1932.

Zhang, C. (2005). Madden-Julian Oscillation. *Reviews of Geophysics*, Vol. 43, pp. 1-36.

6

Climatology of the Northern-Central Adriatic Sea

Aniello Russo[1,2], Sandro Carniel[2,*], Mauro Sclavo[2] and Maja Krzelj[3]
[1]Marche Polytechnic University-DISVA, Ancona
[2]The National Research Council, Institute of Marine Sciences, Venice
[3]University of Split, Center of Marine Studies, Split
[1,2]Italy
[3]Croatia

1. Introduction

It is well know that the ocean processes exert a great influence on global climate as well as affect the local climate of coastal areas (Russo et al., 2002). Within the Mediterranean region (see Fig. 1a), the presence of the Adriatic Sea influences the atmospheric properties of the surrounding regions over long and short time-scales, and has obviously a relevant influence on human activities and ecosystems (Boldrin et al., 2009).

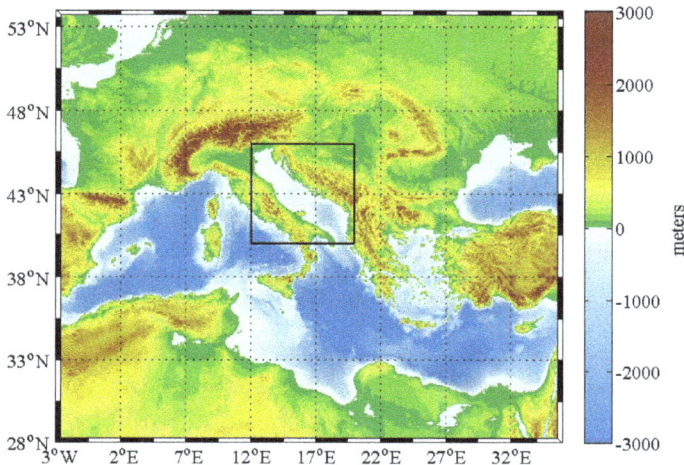

Fig. 1a. Orography and bathymetry of the Mediterranean region. The inset shows the Adriatic basin.

This Chapter will describe the main climatological characteristics of the northern-central Adriatic Sea (see Figure 1b) assessed on a human time-scale, i.e. the last few decades.

---

* Corresponding author

Generally speaking, the Adriatic sea and the above atmosphere are tightly linked and together form a dynamic component of the local climate system, where the relationships are intrinsically two-way (Pullen et al., 2006). For example, short-term fluctuations in wind or temperature (i.e. weather), as well as the atmospheric action of blocking rays, can influence the currents through wind-driven processes (see for instance Carniel et al., 2009) and prevent the warming of the underlying ocean (Boldrin et al., 2009; Dorman et al., 2006).

At the same time, the atmospheric temperature and precipitation rates are relevantly affected by the variation of the sea surface temperature (SST) through processes like evaporation and cyclone development.

Besides, the atmosphere can strongly affect the density differences that are triggering baroclinic Adriatic currents (Vilibic and Supic, 2005), by intense cooling driven by turbulent heat fluxes exchange through the air-sea interface, or by changing surface salinity according to evaporation/precipitation balance. Namely, currents generated during processes typical of the "deep-water formation" are particularly important for climate. For instance, in the relatively shallow northern Adriatic Sea, during winter time, surface cooling due to strong sensible and evaporative heat losses can cause water to become more dense (Bergamasco et al., 1999; Vilibic and Supic, 2005), leading to mixing and convective overturning processes. The resulting dense waters tend to spread out over the whole basin, moving to the southern region (Carniel et al., 2012) and triggering a return circulation from the Ionian and Aegean region that is called into the basin by continuity. Such waters are transported by the Adriatic Sea eastern currents and influence the local climate by transporting relevant amount of heat (Boldrin et al., 2009).

A thorough description of the northern-central Adriatic Sea main climatological characteristics would then help to understand the influence that this sea may have, also through its direct link with the atmosphere (e.g. temperature and precipitation) and marine ecosystems, on relevant aspects and activities of the society, such as tourism, fisheries, navigation, etc.

However, the hydrodynamic behavior of this portion of the Adriatic Sea (see again Figure 1b) is rather complex; from many different points of view, indeed, it can be considered as a reduced-scale oceanic one (Signell et al., 2005). During its annual hydrologic cycle, the basin is actually changing its main hydrodynamic features, from those typical of a homogenous, well-mixed basin, to those of a stratified system. Generally speaking, the northern and shallower portion of the Adriatic Sea is characterized by the sinking of heavier and denser waters (and therefore by a complete mixing) during the winter season, by a rather large surface warming during summer, and by abundant precipitation rates and discharges from rivers (mainly, the Po river) during spring and fall, and by different wind regimes with a prevalence in the cold season of strong dry northeasterly winds (called Bora, see Dorman et al., 2006; Boldrin et al., 2009) while in the warm season moderate southeasterly winds (Scirocco) are more frequent (e.g., Signell et al., 2005). All such processes in turn affect the Adriatic biogeochemical characteristics and its marine ecosystems. In the previous century several interesting studies have been carried out by international scientists dealing with the circulation of the Adriatic sea. Some of them are well condensed in Artegiani et al. (1997a; 1997b) as far as the thermohaline climatological characteristics and circulation are concerned, Poulain (2001) for the surface circulation using lagrangian devices, and particularly Cushman-Roisin et al. (eds. 2001), where a large review of physical oceanography studies updated until the end of the previous century was presented. Concerning bio-chemical properties, Zavatarelli et al. (1998) defined for the first time a climatology of the main characteristics in the Adriatic Sea.

In the following text seasonal climatology of physical and chemical properties of the water masses in the northern and central Adriatic sea are described.

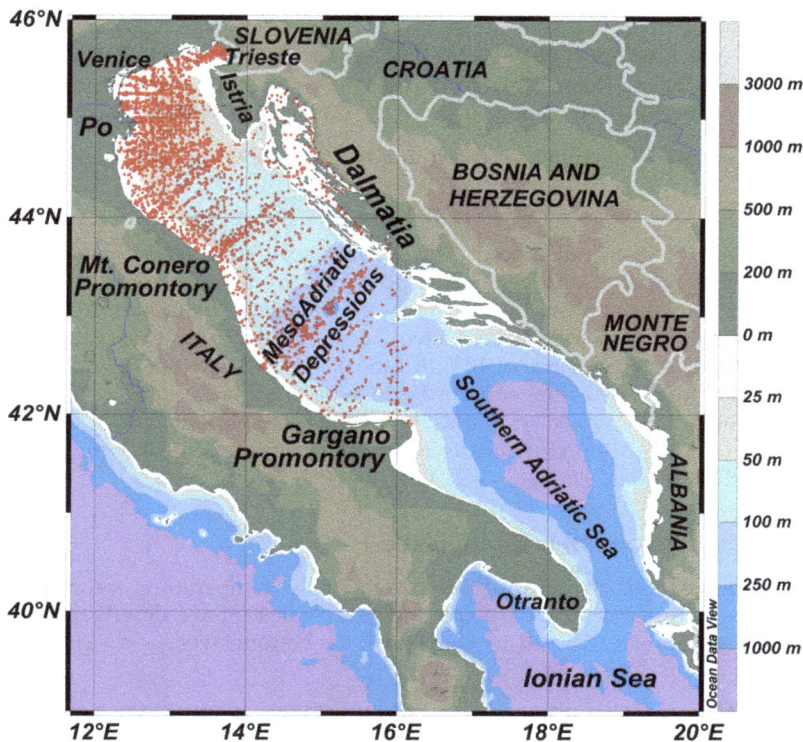

Fig. 1b. Adriatic Sea orography and bathymetry. The dots represent the stations extracted from MEDATLS 2002 archive and used to compute the seasonal climatology of the northern-central portion of the Sea.

2. Materials and methods

First climatological studies of the Adriatic Sea date back to the end of the previous century, namely Russo and Artegiani (1996), Artegiani et al. (1997a and 1997b), Zavatarelli et al. (1998) gathered and analyzed more than 4000 stations distributed in the whole Adriatic Sea.

In the same period some relevant European projects started, with the aim of establishing larger data base focused in the Mediterranean region. The final result was reached in the data base MEDAR/MEDATLAS 2002 (MEDAR Group, 2002). This dataset encompasses more than 8800 quality checked stations in the northern and central basin.

Using the stations located in the northern central Adriatic region, a seasonal climatology has been derived (a monthly one is not achievable, standing the low data frequency during some specific months) for temperature, salinity, dissolved oxygen and nutrients (nitrates, nitrites, ortho-silicates, ortho-phosphate) that will be discussed in the next sections. Seasons are defined as follows: winter season is represented by January-February-March months (JFM), spring by April-May-June (AMJ), summer brackets July-August-September (JAS), while fall includes October-November-December (OND). Data have been selected accordingly to seasons and spatially interpolated by means of an Objective Analysis (OA) technique adopted for ocean data (see Bretherton et al., 1976 and Carter and Robinson, 1987), as applied in Artegiani et al. (1997b) and Russo et al. (2002). The interpolation has been performed according to the optimal linear estimation theory, based on the Gauss-Markov theorem.

The adopted correlation function is isotropic and has the following expression:

$$C(r) = (1 - r^2/a^2) \exp(-r^2/2b^2), \quad r^2 = x^2 + y^2 \tag{1}$$

where a is the zero-crossing distance, b the spatial decay scale. These parameters should meet the condition $a > 2^{1/2} b$ to generate a positive-definite correlation matrix. Since the Chapter is focused on the analysis of climatological fields, large values (a= 100 km and b= 60 km) compared to mesoscale features have been chosen. The observational noise is approximated by an error variance of 10%. All the observations closer to each other than 10 km have been preliminarily averaged in order to avoid very high correlations. The number of influential points is chosen to be 10 and the mapping resolution is 0.05° both in latitude and longitude. The total domain consists of 81×81 points, and areas with error variance greater than 30% have been blanked.

3. Seasonal climatology of water masses

Main thermohaline and chemical properties (namely temperature, salinity, dissolved oxygen, nitrates, nitrites, ortho-phosphate and ortho-silicates) are described seasonally at the surface, at the depth of 10 m, 20 m, 50 m and at the bottom layer.

3.1 Temperature

Sea surface temperature (see Figure 2) shows a strong seasonal cycle, with winter values well below 10 °C along the Italian coastal region till the Gargano and minima in the northern region. Maxima values, above 25 °C, are reached in the south-western area. During winter the mean temperature in the basin is of 11.56 °C (12.23 °C is the median). A thermal gradient is present both along the longitudinal and transversal axis, with mean temperature below 7 °C in the north and above 13 °C in the south-eastern area. This is also due to the cyclonic circulation of the basin, that brings in warmer waters from the southern basin, while the northernmost waters, subjected to large heat losses due to Bora winds (strong katabatic winds coming from NE), are transported to the south-east along the Italian coast (Dorman et al., 2006). Colder waters are also found in the sea region in the proximity of the Dalmatian islands and eastern coast, an area that is directly exposed to the cold Bora wind action. During spring time the situation is more variable. Mean temperatures in the region are comprised between 14-23 °C (mean value 17.3 °C, median 17.07 °C), with maxima found in the western region and minima in the Dalmatian islands area. In the summer time the thermal variability is much lower, with mean values bracketed between 21 and 27 °C (mean 24.47 °C, median 24.44 °C), and maxima in the south-west region, minima in the eastern one (also because of upwelling phenomena). During the fall season the surface waters get cooled down, and this happens faster in the north-east and south-west part of the basin, with values between 13 and 22 °C (averaged values in the area being 18.14 °C, median 18.37 °C).

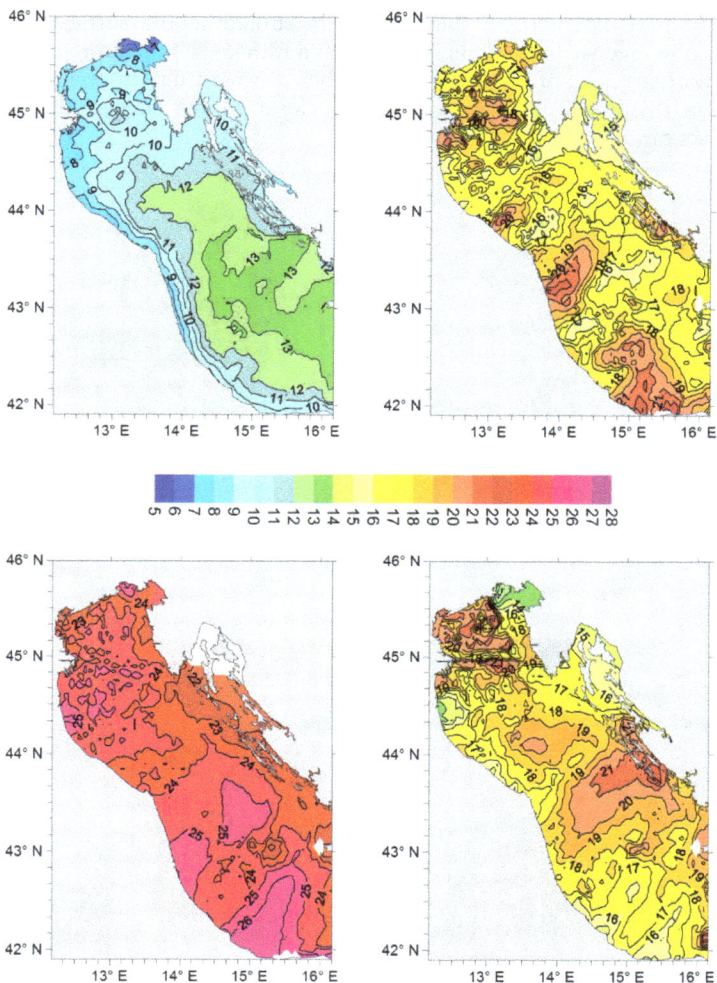

Fig. 2. Seasonal climatological maps (winter top left, spring top right, summer bottom left, autumn bottom right) of surface temperature (°C) in the northern-central Adriatic Sea.

At the depth of 10 m the temperature fields (see Figure 3) exhibit similar trends with respect to the surface ones, even though they are clearly attenuated. Seasons showing more similar fields at surface and 10 m are autumn and winter, mostly because of vertical mixing processes typical of these seasons, characterized also by an intense heat loss. Spring time is the most variable season, even though to a less extent to what depicted for the surface conditions. Temperatures are now bracketed between 12 and 22 °C, with mean at 16.07 °C (median 15.59 °C). Temperatures of the northern area are lower than corresponding surface ones, and this indicates that the solar radiation heat is stored in the very upper meters by the turbid waters brought in by the northern Adriatic rivers. During summer time the temperatures at 10 m depth are between 20 and 27 °C (mean 23.35 °C, median 23.1 °C), and in the southern part of the basin they are again similar to the surface ones, since the thermocline is now deeper than 10 m, until which a rather well mixed region is present. In any case, the northern part of the basin still exhibits a substantial difference, always due to the fact that in this region the thermocline is closer to the surface due to the above mentioned effect of solar radiation absorption.

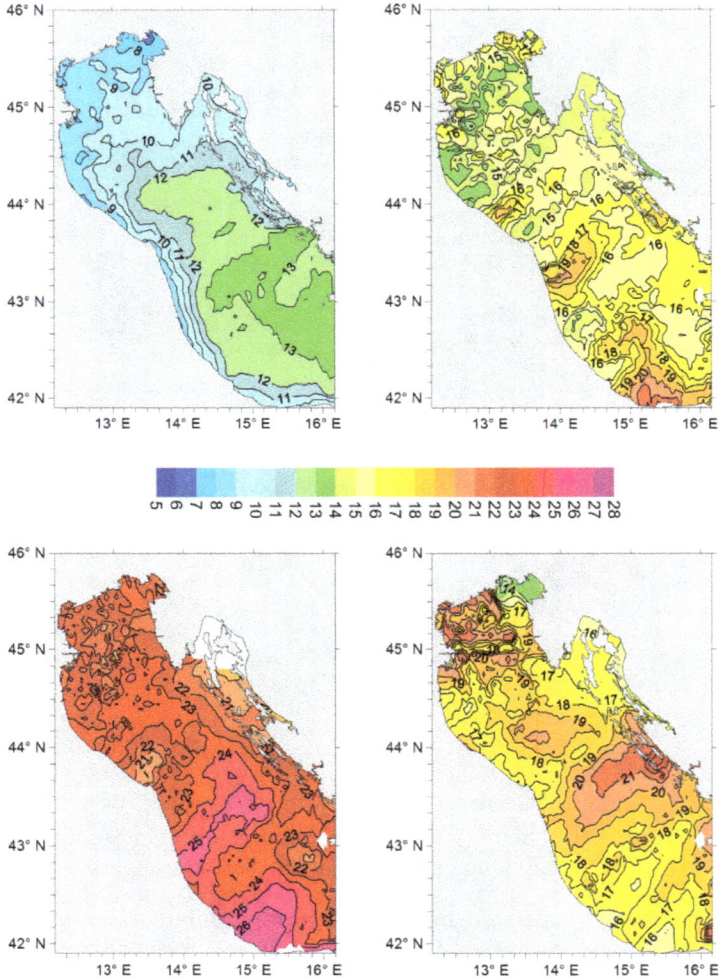

Fig. 3. Seasonal climatological maps (winter top left, spring top right, summer bottom left, autumn bottom right) of 10 m depth temperature (°C) in the northern-central Adriatic Sea.

Moving down to 20 m depth level (see Figure 4), the thermal fields are again resembling those at upper levels during autumn and winter, while increasing differences are evident during spring and summer time. During spring time the mean temperatures are between 10 °C (northern region) and 19 °C (southern area), with a mean value 14.71 °C (median

14.79 °C). A relevant thermal gradient is present in the region between the Po delta-Istria peninsula and the area at its south. Mean summer temperatures are between 15 and 25 °C, with a mean value of 19.29 °C (median 18.85 °C). Minima are observed in front of the Po delta region, where the river waters tend to create a strong thermo-haline stratification that is preventing the vertical mixing of cold water produced during winters.

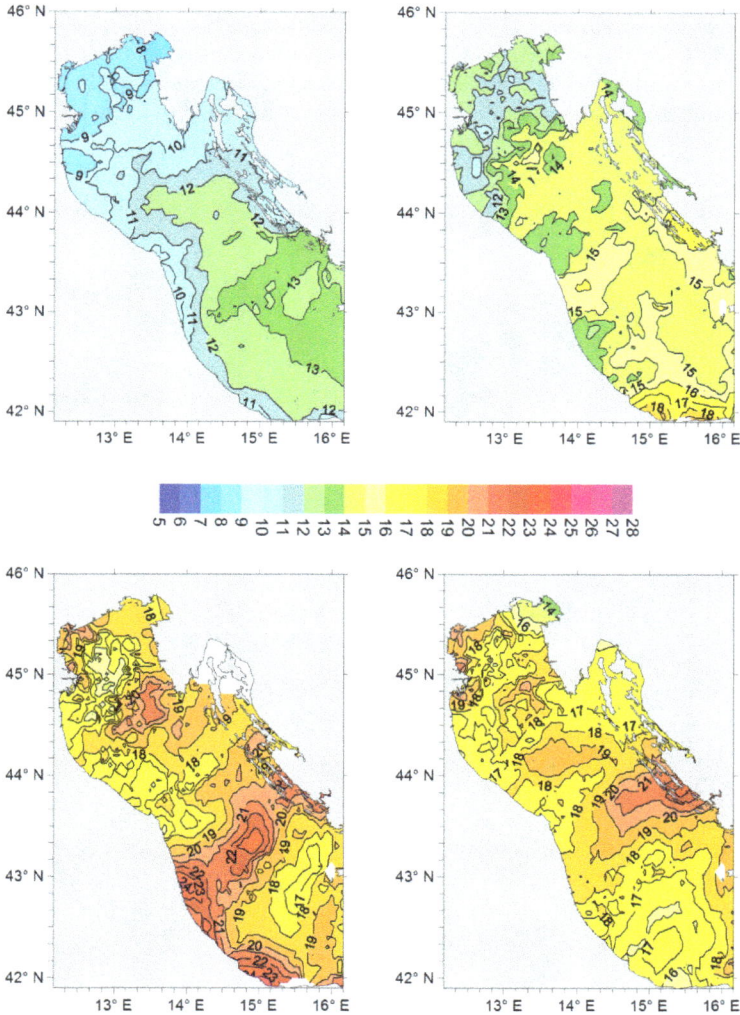

Fig. 4. Seasonal climatological maps (winter top left, spring top right, summer bottom left, autumn bottom right) of 20 m depth temperature (°C) in the northern-central Adriatic Sea (note that a small area close to the Italian coasts is actually shallower than 20 m).

At 50 m depth the thermal variability is very much reduced (see Figure 5), with values between 10 and 18 °C. During winter time typical values are between 10 °C (in the north-west region) and 14 °C (central-east sector), with mean values to 12.69 °C (median 13.00 °C). In spring temperatures are slightly higher, about 0.5 °C, with a mean value of 13.35 °C (median 13.47 °C). Lowest values are observed in the western area (about 12 °C) while the highest ones (about 14 °C) are present in the eastern sector. During summer time the slow warming of the water column keeps continuing because of diffusion and mixing from the upper levels, and the mean and median temperature increase of about 1 °C, reaching respectively 14.29 °C and 14.38 °C. Lowest values (around 13 °C) are present in the northern and western area. Maxima are around 15 °C in the eastern and central area of the basin. Highest temperatures are observed during autumn, since in this season the thermocline is more easily broken by storms and surface cooling processes, and the surface waters are mixed with those below.

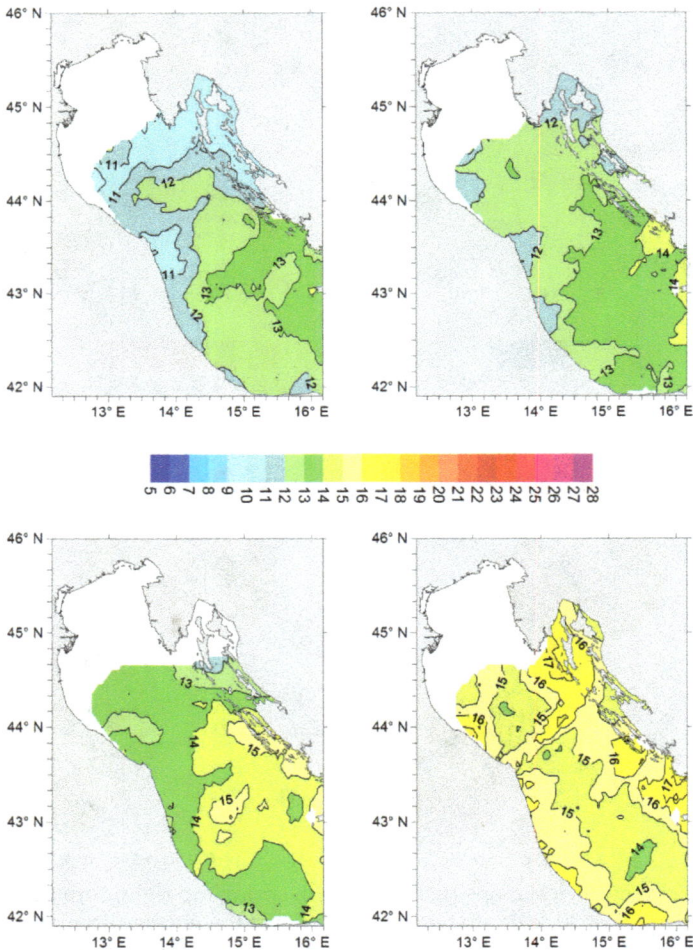

Fig. 5. Seasonal climatological maps (winter top left, spring top right, summer bottom left, autumn bottom right) of 50 m depth temperature (°C) in the northern-central Adriatic Sea (note that areas close to the Italian coasts are actually shallower than 50 m).

Mean temperature and median one are 15.57 and 15.24 °C respectively, with highest values that are aligned almost parallel to the eastern (until 18 °C) and western coasts, while the lowest values (up to 14 °C) are found in the middle of the basin.

In the layer closer to the sea bottom (see Figure 6), the thermal variability during the seasons is once again rather large (from 7 to 25 °C), mostly because of the bathymetric variations.

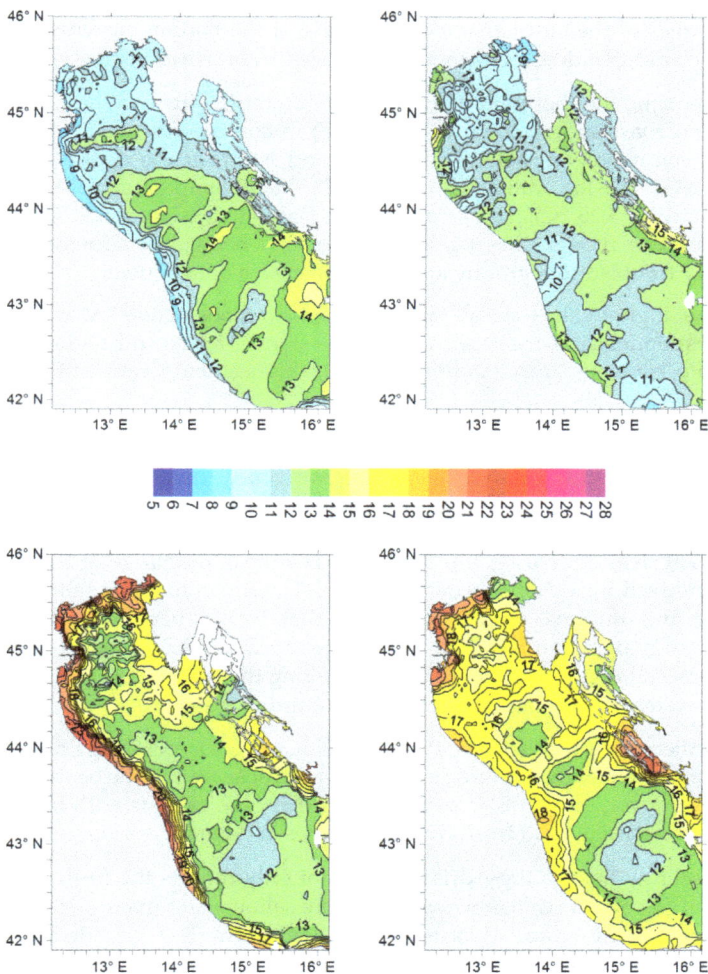

Fig. 6. Seasonal climatological maps (winter top left, spring top right, summer bottom left, autumn bottom right) of bottom temperature (°C) in the northern-central Adriatic Sea.

During winter the mean temperature is about 12.25 °C (median 12.92 °C); closer to the western coast, where the bottom is shallower, cold waters are encountered, with temperatures lower than 9 °C, while maxima (more than 14 °C) are found in the south-eastern region, reflecting the flow of warmer and southern waters. Relative minima are depicted in the central-southern region, in proximity of the meso-Adriatic depressions (see Figure 1b), where the dense and cold northern Adriatic waters can get accumulated without being vertically mixed during the year because of relatively high depths.

During spring time the temperature variations are limited, being the mean and median respectively of 12.73 and 12.78 °C. However, the shallow water regions along the western coasts exhibit now larger temperatures, similar to those of the eastern ones, while the central and southern areas of the basin are colder, because of the bottom currents that flush along south-east the colder and denser northern Adriatic waters formed during winter time.

During summer time the thermal picture is a lot different, with very high temperatures all along the Italian coast and minima (below 12 °C) in correspondence of the meso-Adriatic pits. Even the central region between Po delta and Istria, in the depths around 40 m, is showing now relatively low values (around 13 °C). Mean temperature of the entire domain is 15.52 °C (median 14.26 °C). The area between Dalmatian islands and Croatian coasts is very much variable, though, having warmer regions being interrupted by colder ones, reflecting a very troubled topography and probably also a data paucity.

During fall time higher values are all long the coastal regions, while lower temperatures are found in correspondence of meso-Adriatic depressions. The central regions, including the quite shallow Po-Istria area, reach in this season the highest temperature of the whole year.

3.2 Salinity

Surface salinity (see Figure 7) shows a strong variability (from less than 30.00 to 38.50; salinity is expressed with a non-dimensional number substantially corresponding to grams of salt dissolved in a kg of water), especially if compared to the Mediterranean Sea, since there are relevant river discharges (the Po river has a mean discharge of 1500 m^3/s) and the basin is characterized by a shallow coastal region. However, maxima salinities are reached during winter and minima ones in summer, that is not perfectly matching Po river discharge, whose maxima are reached in spring-autumn and minima in summer. During winter, the mean salinity is 37.40 (median 38.18); along the coastal region there exists a band of low salinity waters, which largest extension is found in front of the Po delta area.

The whole northern area between the Po river delta and Istria peninsula has salinity values lower than 38.00, while higher values (up to 38.40) are observed in the central-eastern region of the basin. Moreover, along the Croatian coasts there are regions with low salinity water masses as well, because of local fresh river discharges and karstic springs.

The resulting salinity field in the Adriatic basin not only reflects the fresh water discharges distribution, but the marine dynamics as well. More saline water masses are brought in from south, transported by the so-called Eastern-Adriatic Current (EAC), while fresher waters are flowing to south-east with the Western Adriatic Current (WAC). Besides, dominant Bora winds from north-east are pushing the fresher waters closer to the western coast. During spring time a generalized decrease of the mean salinity values is depicted, with average of 36.93 (median 37.71); a significant offshore propagation of the area at low salinity is evident, as a consequence of the high discharge rate of the Po river and of the onset of the thermal stratification, that is confining the riverine waters in the upper layers of the water column. Minima are now well below 30.00 in the Po delta region, while values larger than 34.00 are found in small areas of the southern part. Along the Croatian coasts as well the region with low salinity values is now located more offshore. During summertime a slight decrease in the surface salinity values is observed (mean 36.79, median 37.60) together with a further propagation offshore of low salinity waters that reach now the whole northern part. This is most likely due to the strong stratification typical of this season, that is in short confining the relatively low amount of fresh waters brought in to the very upper layers of the sea. Water with relatively high salinity are now evident just in the central-south region, north of Gargano. During autumn, when the atmospheric circulation gets again intense and starts cooling the sea surface, low salinity waters are again confined to the Italian coasts; mean value is now up to 37.05 (median 37.78), since the stratification is now destroyed and the waters are vertically mixing. Together with this, there is a flow of saline waters along the eastern coasts from the southern Adriatic Sea.

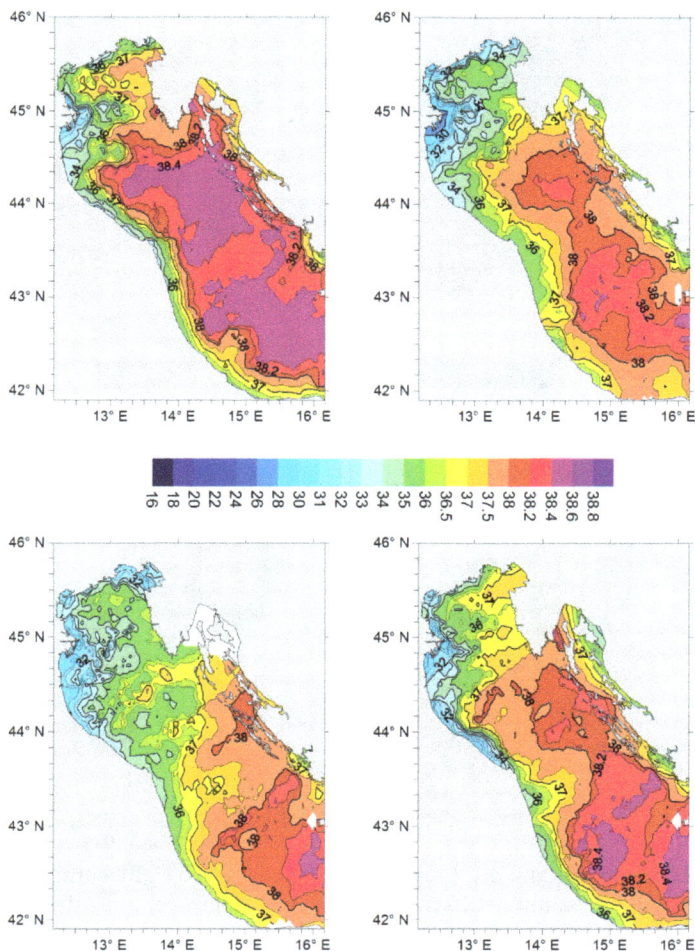

Fig. 7. Seasonal climatological maps (winter top left, spring top right, summer bottom left, autumn bottom right) of surface salinity in the northern-central Adriatic Sea.

Along the Italian coast a a particular case is provided by the coastal areas around the Conero Promontory (see Figure 1b): in the northern part of them, very low salinity waters are present (below 32.00, with a strong gradient coast-offshore), while in the southern region values are beyond 35.00. The Conero Promontory favors an offshore spreading of coastal freshwaters and their dilution with saltier waters. In the corresponding Croatian regions in the eastern part of the basin, salinity is now higher than 38.00 everywhere, except in the region between the Croatian and Dalmatian coasts, where the balance is modified by fresh water discharges from the mainland.

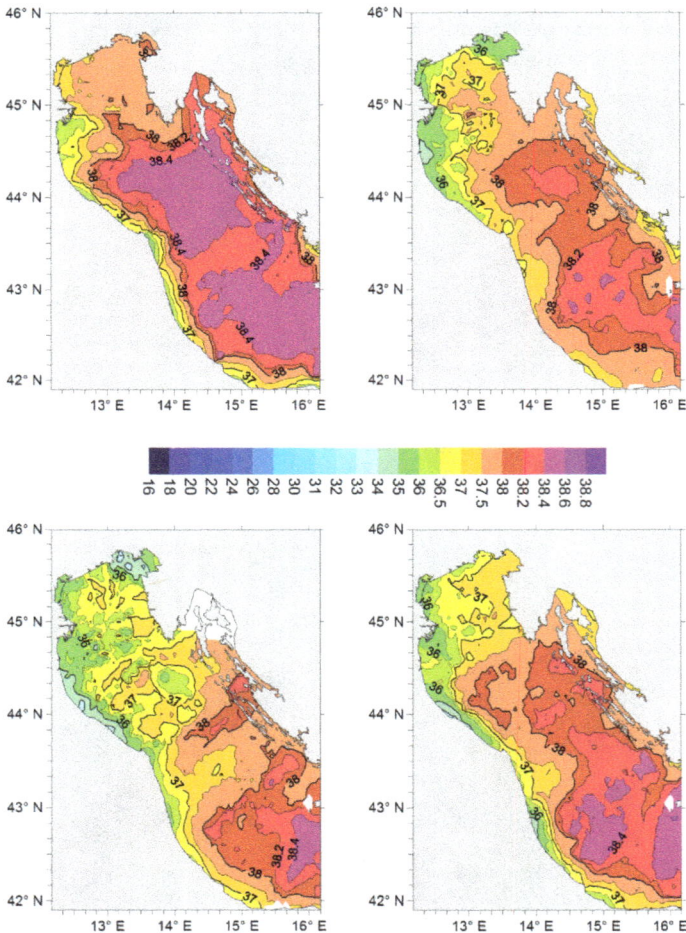

Fig. 8. Seasonal climatological maps (winter top left, spring top right, summer bottom left, autumn bottom right) of salinity at 10 m depth in the northern-central Adriatic Sea.

Analyzing the situation at 10 m depth (see Figure 8), salinity fields exhibit a seasonal variation (from 34 to 38.5) that is significantly lower with respect to the surface. Nevertheless, the trend seems to be similar: maxima during winter, an increasing offshore propagation of low salinity areas from the Italian coast while transitioning from winter to summer. During winter the trend is substantially similar to the surface one, since in this season the water column is very well mixed. The domain mean salinity value is 37.94, median being 38.25.

The northern region is the only one where values that are lower with respect to the previous situation can be seen (most likely due to the Po river flooding events that reach offshore areas before getting mixed). During spring period the mean salinity value reaches 37.71 (median 37.94), and a salinity lowering compared to the previous season is evident in almost the whole basin. The same is evident in summer for most of the basin, with a relevant freshening above all in the northern area, while during autumn salinities begin to increase.

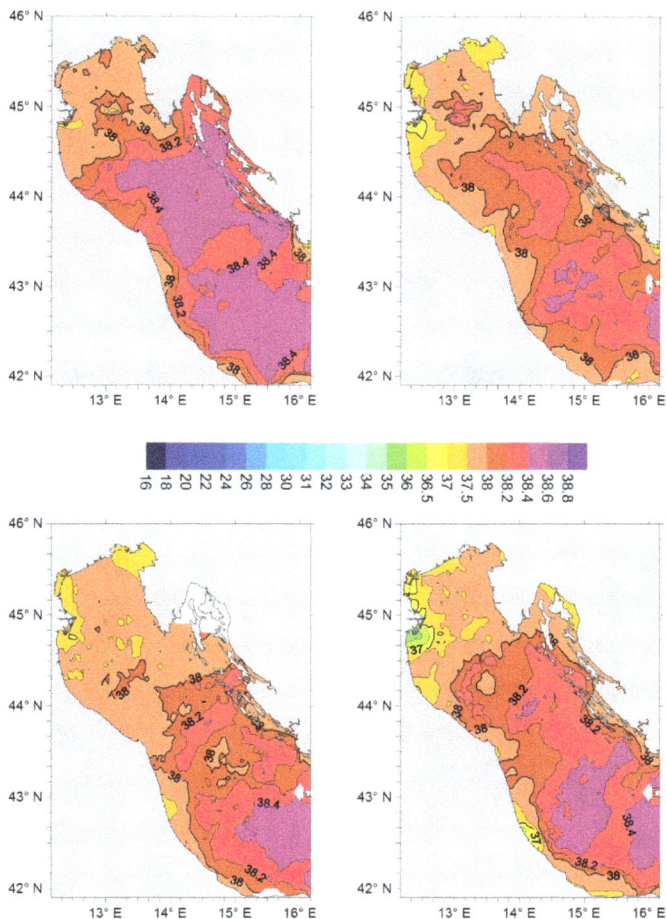

Fig. 9. Seasonal climatological maps (winter top left, spring top right, summer bottom left, autumn bottom right) of salinity at 20 m depth in the northern-central Adriatic Sea.

Salinity fields at 20 m depth (see Figure 9) are resembling those at 10 m depth as far as the central-eastern sectors are concerned, while they change significantly in the north-western area, where the salinity are higher compared to the 10 m depth situation. During winter time the mean salinity is about 38.22 (median 38.34), in spring the mean decreases to 38.07 (median 38.14), in summer similar values are present (38.18 and 38.10), while in the autumn time the mean value is 38.00 (median 38.15). Looking at the transect region between Mt. Conero Promontory and Croatia a salinity gradient is still detectable, 38.00 or less along the Italian coasts, 38.20 or more in the central-eastern region.

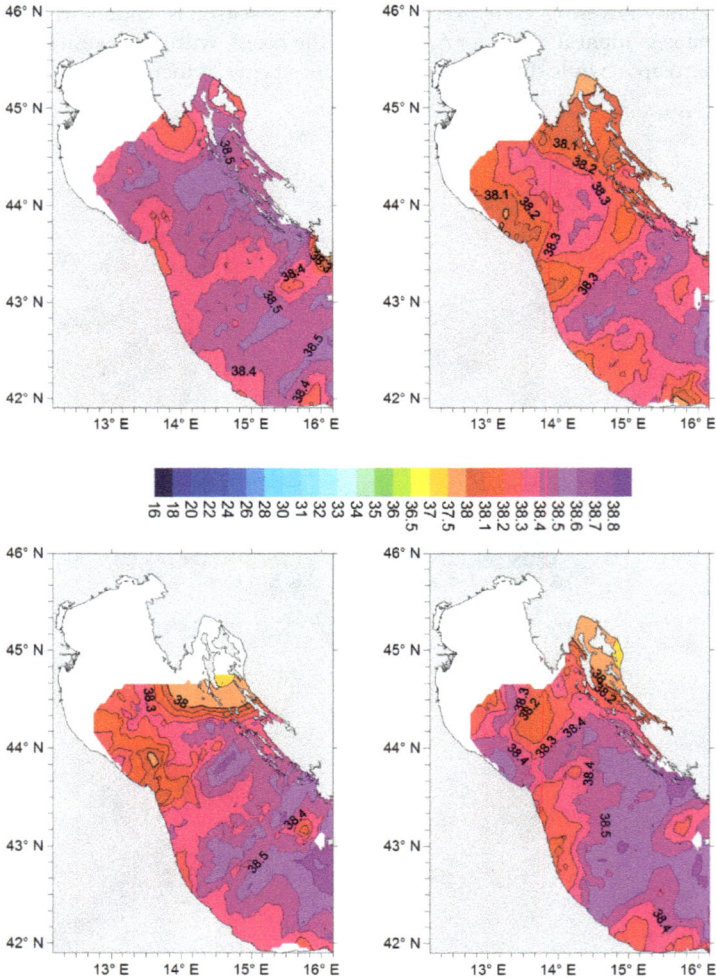

Fig. 10. Seasonal climatological maps (winter top left, spring top right, summer bottom left, autumn bottom right) of salinity at 50 m depth in the northern-central Adriatic Sea.

Moving down to 50 m (see Figure 10) the salinity fields reach rather high values, above 38.00 almost everywhere, but in the Kvarner gulf and the northern part of the area between Dalmatian islands in the summer and autumn seasons. During winter the mean salinity is

38.42 (median 38.45), in spring values are slightly less (38.37 and 38.39), in summer mean salinity is again up to 38.42 (median 38.39), while in autumn it reaches 38.39 (38.45).

Even at this water depth, in the area between Conero Promontory area and Croatia patches with low salinity waters (albeit larger than 38.00) are found in the western part, while maxima (above 38.50) are common in the eastern sector, especially in autumn and winter.

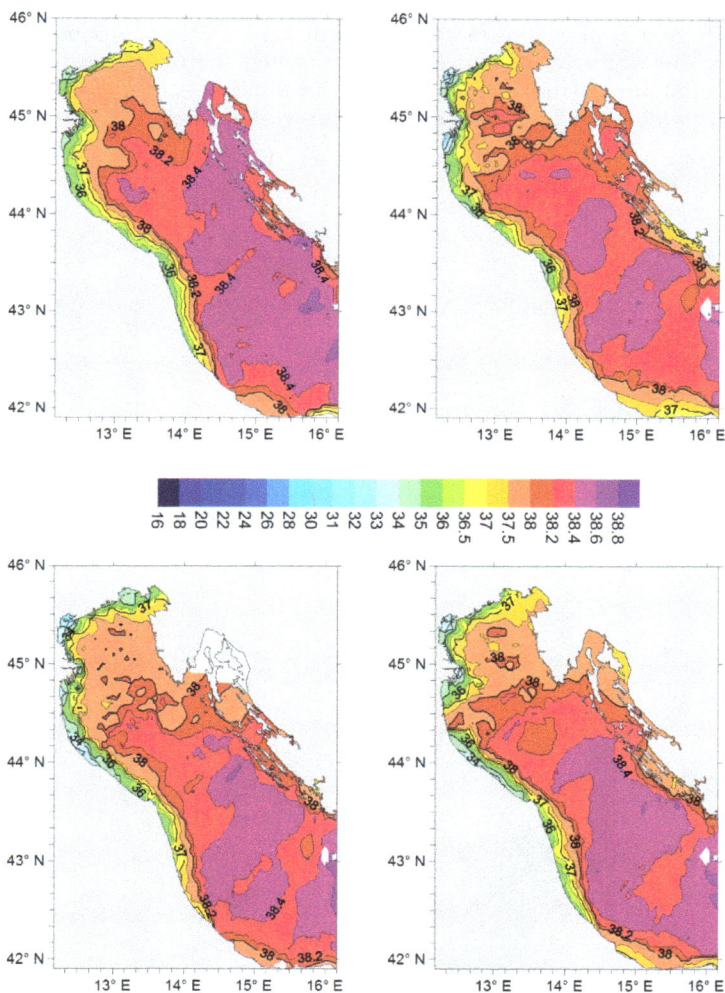

Fig. 11. Seasonal climatological maps (winter top left, spring top right, summer bottom left, autumn bottom right) of salinity at the bottom in the northern-central Adriatic Sea.

In proximity of the bottom (see Figure 11) the salinity fields have a rather similar behavior during all seasons, with very low values along the shallow western coast, from Trieste up to south of the Conero Promontory (see again Figure 1b), and rather high values in the central-eastern regions. The reduced variability throughout the season is also evident in the mean (and median) values: during winter values are up to 37.98 and 38.89 respectively, in spring these values are preserved to 37.90 (38.29); during summer values are 37.87 (38.32) and in autumn 37.84 (38.29).

3.3 Dissolved oxygen

The dissolved oxygen at the surface (see Figure 12) exhibits a seasonal cycle clearly influenced by the water temperature behavior; when the temperature increases, the maximum amount of available oxygen in the water diminishes. Indeed, dissolved oxygen maxima are in winter (up to 8 ml/l in the north-western area, where also temperature

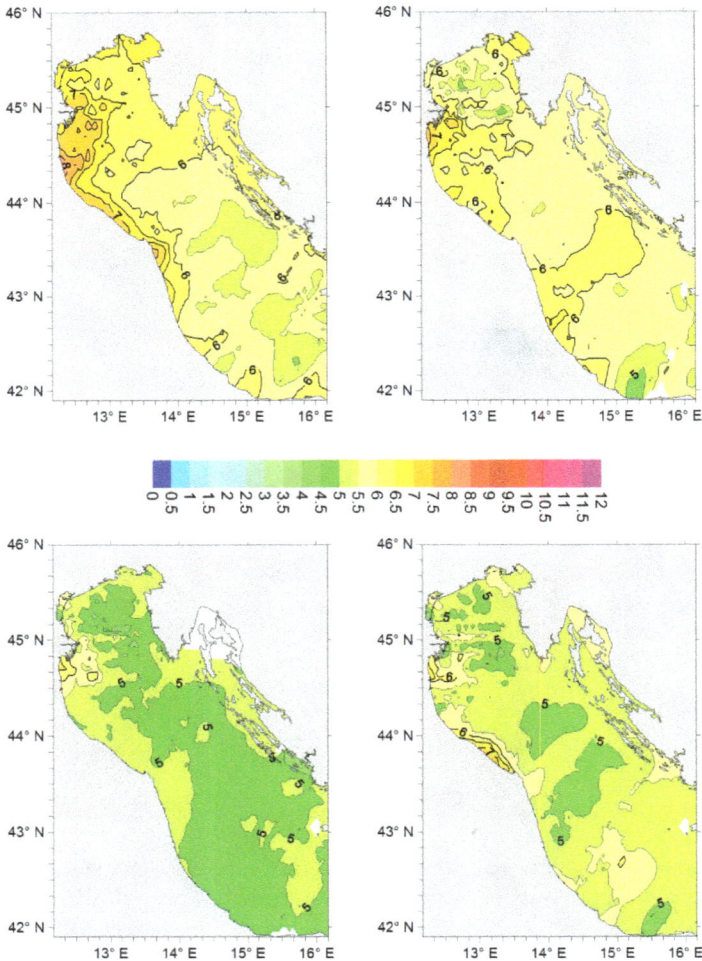

Fig. 12. Seasonal climatological maps (winter top left, spring top right, summer bottom left, autumn bottom right) of surface dissolved oxygen (ml/l) in the northern-central Adriatic Sea.

minima are found), while minima (4.5 ml/l) are measured in summer, in the southern-eastern part of the basin. During all seasons, the amount of oxygen dissolved in the western areas is generally larger than in the corresponding eastern areas.

Dissolved oxygen fields at 10 m depth (not showed) are very similar to the surface ones during all seasons, showing a reduced spatial variability.

At 20 m depth, the dissolved oxygen fields (see Figure 13) are different, mostly during summer time. At this depth, the western regions are characterized by lower values, and minima are present in the Gulf of Trieste and southern of the Po delta region, indicating hypoxic conditions in such shallow regions.

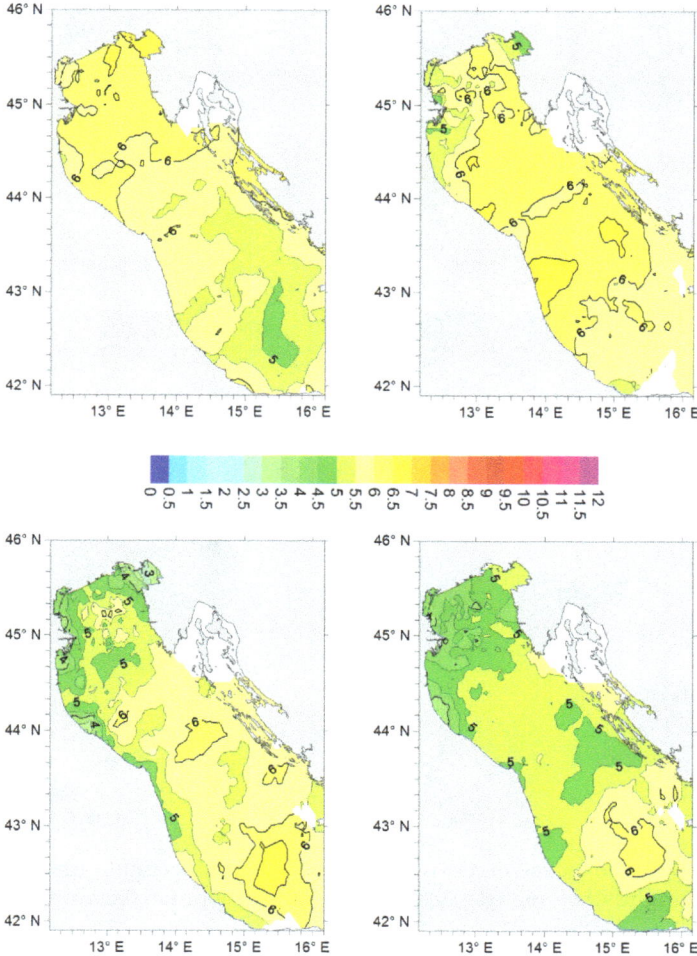

Fig. 13. Seasonal climatological maps (winter top left, spring top right, summer bottom left, autumn bottom right) of dissolved oxygen (ml/l) at 20 m depth in the northern-central Adriatic Sea.

At 50 m depth, the dissolved oxygen fields (see Figure 14) are again rather homogenous both in winter and spring, while in summer and autumn lower values are present in the western region.

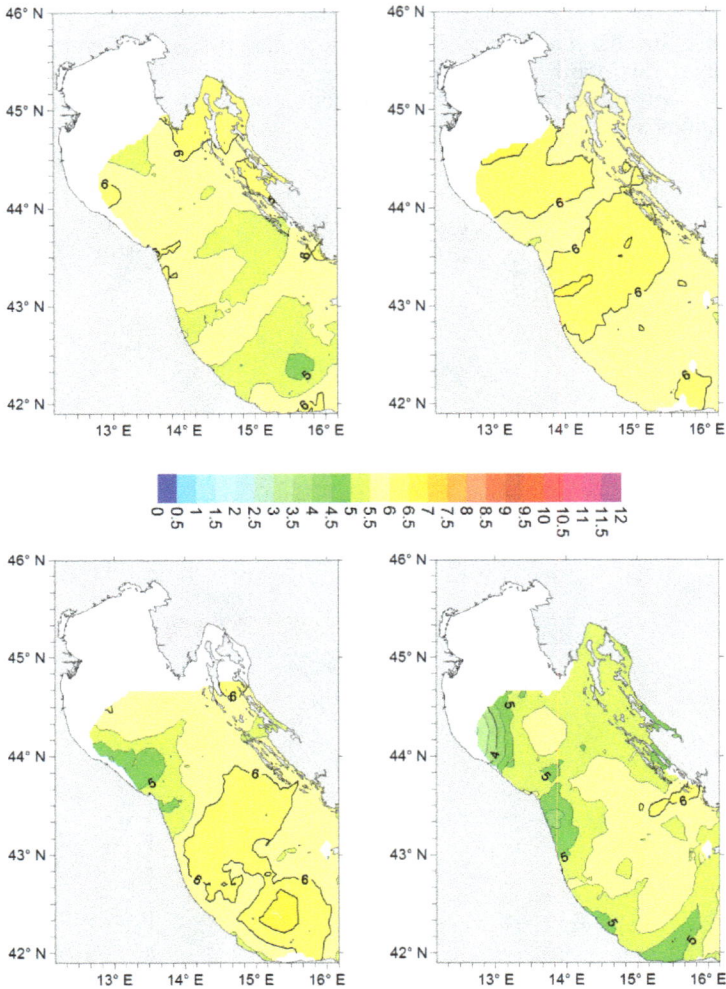

Fig. 14. Seasonal climatological maps (winter top left, spring top right, summer bottom left, autumn bottom right) of dissolved oxygen (ml/l) at 50 m depth in the northern-central Adriatic Sea.

In the bottom layers (see Figure 15) the variations to the content of dissolved oxygen are more evident, namely the low values are present in a large area close to the Po delta region in the summer and autumn seasons, again indicative of hypoxic events.

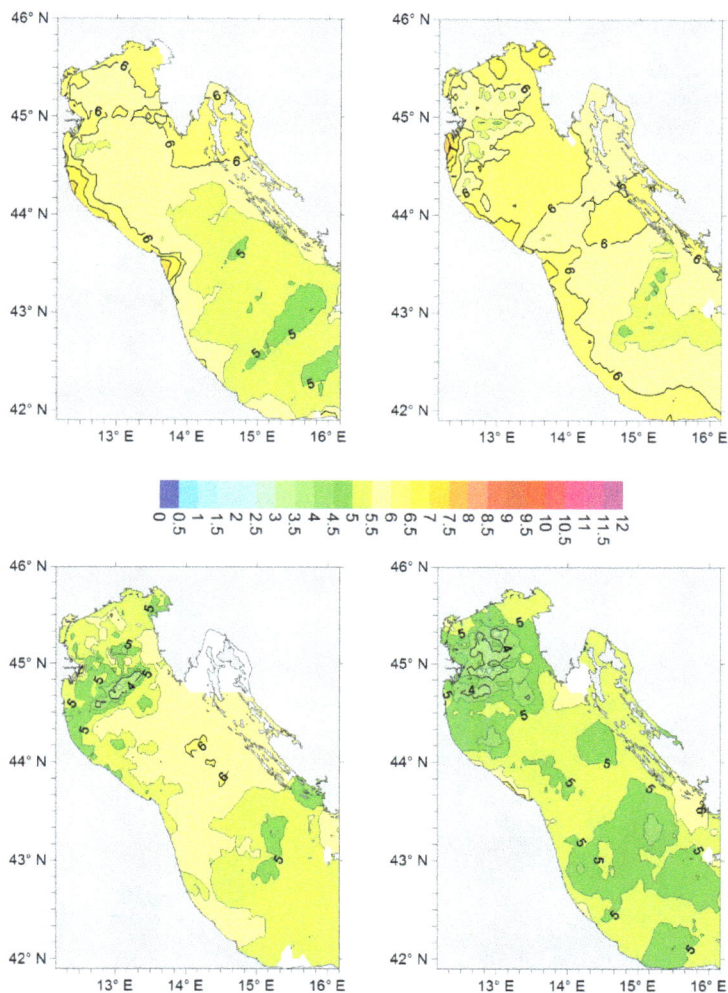

Fig. 15. Seasonal climatological maps (winter top left, spring top right, summer bottom left, autumn bottom right) of dissolved oxygen (ml/l) at the bottom in the northern-central Adriatic Sea.

3.4 Nutrients

Unfortunately, available data for nitrates (NO_3) and nitrites (NO_2), orthophosphate (PO_4) and orthosilicate (SiO_4) are far less than those collected for temperature and salinity. The first consequence is that the obtained climatological fields are less reliable, and often do not cover the whole examined area. The description will therefore be more general and less detailed. On the other hand NH_4 is even less present in the acquired data, and it is not therefore possible to represent it (the total inorganic nitrogen content, TIN, can however be estimated with a good degree of approximation by the NO_3+NO_2 pools).

3.4.1 Nitrates and nitrites

NO_3 is the most abundant nitrogen form almost anywhere in each season, even because autotrophic organisms tend to prefer NH_4, NO_2 and NO_3, in strict order. At the surface (see Figures 16 and 17) both these latter nitrogen forms reach maxima along the western coast, a clear signature of the river discharge. However, nitrates maxima are close to the Po river delta and north of it, while nitrites maxima are present also to the south of the Po river delta. The season with lowest nitrites and nitrates concentration is summer, mainly because of the minimum river discharge brought into the basin.

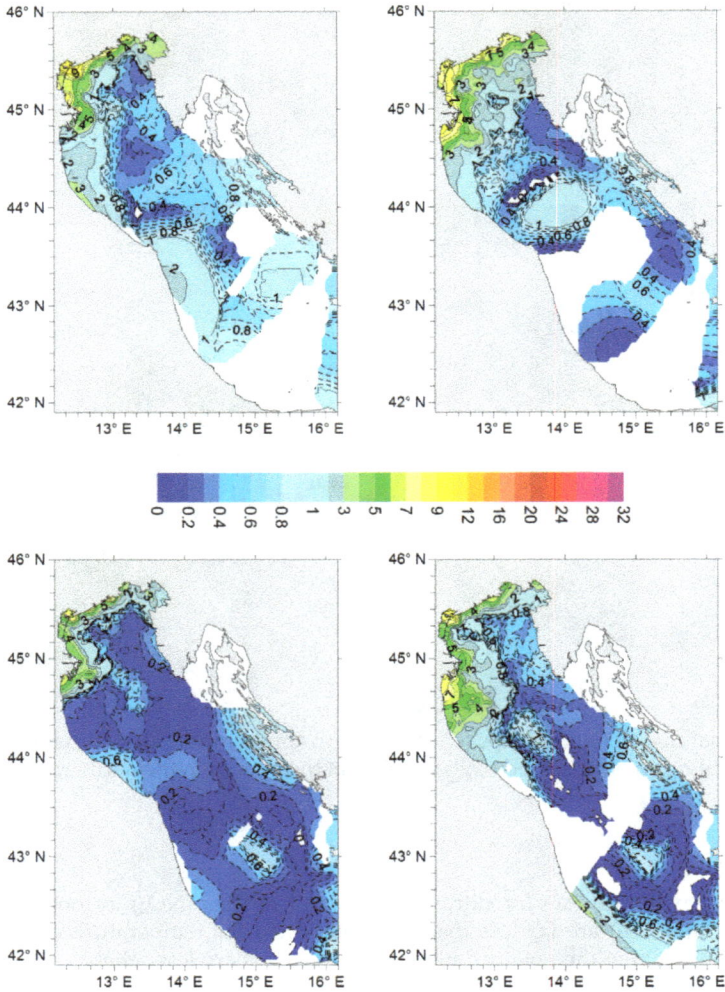

Fig. 16. Seasonal climatological maps (winter top left, spring top right, summer bottom left, autumn bottom right) of N-NO_3 (mM/m³) in the surface waters of the northern-central Adriatic Sea.

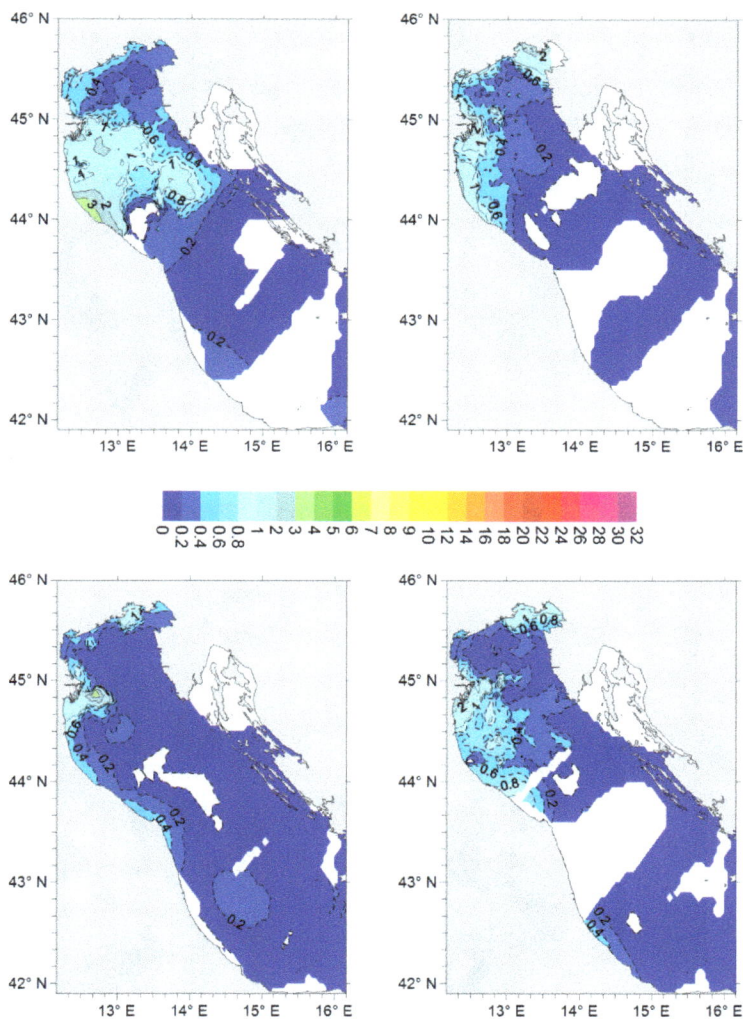

Fig. 17. Seasonal climatological maps (winter top left, spring top right, summer bottom left, autumn bottom right) of N-NO$_2$ (mM/m^3) in the surface waters of the northern-central Adriatic Sea.

At 10 m depth and 20 m depth (not showed) the general trend is close to that shown at the surface, but the absolute values are considerably smaller.

At 50 m depth (see Figures 18 and 19) the concentration increases in the north-western region, probably because in this area the analyzed depth level is close to the bottom and is affected by demineralization processes of the organic matter due to bacteria.

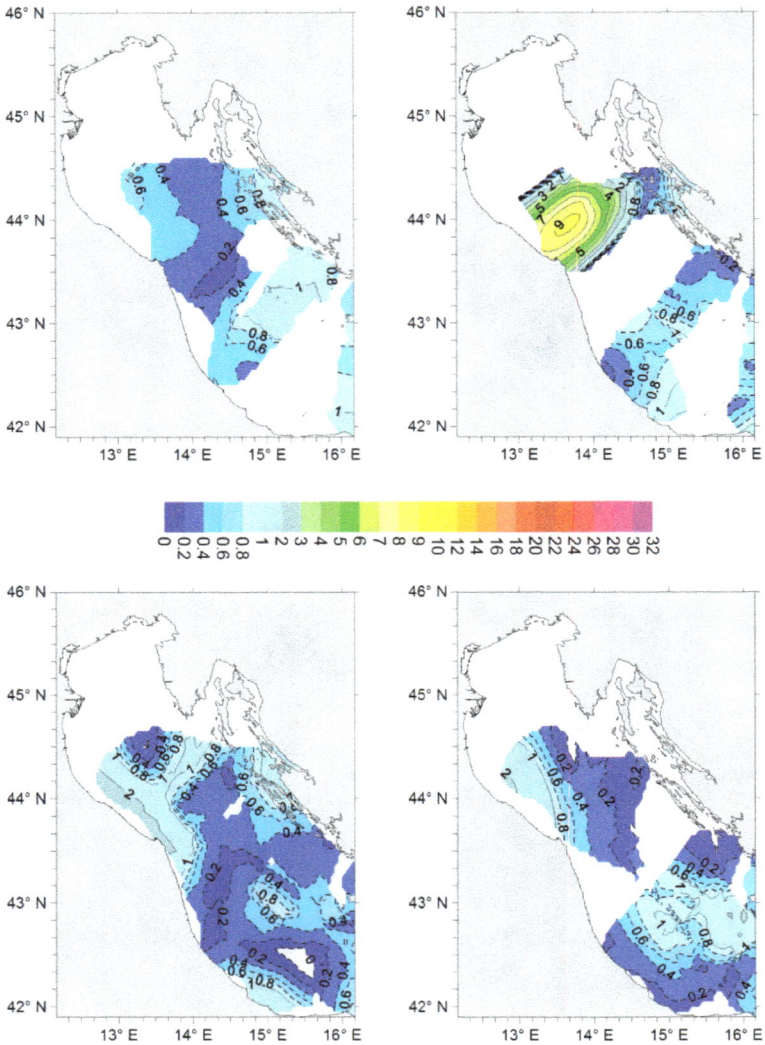

Fig. 18. Seasonal climatological maps (winter top left, spring top right, summer bottom left, autumn bottom right) of N-NO$_3$ (mM/m^3) at 50 m depth in the northern-central Adriatic Sea.

The same line of reasoning can be applied for the rather large values that are observed in the proximity of the bottom (shown in Figures 20 and 21), particularly in the north-western and meso-Adriatic depressions areas. The high values observed offshore in the central area of the domain during spring time are most likely due to data not quality checked that should be rejected.

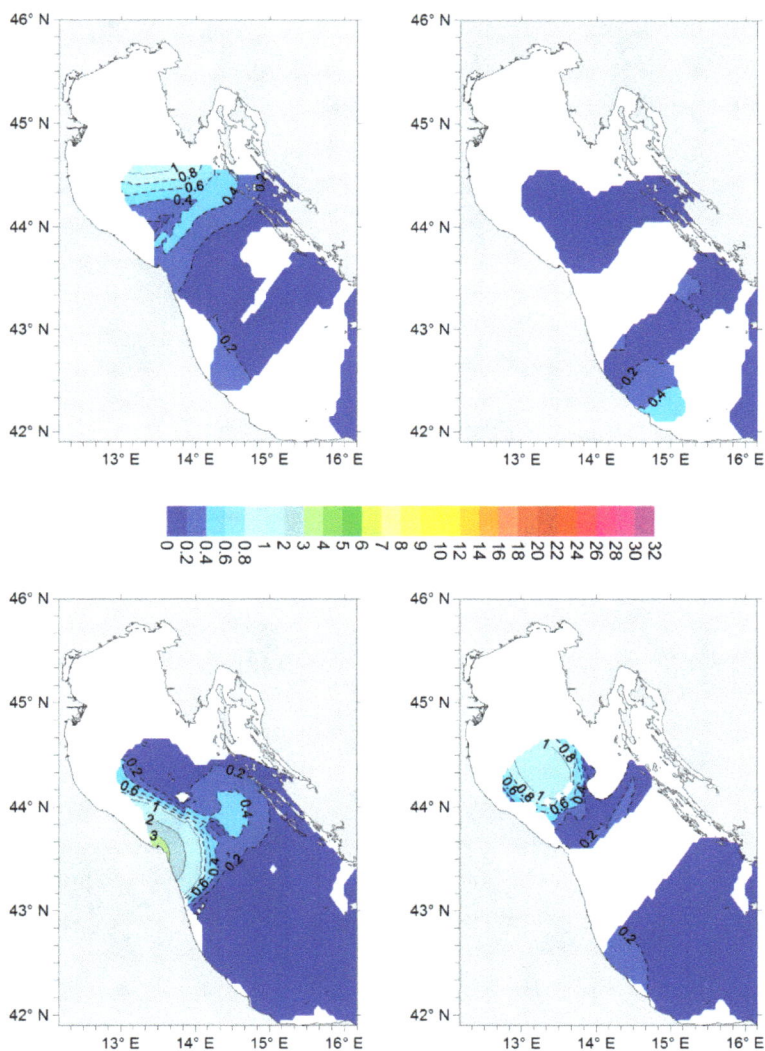

Fig. 19. Seasonal climatological maps (winter top left, spring top right, summer bottom left, autumn bottom right) of N-NO₂ (mM/m³) at 50 m depth in the northern-central Adriatic Sea.

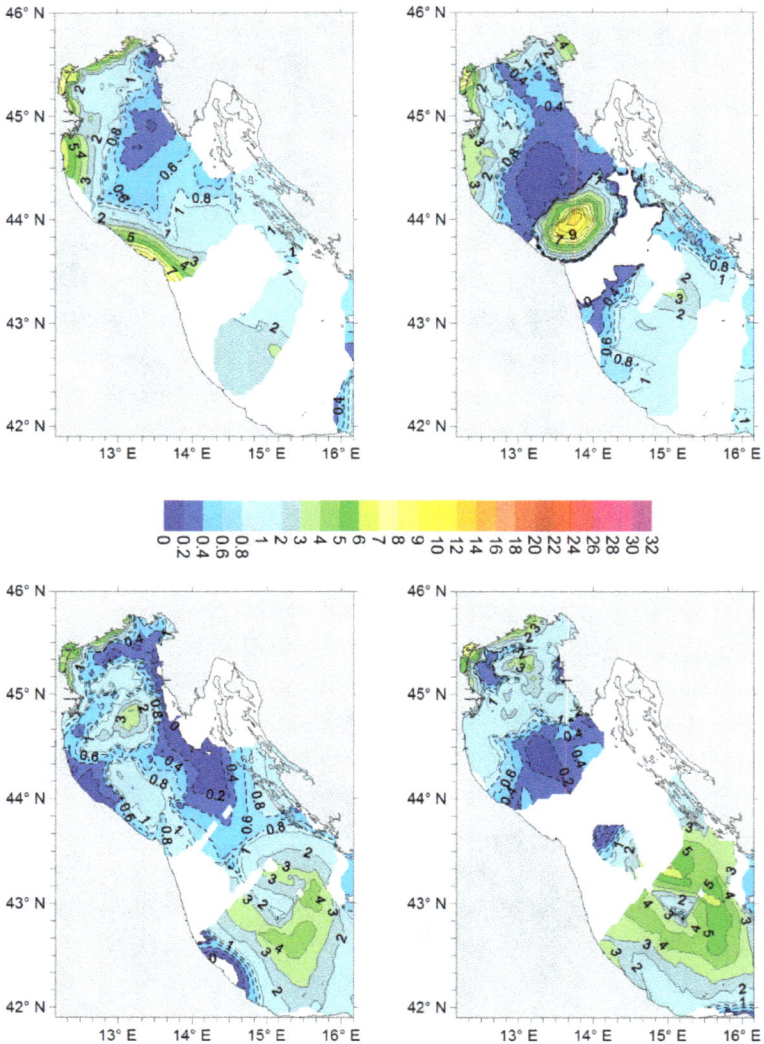

Fig. 20. Seasonal climatological maps (winter top left, spring top right, summer bottom left, autumn bottom right) of N-NO₃ (mM/m³) at the bottom in the northern-central Adriatic Sea.

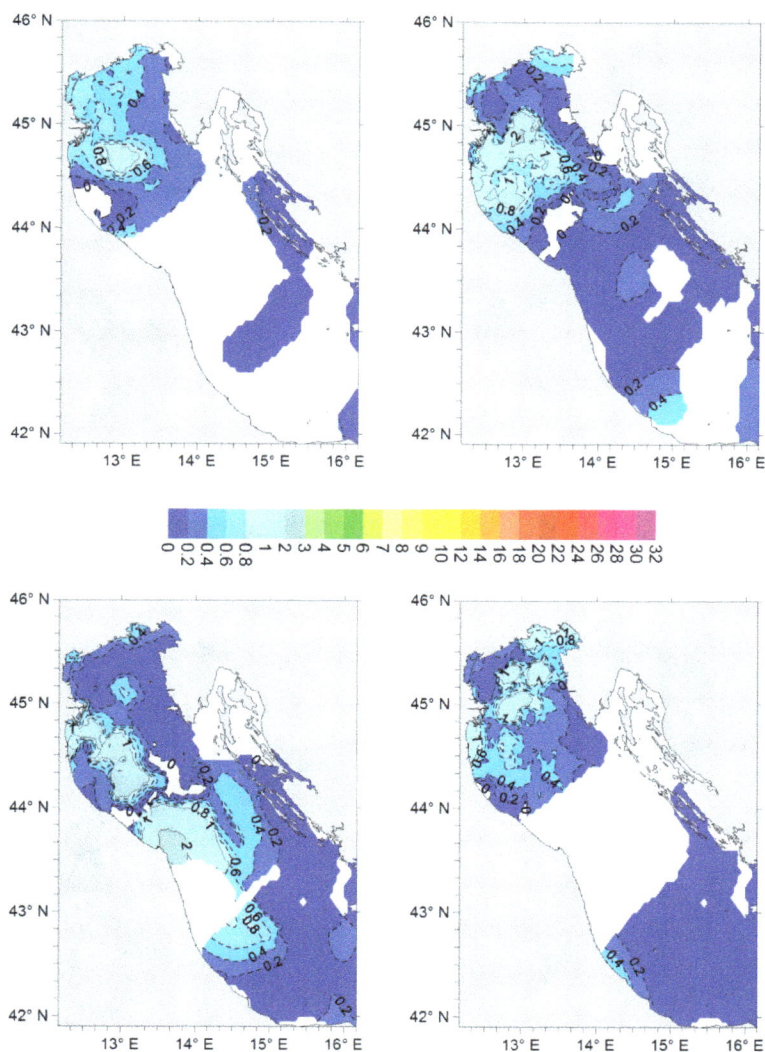

Fig. 21. Seasonal climatological maps (winter top left, spring top right, summer bottom left, autumn bottom right) of N-NO$_2$ (mM/m^3) at the bottom in the northern-central Adriatic Sea.

3.4.2 Orthophosphathes

Values of orthophosphates at the surface (see Figure 22) are typically very low everywhere but in the northern area, especially in the proximity of the river Po, due to the amount of river discharge.

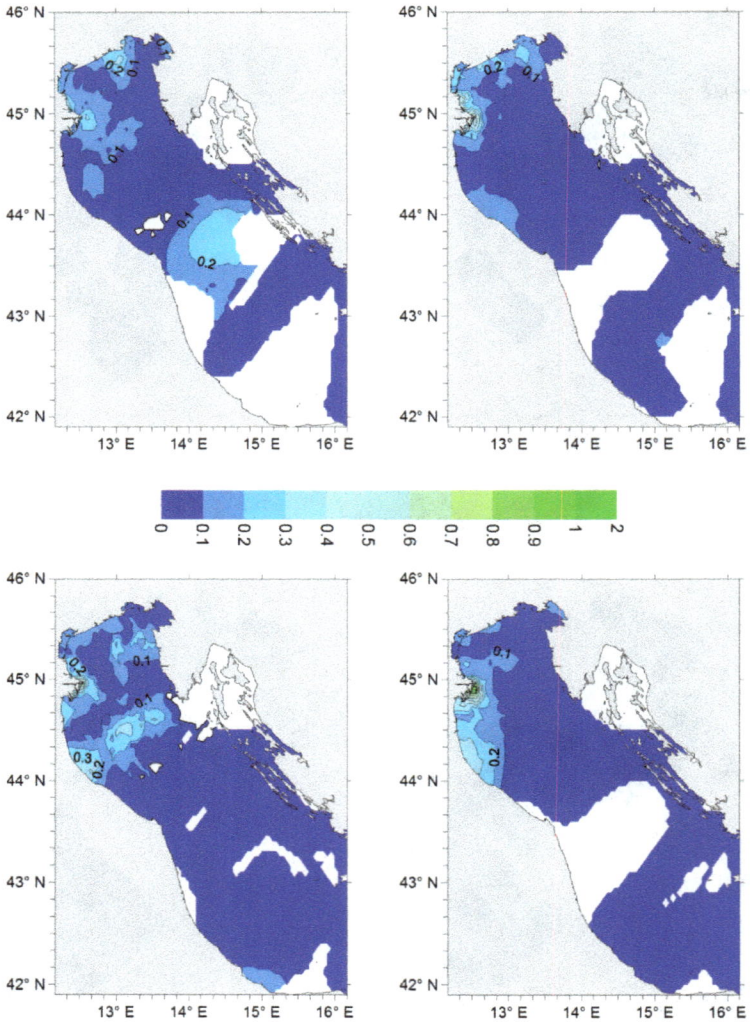

Fig. 22. Seasonal climatological maps (winter top left, spring top right, summer bottom left, autumn bottom right) of P-PO$_4$ (mM/m^3) in the surface waters of the northern-central Adriatic Sea.

The same situation is present at 10 m depth and 20 m as well (not presented), even though with values generally lower with respect to the surface ones.

At 50 m depth, the P-PO4 concentrations (Figure 23) are extremely low anywhere.

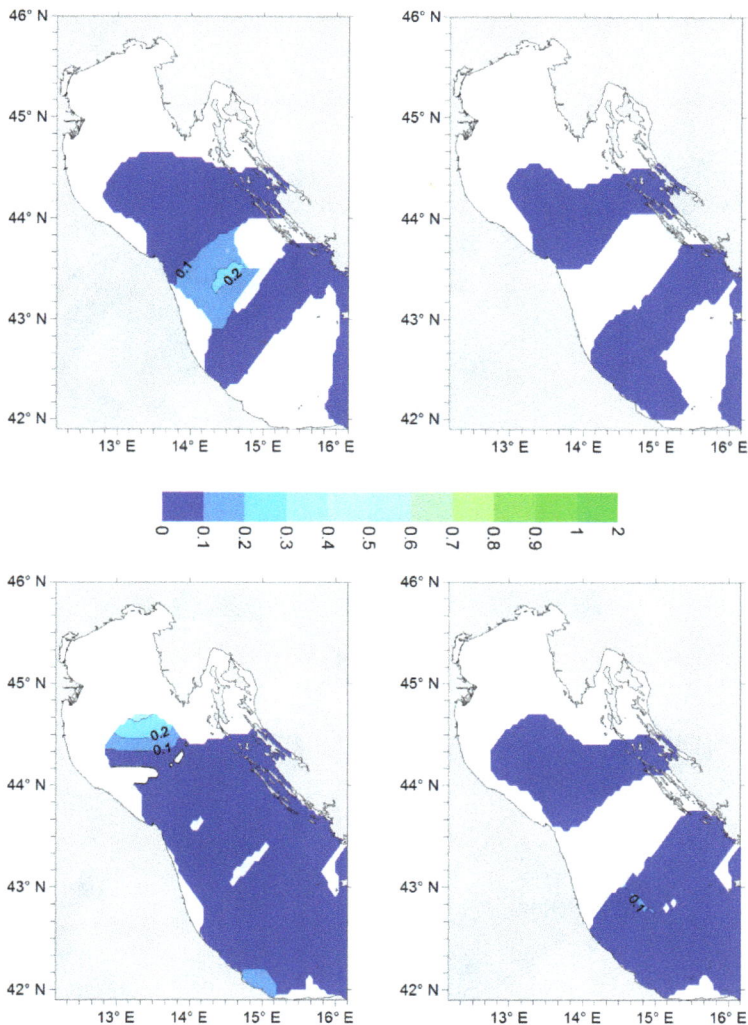

Fig. 23. Seasonal climatological maps (winter top left, spring top right, summer bottom left, autumn bottom right) of P-PO$_4$ (mM/m^3) at 50 m depth in the northern-central Adriatic Sea.

At the sea bottom (Figure 24) the values tend to increase in the northern part and in the area of the meso-Adriatic depressions, as a consequence of mineralization processes of the organic matter.

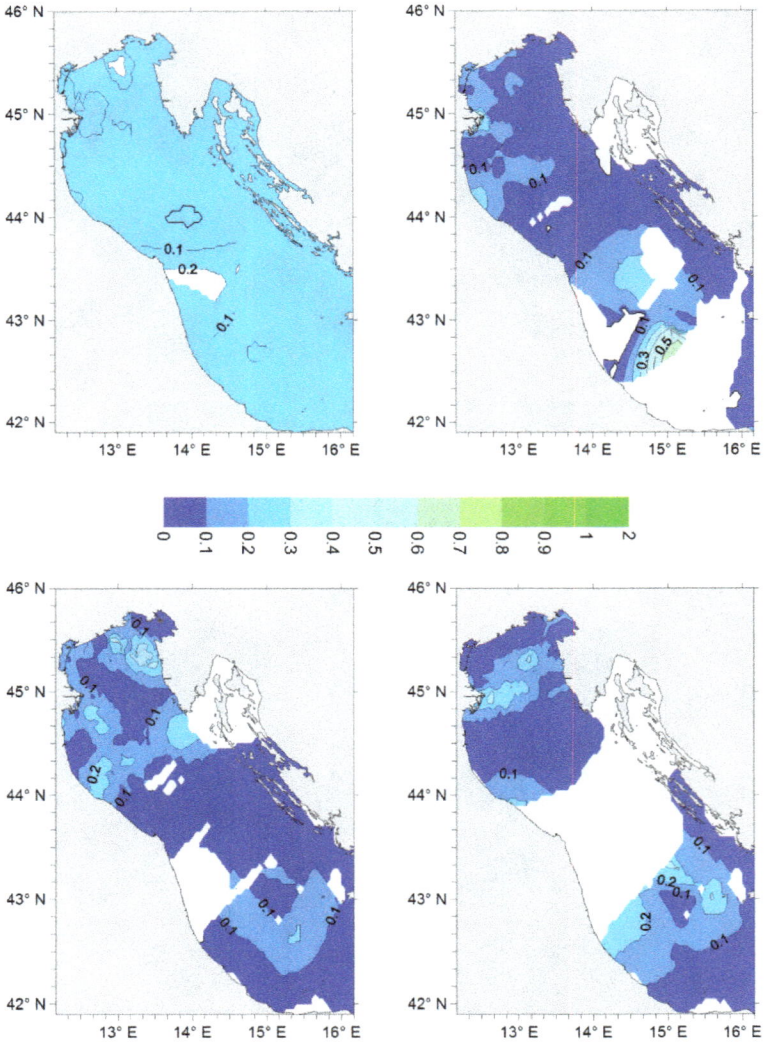

Fig. 24. Seasonal climatological maps (winter top left, spring top right, summer bottom left, autumn bottom right) of P-PO$_4$ (mM/m^3) at the bottom in the northern-central Adriatic Sea.

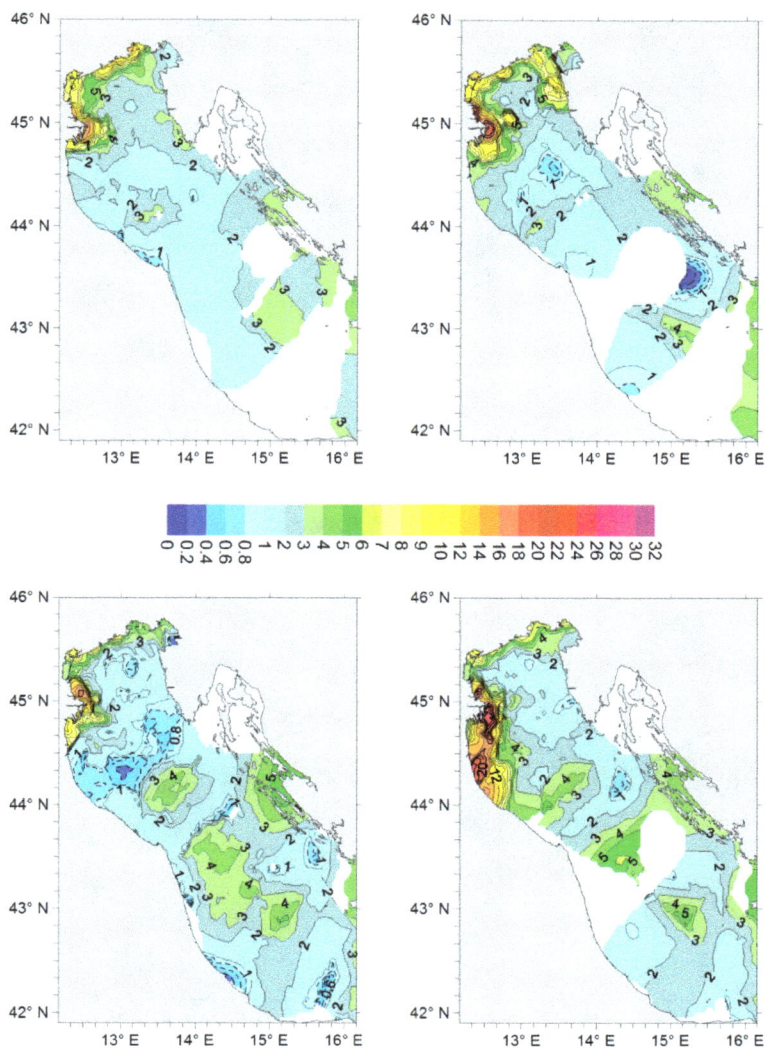

Fig. 25. Seasonal climatological maps (winter top left, spring top right, summer bottom left, autumn bottom right) of surface Si-SiO$_4$ (mM/m^3) in the northern-central Adriatic Sea.

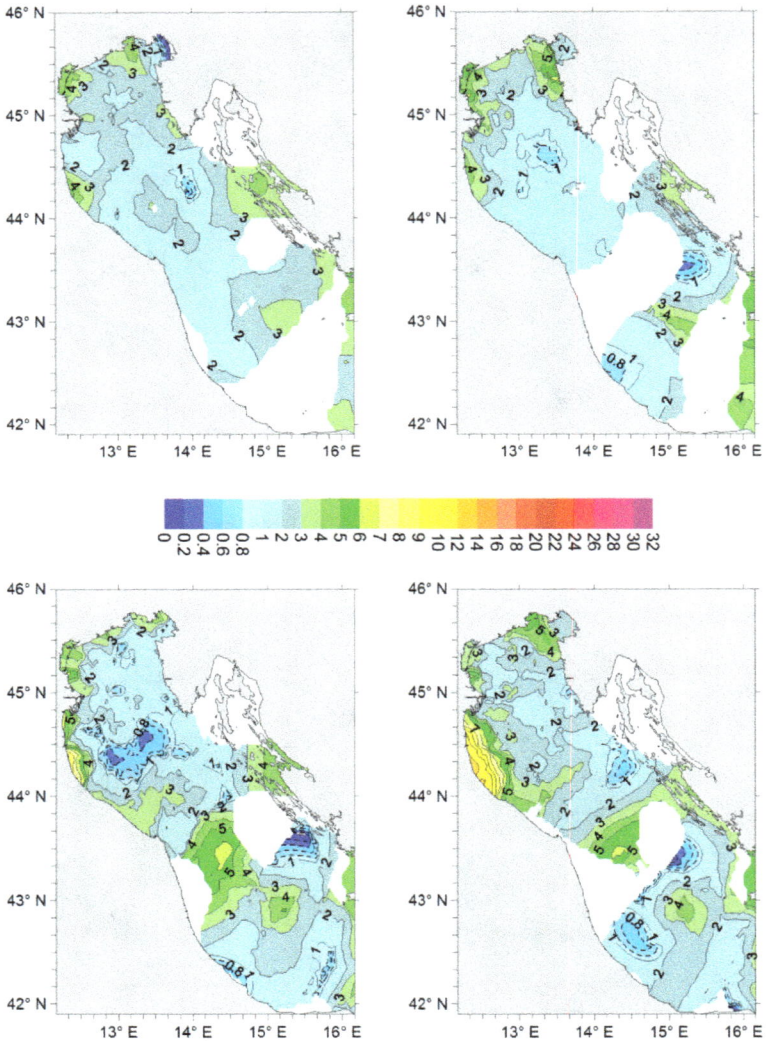

Fig. 26. Seasonal climatological maps (winter top left, spring top right, summer bottom left, autumn bottom right) of Si-SiO$_4$ (mM/m^3) at 10 m depth in the northern-central Adriatic Sea.

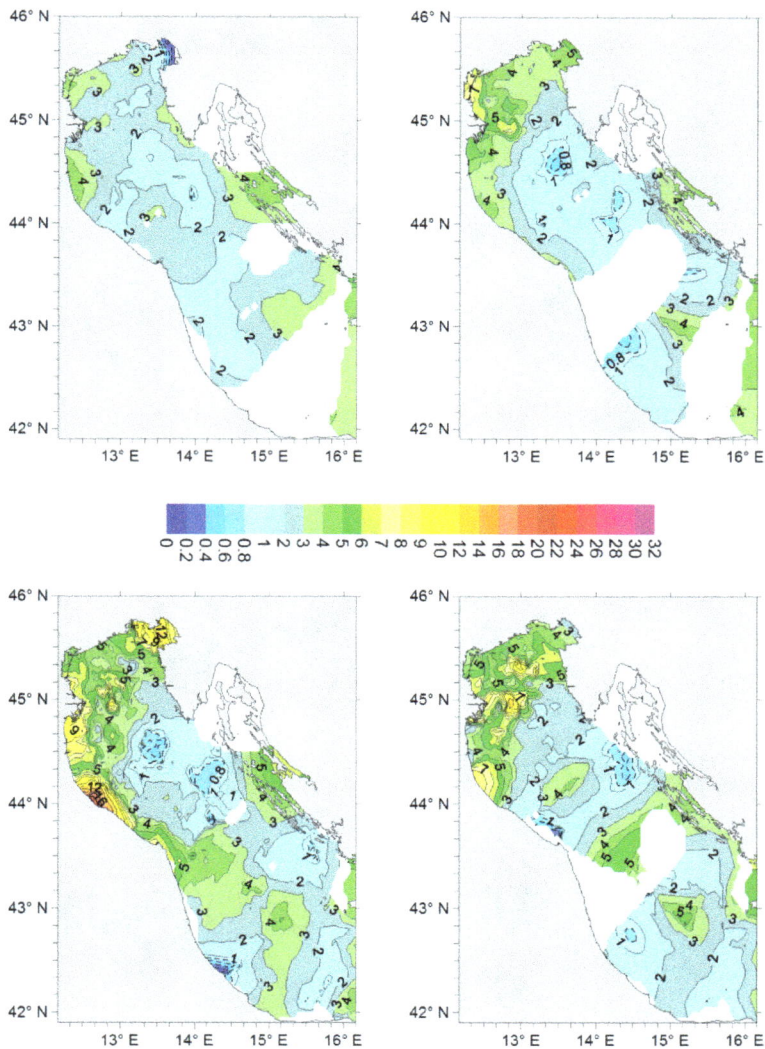

Fig. 27. Seasonal climatological maps (winter top left, spring top right, summer bottom left, autumn bottom right) of Si-SiO$_4$ (mM/m^3) at 20 m depth in the northern-central Adriatic Sea.

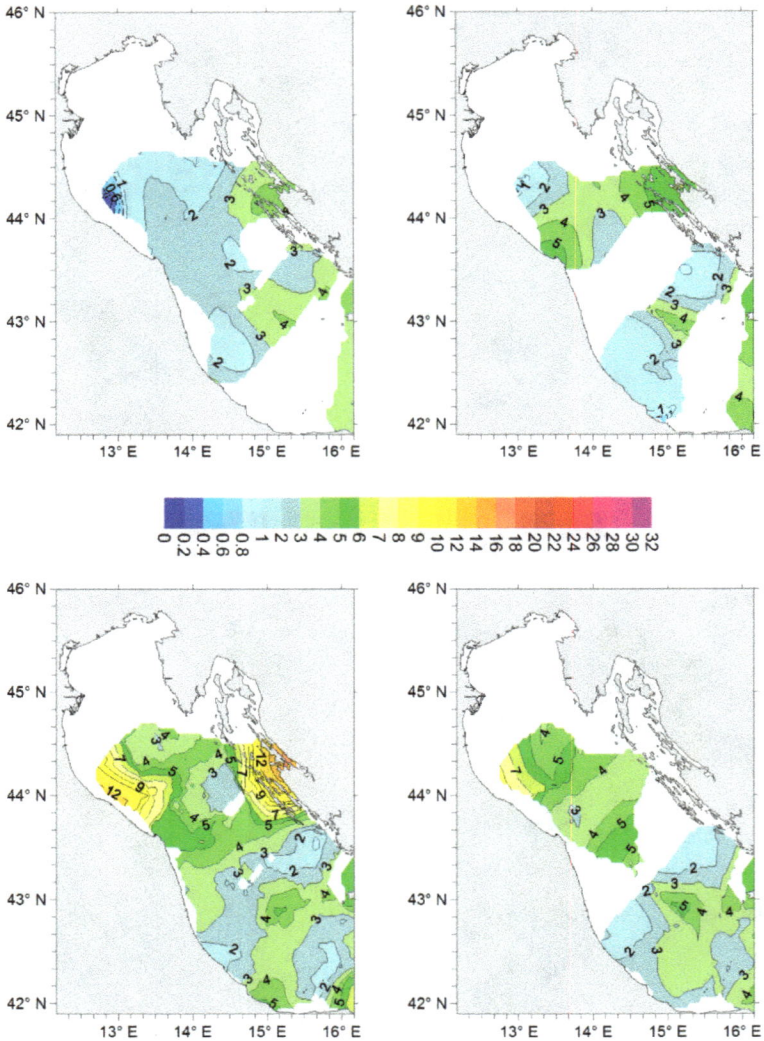

Fig. 28. Seasonal climatological maps (winter top left, spring top right, summer bottom left, autumn bottom right) of Si-SiO₄ (mM/m³) at 50 m depth in the northern-central Adriatic Sea.

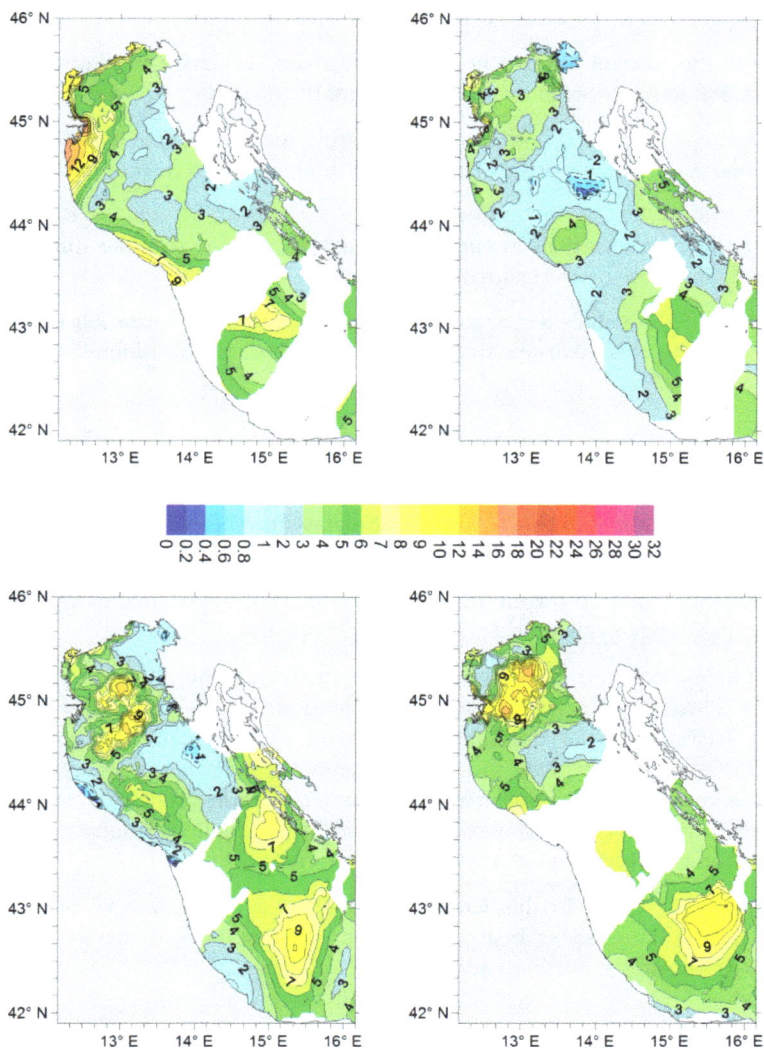

Fig. 29. Seasonal climatological maps (winter top left, spring top right, summer bottom left, autumn bottom right) of Si-SiO4 (mM/m³) at the bottom in the northern-central Adriatic Sea.

3.4.3 Orthosilicates

Orthosilicates concentrations are far more larger than other nutrient salts, reaching rather high values at the surface (see Figure 25) in the western northern Adriatic region, which is more affected by river discharge. It can be seen that maxima concentrations are reached at the Po delta in autumn and spring, i.e. the seasons with the highest Po river runoff.

At 10 m depth the concentrations significantly decrease, as shown in Figure 26, and quite high values are mainly observed in autumn south of the Po delta.

At 20 m depth (see Figure 27) winter and also spring maps resemble the 10 m ones, while during summer and autumn orthosilicates reach higher values in the north-western area of the basin.

At 50 m very low concentrations are again depicted, but during summer time in the north-western region and close by the Dalmatian island and Croatian coast.

At the bottom Si-SiO4 values are again considerably high (see Figure 29) in the northern area, and also in the meso-Adriatic depressions, again because of mineralization processes of the organic matter.

4. Conclusions

A climatology of main thermohaline and chemical properties of the northern and central Adriatic Sea based on the MEDAR/MEDATLAS dataset has been drawn and described.

A large variability of temperature and salinity is evident at horizontal, vertical and seasonal scales. The same can be said for nitrates and orthosilicates, while smaller variations can be observed for dissolved oxygen, nitrites and orthophosphates.

Interesting phenomena can be individuated by analyzing the climatological maps. For example, the dissolved oxygen minima found in summer and autumn at bottom in the central area offshore the Po delta, combined with nutrient salts maxima (particularly nitrates, orthosilicates and nitrites) and low temperatures, indicate that dense waters formed during winter may persist in the area and remineralization processes contribute to the bottom water nutrient enrichment and dissolved oxygen depletion. Similar processes can be evidenced also in the deeper meso-Adriatic depressions.

The climatology presented in this Chapter will serve as "benchmark" for carrying out further analysis of the changes in the oceanographic properties that may have occurred during the last years in the northern and central Adriatic Sea.

5. Acknowledgements

This contribution was supported by the project "MARINA", funded by Regione Veneto within the initiatives of the regional law n.15/2007 and by the project SECURSEA funded by Regione Marche within the EU project SECURSEA (PIC INTERREG IIIA). Partial support was received also from the Italian Ministry of University and Research FIRB Project, code RBFR08D825 (project "DECALOGO") and from the PRIN 2008YNPNT9_005 grant.

6. References

Artegiani, A., Bregant D., Paschini E., Pinardi N., Raicich F. and Russo A., 1997a. The Adriatic Sea General Circulation. Part I: air - sea interactions and water mass structure. *Journal of Physical Oceanography*, 27(8), 1492-1514.

Artegiani, A., Bregant D., Paschini E., Pinardi N., Raicich F. and Russo A., 1997b. The Adriatic Sea General Circulation. Part II: Baroclinic circulation structure. *Journal of Physical Oceanography*, 27(8), 1515-1532.

Bergamasco, A., Oguz T., and Malanotte-Rizzoli P., 1999. Modeling dense water mass formation and winter circulation in the northern and central Adriatic Sea. *Journal of Marine Systems*, 20, 279-300.

Boldrin, A., Carniel S., Giani M., Marini M., Bernardi Aubry F., Campanelli A., Grilli F. and Russo A., 2009. The effect of Bora wind on physical and bio-chemical properties of stratified waters in the Northern Adriatic. *Journal of Geophysical Research – Ocean*, 114, C08S92. DOI:10.1029/2008JC004837

Bretherton, F.P., Davis R.E. and Fandry C.B., 1976. A technique for objective analysis and design of oceanographic experiments applied to MODE - 73. *Deep - Sea Research*, 30, 985-1002.

Carniel, S., Warner J.C., Chiggiato J. and Sclavo M., 2009. Investigating the impact of surface wave breaking on modelling the trajectories of drifters in the Northern Adriatic Sea during a wind-storm event. *Ocean Modeling*, 30, 225-239. DOI:10.1016/j.ocemod.2009.07.001

Carniel, S., Bergamasco A., Book J.W., Hobbs R.W., Sclavo M. and Wood W.T., 2012. Tracking bottom waters in the Southern Adriatic applying Seismic Oceanography techniques. [in press, *Continental Shelf Research*] DOI:10.1016/j.csr.2011.09.0034

Carter, E.F., and Robinson A.R., 1987. Analysis models for the estimation of Oceanic Fields. *J. Atmos. Oceanic Technol.*, 4, 49-74.

Cushman-Roisin, B., Gacic M., Poulain P.-M. and Artegiani (eds.) A., 2001. Physical Oceanography of the Adriatic Sea: Past, Present and Future. Kluwer Academic Publishers, Dordrecht/Boston/London.

Dorman, C.E., Carniel S., Cavaleri, L., Sclavo, M., Chiggiato J. and others, 2006. February 2003 marine atmospheric conditions and the Bora over the Northern Adriatic. *Journal of Geophysical Research–Ocean*, 111, C03S03. DOI: 10.1029/2005JC003134

MEDAR Group, 2002. MEDATLAS/2002 database. In: *Mediterranean and Black Sea database of temperature salinity and bio-chemical parameters*, 30 January 2010, Available from: http://www.ifremer.fr/medar/

Poulain, P.-M., 2001. Adriatic Sea surface circulation as derived from drifter data between 1990 and 1999. *Journal of Marine Systems*, 29, 3-32.

Russo, A., and Artegiani A., 1996. Adriatic Sea hydrography. *Scientia Marina*, 60 (Supl.2) : 33-43.

Russo, A., Rabitti S. and Bastianini M., 2002. Decadal climatic anomalies in the Northern Adriatic Sea inferred from a new oceanographic data set. *P.S.Z.N.: Marine Ecology*, 23, Supplement 1, 340-351.

Signell, R.P., Carniel S., Cavaleri L., Chiggiato J., Doyle J., Pullen J. and M. Sclavo, 2005. Assessment of wind quality for oceanographic modeling in semi-enclosed basins. *Journal of Marine System*, 53(1-4), 217-233. DOI: 10.1016/j.marsys.2004.03.006

Pullen J., Doyle J. and Signell R.P., 2006. Two-way air-sea coupling: A study of the Adriatic. *Monthly Weather Review*, 134(5), 1465–1483.

Vilibic, I. and Supic N., 2005. Dense water generation on a shelf: the case of the Adriatic Sea. *Ocean Dynamics*, 55 (5-6), 403-415.

Zavatarelli, M., Raicich F., Bregant D., Russo A. and Artegiani A., 1998. Climatological biogeochemical characteristics of the Adriatic Sea. *Journal of Marine Systems*, 18, 227-263.

7

Climatology of the U.S. Inter-Mountain West

S. Y. Simon Wang and Robert R. Gillies

Utah Climate Center / Department of Plants, Soil and Climate
Utah State University, Logan UT
USA

1. Introduction

The Inter-Mountain West (IMW) of North America is a region that lies between the Rocky Mountains to the east and the Cascades and Sierra Nevada to the west (**Fig. 1**). The climate of the IMW is generally semi-arid but this varies by location and elevation. An estimated 50-80% of the IMW's streams and rivers are fed by mountain snowpack (Marks and Winstral 2001), while the majority of the streams and rivers flow into desert sinks or closed-basin lakes such as the Great Salt Lake (**Fig. 1**). These streams and rivers create some agriculturally productive areas in the otherwise dry basins and mountain valleys. In particular, the Colorado River supplies water to the population-booming southwestern

Fig. 1. The Inter-Mountain West (IMW) region (yellow dashed outline), the Great Salt Lake (arrow), and the tree ring site for the reconstructed precipitation (red dot) as shown in Fig. 2. Background is terrain (map source: Unidata).

states and cities. Climate in the Colorado River Basin has been a subject of intense research due to its projected drying trend (Barnett and Pierce 2008). Change in winter precipitation regime (i.e. ratio between rainfall and snowfall) is also a subject of interest not only because its role in water resource but also its impact on recreational (ski) industry in the IMW.

Paleoclimate records indicate that the current mean state of climate in the IMW may not be stable. Over most of the past millennium, droughts in the IMW were generally more intense and lasted longer than those experienced in the 20th century, which had included the "severe droughts" of the 1930s and 1950s. **Fig. 2** presents a long-term precipitation proxy based on tree ring data collected near the upstream Colorado River in northeastern Utah (red dot in Fig. 1). It is immediately apparent that even the worst droughts of the 20th century are dwarfed in both magnitude and duration compared to the previous droughts of the 14th-18th centuries, as well as the megadrought of the 13th century (Gray et al. 2004). Moreover, the 20th century has been a relatively wet period when compared to the pre-19th century era. These observations may imply an unstable state of modern-time, relatively wet precipitation regime that carries the potential to resume its pre-18th century climatology with longer, deeper droughts. Previous studies of tree ring-based drought analysis (Cook et al. 1997; Herweijer et al. 2007) have found a similar non-stationarity in the IMW climate change, in the sense that drought frequency has shifted from being centennial and more intense in the early millennium into being multi-decadal and less intense in the later millennium. To what extent such a cyclic feature may change or persist into the future is important information for water management.

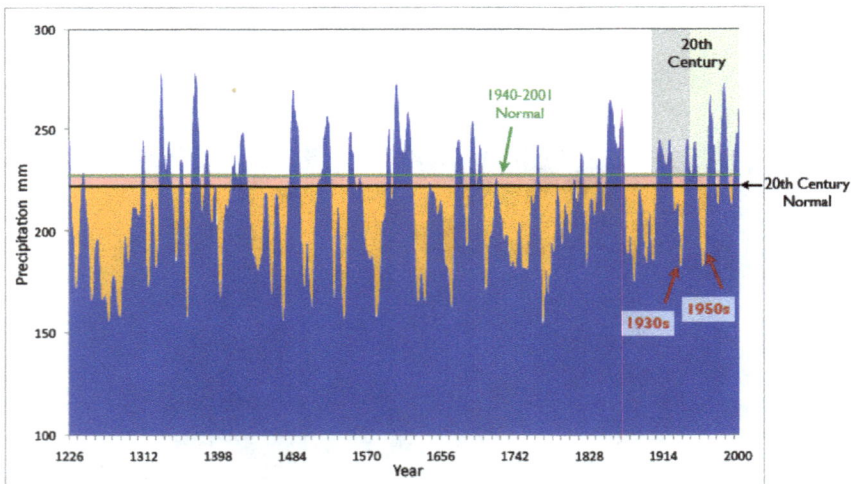

Fig. 2. (a) Tree ring-reconstructed annual precipitation proxy (blue shadings) generated from samples collected in northeastern Utah (red dot in Fig. 1) by Gray et al. (2004). Orange shadings indicate precipitation deficits under the 1940-2001 mean.

It is now well established that quasi-periodic climate modes in the Pacific and Atlantic Oceans modulate the IMW climate. A sizable body of research (Barlow et al. 2001; Schubert et al. 2004, 2009; Seager et al. 2005; among others) has explored the physical linkage between the Pacific sea surface temperature (SST) variations and climate anomalies in North America. Such Pacific SST variations include *(a)* multi-decadal modes like the Pacific Decadal Oscillation (PDO) (Mantua et al. 1997) and the Interdecadal Pacific Oscillation (IPO; Folland et al. 2002), *(b)* decadal modes such as the Pacific quasi-decadal oscillation (QDO) (Tourre et al. 2001; White et al. 2003), and *(c)* interannual modes associated with the El Niño-Southern Oscillation (ENSO). Covariability between these Pacific climate modes and the drought evolution in the IMW has also been reported (e.g., Sangoyomi 1993; Zhang and Mann 2005). Moreover, the IMW's climate regime is further complicated by unique seasonal variations in precipitation, characterized by a combination of annual and semiannual cycles whose timing varies across the region. Such a complex climatology poses a great challenge to climate models and climate prediction.

This chapter discusses the complex climate regimes of the IMW with a focus on four different timescales: (Section 2) seasonal and intraseasonal, section (3) interannual, and section (4) decadal variabilities. These different scales of climate variability interact with each other and this further modulates the IMW climate variability; Section 5 discusses such an interaction. Simulations of the IMW climate by some global and regional climate models are evaluated. Finally, Section 6 provides a summary of the chapter.

2. Seasonal climate variability

a. Annual and semiannual cycles

Pronounced annual and semiannual cycles in precipitation characterize the IMW's climatology. Earlier studies (e.g., Hsu and Wallace 1976) have noted that the phase of the annual cycle changes by six months going from east to west across the Rocky Mountains, while the semiannual cycle changes phase from north to south. The precipitation patterns corresponding to these annual and semiannual cycles are illustrated in **Figs. 3a** and **3b**, respectively, through correlation maps of the first and second principal components (PC) of monthly data of the Climate Prediction Center Merged Analysis of Precipitation (CMAP; Xie and Arkin 1997). In the IMW, the combination of the annual cycle (i.e. east-west) and semiannual cycle (i.e. north-south) forms four seasonal precipitation regimes that meet in the central IMW near Utah. However, atmospheric general circulation models (AGCMs) do not simulate well such seasonal cycles in precipitation, particularly the semiannual cycle (Boyle 1998). In addition, complexity also arises from interaction between the atmospheric circulation and orography encountered in the IMW, amplifying such model deficiencies. For instance, the NCAR Community Climate Model version 3 (CCM3) produced a distorted semiannual precipitation pattern in the IMW, shown in **Fig. 3c**. Wang et al. (2009a) have suggested that the semiannual precipitation cycle in the IMW is likely related to the onset and development of the North American Monsoon (NAM) that features a change in the upper-level circulation regime, evolving from a large-scale trough in spring into a quasi-stationary anticyclone in the monsoon months (Higgins et al. 1997). The circulation regime change is followed by a precipitation phase reversal in the north-south direction, as is shown in **Fig. 3b**. A similar but weaker change in circulation pattern occurs in early spring, when an upper-level trough forms over the southwest U.S. and gradually migrates westward (Wang and Chen 2009). These features pose a great challenge to AGCMs in simulating the precipitation seasonal cycle of the IMW.

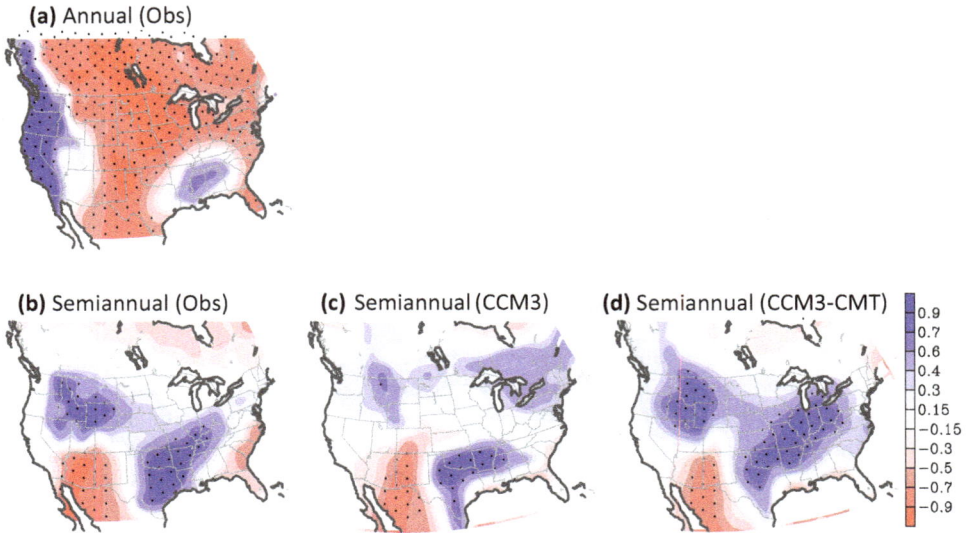

Fig. 3. Correlation maps of monthly precipitation anomalies with respect to (a) the annual cycle and (b) the semiannual cycle using the CMAP (observed) data. (c) and (d) Same as (b) but using the control run and the coupled-CMT run of CCM3, respectively. The CCM3 precipitation was correlated with the PC time series of the CMAP (observed) data.

To further examine the AGCM performance, we adopted a set of CCM3 simulations with (a) the control run as in **Fig. 3c** and (b) the inclusion of a convective momentum transport (CMT) in the convection scheme as in **Fig. 3d**, generated by Wu et al. (2007). The reason here to include the CMT experiment was to evaluate the impact of convective processes on the simulated seasonal cycles, since the semiannual cycle of the IMW precipitation is linked to the NAM onset – i.e. convective rainfall. In AGCMs, CMT is far more intricate to parameterize than thermodynamic transports due to complicated cloud-scale pressure gradients induced by organized convection (Moncrieff 1992; Wu et al. 2007). As shown in **Fig. 3d**, the inclusion of CMT considerably improves the simulation of the semiannual precipitation change. Of note is that the improvement appears not only in the IMW but also in the central and southeast United States. An improvement like this signifies that the semiannual cycle in precipitation is closely related to cloud system feedback, since convective clouds not only release latent heat and redistribute heat/moisture but also transport momentum. Possibly, the convective momentum tendencies adjust the local Hadley circulation across the IMW, while the responses of meridional wind to the more realistic heating improve the secondary meridional circulation associated with the NAM development.

Regional climate models (RCMs) are thought to produce a more realistic precipitation seasonal cycle than AGCMs, particularly in regions having topographically enhanced precipitation (Leung et al. 2006). The IMW's geography is characterized by four major mountain ranges: 1) the Cascade Range, 2) the Bitterroot Range, 3) the Wasatch Range and 4) the Colorado Rockies, denoted in **Fig. 4a** as regions 1-4, while central Arizona is denoted as region 5. In the cold season, precipitation occurs mainly on the windward side of these mountain ranges (**Fig. 4b**).

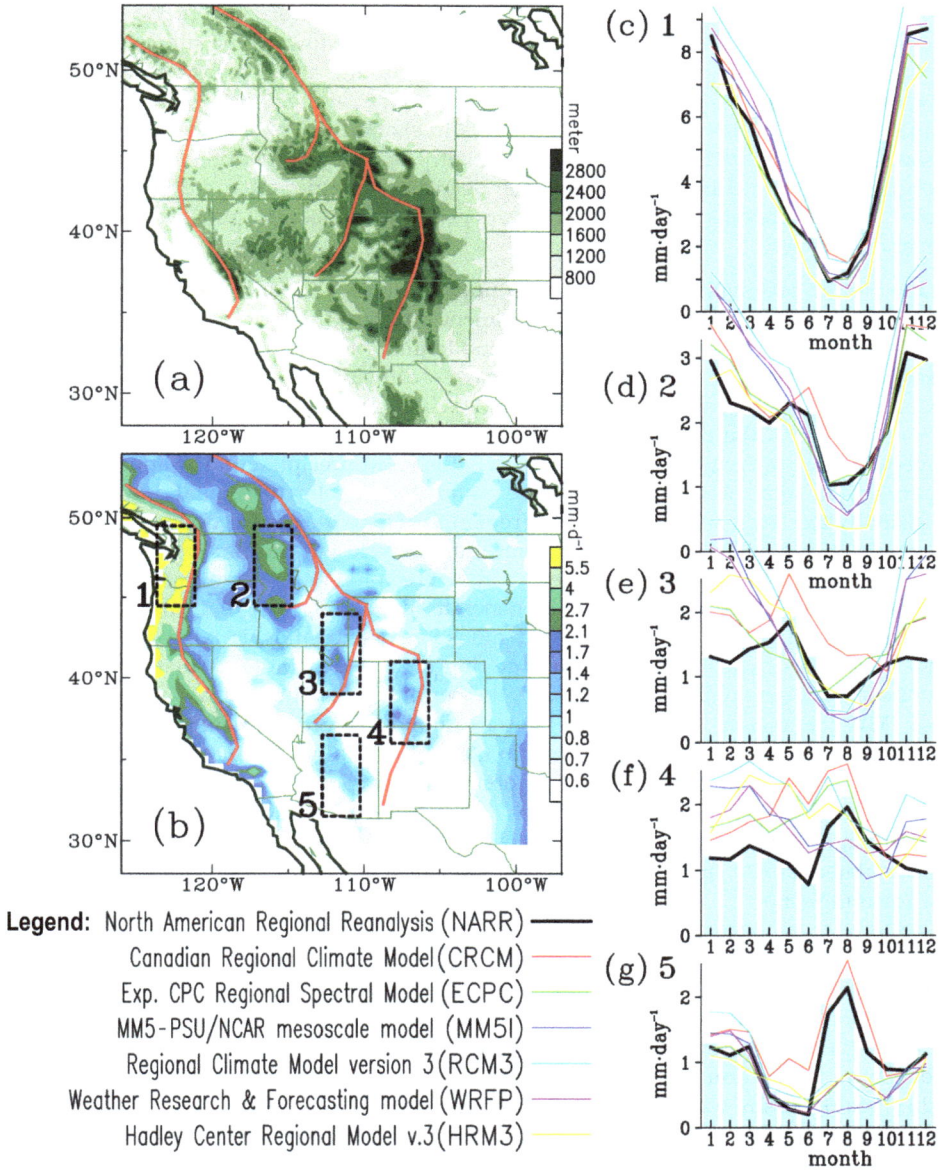

Fig. 4. (Adopted from Wang et al. 2009a) (a) Orography and (b) cold-season rainfall (November-May, UDel) of the IMW, where the red lines outline the major mountain ranges. (c)-(g) Monthly histograms of UDel rainfall averaged from the five regions indicated in (b), superimposed with the corresponding precipitation of NARR (thick black line) and all RCMs (color lines). The abbreviation of the RCMs and their corresponding color codes are given under (b). Note the precipitation scale in (c) is twice of that in (d)-(g).

Wang et al. (2009a) showed that, from the coastal area toward the IMW (i.e. regions 1 through 4) the seasonal cycle evolves from a winter regime toward the summer regime with increasing semiannual variability. This transition of the precipitation seasonal cycle is illustrated in **Figs. 4c-g** by two sets of observation – the North American Regional Reanalysis (NARR) (Mesinger et al. 2006; black line) and the University of Delaware (UDel) (Legates and Willmott 1990; blue bar). Apparently, spring precipitation becomes important in region 2 and peaks in region 3, but then decreases in regions 4 and 5 where a monsoon rainfall regime prevails in summer, corresponding to the combination of the annual and semiannual cycles.

Analyzing the precipitation climatology simulated by six RCMs participating in the North American Regional Climate Change Assessment Program (NARCCAP; Mearns et al. 2009), in which the RCMs were driven by the same boundary conditions of a global reanalysis, Wang et al. (2009a) found that the RCMs have a difficulty in replicating the precipitation seasonal cycle. As revealed in **Figs. 4c-g**, in the Cascade Range (region 1) where the annual cycle is dominant, the phases of the RCM precipitation show marked consistency with the observations. However, beginning in region 2, the RCM precipitation variation increasingly departs from the observation and the simulated winter precipitation amounts are consistently too large. In the inner IMW (region 3), where the annual cycle is weak, four out of the six RCMs still produce a dominant annual cycle and do not capture the elevated spring precipitation. In region 4 where winter precipitation and summer monsoon rains have equal contributions, the RCMs generate mixed signals that disagree with the observations. In region 5, the simulated monsoon precipitation is only captured by one model (red line), which used spectral nudging (i.e., incorporating reanalysis information into the simulation domain). In summary, the performances of the RCMs are weakest in the inner IMW (i.e. regions 2-4) in which the transition of seasonal climate regimes takes place. Due to the overprediction in winter precipitation, most RCMs have a tendency to produce too strong of an annual cycle and this obscures the relatively good performance in the semiannual cycle (Wang et al. 2009a). Such a character differs from most AGCMs that do not simulate the semiannual cycle well. The overprediction of winter precipitation also affects the simulated ENSO impact on the IMW climate; this model deficiency will be further discussed in Section 3.

b. Intraseasonal variation

Winter and spring weather conditions in the IMW are characterized by considerable intraseasonal variations (on the order of 3-8 weeks). These intraseasonal variations are linked to (1) the evolution of free external Rossby waves (Lau and Nath 1999), (2) tropical-midlatitude interaction associated with the Madden-Julian Oscillation (MJO) (Mo 1999), and (3) short Rossby waves propagating through the jet stream waveguide, referred to as the circumglobal teleconnection (CGT) (Branstator 2002). For external Rossby waves, the forcing and dynamics have been studied extensively; i.e. the waves slowly propagate westward in response to the planetary β effect at a speed coincident with the intraseasonal timescale (Branstator 1987; Horel and Mechoso 1988; Lau and Nath 1999). On the other hand, the midlatitude response of the circulation to the tropical MJO forcing normally results in stationary or eastward-moving synoptic-scale waves (Mo and Paegle 2005). For the CGT, upstream transient vorticity forcing, usually occurring in Eurasia and East Asia, produces short-wave response with the energy propagating eastward towards

North America. The CGT pattern is known to change month-by-month (i.e. intraseasonal) (Ding and Wang 2005), yet it has a profound impact on extreme spring weather conditions in the IMW (Wang et al. 2010b).

A recent study (Gillies et al. 2010a) has shown that winter weather – including temperature, precipitation and snowpack – fluctuates in strong association with a "30-day mode" that dominates persistent weather regimes in the IMW. Analyzing the geopotential height sounding data in Salt Lake City during 1980-2008, Gillies et al. (2010a) constructed a composite lifecycle of the 30-day mode (**Fig. 5a**). This 30-day mode strongly affects precipitation and snow depth over the IMW (**Fig. 5b**; centered in Salt Lake City with a 200 km radius), as well as temperature that effects snowmelt. Based on the 8-phase lifecycle of the 30-day mode, Gillies et al. (2010b) constructed the composite patterns of the 200-mb streamfunction and velocity potential (**Fig. 6**). The composite lifecycle of the 30-day mode – i.e. centered in the IMW – depicts a global eastward propagation of the velocity potential with a predominant zonal wave-1 pattern, which resembles the signature MJO structure. Meanwhile, a series of short-wave cells is excited within the eastward propagating velocity potential and propagates towards North America. At phases 2-3 when a stationary ridge prevails over the Great Basin, the associated wave train follows the "great circle" route of the Pacific-North America (PNA) pattern (Horel and Wallace 1981). An oppositely signed circulation anomaly appears at phases 6-7. Despite the propagating feature in the velocity potential, the streamfunction wave trains appear to be quasi-stationary. Such a feature underscores the fact that wintertime stationary waves in North America (and in the IMW) fluctuate in response to the tropical–extratropical linkages of the MJO (e.g., Kushnir 1987; Mo and Paegle 2005). These results indicate that the occurrences of either persistent ridging events or prolonged precipitation spells in the IMW are "phase locked" with the MJO evolution – at least at the higher-frequency end of the MJO spectrum that spans 30-60 days.

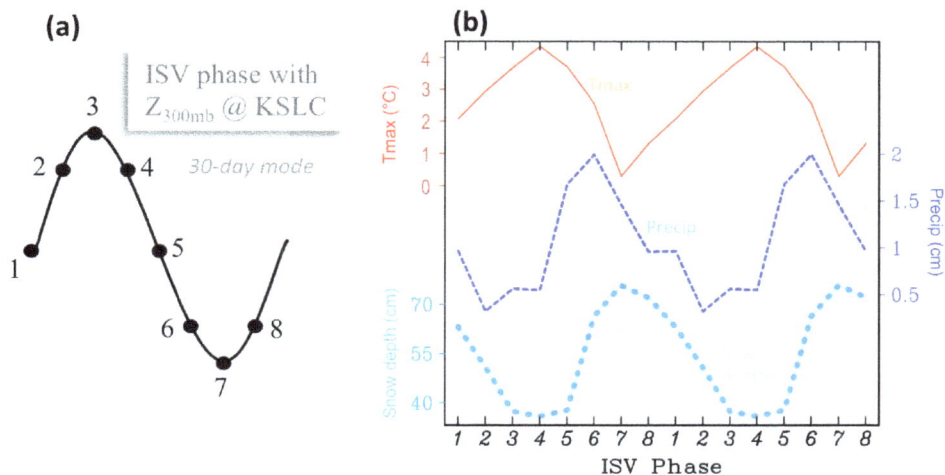

Fig. 5. (a) Lifecycle of the intraseasonal variability (ISV) based on Salt Lake City sounding of geopotential height (300mb), and (b) the 8-phase composites of Tmax (red), precipitation (blue dashed), and snow depth (cyan dotted) in Utah and upper Colorado Basin during 1980-2008 (after *Gillies et al. 2010a*).

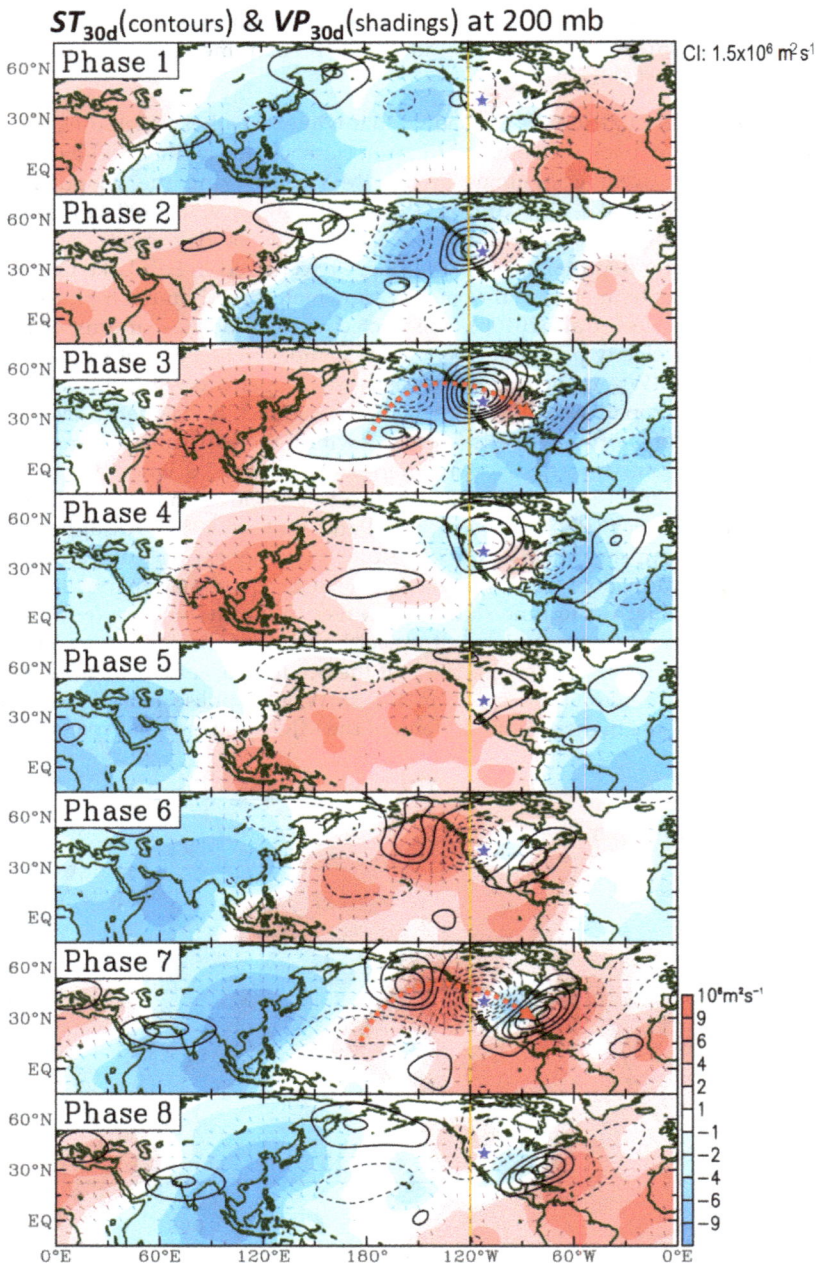

Fig. 6. Eight-phase composites of the 200mb velocity potential (shadings) and streamfunction constructed form the 30-day mode at Salt Lake City. Adopted from *Gillies et al. (2010b)* with permission from the American Meteorological Society.

3. Interannual variability

The impact of ENSO on precipitation and drought anomalies over western North America has been studied extensively. It is now well established that ENSO tends to produce a so-called North American dipole structure in precipitation and temperature, encompassing the Pacific Northwest and the Southwest with opposite polarity (Dettinger et al. 1998; and subsequent works). However, short-term climate forecasts (6 months to 2 years) based on this conceptual model have frequently failed during the recent decade (Wood 2011). Past studies (Rajagopalan and Lall 1998; Zhang and Mann 2005) have noted that the central part of the IMW is shielded from direct influence of ENSO as the region lies between the marginal zone of the north-south dipole pattern. Later studies (Hidalgo and Dracup 2003; Brown 2011) also found that the ENSO-climate connection in the IMW is not stable; rather, it fluctuates following a long-term oscillatory manner. This feature is further discussed in Section 5.

The CGT pattern (cf. Section 2b), which is characterized as a short Rossby wave train along the jet stream waveguide with a zonal wave-5 structure (Branstator 2002; Ding and Wang 2005), has been found to be linked to persistent rainy conditions over the IMW (Wang et al. 2010b). The spatial scale of the CGT wave train is shorter than that of the classic PNA pattern (wave 1-3; Hoskins and Karoly 1981), while its variability is *uncorrelated* with ENSO (Ding and Wang 2005). During spring, the North American circulations undergo a considerable change as the jet stream shifts rapidly northward. From April to July, a synoptic-scale trough develops over the southwest U.S., deepens and migrates to the West Coast and eventually merges with the oceanic trough over the North Pacific (Higgins et al. 1997; Wang and Chen 2009). The development of this trough enhances moisture and facilitates synoptic disturbances toward the IMW during spring, contributing to the rainy season there (cf. Fig. 4e).

Such a seasonal transition takes place through the formation of a standing wave train, manifest as stationary short waves embedded in the climatological jet stream (**Fig. 7a**). The distinct short-wave trough over the West Coast (arrow indicated) reflects the spring trough that enhances precipitation in the IMW. Wang et al. (2010b) found that extremely wet spring seasons in the IMW often occur with the presence of the CGT. Specifically, a *wet* (dry) spring in the IMW occurs when the CGT is *in-phase* (out-of-phase) with the standing wave train. For example, **Fig. 7b** shows the record precipitation that occurred in June 2009 as the climatological trough considerably deepened and was embedded in a series of short waves. Here, the CGT is further illustrated by the filtered geopotential height in the zonal wave-5 regime using harmonic analysis. Visual comparison between **Figs. 7a** and **7b** indicates that the 2009 wave train is in-phase with the standing wave train. Analyzing a 50-year period of two reanalysis datasets (NCEP and ERA; see caption) through the empirical orthogonal function (EOF) analysis, Wang et al. (2010b) confirmed the CGT linkage with the spring precipitation anomalies in the IMW. As shown in **Fig. 7c**, the first leading mode (26%) depicts the CGT as an amplification effect of the standing wave train (due to their coincident phases). Comparison of the EOF time series with the IMW precipitation anomalies in **Fig. 7d** further reveals that precipitation anomalies were clearly dependent with the phasing of the CGT.

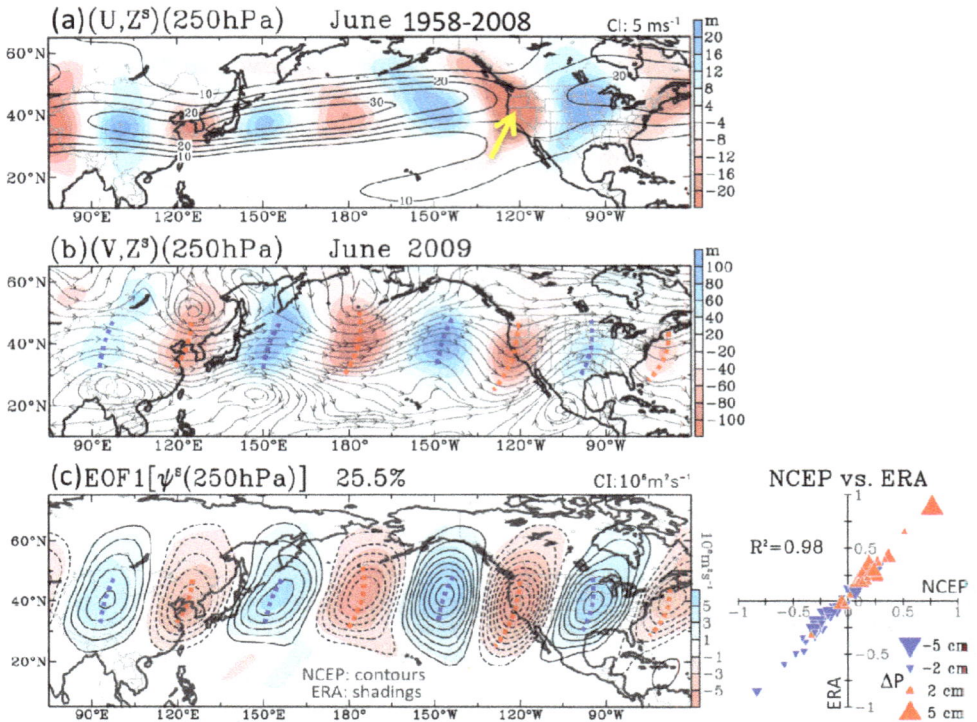

Fig. 7. (a) long-term 250hPa zonal wind (contours) and the shortwave-regime geopotential height (shadings; zonal wave-5) in June; yellow arrow indicates the spring trough. (b) The 250hPa streamlines and shortwave-regime geopotential height (shadings) in June 2009, marked with trough (red) and ridge (blue) lines. Note the consistent phases of geopotential height between (a) and (b). (c) First EOF of shortwave-regime streamfunction at 250 hPa from two reanalyses: NCEP (contours; Kalnay et al. 1996) and ERA (shadings, combined from ERA40 and ERA-Interim; Simmons et al. 2007). (d) Relationship between the first EOFs of NCEP1 and ERA, with the scatters represented by the IMW precipitation anomalies (ΔP; Legates and Willmott 1990) in terms of sign and size (After Wang et al. 2010b).

As mentioned in Section 2, Wang et al. (2009a) found that precipitation simulations by the NARCCAP RCMs (Section 2a) exhibit a "false" ENSO signal in the central IMW, likely due to their general tendency to overpredict winter precipitation. By examining the ENSO precipitation pattern constructed from the composite differences between six El Niño (1982/83, 87/88, 91/92, 94/95, 97/98, 2002/03) and four La Niña (1984/85, 88/89, 95/96, 98-2001) winters, which was based on the NOAA Climate Prediction Center[1], **Fig. 8a** reveals the typical north-south precipitation dipole over the western United States. This precipitation dipole is reasonably simulated by the NARCCAP models. However, in the central IMW (i.e. regions 3 and 4 in Fig. 4b indicated by yellow arrows), the RCM precipitation anomalies are uniformly too large compared to the observations. It is known

---

[1] http://www.cpc.noaa.gov/products/analysis_monitoring/ensostuff/ensoyears.ERSST.v3.shtml

that precipitation variations in the central IMW are not exactly in-phase with ENSO; instead, precipitation variations lag ENSO by a quarter-phase in both the 3-6 year frequency (Rajagopalan and Lall 1998) and the decadal frequency (Wang et al. 2009b). This feature is manifest by the weak winter precipitation anomalies in the observations near regions 3 and 4 (**Fig. 8a**), where the precipitation dipole changes sign. In the RCMs, however, significant ENSO signals remain in the central IMW extending from the Pacific Northwest, suggesting that most RCMs simulate the ENSO impacts too far inland. On the other hand, the ENSO composites made for the subsequent spring seasons (**Fig. 8b**) do not reveal such a systematic bias; this suggests that overprediction of winter precipitation over the IMW may be the cause of such a false ENSO signal.

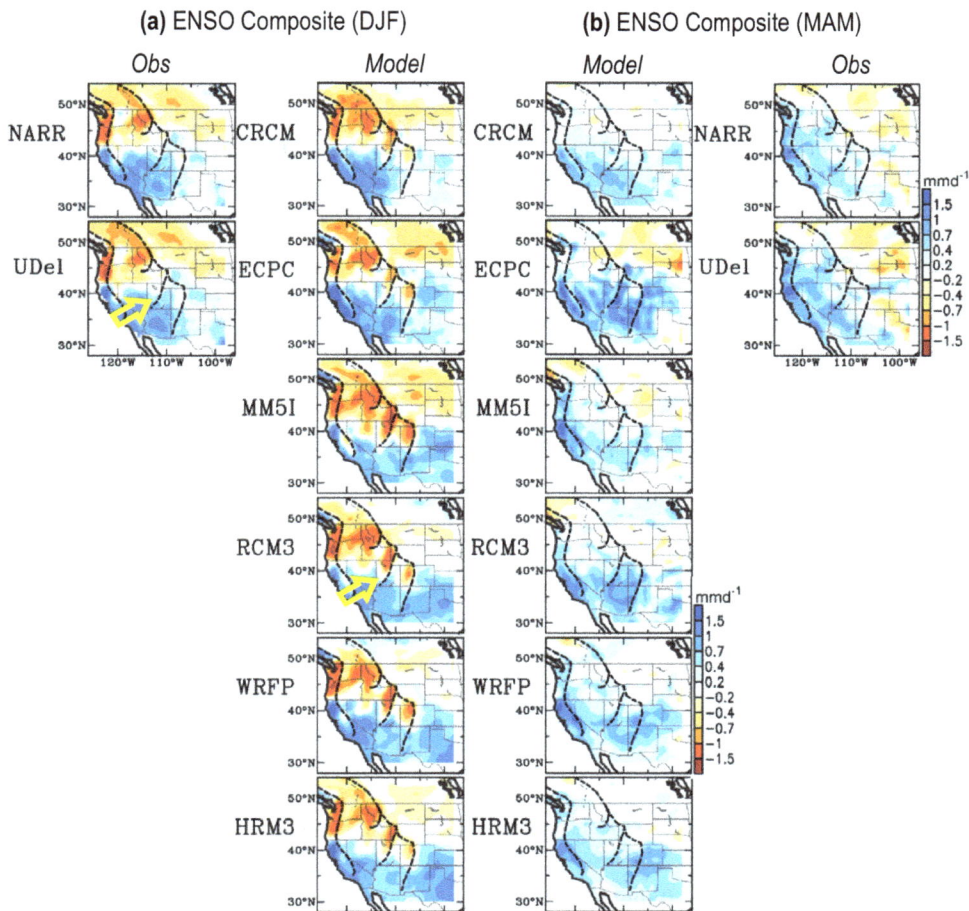

Fig. 8. (a) Differences of precipitation composites between El Niño and La Niña winters (December-February). (b) Same as (a) but for the subsequent springs (March-May). The major mountain ranges are outlined by black dashed lines. The interval of shaded contours is given in the lower right. Adopted from Wang et al. (2009a).

4. Decadal variability

a. Multi-decadal cycle

A wealth of research (Gray et al. 2003; Seager et al. 2005; Herweijer et al. 2007; and others) has established a link between low-frequency climate variability in the IMW and the Pacific climate oscillations. Recent studies focusing on the decadal-scale climate variability have turned attention to the Great Salt Lake (GSL), a large closed-basin lake located in the heart of the IMW (**Fig. 1**). As a pluvial lake, the GSL integrates hydrological forcings over a substantial watershed. When coupled with the lake's shallowness, the accumulated water results in extensive fluctuations in the lake elevation. The large drainage area of the GSL dampens out high-frequency variability and therefore is more responsive to climatic variabilities at longer timescales (Lall and Mann 1995). Consequently, the long-term change in the GSL elevation reflects the persistent wet/dry periods in the IMW. This feature is revealed in **Fig. 9a**, the GSL's 150 years of elevation record (shaded curve) superimposed with the lake elevation tendency (ΔGSL; dotted line). Applied with the 20-year lowpass filter, the ΔGSL exhibits a marked multi-decadal variability (solid line) that is coherent with the low-frequency variation of the Palmer Drought Severity Index (PDSI; red dashed line), reconstructed from tree rings near the GSL (Cook and Krusic 2004). Such a result supports the notion that any long-term changes in the GSL elevation respond to sustained wet and dry periods of the surrounding region. A recent study by Wang et al. (2011a) noticed that a multi-decadal (~30 year) and a quasi-decadal (10-15 year) frequency bands stand out significantly in the GSL elevation spectrum; this feature is shown in **Fig. 9b**. These two frequency bands are also the leading timescales of the drought variability in the IMW (Herweijer et al. 2007).

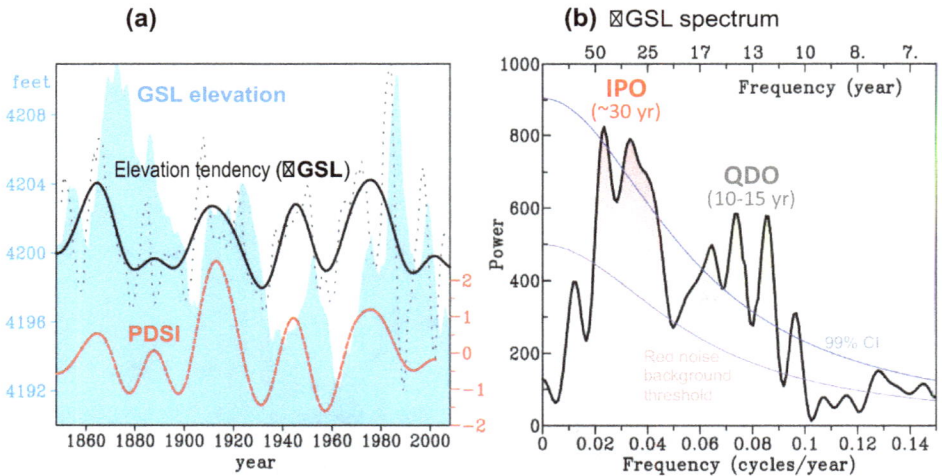

Fig. 9. (a) The GSL elevation (blue shaded graph) overlaid with the elevation tendency (ΔGSL; black line) and the PDSI (red dashed line) at the nearby grid points, after a 20-year lowpass filter. The unfiltered ΔGSL is shown with a gray dotted line for reference. (b) The Multi-Taper Method (MTM) spectral analysis of the ΔGSL during the period 1848-2008 with 2π tapper, overlaid with the 99% confidence interval (upper blue line) and the red noise background threshold (lower blue line). Adopted from Wang et al. (2011a).

The analysis of Wang et al. (2011a) revealed that hydrological factors controlling the multi-decadal variations of the GSL elevation respond to a particular teleconnection that is induced at the transition point of the IPO, that is, a basin-scale interdecadal variability that exhibits a loading pattern in the tropical Pacific SSTs (Folland et al. 2002). The transition lies approximately halfway between the warmest and coldest tropical SST anomalies in the central Pacific (corresponding to the extreme IPO phases). A distinctive teleconnection pattern develops during such transition points, as was argued by Wang et al. (2011a) and delineated in **Fig. 10**. Using a 700-year record (1300-2003) of the tree ring-reconstructed Niño 3.4 index (Cook et al. 2009), denoted as CNiño3.4, as well as the Twentieth Century Reanalysis (20CR) Version 2 during 1871-2010 (Compo et al. 2011), the 250hPa streamfunction and SSTs were regressed upon the CNiño3.4 index. All the variables were bandpass filtered with 20-45 years to reflect the IPO. At year zero (yr+0),

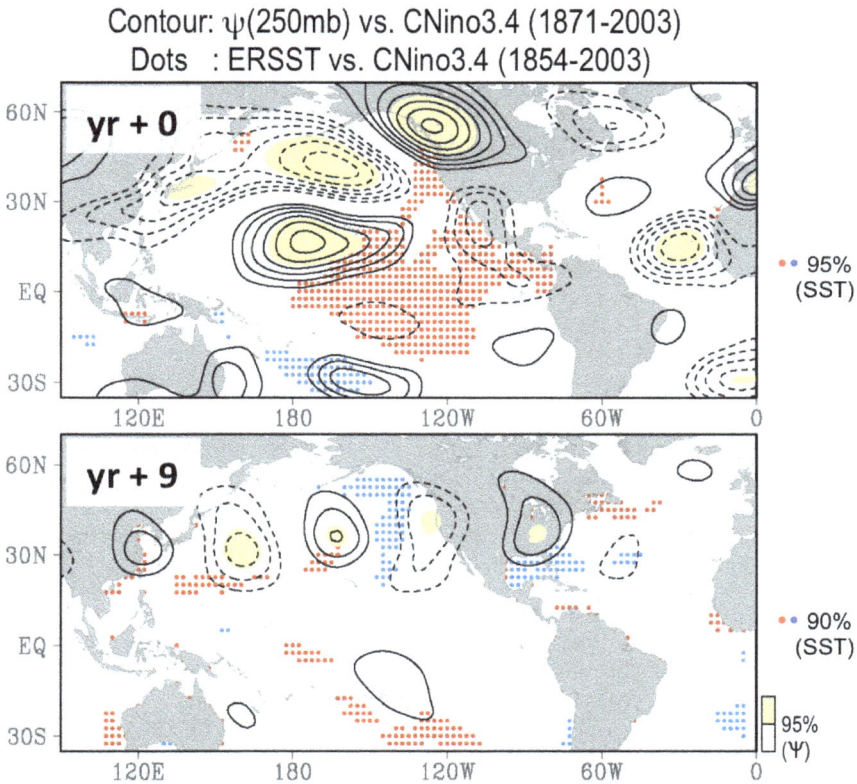

Fig. 10. (a) Patterns of the 250mb streamfunction (ψ; contours) and SSTs (dotted) regressed upon the CNiño3.4 index at different lags. The contour interval is 10-6 m2 s-1 omitting zeros. All data were bandpass filtered with 20-50 years. Values at the 95% confidence interval are indicated by yellow shadings for ψ and by dots for the SSTs, based on Student's t-test.

the SST response to the CNiño3.4 depicts the classic IPO "horseshoe pattern" (Zhang and Delworth 2007) consisting of a widespread eastern tropical warming surrounded by midlatitude cooling. A clear PNA pattern of the streamfunction field emerges in response to such a tropical Pacific warming (**Fig. 10a**). Such SST and circulation patterns are known to produce the North American dipole that juxtaposes the GSL area with wet conditions the SST response to the CNiño3.4 depicts the classic IPO "horseshoe pattern" (Zhang and Delworth 2007) consisting of a widespread eastern tropical warming surrounded by midlatitude cooling. A clear PNA pattern of the streamfunction field emerges in response to the north and dry conditions to the south (Gershunov and Barnett 1998); however it is not the modulating force for the GSL change (Wang et al. 2009b). At yr+9, the basin-wide SST pattern reveals weak anomalies in the central equatorial Pacific with noticeable cooling in the northeastern Pacific (**Fig. 10b**). Meanwhile, a cyclonic cell develops over western North America and is embedded in a zonal wave train. Such a teleconnection wave train appears to be a Rossby wave response to upstream forcings (similar to the situation in **Fig. 7**), possibly induced by warm SST and/or convection anomalies in the western North Pacific (Lau and Weng 2002; Wang et al. 2011b). The resulting cyclonic circulation over the West Coast influences precipitation in the interior West (Barlow et al. 2001). The results presented here therefore suggest that the GSL's multi-decadal variability (and associated local wet/dry cycle) is modulated by the IPO's *transition phases* between the warm and cold, rather than by its extreme phases.

b. Quasi-decadal cycle

The marked quasi-decadal variability (10-15 year) revealed form the IMW precipitation variations, the GSL elevation change (cf. Fig. 9b), and the Pacific SSTs indicates a co-variation between the GSL elevation and the Pacific QDO. The Pacific QDO has a substantial SST variation in the central tropical Pacific near the NINO4 region (Allan 2000). At the quasi-decadal timescale, the SST anomalies in the NINO4 region [ΔSST(NINO4)] exhibit a significant, yet inverse coherence with the GSL elevation (**Fig. 11**). However, such a coherence denies any direct association (either in-phase or out-of-phase) between the precipitation anomalies in the GSL watershed and the Pacific QDO because, in a given frequency, the precipitation variations always lead the GSL elevation variations (Lall and Mann 1995). In other words, only an indirect link or a coincidence can explain such an association between the Pacific QDO and the precipitation source of the GSL.

Wang et al. (2010a) investigated this phenomenon and found that the quasi-decadal variation in the IMW precipitation consistently lags the Pacific QDO by a quarter-phase, i.e. 3 years after the peak of the warm-phase Pacific QDO occurs, an anomalous trough develops over the Gulf of Alaska and enhances the IMW precipitation (similar to the situation in **Fig. 10**). For the opposite, i.e. the cool-phase Pacific QDO, 3 years later an anomalous ridge forms in the same location and thus reduces the IMW precipitation. These findings describes a process that the quasi-decadal coherence between ΔSST(NINO4) and the GSL elevation reflects a sequential process that begins with the warm/cool phase of the Pacific QDO and ultimately affects the GSL elevation, through modulations of the quadrature amplitude modulation of the Pacific QDO. This process is illustrated in **Fig. 11** with bandpass filtered time series of ΔSST(NINO4), the IMW precipitation (P), and the GSL elevation. The phase shift creates consistent time lags between the GSL elevation and the

Pacific QDO, leading to a phase lag of ~3 years in the quasi-decadal frequency (**Fig. 11**). These processes attribute the phase lags in the occurrence of the IMW droughts/pluvials in the IMW, to the lifecycle of the decadal-scale Pacific climate oscillations.

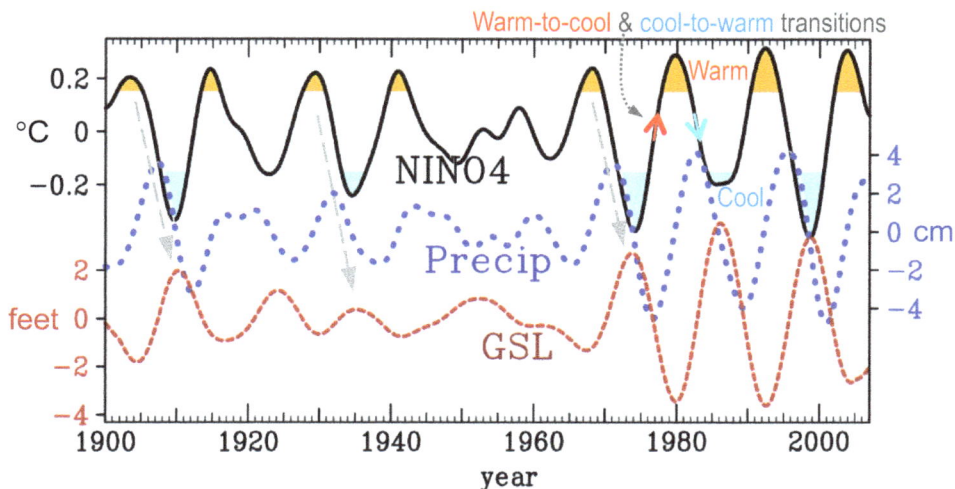

Fig. 11. Time series of the 10-15 year bandpassed ΔSST(NINO4) (black line), precipitation in the GSL watershed (blue dotted line), and the GSL elevation (red dashed line). The GSL elevation lags the precipitation by 3 years while the precipitation lags the Pacific QDO by another 3 years. (Adopted from Wang et al. 2010a) with permission of the American Meteorological Society

c. Paleoclimate evidence

Because instrumental precipitation data in the IMW only date back to the 1890s, and the length of reliable atmospheric data is considerably shorter, proxy precipitation constructed by tree-ring chronology was examined. Compiling the tree ring records in northeastern Utah, Gray et al. (2004) constructed a 776 year precipitation record from 1226 to 2001 – this data has been shown in **Fig. 1**. The power spectra of Gray et al.'s tree-ring precipitation (**Fig. 12a**) depict three significant modes, with one covering the 150-200 year frequency and the other two corresponding to the 30 year and 10-15 year cycles of the GSL elevation (cf. **Fig. 9b**). The 150-200 year mode echoes the "secular mode" of GSL that was pointed out by previous studies (Lall and Mann 1995). As shown in **Fig. 1**, its uptrend during the 20th century coincides with the climate regime shift observed in the Colorado River Basin during the 1970s (Hidalgo and Dracup 2003). The 30-year cycle of the tree-ring precipitation is most pronounced between 1500 and 1650 and has weakened since starting around 1650. Recent tree-ring chronologies have been developed into proxies for atmospheric circulation patterns. Examining the proxy index of the Pacific Decadal Oscillation (PDO) constructed by Biondi et al. (2001) for the 1661-1991 period and comparing it with Gray et al.'s tree-ring precipitation, Wang et al. (2010a) found three peaks in the 30 year, 10-15 year, and 4-5 year frequency bands that are significant above the 90% confidence level in the spectral coherence analysis (**Fig. 12b**). It is known that the PDO (which covers the North Pacific north of 20°N, in contrast to the IPO that covers the entire Pacific Basin) contains both the interdecadal and interannual signals in the tropical Pacific (Zhang et al. 1997). Therefore, the three coherence zones in **Fig. 12b** appear to be responses of the PDO, QDO, and ENSO modes, respectively. These modes not only affect the IMW climate individually but also collectively; their interplay is discussed next.

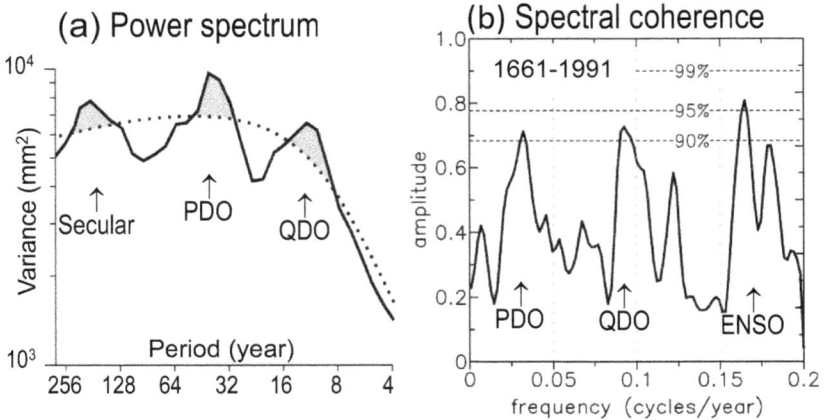

Fig. 12. (a) The global wavelet power spectrum of the tree-ring reconstructed precipitation over northeast Utah (*Gray et al. 2004*) from 1226 to 2001 superimposed with the red-noise 99% significance level (dotted curve). (b) MTM coherence between the tree-ring precipitation and the proxy PDO index constructed by Biondi et al. 2001 for 1661-1991. The three frequency bands corresponding to the IPO, QDO, and ENSO are indicated. (Adopted from Wang et al. 2010a)

5. Scale interaction between different climate modes

a. Interdecadal vs. interannual

Focusing on timescales longer than ENSO, previous studies (e.g., Gershunov and Barnett 1998; Brown and Comrie 2004) have identified a pronounced interaction between interannual and decadal climate variabilities originating from the Pacific. Those studies concluded that constructive and destructive superposition between ENSO and the PDO can respectively strengthen and weaken the North American dipole structure in precipitation and temperature. Recent studies (e.g., Brown 2011) continued to explore such a multi-decadal, ENSO-related variability over the western United States. But to date, how these different scales of natural variability interact and how such interactions influence the IMW climate remains unclear. Although ENSO is known to fluctuate within a broad frequency band of 2-7 years, several studies (e.g., Allan 2000; Tourre et al. 2001) have argued that there are two distinct modes of ENSO – a 2-3 year "quasi-biennual" oscillation and a 3-6 year interannual oscillation. These two ENSO modes are associated with unique atmosphere-ocean interactions and are linked to different regional climate patterns (Mo 2010).

To substantiate this character, time series of precipitation averaged in four western states of Nevada, Utah, Idaho and Wyoming (derived from the US Historical Climatology Network data), as well as the Nino3.4 SST index, are displayed in **Fig. 13a**. Here, both time series were bandpass filtered with 3-6 years to focus on the interannual mode, following Rajagopalan and Lall (1998). Visual inspection finds that the two variables are not consistently in-phase or out-of-phase; instead, their relationship fluctuates. By computing a running correlation analysis of every 15 years, centered at the 7th year, the relationship between the precipitation

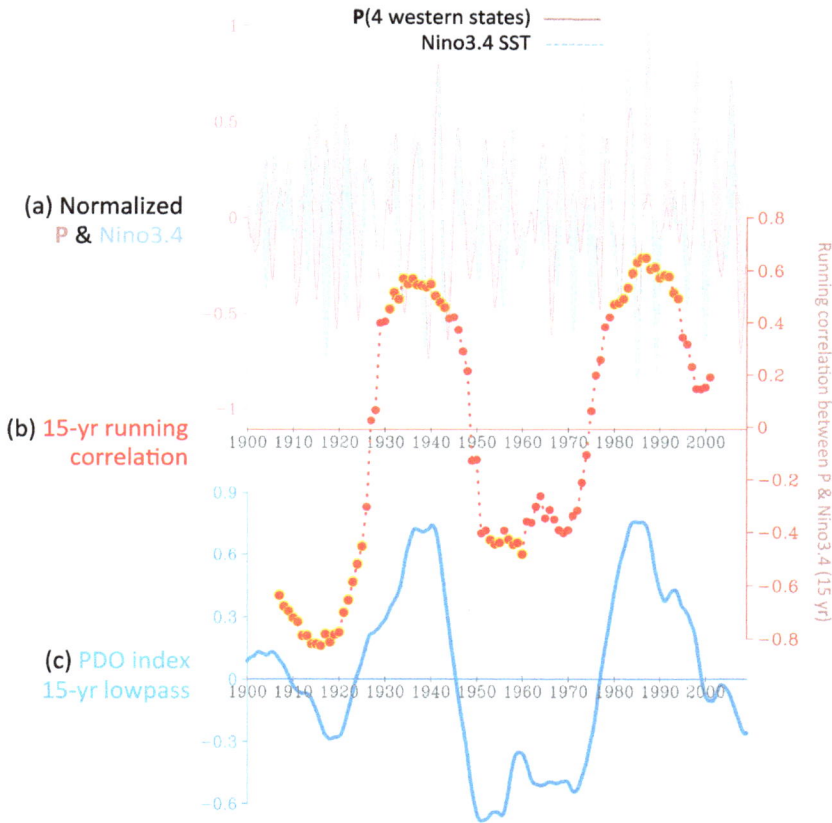

Fig. 13. (a) Time series of precipitation (P) averaged within the 4-state region (see text) and the Nino3.4 SST index. (b) Running correlation coefficients of a 15 year period between P and Nino3.4, with significant correlations outlined by yellow circles. (c) The PDO index with 15-year lowpass filter.

and Nino3.4 SST anomalies reveals a low-frequency oscillation (**Fig. 13b**), with some 20 years featuring negative correlations and the next 20 years showing positive correlations, only to transition to negative again. Superimposing the PDO index[2] smoothed by a 15-year lowpass filter (to highlight its low-frequency signal, **Fig. 13c**), it becomes clear that the alternating correlation between ENSO and the precipitation anomalies is a modulation from

[2] The PDO index was obtained from the University of Washington at
http://jisao.washington.edu/pdo/PDO.latest.

the PDO. Apparently, positive PDO phases shift the ENSO-induced North American dipole northward, causing the central IMW to respond positively to ENSO. Likewise, negative PDO phases shift the dipole southward, thereby leading the central IMW to react negatively with ENSO. On the other hand, correlation analysis between the Nino3.4 SST anomalies and precipitation in Arizona and New Mexico (not shown) does not reveal any discernable change with respect to the PDO phasing. This finding result suggests that the PDO modulation on ENSO's climate impact is only effective around the transition zone of the precipitation dipole in the central IMW. This result echoes the observation by Brown (2011) that the relationship between cold-season ENSO conditions and circulation anomalies over the western United States varies with phasing of the PDO. The results presented here and from Brown (2011) therefore suggest uncertainty on decadal time scales for using ENSO conditions as a seasonal climate forecast tool.

b. Interannual vs. intraseasonal

We have observed that the intraseasonal variability of the IMW climate also undergoes modulation from certain interannual modes. Analyzing episodes of persistent temperature inversions in Salt Lake Valley and Cache Valley (80 miles northeast of Salt Lake City), Gillies et al. (2010a) noticed a tendency for those persistent inversion episodes to delay their occurrence by about 5 days each year. This intriguing feature is illustrated in **Fig. 14a**, which

shows the temperature lapse rate at the Salt Lake City International Airport (KSLC; shadings) overlaid with the PM2.5 concentrations in Salt Lake City (blue line), for winters 1998/99 to 2007/08. Throughout this decade, the occurrences of persistent temperature inversions (i.e. events lasting longer than 7 days) began in mid-December 1998, shifted to late December in 1999 and then continued to "delay" through late January 2007. Because such prolonged temperature inversion events are linked to episodes of semi-stationary ridge developed over the IMW, we analyzed the NCEP/NCAR Reanalysis (Kalnay et al. 1996) using the 700hPa geopotential height at the nearest grid point of KSLC from December 1948 to February 2010. The geopotential height was bandpass filtered with 20-40 days based on the 30-day mode as identified in Gillies et al. (2010a). The analysis of 62 winters (**Fig. 14b**) reveals episodes of prolonged ridge events (dark gray) that seem to delay each year by one week, as indicated by yellow arrows. This "migration" in the timing of the ridging episodes reappears in early December about every 5-6 years. This timescale coincides with that of the interannual ENSO cycle. It is possible that persistent ridging and inversion episodes in the IMW undergo an external forcing that modulates the seasonal timing of their occurrence.

To examine the possible interannual forcing on the intraseasonal variability of the persistent ridging events principal component (PC) analysis was carried out using the bandpass filtered 700hPa geopotential height as that shown in Fig. 14b. The first two leading modes (PC1 and PC2; **Fig. 15a**), accounting for 27% and 19% of the variance, depict a quadrature phase shift in the timing of the intraseasonal fluctuation. The year-to-year coefficients of PC1 and PC2 are significantly correlated at a 1-year lag (r=0.48), suggesting that PC1 tends to transitions into PC2 a year after. Meanwhile, PC3 shows weaker amplitude in the middle of the winter (9% of the variance). The regression patterns of winter streamfunction at 300 hPa and SSTs with each PC coefficient are shown in **Fig. 15b**. The regression with PC1 reveals a La Nina type of SST and circulation structure characterized by cold SST anomalies in the central equatorial Pacific associated with a PNA-like teleconnection pattern. The regression with PC2 depicts weak and disorganized SST anomalies accompanied by a short-wave circulation pattern. Noteworthy is that this short-wave pattern is in-phase with the winter standing wave train, similar to that presented in Fig. 7a, but at a slightly different latitude with respect to the winter jet stream. The troughs (red) and ridges (blue) of the winter standing wave train are indicated in **Fig. 15b**. Likely, PC2 is at the transition point between the La Nina and El Nino phases, where short-wave circulation response is a dominant feature (Wang et al. 2011b). The regression with PC3 outlines the SST and circulation patterns very similar to the North Atlantic Oscillation (Hurrell 2003), with a seesawing dipole between the Icelandic low and the Bermuda high. This mode (PC3) appears to modify the transition between PC1 and PC2 and further change the occurrence timing. These observations require further examination through GCM experiments prescribed with SST forcing conditions as those in **Fig. 15**.

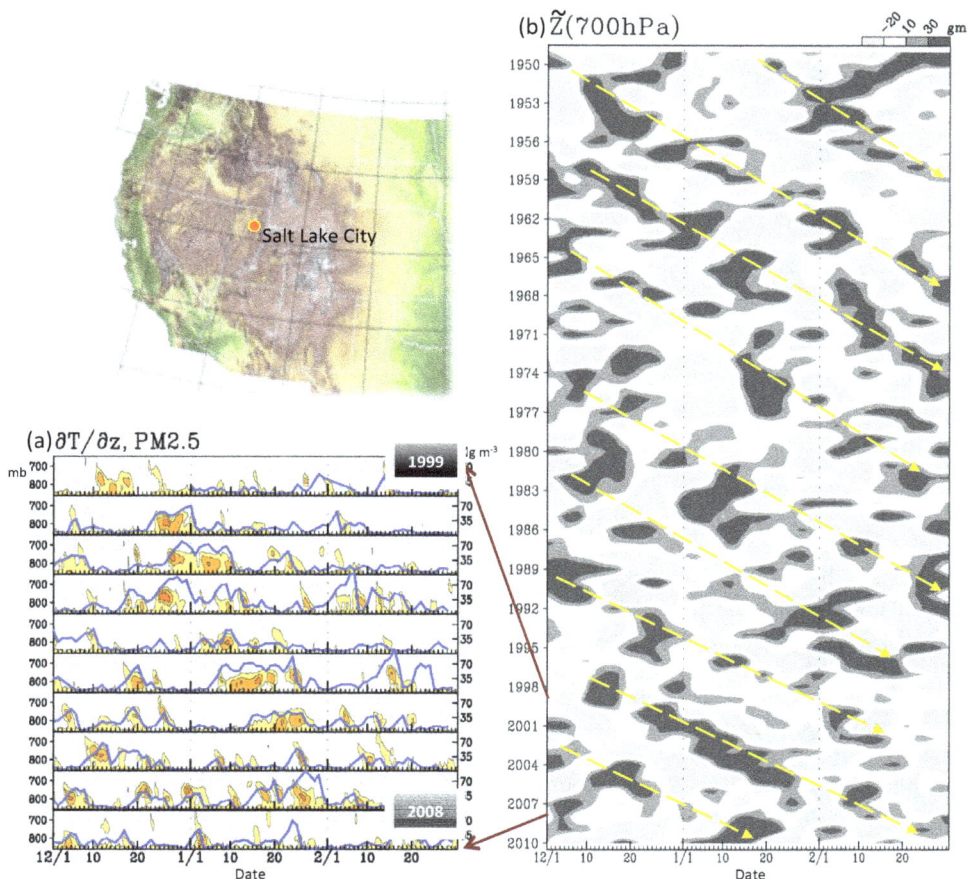

Fig. 14. (a) The 00Z temperature lapse rate at KSLC (orange contours) overlaid with PM2.5 concentrations within Salt Lake Valley for the winters of 1999-2008. (b) Bandpass filtered geopotential height at 700mb with 20-40 days interpolated onto the KSLC location derived from NCEP reanalysis; yellow arrows indicate the time shift in persistent ridging events. The geographical location and terrain are given at top left.

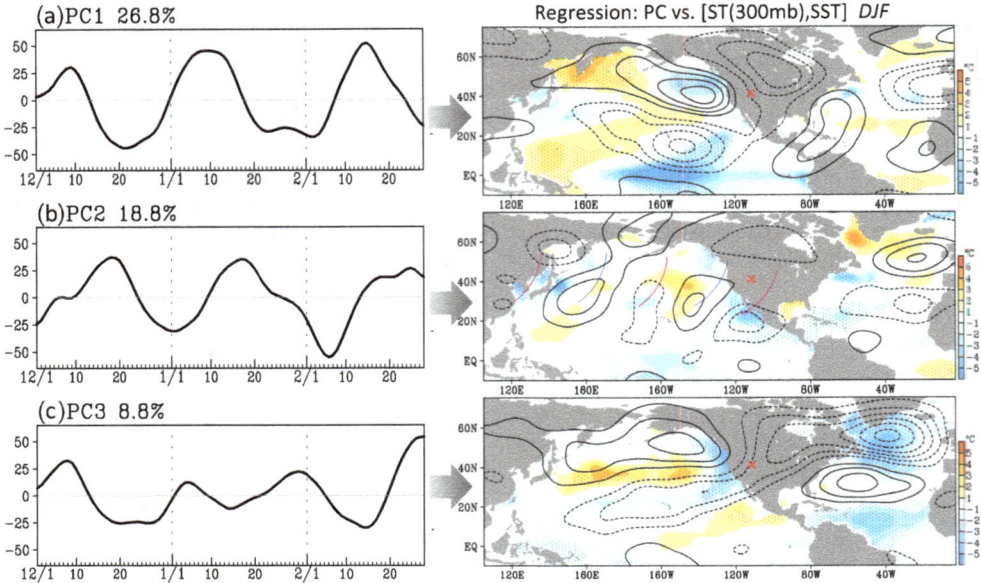

Fig. 15. (a) First 3 leading PCs of the bandpass filtered 700mb geopotential height at KSLC over the 1980-2008 period, and (b) corresponding patterns of 300mb streamfunction (contours) and SSTs (shading) regressed upon each PC time series. KSLC is indicated by a red cross.

6. Summary

The IMW is situated in the marginal zone of different climate regimes between (a) the annual and semiannual cycles, (b) the ENSO-induced north-south dipole, and (c) the PDO-related north-south seesaw pattern. The IMW climate variability undergoes robust modulations by the Pacific teleconnection. However, past research has almost exclusively focused on the extreme phases of the Pacific oscillatory modes (such as ENSO and the PDO/IPO), even though those modes feature a well-defined lifecycle. In this chapter we demonstrated that the Pacific oscillatory modes induce different types of teleconnection during their transition between the extreme warm and cold phases. Such teleconnection processes are subtle and may be difficult to monitor, but they have been found to control the climate variability of the IMW more profoundly than the warm/cold phases of the Pacific oscillatory modes. Moreover, the circulation anomalies affecting the IMW are more responsive to midlatitude short-wave trains rather than the long-wave PNA pattern. This feature poses a challenge to climate modeling and seasonal climate prediction for the IMW, especially the upper Colorado River Basin. The interplay between the various climate oscillations and their collective effects further complicate the climate variability of the IMW. Current climate forecast models have a difficulty in realistically depicting the IMW's climate regime and the transition-phase teleconnections associated with ENSO (and /IPO).

In the changing climate, future water resources of the IMW may be strained and threatened due to the rapid increase in water demand and the projected decrease in precipitation.

Historical climate data indicate that episodic events of extreme drought will compound the problem of increased demand and that this may interact with projected climate change in unforeseen and potentially worrisome ways. This is particularly a concern for water management in the IMW, which supplies the headwater of the Colorado River. Further diagnostics and modeling studies are required to isolate the effect of individual climate oscillatory modes in order to improve climate prediction for the IMW.

7. Acknowledgements

We are indebted to Dr. Xiaoqing Wu at Iowa State University for sharing the CCM3 simulation data. Insightful comments from Dr. Huug van dan Dool of the Climate Prediction Center are highly appreciated. This study was supported by the Utah State University Agricultural Experiment Station and approved as paper number 8359, as well as the Bureau of Reclamation Grant R11AC81456.

8. References

Allan, R., 2000: ENSO and climatic variability in the last 150 years. *El Niño and the Southern Oscillation: Multiscale Variability, Global and Regional Impacts*, H. F. Diaz, and V. Markgrav, Eds., Cambridge Univ. Press, 3–56.

Barlow, M., S. Nigam, and E. H. Berbery, 2001: ENSO, Pacific Decadal Variability, and U.S. Summertime Precipitation, Drought, and Stream Flow. *J. Climate*, 14, 2105-2128.

Barnett, T. P., and D. W. Pierce, 2008: When will Lake Mead go dry? *Water Resour. Res.*, 44, W03201.

Biondi, F., A. Gershunov, and D. R. Cayan, 2001: North Pacific Decadal Climate Variability since 1661. *J. Climate*, 14, 5-10.

Boyle, J. S., 1998: Evaluation of the Annual Cycle of Precipitation over the United States in GCMs: AMIP Simulations. *J. Climate*, 11, 1041-1055.

Branstator, G., 1987: A Striking Example of the Atmosphere's Leading Traveling Pattern. *J. Atmos. Sci.*, 44, 2310-2323.

— —, 2002: Circumglobal Teleconnections, the Jet Stream Waveguide, and the North Atlantic Oscillation. *J. Climate*, 15, 1893-1910.

Brown, D. P., 2011: Winter Circulation Anomalies in the Western United States Associated with Antecedent and Decadal ENSO Variability. *Earth Interactions*, 15, 1-12.

Brown, D. P., and A. C. Comrie, 2004: A winter precipitation "dipole" in the western United States associated with multidecadal ENSO variability. *Geophys. Res. Lett.*, 31, L09203.

Compo, G. P., and Coauthors, 2011: The Twentieth Century Reanalysis Project. *Q. J. Roy. Meteor. Soc.*, 137, 1-28.

Cook, E. R., and P. J. Krusic, 2004: North American summer PDSI reconstructions, 24 pp.

Cook, E. R., D. M. Meko, and C. W. Stockton, 1997: A New Assessment of Possible Solar and Lunar Forcing of the Bidecadal Drought Rhythm in the Western United States. *J. Climate*, 10, 1343-1356.

Cook, E. R., R. D. D'Arrigo, and K. J. Anchukaitis, 2009: Tree Ring 500 Year ENSO Index Reconstructions# 2009-105.

Dettinger, M. D., D. R. Cayan, H. F. Diaz, and D. M. Meko, 1998: North–South Precipitation Patterns in Western North America on Interannual-to-Decadal Timescales. *J. Climate*, 11, 3095-3111.

Ding, Q., and B. Wang, 2005: Circumglobal Teleconnection in the Northern Hemisphere Summer*. *J. Climate*, 18, 3483-3505.

Folland, C. K., J. A. Renwick, M. J. Salinger, and A. B. Mullan, 2002: Relative influences of the Interdecadal Pacific Oscillation and ENSO on the South Pacific Convergence Zone. *Geophys. Res. Lett.*, 29, 1643.

Gershunov, A., and T. P. Barnett, 1998: Interdecadal Modulation of ENSO Teleconnections. *Bull. Amer. Meteor. Soc.*, 79, 2715-2725.

Gillies, R. R., S.-Y. Wang, and M. R. Booth, 2010a: Atmospheric scale interactions on wintertime Intermountain West inversions. *Wea. Forecasting*, 25, 1196–1210.

Gillies, R. R., S.-Y. Wang, J.-H. Yoon, and S. Weaver, 2010b: CFS Prediction of Winter Persistent Inversions in the Intermountain Region. *Wea. Forecasting*, 25, 1211–1218.

Gray, S. T., S. T. Jackson, and J. L. Betancourt, 2004: Tree-ring based reconstructions of interannual to decadal-scale precipitation variability for northeastern Utah. *J. Amer. Water Resour. Assoc*, 40, 947–960.

Gray, S. T., J. L. Betancourt, C. L. Fastie, and S. T. Jackson, 2003: Patterns and sources of multidecadal oscillations in drought-sensitive tree-ring records from the central and southern Rocky Mountains. *Geophys. Res. Lett.*, 30, 49-41 - 49-44.

Herweijer, C., R. Seager, E. R. Cook, and J. Emile-Geay, 2007: North American Droughts of the Last Millennium from a Gridded Network of Tree-Ring Data. *J. Climate*, 20, 1353-1376.

Hidalgo, H. G., and J. A. Dracup, 2003: ENSO and PDO Effects on Hydroclimatic Variations of the Upper Colorado River Basin. *Journal of Hydrometeorology*, 4, 5-23.

Higgins, R. W., Y. Yao, and X. L. Wang, 1997: Influence of the North American Monsoon System on the U.S. Summer Precipitation Regime. *J. Climate*, 10, 2600-2622.

Horel, J. D., and J. M. Wallace, 1981: Planetary-Scale Atmospheric Phenomena Associated with the Southern Oscillation. *Mon. Wea. Rev.*, 109, 813-829.

Horel, J. D., and C. R. Mechoso, 1988: Observed and Simulated Intraseasonal Variability of the Wintertime Planetary Circulation. *J. Climate*, 1, 582-599.

Hoskins, B. J., and D. J. Karoly, 1981: The Steady Linear Response of a Spherical Atmosphere to Thermal and Orographic Forcing. *J. Atmos. Sci.*, 38, 1179-1196.

Hsu, C.-P. F., and J. M. Wallace, 1976: The Global Distribution of the Annual and Semiannual Cycles in Precipitation. *Mon. Wea. Rev.*, 104, 1093-1101.

Hurrell, J. W., 2003: *The North Atlantic oscillation: climatic significance and environmental impact.* Vol. 134, American Geophysical Union, 279 pp.

Kalnay, E., and Coauthors, 1996: The NCEP/NCAR 40-Year Reanalysis Project. *Bull. Amer. Meteor. Soc.*, 77, 437-471.

Kushnir, Y., 1987: Retrograding Wintertime Low-Frequency Disturbances over the North Pacific Ocean. *J. Atmos. Sci.*, 44, 2727-2742.

Lall, U., and M. E. Mann, 1995: The Great Salt Lake: A barometer of low-frequency climatic variability. *Water Resour. Res.*, 31, 2503-2515.

Lau, K. M., and H. Weng, 2002: Recurrent Teleconnection Patterns Linking Summertime Precipitation Variability over East Asia and North America. *Journal of the Meteorological Society of Japan*, 80, 1309-1324.

Lau, N.-C., and M. J. Nath, 1999: Observed and GCM-Simulated Westward-Propagating, Planetary-Scale Fluctuations with Approximately Three-Week Periods. *Mon. Wea. Rev.*, 127, 2324-2345.

Legates, D. R., and C. J. Willmott, 1990: Mean seasonal and spatial variability in gauge-corrected, global precipitation. *Int. J. Climatol.*, 10, 111-127.

Leung, L. R., Y.H. Kuo, and J. Tribbia, 2006: Research needs and directions of regional climate modeling using WRF and CCSM. *Bull. Amer. Meteor. Soc.*, 87, 1747–1751.

Mantua, N. J., S. R. Hare, Y. Zhang, J. M. Wallace, and R. C. Francis, 1997: A Pacific Interdecadal Climate Oscillation with Impacts on Salmon Production. *Bull. Amer. Meteor. Soc.*, 78, 1069-1079.

Marks, D., and A. Winstral, 2001: Comparison of Snow Deposition, the Snow Cover Energy Balance, and Snowmelt at Two Sites in a Semiarid Mountain Basin. *Journal of Hydrometeorology*, 2, 213-227.

Mearns, L., W. J. J. Gutowski, R. Jones, R. Leung, S. McGinnis, A. Nunes, and Y. Qian, 2009: A Regional Climate Change Assessment Program for North America. *Eos Trans. AGU*, 90, 311-312.

Mesinger, F., and Coauthors, 2006: North American Regional Reanalysis. *Bull. Amer. Meteor. Soc.*, 87, 343-360.

Mo, K., and J. Paegle, 2005: Pan-America. *Intraseasonal variability in the atmosphere-ocean climate system*, K.-M. L. a. D. E. Waliser, Ed., Springer, 95-124.

Mo, K. C., 1999: Alternating Wet and Dry Episodes over California and Intraseasonal Oscillations. *Mon. Wea. Rev.*, 127, 2759-2776.

— —, 2010: Interdecadal Modulation of the Impact of ENSO on Precipitation and Temperature over the United States. *J. Climate*, 23, 3639-3656.

Moncrieff, M. W., 1992: Organized Convective Systems: Archetypal Dynamical Models, Mass and Momentum Flux Theory, and Parametrization. *Quarterly Journal of the Royal Meteorological Society*, 118, 819-850.

Rajagopalan, B., and U. Lall, 1998: Interannual variability in western US precipitation. *Journal of Hydrology*, 210, 51-67.

Sangoyomi, T. B., 1993: Climatic variability and dynamics of Great Salt Lake hydrology, PhD, Utah State University, 247 pp.

Schubert, S., and Coauthors, 2009: A U.S. CLIVAR Project to Assess and Compare the Responses of Global Climate Models to Drought-Related SST Forcing Patterns: Overview and Results. *J. Climate*, 22, 5251-5272.

Schubert, S. D., M. J. Suarez, P. J. Pegion, R. D. Koster, and J. T. Bacmeister, 2004: On the Cause of the 1930s Dust Bowl. *Science*, 303, 1855-1859.

Seager, R., Y. Kushnir, C. Herweijer, N. Naik, and J. Velez, 2005: Modeling of Tropical Forcing of Persistent Droughts and Pluvials over Western North America: 1856–2000*. *J. Climate*, 18, 4065-4088.

Simmons, A. S., D. D. Uppala, and S. Kobayashi, 2007: ERA-interim: new ECMWF reanalysis products from 1989 onwards. *CMWF Newsl* 110, 29-35.

Tourre, Y., B. Rajagopalan, Y. Kushnir, M. Barlow, and W. White, 2001: Patterns of Coherent Decadal and Interdecadal Climate Signals in the Pacific Basin During the 20 th Century. *Geophys. Res. Lett.*, 28, 2069-2072.

Wang, S.-Y., and T.-C. Chen, 2009: The Late-Spring Maximum of Rainfall over the U.S. Central Plains and the Role of the Low-Level Jet. *J. Climate*, 22, 4696-4709.

Wang, S.-Y., R. R. Gillies, and T. Reichler, 2011a: Multi-decadal drought cycles in the Great Basin recorded by the Great Salt Lake: Modulation from a transition-phase teleconnection. *J. Climate*. doi: http://dx.doi.org/10.1175/2011JCLI4225.1

Wang, S.-Y., R. R. Gillies, E. S. Takle, and W. J. Gutowski Jr., 2009a: Evaluation of precipitation in the Intermountain Region as simulated by the NARCCAP regional climate models. *Geophys. Res. Lett.*, 36, L11704.

Wang, S.-Y., R. R. Gillies, J. Jin, and L. E. Hipps, 2009b: Recent rainfall cycle in the Intermountain Region as a quadrature amplitude modulation from the Pacific decadal oscillation. *Geophys. Res. Lett.*, 36, L02705.

Wang, S.-Y., R. R. Gillies, J. Jin, and L. E. Hipps, 2010a: Coherence between the Great Salt Lake Level and the Pacific Quasi-Decadal Oscillation. *J. Climate*, 23, 2161-2177.

Wang, S.-Y., R. R. Gillies, L. E. Hipps, and J. Jin, 2011b: A transition-phase teleconnection of the Pacific quasi-decadal oscillation. *Clim. Dynamics*, 36, 681-693.

Wang, S.-Y., L. E. Hipps, R. R. Gillies, X. Jiang, and A. L. Moller, 2010b: Circumglobal teleconnection and early summer rainfall in the US Intermountain West. *Theor. Appl. Climatol.*, 102, 245-252.

White, W. B., and Z. Liu, 2008: Resonant excitation of the quasi-decadal oscillation by the 11-year signal in the Sun's irradiance. *J. Geophys. Res.*, 113, 1-16.

White, W. B., Y. M. Tourre, M. Barlow, and M. Dettinger, 2003: A delayed action oscillator shared by biennial, interannual, and decadal signals in the Pacific Basin. *J. Geophys. Res.*, 108, 15-11 - 15-18.

Wood, A. W., 2011: Development of a Seasonal Climate and Stream Flow Forecasting Test Bed for the Colorado River Basin. *36th Climate Diagnostic and Prediction Workshop*.

Wu, X., L. Deng, X. Song, and G. J. Zhang, 2007: Coupling of Convective Momentum Transport with Convective Heating in Global Climate Simulations. *J. Atmos. Sci.*, 64, 1334-1349.

Xie, P., and P. A. Arkin, 1997: Global Precipitation: A 17-Year Monthly Analysis Based on Gauge Observations, Satellite Estimates, and Numerical Model Outputs. *Bull. Amer. Meteor. Soc.*, 78, 2539-2558.

Zhang, R., and T. L. Delworth, 2007: Impact of the Atlantic Multidecadal Oscillation on North Pacific climate variability. *Geophys. Res. Lett.*, 34, L23708.

Zhang, Y., J. M. Wallace, and D. S. Battisti, 1997: ENSO-like Interdecadal Variability: 1900–93. *J. Climate*, 10, 1004-1020.

Zhang, Z., and M. E. Mann, 2005: Coupled patterns of spatiotemporal variability in Northern Hemisphere sea level pressure and conterminous U.S. drought. *J. Geophy. Res.*, 110, D03108.

8

Spatial and Temporal Variability of Sea Surface Temperature in the Yellow Sea and East China Sea over the Past 141 Years

Daji Huang[1,2], Xiaobo Ni[1], Qisheng Tang[3],
Xiaohua Zhu[1,2] and Dongfeng Xu[1,2]
[1]State Key Laboratory of Satellite Ocean Environment Dynamics
Second Institute of Oceanography, State Oceanic Administration
[2]Department of Ocean Science and Engineering, Zhejiang University
[3]Yellow Sea Fisheries Research Institute, Chinese Academy of Fishery Sciences
China

1. Introduction

The Yellow Sea and East China Sea (YES) are marginal seas in the northwest Pacific. There is in fact a smaller sea, the Bohai Sea, to the north of the Yellow Sea. For most discussions in the chapter, we shall treat the Bohai Sea as part of the Yellow Sea. The YES is one of the mostly intensively utilized sea in the world, for example, heavy fishery and marine aquaculture. The use of the YES is closely related to its climate variability, though it is not well-know because until now there has been a lack of adequate observational data. To know the climatology of sea surface temperature (SST, all the acronyms used in the chapter are listed in Table 1) in the YES and their relationship with regional and global climate have both scientific and social importance.

There have been recent studies, some associated with marine ecosystem, on the long-term temperature variation in the YES. The results indicate that SST has risen significantly in the 20th century. The observed annual mean SST in the Bohai Sea increased by 0.42°C from 1960 to 1997 (Lin et al., 2001), and the observed annual mean of water-column average temperature along the 36°N section between 120°45′E and 124°30′E in the Yellow Sea increased by 1.7℃ from 1976 to 2000 (Lin et al., 2005). On the eastern side of the Yellow Sea, there has been an increase of 1.8℃ and 1.0℃, respectively, in water temperature in February and August over the past 100 years (Hahn, 1994). Using the Hadley SST data from 1901 to 2004, Zhang et al. (2005) found that, in the YES, the annual mean SST was cold from 1900s to 1930s, warm in the 1950s, slightly cold in the 1960s and warming again from 1980s.

Until now, the inter-relation between the SST in the YES with the regional climate is not well documented, though their interaction is rather distinct, for instance, the variability of the land surface air temperature over China affects the SST in the YES particularly in winter, and the SST in the YES also have some influence on the air temperature, fog and precipitation over China, especially along the coastal area. However these have not been well studied, especially in climatological prespective due to the lack of long time dataset.

For all study of the spatial and temporal variability of SST in the YES, the annual mean data is used, while seasonal variability is filtered out. In this study, we use the Met Office Centre's Hadley SST data (Rayner et al., 2003) to investigate the seasonal variability, inter-annual to decadal variability and long-term trend of SST in the YES.

Acronym	Expanded form
AC	annual component
AM	annual mean component
AR	annual range
CC	cold regime with a cold trend,
CW	cold regime with warm trend
EASM	the East Asian summer monsoon
EMD	Empirical Mode Decomposition
ENSO	El Nino–Southern Oscillation
EOF	Empirical Orthogonal Function
JJA	June, July and August
JMA	the Japan Meteorological Agency
PAC	the normalized annual precipitation anomaly over China
PDO	Pacific Decadal Oscillation
RMSE	root mean squared error
STD	standard deviation
SST	sea surface temperature
TAC	annual surface air temperature anomaly over China
WC	warm regime with cold trend
WW	warm regime with warm trend
YES	the Yellow Sea and East China Sea

Table 1. List of the acronyms used in the chapter

2. Data and methods

2.1 Data

The monthly SST in the YES is extracted from HadISST1 SST dataset for the period from 1870 to 2010, i.e., 141 years. There are 188 grid points in the YES (Fig. 1). HadISST1 SST data set, produced by the Met Office Hadley Centre, is a monthly global 1° latitude-longitude grid data start from 1870 till present. HadISST1 temperatures are reconstructed using a two-stage reduced-space optimal interpolation procedure, followed by superposition of quality-improved gridded observations onto the reconstructions to restore local detail. HadISST1 compares well with other published analyses, capturing trends in global, hemispheric, and regional SST well, containing SST fields with more uniform variance through time and better month-to-month persistence than those in global SST (Rayner et al., 2003). HadISST1 SST dataset is available at web site http://www.metoffice.gov.uk/hadobs/hadisst/data/download.html.

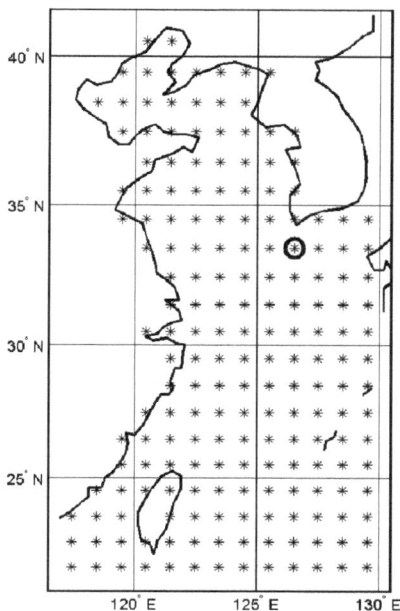

Fig. 1. Study area of the YES, and location of data grid points extracted from HadSST1 SST dataset. A bold cycle indicates the specific data points at (126.5°E, 33.5°N), which is used as a template.

The spatial and temporal variability of SST in the YES is related to the regional and global climate. One climatic effect of the SST in the YES is readily illustrated by much warmer winter temperatures on its east coast than that on the west coast. For example, January temperature is 6.4°C in Nagasaki (32.4°N), Japan, but only 3.7°C in Shanghai (31.3°N), China (Xie et al., 2002). Both the satellite observation and numerical model results show (Xie et al., 2002; Chen et al., 2003) that SST front in the YES plays a significant role on enhancing wind speed and raining cloud above the region. The SST in the YES (a marginal sea between the largest continent Eurasia and the largest ocean Pacific) is a part of the global climate and is closely linked to the Pacific and East Asian climate. The East Asian monsoon system is one of the most active components of the global climate system. El Nino–Southern Oscillation (ENSO) exhibits the greatest influence on the interannual variability of the global climate (Webster et al., 1998). The mature phase of ENSO often occurs in boreal winter and is normally accompanied by a weaker than normal winter monsoon along the East Asian coast (Wang et al., 2000). Consequently, the climate in south-eastern China and Korea is warmer and wetter than normal during ENSO winter and the following spring (Tao & Zhang, 1998; Kang & Jeong, 1996).

In order to investigate the relationship of the variability of SST in the YES with the regional and global climate, the following time series are used.

The annual surface air temperature anomaly over China (TAC), the normalized annual precipitation anomaly over China (PAC) and the East Asian summer monsoon (EASM) index are used to represent the regional climate. The TAC is reconstructed by Tang & Ren

(2005) for 1905 to 2001 and extended by Ding & Ren (2008) to 2005. A monthly mean temperature data obtained by averaging monthly mean maximum and minimum temperatures is used to avoid the inhomogeneity problems with data induced by different observation times and statistic methods between early and late 20th century. The PAC is reconstructed by Ding & Ren (2008), and is normalized with respect to its 30 years (1971-2000) standard deviation (STD). Both TAC and PAC are available from 1905 to 2005 and these time series are digitalized from their published figures (Ding & Ren, 2008).

The EASM index is defined as an area-averaged seasonally (June, July and August, JJA) dynamical normalized seasonality at 850 hPa within the East Asian monsoon domain (10°N-40°N, 110°E-140°E) (Li & Zeng, 2003). There is an apparent negative correlation between the EASM index and summer (JJA) rainfall in the middle and lower reaches of the Yangtze River in China, indicating drought years over the valley are associated with the strong EASM and flood years with the weak EASM. The annual ESAM index is available from 1948 to 2010 and is downloaded from http://web.lasg.ac.cn/staff/ljp/data-monsoon/EASMI.htm.

Both ENSO and Pacific Decadal Oscillation (PDO) indexes are used to represent the global climate. The ENSO index used in the chapter is produced by the Japan Meteorological Agency (JMA). It is the monthly SST anomalies averaged for the area 4°N-4°S and 150°W-90°W. This ENSO index is the JMA index based on reconstructed monthly mean SST fields for the period Jan 1868 to Feb 1949, and on observed JMA SST index for March 1949 to present (Meyers et al., 1999). The monthly ENSO index data file (jmasst1868-today.filter-5) is available from 1868 to 2010 and is downloaded from http://coaps.fsu.edu/pub/JMA_SST_Index/.

The PDO index used in the chapter is updated standardized values for the PDO index, derived as the leading principal component of monthly SST anomalies in the North Pacific Ocean, poleward of 20°N (Zhang, et al., 1997; Mantua et al., 1997). The monthly mean global average SST anomalies are removed to separate this pattern of variability from any "global warming" signal that may be present in the data. The monthly PDO index is available from 1900 to present and is downloaded from http://jisao.washington.edu/pdo/PDO.latest.

All the above mentioned five time series are shown in Fig. 2.

2.2 Data analysis methods

The monthly SST of the last 141 years in the YES contains both spatial and temporal variability. The temporal variability is primarily contributed by seasonal signal for overall variability, and is contributed by inter-annual to decadal signals for the annual mean variability. For long-term variability analysis, it is a common way to remove seasonal signal, and explore only on low frequency variability, i.e., inter-annual to decadal scales and long-term trend. Here, in addition to the common way for analyzing low frequency variability, we shall also investigate seasonal signal to know its spatial and temporal variability. Hereafter, we shall use the term "annual component" (AC) instead of seasonal signal, and "annual mean component" (AM) instead of inter-annual to decadal scales and long-term signal.

The AC as well as AM is spatially interrelated with specific spatial patterns. We shall firstly separate AC and AM from the monthly SST at each grid point. Then, the pattern recognition is used for AC to identify the normalized annual pattern and the time-dependent annual range (AR). The normalized annual pattern is fitted with annual and semi-annual sinusoidal functions to get their amplitudes and time lags. Both AR and AM are further analyzed with

Empirical Orthogonal Function (EOF) methods (Emery & Thomson, 2001) to explore their spatial and temporal variability.

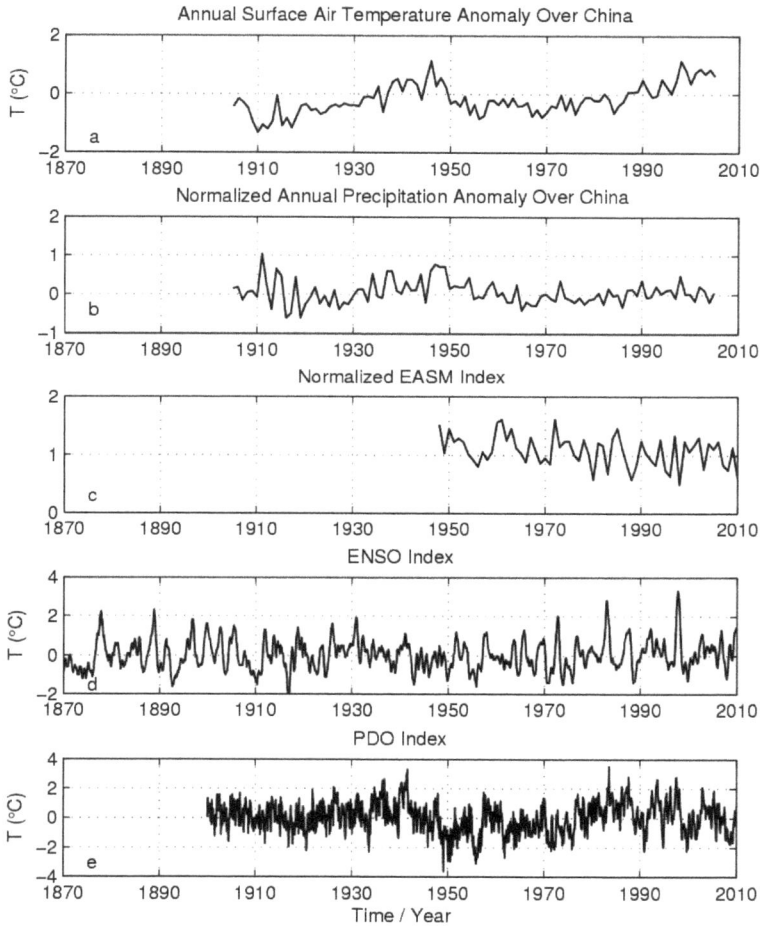

Fig. 2. Time series used to represent regional and global climate. a) Annual surface air temperature anomaly over China (TAC). b) Normalized annual precipitation anomaly over China (PAC). c) EASM Index. d) ENSO index. e) PDO index.

2.2.1 Partition SST into AC and AM

The SST at each grid point is partitioned into AC and AM by applying twice a 12 points moving average on the SST, namely, AM is obtained by moving average and AC is derived by subtracting AM from the SST. For instance, the SST at a specific location (126.5°E, 33.5°N), and the partitioned AC and AM are shown in Fig. 3. The much larger AR of AC for 17°C than the range of AM for 3°Cmeans that seasonal variability is much stronger than low frequency variability on inter-annual to decadal scales.

2.2.2 Harmonic analysis of AC

Fig. 3b shows that AC neither is a sinusoidal function, nor has time independent amplitude. The great similarity of AC suggests that AC at each grid point can be expressed with an annual pattern, especially with a normalized annual pattern, multiply by a time varying AR.

$$SST_{AC} = \tfrac{1}{2}\, AR^*T_{Norm} \tag{1}$$

Where SST_{AC} is AC of SST at a grid point, T_{Norm} is annual pattern of AC normalized by half AR. T_{Norm} is further fitted with annual and semi-annual sinusoidal harmonic functions, which are expressed by their amplitudes and time lags as follows.

$$T_{Norm} = h_1 \cos[\tfrac{2\pi}{12}(t\text{-}t_1)] + h_2 \cos[\tfrac{4\pi}{12}(t\text{-}t_2)] + \Delta T \tag{2}$$

Where, t is time in month and increases from 1 to 12 for January to December. h_1 and h_2 are amplitudes of the annual and semi-annual harmonic functions. t_1 and t_2 are time lags in month of the annual and semi-annual harmonic functions.

Fig. 4 shows the AC of SST at a specific location (126.5°E, 33.5°N) and the validation of above expressions (1) and (2). Fig. 4a and Fig. 4c demonstrate that AC has similar annual pattern with maximum and minimum SST in August and February. The annual maximum and minimum SSTs vary very significantly from year to year, and their temporal variability is reflected by AR (Fig. 4b). The much reduced dispersion and deviation of normalized annual pattern in Fig. 4d mean that normalized annual pattern is a better representation than the un-normalized annual pattern for pattern recognition. The derived annual and semi-annual harmonic functions, namely c_1 and c_2, are shown in Fig. 4e. The small difference (ΔT) between the SST pattern and sum of annual and semi-annual harmonic functions ($c_1 + c_2$) suggests that annual pattern can well be described by equation (2).

2.2.3 Analysis of AR and AM

The time varying AR and AM are further analyzed with EOF method. Firstly, both AR and AM are partitioned into their time independent mean and time dependent anomaly expressed statistically by STD (Fig. 6, Fig. 8). Then, their anomalies are analyzed with EOF method, by decomposing their spatial-temporal anomaly to coherent spatial modes and corresponding temporal modes (Fig. 7, Fig. 9). The larger variance explained by the first two leading EOF modes in general, and the dominant contribution of the first mode in particular, means that the spatial and temporal variability of AR and AM can well be described by their first EOF mode. Conventionally, the spatial mode is normalized to be a unit vector, and the magnitude of variability is expressed in the corresponding temporal mode. This expression is more mathematical than physical, as it is hard to extract the actual variability at a specific time and place by combine the spatial and temporal value. In present chapter, we shall normalize the temporal mode by its maximum, and the magnitude of variability is reflected by the corresponding spatial mode. And the spatial mode is actually the maximum variability occurred when the temporal mode equals one. In this way, we can easily estimate the actual variability at a specific time and place, just by multiply the spatial mode values at that place by the temporal mode value at that time.

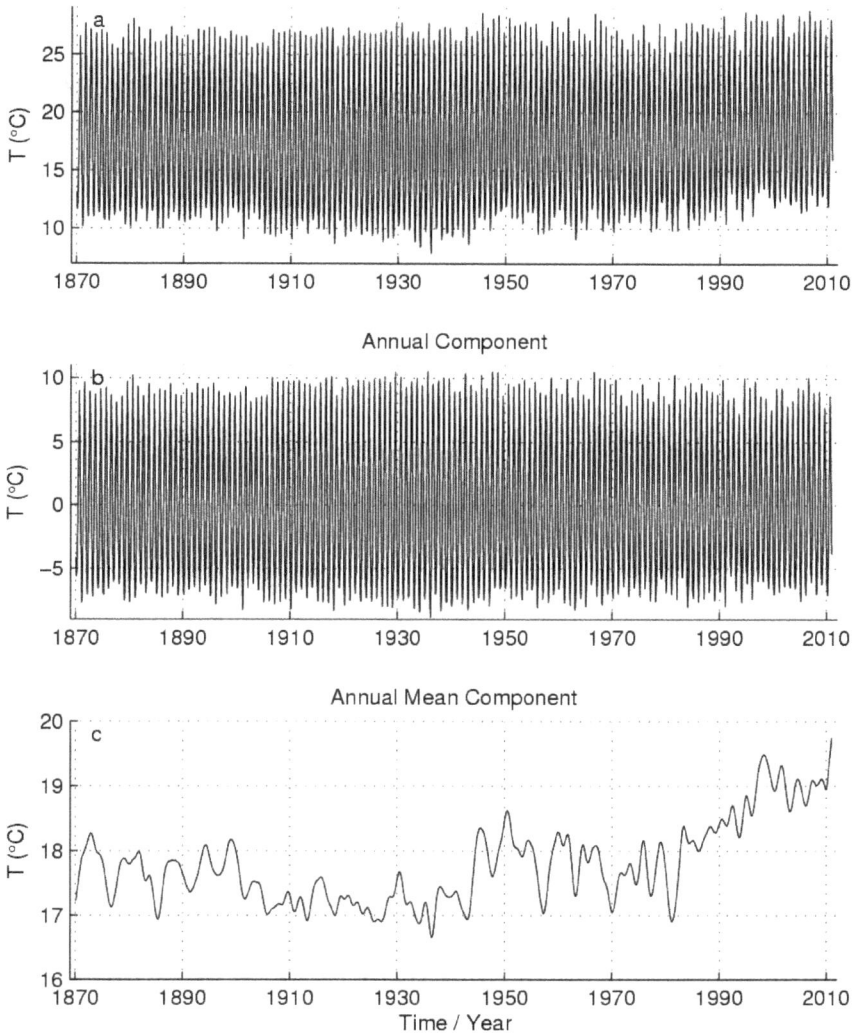

Fig. 3. Partition of SST into AC and AM. a) SST at specific location (126.5°E, 33.5°N) as indicated by a bold circle in Fig. 1. b) Decomposed AC with zero annual mean. c) Derived low frequency de-annual component, AM.

2.2.4 Analysis of regime shifts

The first temporal EOF mode of AM is further analyzed with Empirical Mode Decomposition (EMD) methods (Huang et al., 1998; Huang et al., 2003; Huang et al., 2008). Regime shifts are identified by the decomposed components (Fig. 10). The increase of SST over the latest regime, namely from 1977 to 2010, is fitted with linear regression to obtain the degree of warming of SST in the YES (Fig. 11).

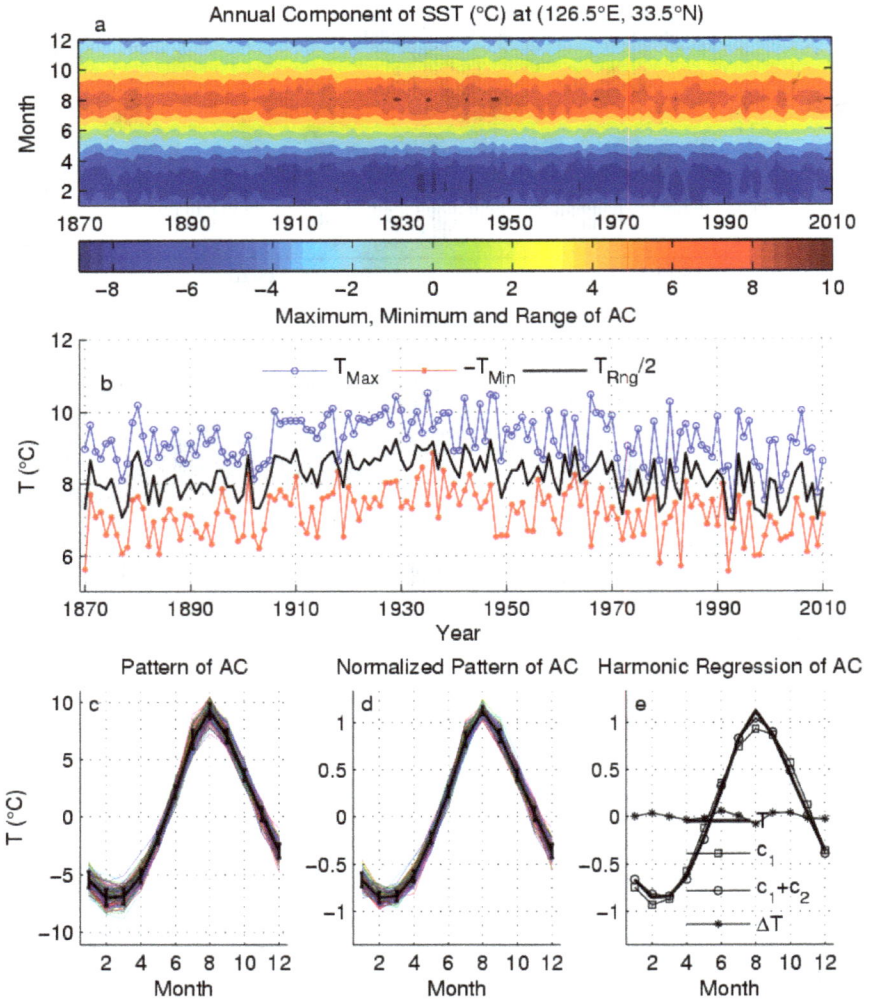

Fig. 4. AC of SST at a specific location (126.5°E, 33.5°N). a) Its contour map. b) The maximum, minimum and half range of AC. c) Annual pattern for 141 years. d) Annual pattern normalized by half AR. e) Harmonic regression of mean normalized annual pattern with annual and semi-annual sinusoidal functions, c_1 and c_2. ΔT is the difference between T and their sum c_1+c_2.

3. Results

We shall present the obtained results from high frequency to low frequency. The spatial and temporal variability of AC will be presented first in terms of normalized annual pattern and AR, followed by AM. The regime shift of AM is presented in conjunction with AR. Finally, we present the correlation between AR and AM with regional and global climate.

3.1 Normalized annual pattern

The fitting of the mean normalized annual pattern with annual and semi-annual sinusoidal functions is validated by root mean squared error (RMSE), which is shown in Fig. 5f. The RMSE is less than 0.04°C in general, comparing with the range of mean normalized annual pattern of 2°C, which means that the mean normalized annual pattern can well be represented by annual and semi-annual sinusoidal functions.

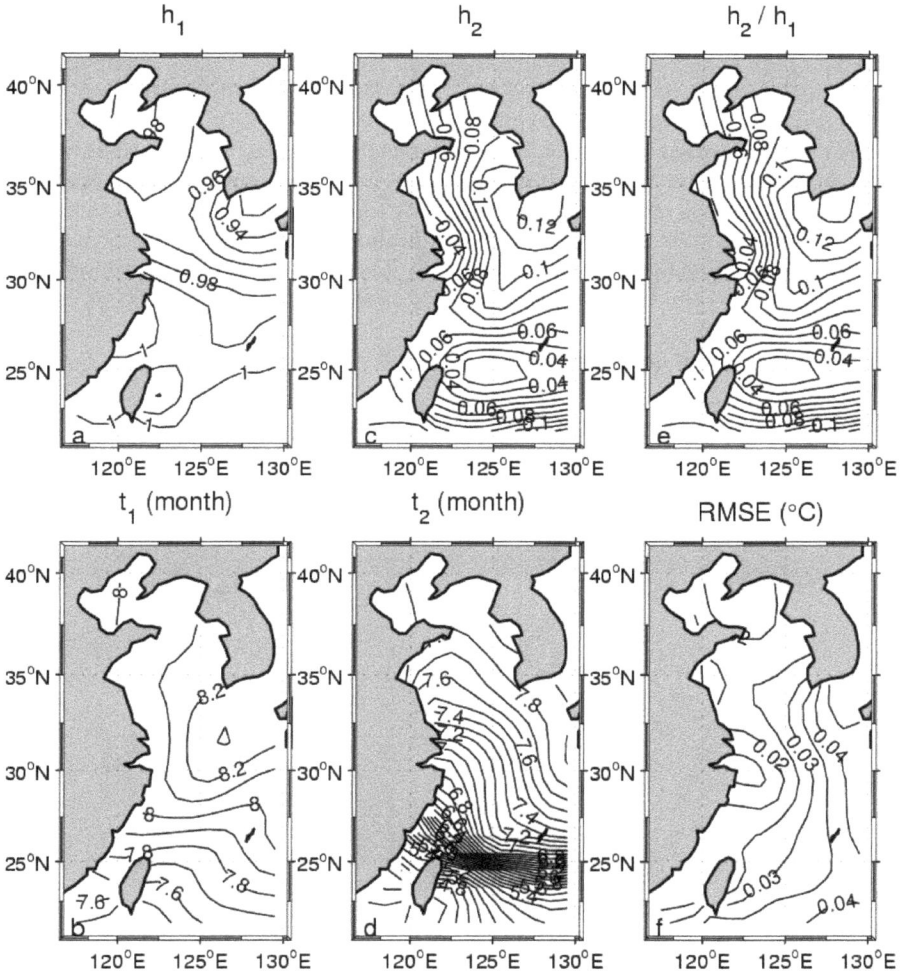

Fig. 5. Harmonic regression of the mean normalized annual pattern with sinusoidal functions. a) h_1, amplitude of annual sinusoidal component. b) t_1, time lag of annual sinusoidal component. c) h_2, amplitude of semi-annual sinusoidal component. d) t_2, time lag of semi-annual sinusoidal component. e) h_1/h_2, ratio of the amplitude of semi-annual to annual components. f) RMSE of harmonic regression of the mean normalized annual pattern with annual and semi-annual sinusoidal functions.

The annual and semi-annual sinusoidal functions are described simply by their harmonic constants, i.e., amplitudes and time lags. The large and close to 1 amplitude ($h_1 > 0.94°C$, Fig. 5a) of annual sinusoidal function means that normalized annual pattern is over dominant by pure annual cycle. The small amplitude of semi-annual sinusoidal function ($h_2 < 0.12°C$, Fig. 5c) means that annual cycle is modified by semi-annual fluctuation, which is most significant to the south of Korean Peninsula. The small ratio of h_2/h_1 (Fig. 5e), which has a very similar spatial pattern to h_2 due to nearly one value of h_1, confirms that annual cycle is much more important than semi-annual cycle.

The time lag of the annual cycle, t_1, is shown in Fig. 5b. From Fig. 5b, the maximum annual SST occurs in later July in the south of YES, and delays gradually northward by half a month to early August in the central YES. In the north part of YES, the maximum annual SST occurs at the beginning of August. Fig. 5d shows that time lag of the semi-annual cycle, t_2, which means that the first maximum semi-annual SST occurs in later April in the southern YES, and delays rapidly north-eastward by three months to later July in the central YES. The semi-annual cycle plays a primary modification on annual cycle, particularly in the area south to Korean Peninsula, where h_2 is largest. The effects of this modification lead to a warmer SST in August and February, which can be clearly identified from Fig. 4e.

3.2 Spatial and temporal variability of AR

Fig. 6a and 6b show the mean and STD of AR. The mean AR increases from $6°C$ in south to $24°C$ in the Bohai Sea. The lower AR in the southern YES means that SST has less annual variability, and the higher AR in the northern YES means that SST has larger annual

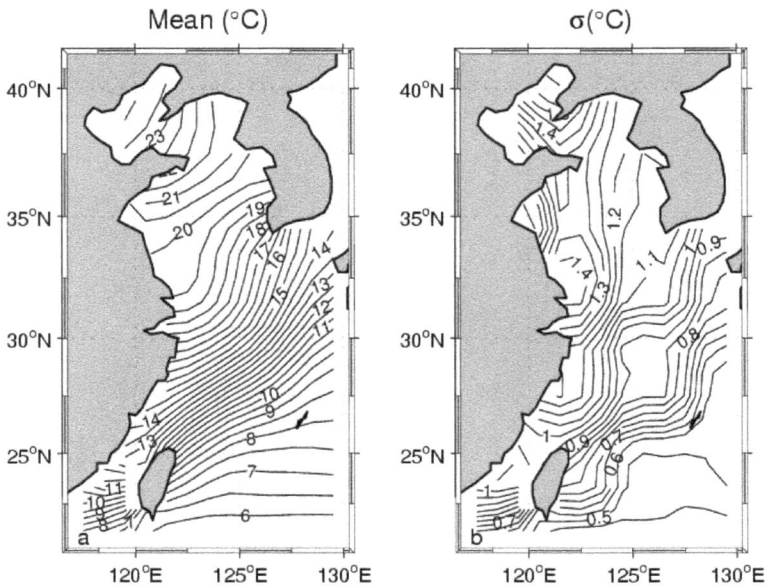

Fig. 6. Mean and STD of AR. a) mean. b) STD.

variability. There is a band of rapid change of mean AR in the continental margin of YES, where, the mean AR increases from 9℃ in the southeast YES to 19℃ in the central Yellow Sea. This band marks a transition zone between oceanic to continental dominant climate.

The STD of AR shows a similar spatial tendency as its mean, i.e., STD increases from 0.5℃ in south to 1.6℃ in the Bohai Sea. There is a band of relatively larger STD oriented primarily in south-north direction in the western YES. Corresponding to the band of rapid change of mean AR, there is also a zone of rapid change of STD, where STD increases from 0.6℃ in the southeast of YES to 1.2℃ in the central Yellow Sea.

The AR as well as its variability increases from south to north in the YES. Both smaller mean and STD of AR in the south means that SST is much more stationary in the southern YES. While, in the northern Bohai Sea, AR and its variability are very larger, show significant annual variation of SST.

The spatial and temporal variability of AR, as expressed by its variance STD in Fig. 6b, is further investigated with EOF method. The first two leading EOF modes explain 84% of total variance (Fig. 7), and the first mode contributes 69% in particular, mean that the spatial and temporal variability of AR can well be described by the first EOF mode.

The first EOF spatial mode shows a spatially coherent in–phase pattern with its amplitude of less than 1℃ in the south increases to greater than 3℃ in the western YES (Fig. 7a). This spatial pattern is very similar to the spatial pattern of STD, as supported by large contribution of 69%. The corresponding temporal mode shows a very distinct inter-annual to decadal variability (Fig. 7c). The larger positive values of about 0.9 in the temporal mode during 1940s mean that AR is much larger during 1940s, which is about 1℃ larger than mean AR in the southern YES and increases by 3℃ in the western YES (Fig. 7c). AR from 1990 to present is reduced about 0.3℃ in the south to 1℃ in the western YES as indicated by larger negative values of about 0.3 in the temporal mode.

3.3 Spatial and temporal variability of AM

Fig. 8a and 8b show the mean and STD of AM. The mean AM decreases from 26℃ in the south to 13℃ in the north of YES. There is a band of rapid change of mean AM in the continental margin of YES, where, the mean AM decreases from 24℃ in the southeast to 17℃ in the central Yellow Sea. This band coincides with that of AR confirms the transition zone between oceanic to continental dominant climate.

The STD of AM shows a relatively uniform spatial pattern with 0.6°C over the entire YES, except in the southeast where STD is much reduced due to the oceanic effects and relatively larger STD in the central YES. There is also a zone of rapid change of STD in the southeast YES, but it is shifted from continental margin to Okinawa trench.

The mean AM decreases from south to north, while its variability is almost same in the entire YES. Larger mean AM and small STD in the south means that climatologically mean SST is high and stationary in the southern YES.

The spatial and temporal variability of AM, as expressed by its variance STD shown in Fig. 8b, is further investigated with EOF method. The first two leading EOF modes explain 94% of total variance (Fig. 9), and the first mode contributes 85% in particular, mean that the spatial and temporal variability can well be described by the first EOF mode.

The first EOF spatial mode shows a spatially coherent in-phase pattern with its amplitude of less than 1.2℃ in south increases to greater than 2.0℃ in the central and south-western YES (Fig. 9a). This spatial pattern is very similar to the spatial pattern of STD, as supported by large contribution of 85% to total variance. The corresponding temporal mode shows very distinct inter-annual to decadal variability (Fig. 9c). The larger positive values of about 0.5 in the temporal mode during the last decade mean that AM is much warmer. Particularly in 1998, the AM is warmer than its mean AM about 1.2℃ in south increases to greater than 2.0℃ in the central and south-western YES. Before 1940s, the AM is generally cold than usual, especially in 1920s.

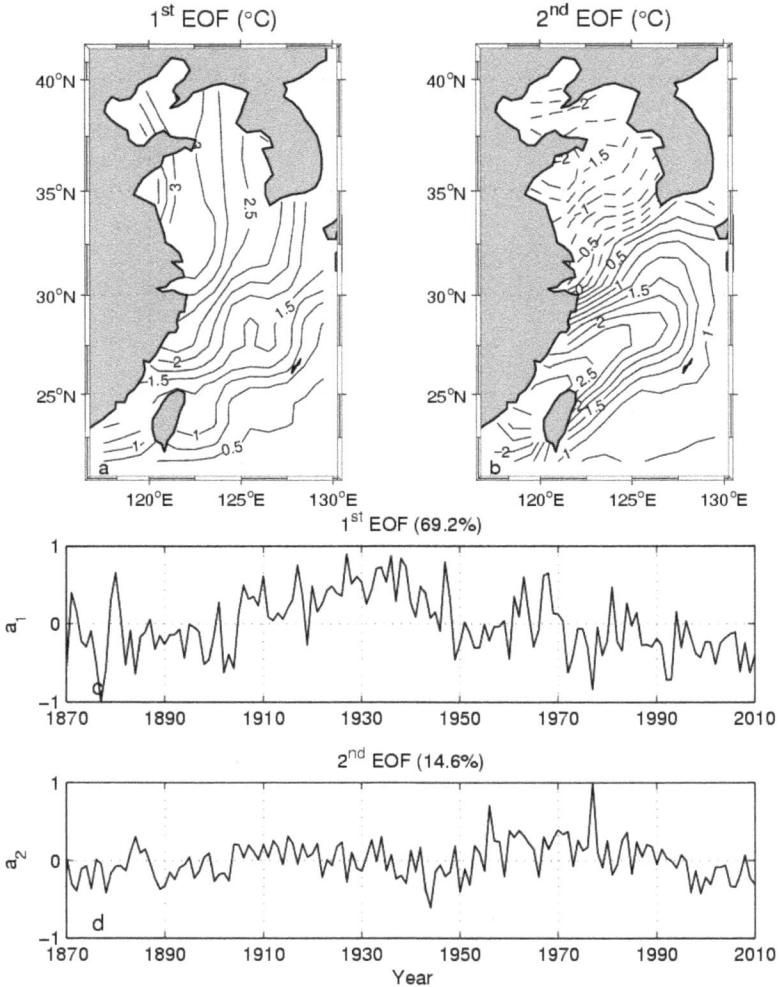

Fig. 7. The spatial and temporal EOF modes of AR. a) The first spatial mode. b) The second spatial mode. c) The first temporal mode. d) The second temporal mode.

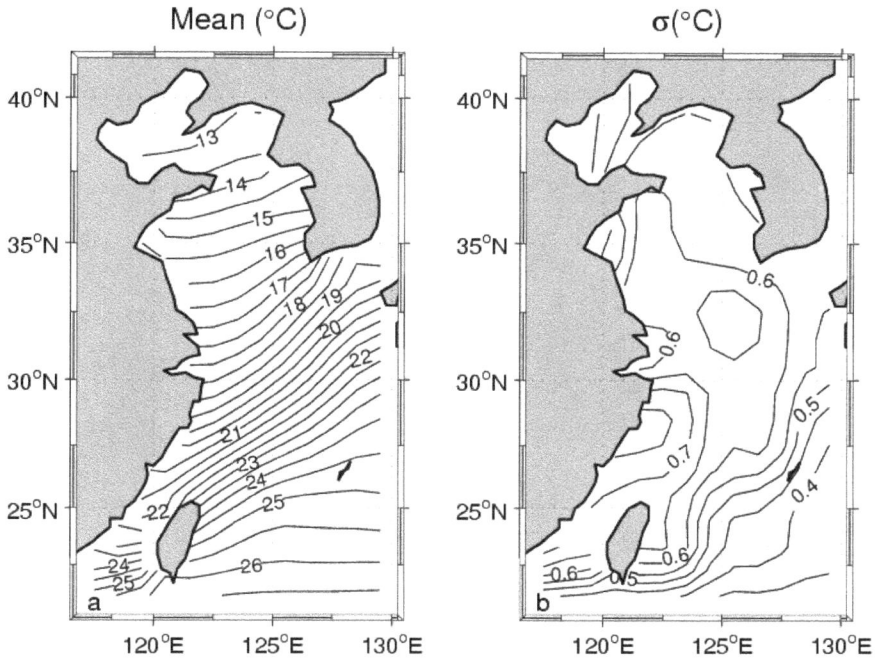

Fig. 8. Mean and STD of AM. a) mean. b) STD.

3.4 Regime shift of AM and AR

Based on EMD analysis of AM and the northern hemisphere air temperature of Asia
(Jones & Moberg, 2003), the variability SST in the YES over the last 141 years is classified
into four regimes (Fig. 10). Namely, Cold regime with a cooling trend (CC) from 1870 to
1900, the mean SST is a slightly cold than usual and AR reduced by 0.5 STD units. Cold
regime with a warming trend (CW) from 1901 to 1944, the mean SST is coldest and is
reduced by 1 STD unit than usual mean SST, and AR is largest and is increased by 1 STD
unit. The third regime is from 1945 to 1976, which is a warm regime with a cooling trend
(WC). The mean SST is slightly warmer than usual and AR is generally in normal. The
fourth regime is most obvious; it is a warm regime with a larger warming trend (WW)
from 1977 to present. During this warmest regime, AR is reduced about 0.7 STD units
with a decrease trend, which means that SST in the YES in getting warmer than ever and
with a much reduced annual range. Consequently, the winter SST increase in the YES is
significantly amplified than other seasons.

AR and AM are significantly negative correlated as show in Fig. 10c. There are three
distinct peaks in the correlation coefficient, with 0, 4 and 10 years time lag of AR with
respect to AM. AR lags AM mean that the variability of AM might affect the variability of
AR for the lagged time interval. The zero time lag means that, in the year when AM is
higher (lower), the corresponding AR is lower (higher) and will have a much warmer
(colder) winter SST than usual in the YES.

The warming trend is further explored with its AM by a linear regression at each grid points from 1977 to 2010. The increment of AM from 1977 to 2010 is shown in Fig. 11. The AM has increased from 0.5℃ to the east of Taiwan to more than 1.6℃ in the central YES, especially to 2℃ to the south of Korean Peninsula. The increment of AM is much larger in the mid-shelf than near shelf and Kuroshio regions.

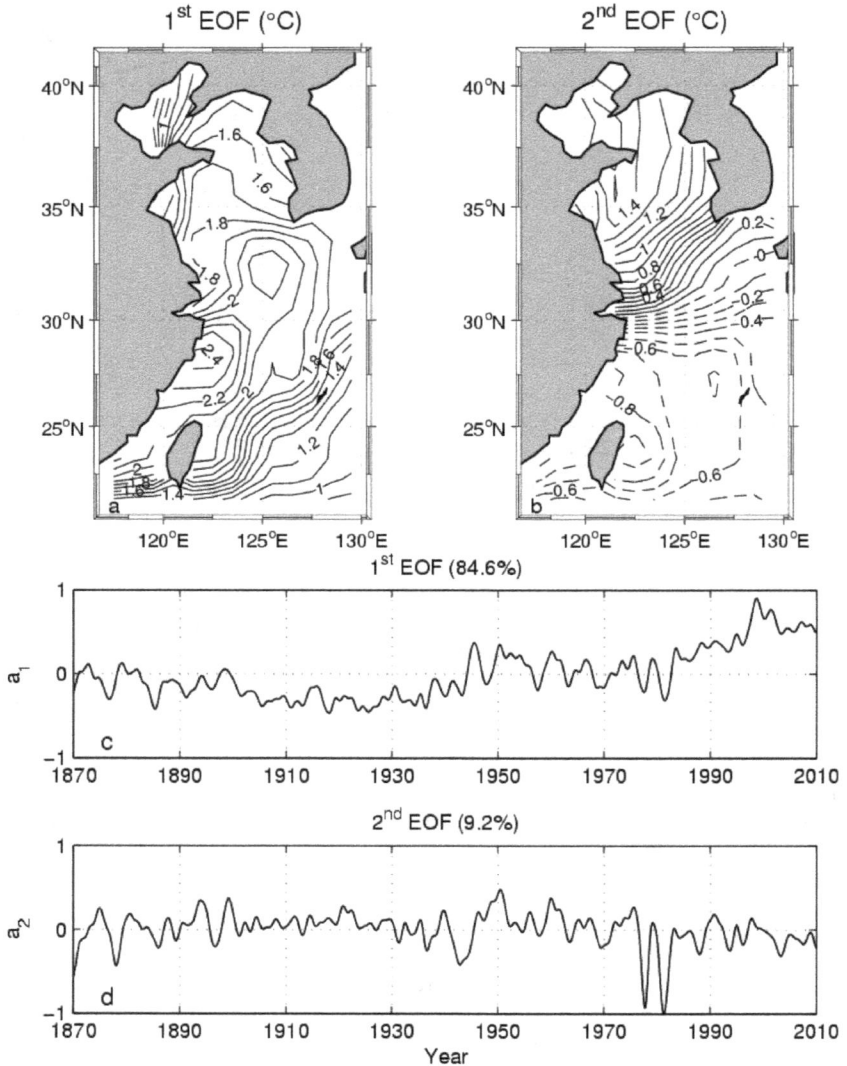

Fig. 9. The spatial and temporal EOF modes of AM. a) The first spatial mode. b) The second spatial mode. c) The first temporal mode. d) The second temporal mode.

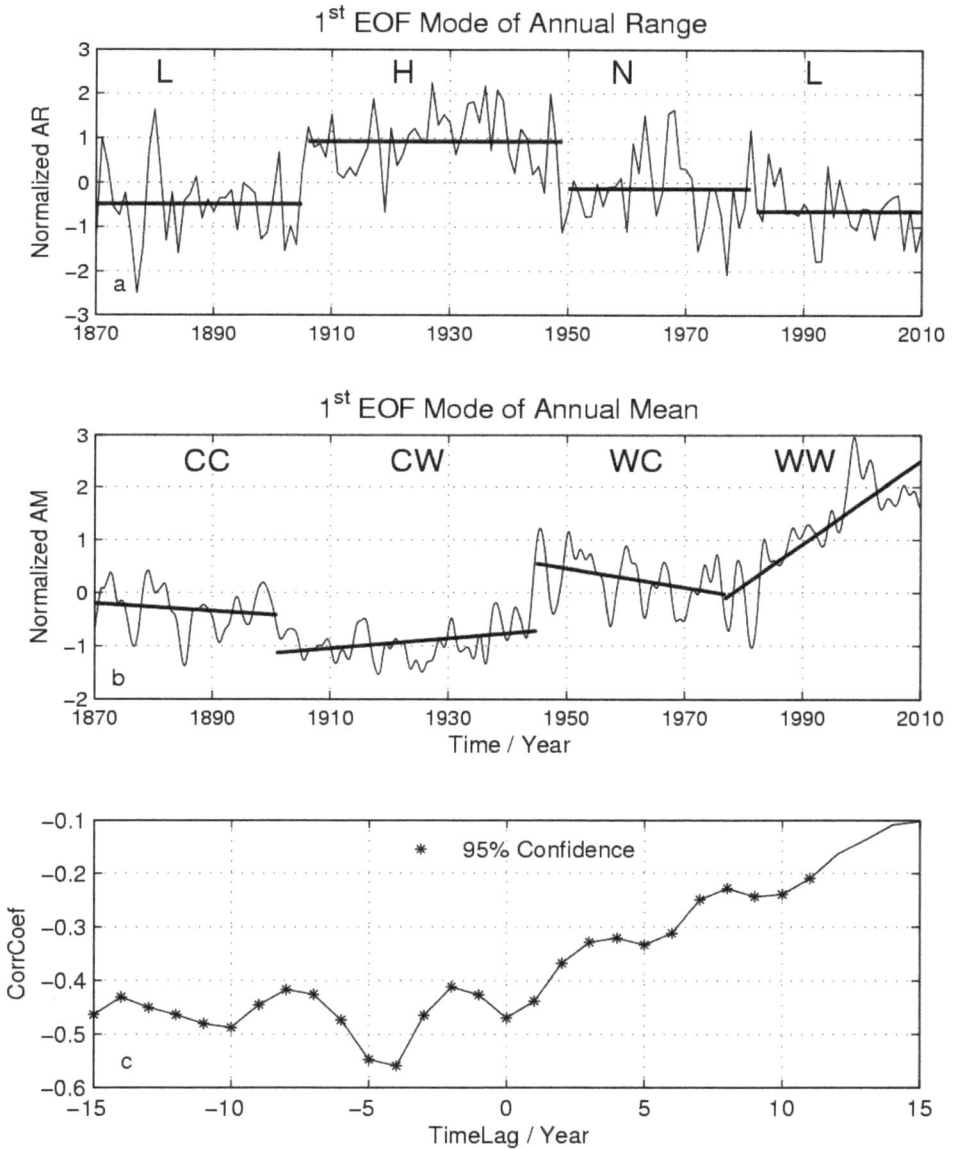

Fig. 10. The regime shifts of SST in the YES. a) The four regimes of in the first EOF mode of
AR, L, H and N stand for the low, high and normal variability of AR. b) The four regimes of
the first EOF mode of AM, CC, CW, WC and WW stand for the cold regime with a cold
trend, cold regime with warm trend, warm regime with cold trend and warm regime with
warm trend. c) Correlation coefficient between AR and AM, negative (positive) time lag
means that AR lags (leads) AM.

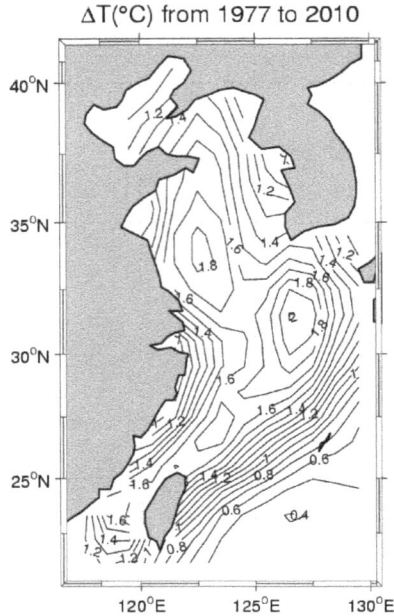

Fig. 11. The increment of AM in the YES over the latest warm regime from 1977 to 2010.

3.5 Relationship between AR and regional and global climate

The variability of AR and AM are closely related to regional and global climate. Their relationships are explored with the correlation between the first EOF temporal mode of AR and AM with TAC, PAC and EASM index which represent regional climate, and with ENSO and PDO indexes which represent global climate.

The time lagged correlation coefficients between the first EOF temporal mode of AR with TAC, PAC, EASM index, ENSO index, and PDO index are shown in the left column of Fig. 12 from top to bottom.

AR is significantly correlated with regional climate, as shown in Figs. 12a to 12c. The peak in Fig. 12a shows that AR and TAC are significantly (95% confidence interval, same hereafter) negative correlated with correlation coefficient of -0.5 for a 10a time ahead for AR against TAC. This correlation coefficient suggests that AR is significantly related with the regional land air temperature anomaly, larger (smaller) AR corresponds to negative (positive) TAC, and the larger AR is generally related to the colder surface air temperature over China 10 years later.

AR and PAC are significantly positive correlated with correlation coefficient of 0.35 for 12a ahead for PAC against AR. AR and EASM index are significantly positive correlated with correlation coefficient of 0.34 for zero time lag, 0.32 for 5a time ahead, 0.33 for 15a time ahead, and 0.32 for 23a time ahead for EASM against AR. As shown in Fig. 12c, strong EASM is generally related to a higher AR of SST in the YES for the corresponding year, 5a, 15a and 23a later.

Spatial and Temporal Variability of Sea Surface Temperature in the Yellow Sea and East China
Sea over the Past 141 Years

229

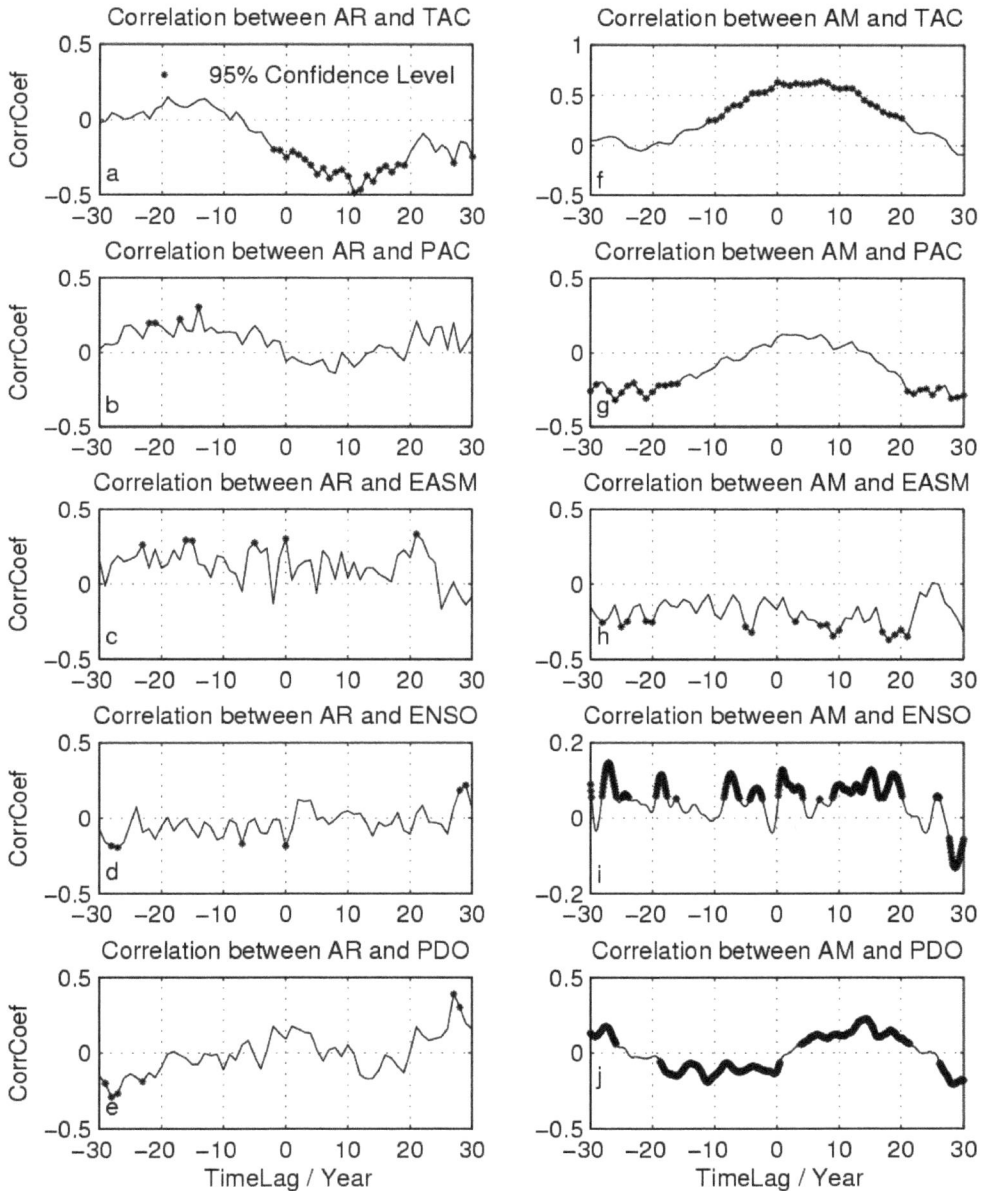

Fig. 12. Relationship between AR and AM with TAC, PAC, and EASM, ENSO and PDO indexes from left to right and top to bottom. Star on line indicates that the corresponding correlation is significant on 95% confidence interval.

AR is also significantly related to the global climate, as shown in Fig. 12d and 12e. AR and ENSO index are significantly negative correlated with correlation coefficient of -0.19 for zero time lags, -0.18 for 7a time ahead, and -0.20 for 27a time ahead for ENSO against AR. These correlation coefficients suggest that AR is significant but with small negative correlation with the global tropical climate. El Nino year is generally related to a lower AR in the YES for the corresponding year, 7a and 27a later. There is also a significant positive correlation with a coefficient of 0.25 for a 29a time ahead for AR against ENSO.

AR and PDO index are most significantly negative correlated with correlation coefficient of -0.30 for 28a time ahead. The correlation coefficient suggests that AR has a significant correlation with the North Pacific climate with a time of 27a later. Positive (negative) phase of PDO is related to a smaller (larger) AR of SST in the YES 27a later. There is also a very significant positive correlation with a coefficient of 0.41 for a 27a time ahead for AR against PDO.

3.6 Relationship between AM with regional and global climate

The time lagged correlation coefficients between the first EOF temporal mode of AM with TAC, PAC, EASM index, ENSO index, and PDO index are shown in the right column of Fig. 12 from top to bottom.

AR is significantly correlated with regional climate, as shown in Figs. 12f to 12h. The large and board peak in Fig. 12f shows that AM and TAC are significantly positive correlated with correlation coefficient of 0.6 for a few years time ahead for AM against TAC. This correlation coefficient suggests that AM is significantly related with the regional land air temperature anomaly, higher (lower) AM corresponds to positive (negative) TAC, and the higher AM is generally related to warm surface air temperature over China from corresponding year to a few years later.

AM and PAC are significantly negative correlation at with correlation coefficient of -0.3 for 21a and 26a ahead for PAC against AM, and 28a ahead for AM against PAC (Fig. 12g). Higher (lower) AM is related with less (more) precipitation on bi-decadal to tri-decadal time lags.

AM and EASM index are significantly negative correlated with correlation coefficient of about -0.35 for 5a time ahead and 9a and 18a time lag of EASM against AM (Fig. 12h). The correlation coefficient suggests that strong (weak) EASM is related to a lower (higher) AM in the YES for 5a later, and lower (higher) AM is related to a strong (weak) EASM 9a and 18a later.

AM and ENSO index are significantly positive correlated with small correlation coefficients of about 0.15 for 3a, 7a, 19a, and 27a time ahead and 1a, 16a and 19a time lag for ENSO against AM (Fig. 12i). These correlation coefficients suggest that El Nino year is generally related to a higher AM in the YES for 3a, 7a, 19a, or 27a later. There is also a significant negative correlation with a coefficient of 0.15 for a 29a time ahead for AM against ENSO.

AM and PDO index are significantly negative correlated with correlation coefficient of -0.20 for 10a time ahead and positively correlated with a correlation coefficient of 0.18 for 27a time ahead for PDO against AM (Fig. 12j). The correlation coefficient suggests positive (negative) phase of PDO is related to a lower (higher) AM in the YES 10a later and to higher (lower) AM in the YES 27a later. There are also significant correlation for AM ahead against PDO with a positive correlation coefficient of 0.25 and a 14a time lead, and a negative correlation coefficient of 0.25 and a 28a time lead.

4. Conclusions and discussions

The SST in the YES over the last 141 years (1870-2010) is partitioned into an AC (seasonal signal with zero mean) and an AM (inter-annual, decadal and long-term trend). The spatial and temporal variability of the AC and AM and their relationship are analyzed with pattern fitting and EOF method. The possible linkage between the identified variability with the known regional and global climate (i.e., TAC, PAC, EASM, ENSO and PDO) is also explored.

AC is represented by a mean normalized annual pattern and a time-varying AR. The mean normalized annual pattern fits well with annual and semi-annual sinusoidal signals with less than 0.04°C RMSE. The annual sinusoidal signal dominates semi-annual signal by contributing greater than 94% to AR in general; the semi-annual signal contributes to more than 10% to AR in the eastern Yellow Sea, particularly to the area south of Korean Peninsula. The annual cycle of SST reaches its highest SST from mid July in the south to early August in the north and central of YES. The mean AR increases from 6°C in the south and east of Taiwan to 24°C in the northern Bohai Sea. The STD of AR is about 0.5°C to the east of Taiwan, and increases from both sides of south and north to the western YES to a value of 1.4°C. This variance is mostly explained by the first EOF mode (69%), which has a coherent in-phase spatial pattern with maximum amplitude in the western YES.

The AM has a mean SST of 26°C in the south of YES, which decreases northward, reaches a minimum mean SST of 13°C in the northern Bohai Sea. The STD of AM has a relatively uniform spatial pattern with maximum variance of 0.7°C in the central to south-western YES. This variance is explained mostly by the first EOF mode (85%), which has a coherent in-phase spatial pattern with its maximum amplitude in the central to south-western YES.

Both AR and AM vary on inter-annual to decadal time scales, and they are significantly negative correlated with a zero and 5a time lag for AR against AM. This correlation suggest a higher (lower) AM is associated with a smaller (larger) AR for the corresponding year and 5a later. Therefore, in the years with higher (lower) AM, often have smaller (larger) AR, and consequently, experience much warmer (colder) than usual winter SST. The variability of winter SST is most significant in four seasons, and the summer SST is the least variable.

Over the last 141 years, the AM in the YES has experienced four regimes, namely, CC from 1970 to 1900, CW from 1901 to 1944, WC from 1945 to 1976, and WW from 1977 to 2010. Corresponding to these four regimes, the AC experienced a smaller, higher, normal and the smallest AR, respectively.

The SST in the YES over the last WW regime (from 1977 to 2010) increases from 0.6°C in the southeast to 2.0°C in the central of YES. During this period, the AR is significantly reduced. The combination of the warming of AM and the reduction of AR leads to a much larger SST increase in winter, i.e., the warming in winter is much more significant than in other seasons in the last WW regime, and particularly in 1998.

Both the AR and AM of SST in the YES is related to the regional and global climate. TAC is positively associated with AM and negatively associated with AR. The warmer (colder) surface air temperature over China is associated with a warmer (colder) SST in the YES, particularly in winter, through the higher (lower) AM and smaller (larger) AR.

Since the precipitation over China is primarily controlled by monsoon. Both PAC and EASM index are positively correlated with AR, and negatively correlated with AM. Both inter-decadal time lagged and leaded significant correlations between AM and PAC suggest that AM and PAC might have some interaction on inter-decadal time scales. Both inter-annual to inter-decadal time lagged and leaded significant correlation between AM and EASM index suggest that AM and EASM might have some interaction on these time scales.

Both the small coefficient between AM and AR with ENSO and PDO indexes mean that AM and AR are definitely related to the variability of the large scale Tropic Ocean and North Pacific Ocean climate, but the contribution from the global climate to the variability of AM and AR in the YES is only about 15-20%. The relatively larger correlation coefficient of between AM and AR with TAC, PAC and EASM index mean that regional climate is more closely related to the variability of SST in the YES.

5. Acknowledgment

HadISST1 data is provided by the Met Office Hadley Centre, UK, dataset is downloaded from http://www.metoffice.gov.uk/hadobs/hadisst/data/download.html. EASM index is provided by Jianping Li at State Key Laboratory of Numerical Modeling for Atmospheric Sciences and Geophysical Fluid Dynamics, Institute of Atmospheric Physics, Chinese Academy of Sciences, and is downloaded from http://web.lasg.ac.cn/ staff/ ljp/ data-monsoon/ EASMI.htm. ENSO index is provided by JMA and is downloaded from http://coaps.fsu.edu/pub/JMA_SST_Index/. PDO index is provided by Nathan Mantua at Joint Institute for the Study of the Atmosphere and Ocean (JISAO), Washington University, dataset is downloaded from http://jisao.washington.edu/ pdo/PDO.latest. This research was supported by the National Basic Research Program of China under Grant No. 2006CB400603 and 2011CB409803, the Zhejiang Provincial Natural Science Foundation of China under Grant No. R504040, the Natural Science Foundation of China under Grant No. 41176021, and the China 908-Project under Grant No. 908-ZC-II-05 and 908-ZC-I-13.

6. References

Chen, D.; Liu, W.T.; Tang, W. & Wang, Z. (2003). Air-sea interaction at an oceanic front: Implications for frontogenesis and primary production, *Geophysical Research Letter*, Vol. 30, No. 14, 1745, doi:10.1029/2003GL017536

Ding, Y. & Ren, Y. (Eds). (2008). *Introduction to climate change in China*, China Meteorological Press, ISBN 978-7-5029-4364-6, Beijing. (In Chinese)

Emery, W.J. & Thomson, R.E. (Eds). (2001). *Data analysis methods in physical oceanography*, Second and revised edition, Elsevier Science B.V., ISBN 0-444-50757-4, Amsterdam, The Netherlands.

Hahn, S.D. (1994). SST warming of Korean coastal waters during 1881-1900, KODC Newsletter 24, pp. 29-37.

Huang, D.; Zhao, J. & Su, J. (2003). Practical implementation of Hilbert-Huang Transform algorithm. *Acta Oceanologica Sinica*, Vol. 22, No. 1, pp. 1-14.

Spatial and Temporal Variability of Sea Surface Temperature in the Yellow Sea and East China
Sea over the Past 141 Years
233

Huang, N.E.; Shen, Z.; Long, S.R.; Wu, M.C.; Shih, H.H.; Zheng, Q.; Yen, N.-C.; Tung, C.C. & Liu, H.H. (1998). The empirical mode decomposition and the Hilbert spectrum for nonlinear and non-stationary time series analysis. *Proceedings of Royal Society of London A*, Vol. 454, No. 1971, pp. 903-995.

Huang, N.E. & Wu, Z. (2008). A review on Hilbert-Huang transform: Method and its applications to geophysical studies. *Review of Geophysics*, Vol. 46, RG2006, doi:10.1029/2007RG000228.

Kang, I. & Jeong, Y. (1996). Association of interannual variations of temperature and precipitation in Seoul with principal modes of Pacific SST, *Journal of the Korean Meteorological Society*, Vol. 32, pp. 339–345

Jones, P.D. & Moberg, A. (2003). Hemispheric and Large-Scale Surface Air Temperature Variations: An Extensive Revision and an Update to 2001. *Journal of Climate*, Vol. 16, No. 2, pp. 206-223.

Li, J. & Zeng, Q. (2003). A new monsoon index and the geographical distribution of the global monsoons. *Advance in Atmospheric Sciences*, Vol. 20, No. 2, pp. 299-302.

Lin, C.; Su, J.; Xu, B. & Tang, Q. (2001). Long-term variations of temperature and salinity of the Bohai Sea and their influence on its ecosystem. *Progress in Oceanography*, Vol. 49, pp. 7–19.

Lin, C.; Ning, X.; Su, J.; Lin, Y. & Xu, B. (2005). Environmental changes and the responses of the ecosystems of the Yellow Sea during 1976–2000. *Journal of Marine Systems*, Vol. 55, pp. 223–234.

Mantua, N.J.; Hare, S.R.; Zhang, Y.; Wallace, J.M. & Francis, R.C. (1997). A Pacific interdecadal climate oscillation with impacts on salmon production. *Bulletin of the American Meteorological Society*, Vol. 78, No. 6, pp. 1069-1079

Meyers, S.D., O'Brien, J.J. & Thelin, E. (1999): Reconstruction of monthly SST in the Tropical Pacific Ocean during 1868-1993 using adaptive climate basis functions. *J. Climate*, Vol. 127, No. 7, pp. 1599-1612

Rayner, N.A.; Parker, D.E.; Horton, E.B.; Folland, C.K.; Alexander, L.V.; Rowell, D.P.; Kent, E.C. & Kaplan, A. (2003). Global analyses of sea surface temperature, sea ice, and night marine air temperature since the late nineteenth century. *Journal of Geophysical Research*, Vol. 108, No. D14, 4407, doi:10.1029/2002JD002670

Tang, G. & Ren, G. (2005). Reanalysis of surface air temperature change of the last 100 Years over china, *Climatic and Environmental Research*, Vol. 10, No. 4, pp. 791-798. (In Chinese)

Tao, S. & Zhang, Q. (1998). Response of the East Asian winter and summer monsoon to ENSO events, *Scientia Atmosphetica Sinica*, Vol. 22, pp. 399–407. (In Chinese)

Wang, B.; Wu, R. & Fu, X. (2000). Pacific–East Asian Teleconnection: How Does ENSO Affect East Asian Climate? *Journal of Climate*, Vol. 13, No. 9, pp. 1517–1536

Webster, P.J.; Magana, V.O.; Palmer, T. N.; Tomas, T. A.; Yanai, M. & Yasunari, T. (1998). Monsoons: Processes, predictability, and prospects for prediction, *Journal of Geophysical Research*, Vol. 103, No. C7, pp. 14451–14510

Xie, S.; Hafner, J.; Tanimoto, Y.; Liu, W.T.; Tokinaga, H. & Xu, H. (2002). Bathymetric effect on the winter sea surface temperature and climate of the Yellow and East China Seas, *Geophysical Research Letter*, Vol. 29, No. 24, 2228, doi:10.1029/2002GL015884

Zhang X.Z.; Qiu, Y.F. & Wu, X.Y. (2005). The Long-term change for sea surface temperature in the last 100 Years in the offshore sea of China. *Climatic and Environmental Research*, Vol. 10, No. 4, pp. 709-807. (In Chinese with English abstract)

Zhang, Y.; Wallace, J.M. & Battisti, D.S. (1997). ENSO-like interdecadal variability: 1900-93. *Journal of Climate*, Vol. 10, No. 5, pp. 1004-1020

Part 3

The Changing Climate

9

Cenozoic Climatic Record for Monsoonal Rainfall over the Indian Region

Mohan Kuppusamy and Prosenjit Ghosh
Centre for Earth Science, IISc, Bangalore
India

1. Introduction

Atmospheric carbon dioxide level is one of the major drivers responsible for the global temperature change (Lacis et al., 2010). The role of carbon dioxide as an important greenhouse gas, and its contribution towards regulation of global surface temperature has been recognized for over a century (Arrhenius, 1896; Chamberlin, 1899; Royer, 2006). The ice core records along with other proxy based records provides an evidence signifying a strong coupling between CO_2 and global temperature for at least the last ~65 m.y. (million years) (Petit et al., 1999; Siegenthaler et al., 2005, Zachos et al., 2001). The intensification of convective hydrological cycle inducing heavy rainfall during high pCO_2 condition is both simulated and estimated from General Circulation Models (GCM) and geochemical analyses of fossil record respectively (Kutzbach & Gallimore, 1989). The evidences of intensification of monsoon, which refer to the rainfall due to seasonal reversal of the wind direction along the shore of the Indian Ocean especially in the Arabian Sea and surrounding regions, are preserved in the sedimentary records from continental and oceanic region (Fig.1). The other factor which affected the regional hydrological cycle apart from the concentration of CO_2 in the atmosphere is tectonic rise of Himalayan mountain. Proxy record based on parameters like stomata index, alkenones and boron isotopes clearly suggested high concentration of CO_2 in the atmosphere (~400 ppm) during Miocene time. The estimated concentration of CO_2 observed in the atmosphere was rather similar to the concentration of CO_2 in the atmosphere measured in the recent years at Mauna Loa (Thoning et al., 1989). The effect of such high CO_2 concentration is seen to have significantly modulated and altered the pattern of rainfall distribution, intensity and its spatial variability. Record from sedimentary archives from the continental and marine sites over the Indian region yielded evidence of warmer, wetter and higher temperature seasonal climate for the Miocene period. A similarity of signature both from continental region and the marine archives support the argument for the change in hydrological condition during last 20 m.y. The marine records are only a few but the largely scattered along the continental margin and central Indian Ocean. A more recent study of such sedimentary sequences lying on the western and eastern India provided glimpses of spatial variability of regional climate. The chapter will narrate the long term variation in Miocene monsoonal rainfall and its spatial pattern using large set of available observations from the palaeo record. This narration will include

- Large synthesis of proxy based parameter from of geological, palentological, biogeochemical and isotopic proxies.

- Implication of upliftment rate of Himalaya and formation of Tibetan Plateau on monsoonal intensity.

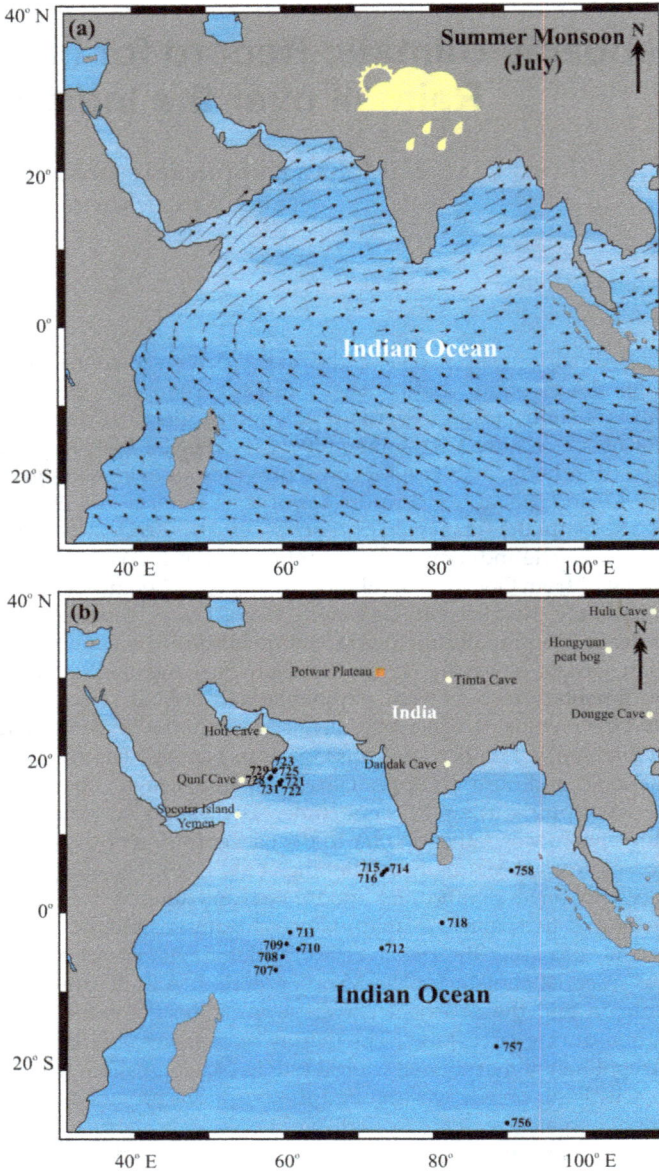

Fig. 1. (a) Climatology of the summer monsoon circulation (July) marked by surface winds (Figure after Wang et al., 2005) and (b) Indian Ocean along with Ocean Drilling Program (ODP) Holes and Cave deposits from south Asia.

The scenario is complex but possible to model with simultaneous observation on rate of upliftment of Tibetan Plateau and pCO_2 level both playing role in influencing the land-sea thermal contrast and strengthening of wind (Hahn & Manabe, 1975). The dual reason provided here will cause strong intensification of the Indian monsoons, with large increment in the summer precipitation (summer monsoon) over the Indian subcontinent and a cold dry winter season (Ruddiman & Kutzbach, 1989; Hastenrath, 1991). According to the tectonic upliftment hypothesis, monsoonal intensification due to changes in the elevation of the Himalayan-Tibetan region had greater role driving the climate in Indian region (Molnar et al., 1993; Clift et al., 2002).

2. Paleoclimate proxies

The evidence of the monsoon rainfall is registered in variety of proxies both over land and oceanic region, arguably the most reliable continuous record of the monsoonal variability comes from the sediments deposited on the ocean floor, from the Arabian Sea region. We cannot directly observe the climates of the geological past in the way that we do record climate parameters since the modern centuries with instrumental record (Plaut et al., 1995). The key variables measured are temperature and humidity. The indirect mean to retrieve such information from sedimentary archives is through analyses of physical and chemical parameter in the sedimentary deposits and these includes physical proxies like clay mineralogy, grain size distribution, heavy minerals, organic bio markers, chemical and isotopic proxies like isotopes in ice core, alkenones, sedimentary organic fraction, inorganic carbonates and organic deposits. Isotopic compositions provide direct relationship allowing understanding of temperature, atmospheric composition and seawater salinity respectively. Along with this multi parameter approach, the importance of distribution of microfossil assemblages and abundances of particular species from marine platform provide direct measure on the response of physical and chemical factors on the biota.

2.1 Paleontological proxies

The empirical use of fossils and fossil assemblages as palaeoclimate indicators can be traced back at least as far as Lyell (1830) and their value was even recognised in classical biostratigraphic literatures (Imbrie & Newell, 1964). The oceanic sedimentary records have been successfully used for reconstruction of monsoon intensity through time (Kroon et al., 1991; Nigrini, 1991; Prell et al., 1992; Prell & Kutzbach, 1992). The biological, chemical and sedimentological parameters indentified as important for understanding the intensity of upwelling phenomena having direct linkage with the strength of the monsoonal winds. Monsoonal variability can therefore be constrained in terms of wind strength through time. The intense seasonal upwelling is induced by the south-westerly monsoon winds from Arabian Sea (Anderson & Prell, 1993). Sediments in the Arabian Sea exhibit a characteristic fauna (foraminiferal species) that are endemic to areas of upwelling. Majority of these species are encountered only in cool temperate water, and therefore, their appearance and abundance in sediment record indicate parameter for upwelling (index). Cullen and Prell (1984) provided the high-quality core-top data banks to indentify, establish and evaluate proxies for monsoonal upwelling. The northern Indian Ocean based on the analysis of 251 core-top samples shows the distribution of planktic foraminiferal dwelling on the surface water and provide tool for paleoceanographic reconstructions. Together with isotopic ratios and trace element concentration, the assemblage of foraminiferal species capture changes in surface water hydrographic condition (e.g. Wefer et al., 1999; Thunell & Sautter, 1992; Kennett & Srinivasan, 1983; Kennett, et al., 1985). The oxygen and carbon isotope composition in shells of planktic foraminifera ($\delta^{18}O_{shell}$, $\delta^{13}C_{shell}$) provides one of the most widely used tools for reconstructing past changes in sea surface temperature and salinity (King & Howard, 2005). Also, high concentration of certain species like *Globigerina bulloides* mirror upwelling conditions in the tropical ocean (Prell et al., 1992; Overpeck et al., 1996; Peeters et al., 2002; Gupta et al., 2003). The $\delta^{18}O$ of planktic foraminiferal calcite records enable reconstruction of global surface temperatures, however suffers from effects like depth habitat reorganization due to climatic evolution (MacLeod et al., 2005). *Neogloboquadrina dutertri* is a significant component of the planktonic foraminiferal group in the Arabian Sea upwelling area (Be & Hutson, 1977; Kroon, 1988) associated with upwelling phenomena together with *G. bulloides* (Auras et al., 1989). The Globigerinoides species (*Globigerinoides sacculifer, Globigerinoides ruber, etc.,*) have occurrence subtropical- tropical surface waters with an oligotrophic character (Be & Hutson, 1977). Intervals with higher frequencies of Globigerinoides are characterized by surface waters condition different from the upwelling situations (Kroon et al., 1991). Paleotemperature reconstruction using benthic foramifera from different sites (Pacific, Atlantic and Indian Ocean) was achieved measuring parameter like $\delta^{18}O$ reflecting either temperature or ice volume effect (Zachos et al., 2001). Benthic foraminifera have ability to adapt changing environmental conditions, enabling them to survive in a wide range of marine environments. Their calcium carbonate shells represent a major and globally significant sink for carbon. Also their calcareous or agglutinated tests lend them good fossilization potential. Thus benthic foraminifera provide important tool for reconstruction of deep-sea paleoceanography and paleoclimatology, based on observations from modern day environmental niche (Gupta & Srinivasan, 1992; Sen Gupta & Machain-Castillo, 1993; Wells et al., 1994; Thomas et al., 1995; Gupta & Thomas, 1999, 2003).

2.2 Clay mineralogy

Composition of clay minerals in marine and terrestrial record reflects weathering processes in the catchment area and basin, adjoining depositional setup (Chamley, 1989; Sellwood et al., 1993; Chamley, 1998; Thiry, 2000). However, for example, the study conducted on the marine sedimentary record from southern latitudes documenting trend in the abundances of smectite, chlorite, and illite. A declining proportion of crystallized smectite and chlorite, and increasing illite in sediments was used to trace the transition from humid to sub-polar and polar conditions (Ehrmann et al., 2005). Ballantyne (1994) and Ballantyne et al. (2006) suggested that low gibbsite concentration in soils deposited in glacial terrain, which can also be used for dating the deposit involving cosmogenic isotopes.

2.3 Palaeosols

Stable isotopic (oxygen and carbon) ratios in paleosol carbonate nodules (Ghosh et al., 2004) and associated organic matter have been used as a proxy for vegetation and climatic changes. While oxygen isotopic composition of carbonate nodules act as an indicator of temperature and salinity condition of environmental water, carbon isotopic composition signifies presence of C_3, C_4 and CAM vegetation types (Cerling et al., 1989). Sellwood et al. (1993) and Retallack (2001) suggested palaeosols are one of the important palaeoclimatic proxy, in particular to estimate the palaeo-precipitation. Presence of palaeosols in a sedimentary record represents intervals with low or no sedimentation. Their mineral and chemical composition reflects the interaction between their source terrigenous clastic sediments and the processes of weathering, which can be physical, chemical and biological.

2.4 pH

The $\delta^{11}B$ of marine carbonate analyzed by Pearson and Palmer (1999) and Pagani et al. (2005a) to deduce pCO_2 estimation relies on the fact that a rise in the atmospheric concentration will mean that more CO_2 is dissolved in the surface ocean, causing a reduction in ocean pH. Sanyal et al. (1995) suggested the oceanic pH provides insights into how the carbonate chemistry of the oceans, including depth to lysocline, has changed through time. Royer et al. (2004) examined palaeo-pH of entire Phanerozoic period has also been estimated based on the calcium-ion concentration of seawater and modeled atmospheric CO_2 concentrations.

2.5 Atmospheric CO_2

Ice core data provided the atmospheric CO_2 concentrations that pre-date ~800 k.y. direct record (Siegenthaler et al., 2005) are of extreme importance for palaeoenvironmental reconstruction (Vaughan, 2007). Pagani (2002) suggested the oceanic proxies, $\delta^{13}C$ of organic materials has been particularly successful in pCO_2 determination, and has recently been refined with studies that focus on $\delta^{13}C$ derived from a single group of long chain hydrocarbons, for example, alkenones and rather than total marine organic material (Henderson, 2002). Pagani et al. (2005b) provided the alkenones record extends back to the Eocene–Oligocene boundary. $\delta^{13}C$ of carbonate, including both marine (Buggisch, 1991), freshwater (Yemane & Kelts, 1996) and pedogenic (Royer, 2001), has also been used as a proxy for atmospheric and oceanic carbon source and for rates of carbon burial (Schouten et al. 2000; Strauss & Peters-Kottig, 2003). Henderson (2002) and Pearson & Palmer (2002) recorded the oceanic pH has also been used as a proxy for atmospheric CO_2 concentration, although this requires assumptions to be made about the equilibrium condition between dissolved inorganic carbon and atmospheric CO_2. By this method, they have reconstructed atmospheric CO_2 concentration back to 60 m.y. ago (Pearson & Palmer, 2000; Demicco et al. 2003). Kasemann et al. (2005) recently extended the technique to deduce CO_2 from Neoproterozoic carbonates. Also GCM studies of monsoon climates have progressed extensively and contributed to our understanding of how the monsoon system evolved to various large scales forcing, including tectonics, CO_2, and orbital variations in incident solar radiation (Kutzbach et al., 1993; Wright et al., 1993, Wang et al., 2005). Atmospheric pCO_2 determination using parameters like stomatal index and carbon isotopic composition of nodules within soil carbonates provides efficient tool to estimate concentration of CO_2 in the atmosphere of geological past (Cerling, 1991; Yapp & Poths, 1992; van der Burgh et al., 1993; McElwain & Chaloner, 1995).

Chemistry-based proxies are the most direct measure and tend to focus on the specific chemical or isotopic system. The proxies are grouped by parameter estimated. Lithological proxies are directly related to the parameter estimated and are grouped by rock type, mineralogy or facies interpreted. Palaeontological proxies are subdivided by whether they use taxonomic methods or focus on some morphological aspect of a group of organisms. There are several proxies which were introduced over the past two decades (Royer et al.,

2001, 2006); the $\delta^{13}C$ of pedogenic minerals (Cerling, 1991; Yapp & Poths, 1992); the stomatal densities and indices in plants (van der Burgh et al., 1993; McElwain & Chaloner, 1995); the $\delta^{13}C$ of long-chained alkenones in haptophytic algae (Pagani et al., 1999); the $\delta^{11}B$ of marine carbonate (Pearson & Palmer, 1999; Pagani et al., 2005b); and the $\delta^{13}C$ of liverworts (Fletcher et al., 2005). Some of the proxy records including soil carbonates (Quade et al., 1989; Sanyal et al., 2004), palaeosols (Rettallack, 1995; Thomas et al., 2002, Ghosh et al., 2004), microfossils (Phadtare et al., 1994), pollens (Hoorn et al., 2000), palaeomagnetic record (Sangode & Bloemendal, 2004) and general sedimentation parameter where precipitation pattern have been used to decipher the changes in the Indian monsoon strength. In the following sections we have provided overviews of monsoon variability.

3. Debate on impact of Cenozoic CO₂ and tectonic process on monsoon

The role of Cenozoic pCO_2 and its effect on monsoonal intensity, including rapid change in the precipitation cycles was discussed in the community through proxy based studies (McGowran, 1989; Brady, 1991; Worsley et al., 1994; Kerrick & Calderira, 1998; Beck et al., 1998; Pearson & Palmer, 2000). During this debate some researcher have stressed the importance of the changing inputs to the atmosphere such as volcanic and hydrothermal outgassing (Owen & Rea, 1985; Berner et al., 1993; Ghosh et al., 2001) or metamorphic decarbonation reactions (Kerrick & Calderira, 1998), while others have focused on outputs such as weathering of silicate minerals and limestone formation (Brady, 1991; Raymo & Ruddiman, 1992; Worsley et al., 1994) or organic carbon burial (Berger & Vincent, 1986; McGowran, 1989; Beck et al., 1998). Cenozoic time CO_2 concentration has gradually decreased from >1000 ppmv during the Paleocene and the beginning of the Eocene period to <300 ppmv during the Pleistocene. This long-term decrease is partly due to reduction in volcanic emissions, which were particularly large during the Paleocene and Eocene epochs, but which have diminished since then, and to changes in the rate of weathering of silicate rocks. Figure 2 shows the sharp decline in the CO_2 concentration is associated with a cooling from the warm conditions of the early Eocene climatic optimum between ~52-50 Ma. This shift is often referred to as a transition from a greenhouse climate to an icehouse, in which permanent ice sheets formation were initialized over Antarctica (starting ~35 Ma) and Arctic regions (around ~3 Ma at Greenland). The warming during the Paleocene Eocene Thermal Maximum (PETM) around ~55 Ma ago, which also had a major impact on life on Earth, is better documented (Fig. 2). During this event which lasted less than ~1.7 Ma, the global temperature increased by more than 5°C in less than 10,000 years. This period is also characterised by a massive injection of carbon into the atmosphere-ocean system as recorded by the variations in the $\delta^{13}C$ measured in the sediments. An apparent discrepancy between oxygen isotope data and other paleoclimate proxies for the span from ~26-16 Ma was resolved by comparison against eustatic estimates from continental margin stratigraphy. Ice-volume estimates from oxygen isotope data compare favorably with stratigraphic and palynological data from Antarctica, and with estimates of atmospheric carbon dioxide for the early Oligocene through early Miocene (34-16 Ma). These isotopic data suggest that the East Antarctic Ice Sheet grew to as much as 30% greater than the present-day ice volume at glacial maxima. This conclusion is supported by data on seismic reflection and stratigraphic data from the Antarctic margin that suggest that the ice sheet may have covered much of the continental shelf region at Oligocene and early Miocene glacial maxima. Palynological data suggest long-term cooling during the Oligocene, with near tundra environments developing

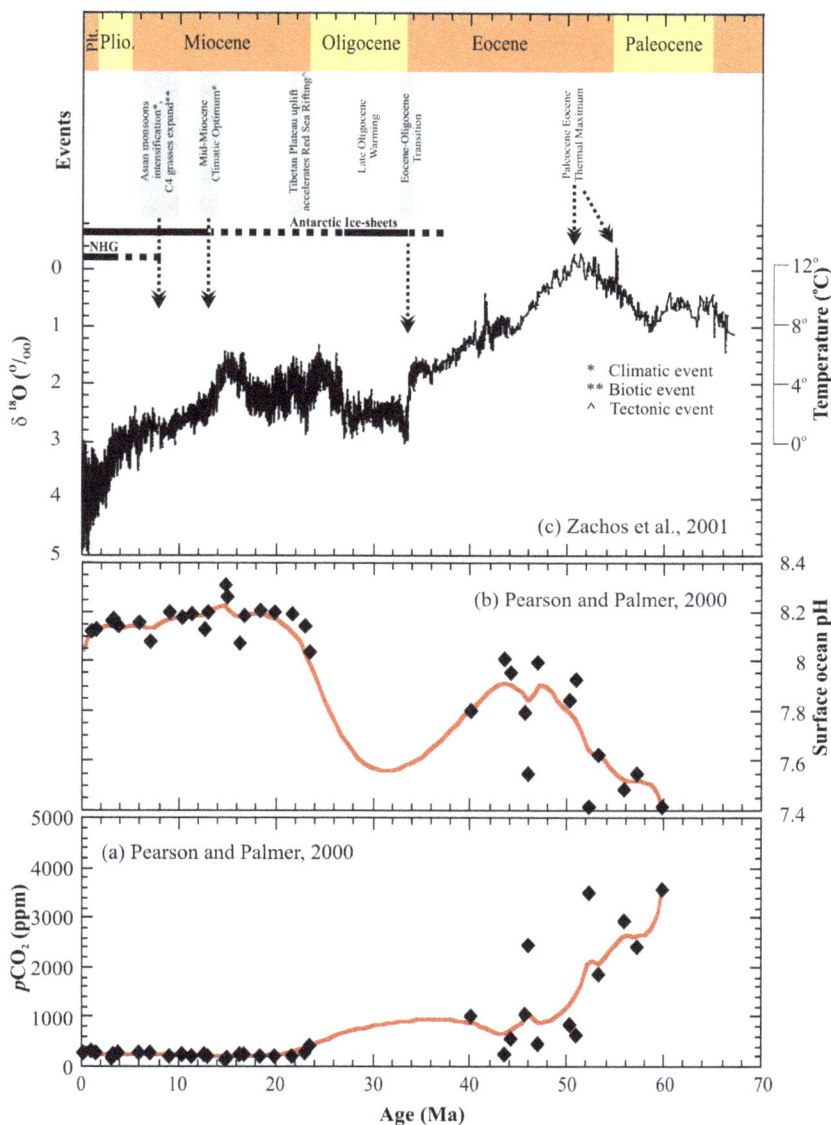

Fig. 2. The global climate over the past ~65 million years based on deep-sea oxygen-isotope measurements in the shell of benthic foraminifera. The $\delta^{18}O$ temperature scale, on the right axis, is only valid for an ice-free ocean. It therefore applies only to the time preceding the onset of large-scale glaciation in Antarctica (about 35 million years ago, see inset in the upper left corner), Figure from Zachos et al. (2008). Estimates of Sea surface pH and pCO_2 levels in Cenozoic climate change. The cure spanning most of the Cenozoic is estimated from the surface ocean pH as derived from the $\delta^{11}B$ ratio of planktonic foraminifers (Pearson & Palmer, 2000).

along the coast at glacial minima by the late Oligocene. Crowley (2000) has reported concentration of CO_2 changes in the atmosphere commonly regarded as a likely forcing mechanism on global climate over geological time scale because of its large and predictable effect on global temperature. An important factor of tectonic forcing associated with the reorganization of topography and elevation of the mountainous regions directly affect the atmosphere circulation and the wind prevailing over the region. Tectonic reorganizations by themselves were not the dominant cause of the warm climates of the Cenozoic, but they affected climate by controlling the processes that control ocean circulation, transport or trap solar heat, and maintain greenhouse gas levels in the atmosphere (Lyle et al., 2008). There was considerable fluctuation in these variables in the Eocene and Oligocene time period, but since the earliest Miocene the system seems to have been much steadier and more closely comparable to the present, despite continuing climate cooling. This suggests that other factors, such as complex feedbacks initiated by tectonic alteration of the ocean basins, were also important in determining global climate change (Pearson & Palmer, 2000).

3.1 Linkage between Indian monsoon system and uplift of the Tibetan Plateau

The major causes for the origin and intensification of the Indian monsoon system can be understood based on the upliftment rates recorded the Himalayas and Tibetan Plateau, because mountains modulate the land-sea thermal gradient. The growth of the Tibetan Plateau has been cited as being a triggering mechanism for a much stronger summer monsoon than might otherwise be predicted (Ruddiman & Kutzbach, 1989; Molner et al., 1993) and areas of rapid exhumation in the Himalayas have been correlated with zones of the most intense of summer monsoon (Thiede et al., 2004). The consequence of this was erosion which not only affected the mountain topography, but its transported large volumes of the sediments to the ocean. Due to this chemical reaction between silicate and acidic water was held responsible for the consumption of the atmospheric CO_2, a greenhouse gas, which in turn can drive long-term global cooling (Raymo & Ruddiman, 1992). Unfortunately, the timing of the tectonic events is poorly constrained relative to the timing of climate changes (Gupta et al., 2004) and Changes in continental position and height have frequently been invoked as causes of large-scale climatic shift during Cenozoic (Lyle et al., 2008). Data compilation of Aeolian records spanning early Miocene to late Pleistocene (central China) were used to infer aridity of the mid-continent was caused by global cooling or topographic uplift of the Tibetan Plateau, the latter might caused a rain shadow effect (Lu et al., 2004). A close association of drying proxies with global cooling and suggest a reduced role for topographic growth on climate change (Lu et al., 2004). The time between Miocene to recent was examined by Zheng et al. (2004), where sediments along the northern edge of the Tibetan Plateau in the Tarim Basin were investigated. The sections dated as ~8 Ma at the base grade upwards to a finer grade and more distal clastic sediments occurrences in a coarse alluvial fan deposits related to the uplift of northern Tibet were documented. At the same time aeolian dunes formation and Playa lakes started form after ~8 Ma, suggesting an enclosed desert basin from that time, suggesting that plateau uplift may be causing the regional climatic drying (Zheng et al. 2004). Gupta et al. (2004) examined the deep-sea benthic foraminiferal census data combined with published data shows that a major increase in biogenic productivity at ~10–8 Ma throughout the Indian Ocean, the equatorial Pacific, and southern Atlantic. The authors suggest that a change has been linked to initiation or intensification of the Indian Ocean monsoons. They argued, however, that the oceanic productivity was changed in a larger region than that affected by the monsoons, and secondly the effects of mountain uplift on global climate are not well documented (Hay et al., 2002). More recent workers have tried to estimates the rate of upliftment of mountains during Cenozoic time using soil carbonates, clays and organic component in the sedimentary deposits (Garzione et al., 2004; Chamberlin, 2006). Hoang et al.(2010) in the Gulf of Tonkin in the South China Sea, their study was based on a seismic stratigraphic analysis of sediments in the Song Hong-Yinggehai Basin formed as a pull-apart basin at the southern end of the continental Red river fault zone. During ~4-3 Ma the basin was filled by the sediments transported from the Red river, allowing periods of faster erosion and between ~15 and 10 Ma to be identified and linked to times of strong summer monsoon. However, the earliest pulse of fast erosion at 29.5-21 Ma may have been triggered by tectonic rock uplift along the Red river fault zone. The chemical weathering has gradually decreased in SE Asia after ~25 Ma, probably because of global cooling, whereas physical erosion became stronger, especially after ~12 Ma (Hoang et al., 2010). Kitoh, (2004) also studied the impact of Tibetan plateau uplift using by a GCM. Their model suggests that when there is no plateau present monsoon precipitation is limited in the deep tropics during northern hemispheric summer. However, as the Tibetan Plateau was uplifted, rainfall increases and affected climate over inland region over the south-eastern Tibetan Plateau. The Indian monsoon system got more intensified with upliftment of Tibet which commenced ~15 Ma. A rapid rate of upliftment during 10-8 Ma (1000-2000 m), coincides with the period of climate change or monsoon intensification (Harrison et al., 1992; Prell et al., 1992; Molnar et al., 1993). The Himalayas associated with Tibetan uplift and climate change connection is the classic example demonstrating a linkage between mountain growth and climate, yet despite significant attention and study a consensus history of the growth of the Himalayas and associated Tibetan Plateau has been elusive (Molnar et al., 1993; Copeland, 1997; Lyle et al., 2008).

4. Cenozoic climate record for estimated correlations: CO_2 and temperature (Greenhouse climates)

The link between the level of atmospheric CO_2 and global surface temperature is intensely important to understand. Solomon et al. (2007) suggested that implicated as the predominant cause of recent global climate change is release of fossil fuel CO_2 to the atmosphere by human activity. Zachos et al. (2001) has developed another database of $\delta^{18}O$ isotopes that cover the past 65 m.y., the time period when Antarctica glaciated over is clearly evident and based on which he showed that the trend in the $\delta^{18}O$ record suggesting a increasing and decreasing features reflect periods of global warming and cooling, and ice sheet growth and decay (Fig. 2). During Cenozoic period around ~59 Ma and 52 Ma (mid-Paleocene to early Eocene), the most distinct warming epoch, as expressed by enriched $\delta^{18}O$ composition was peaked at the PETM at ~52-50 Ma. The global average temperature exceeded by 6°C during the PETM and the periods witnessed deglaciation of Antarctica and Greenland glaciers started rebuilding at ~14 Ma. Zachos et al. (2001) showed that although having a high CO_2 concentration and high temperature, the Paleocene to early Eocene temperature record is a climate puzzle: globally averaged surface temperatures were significantly warmer than the present day and there is no convincing evidence for ice

(Frakes et al., 1992), but CO_2 estimates range from <300 to >2000 ppm (Fig. 3). Over the last 65 m.y., the CO_2 concentration has gradually decreased from more than 1000 ppmv during the Paleocene and the beginning of the Eocene epochs to less than 300 ppmv during the Pleistocene and the Cenozoic climate trend is characterized by a deep-sea cooling of approximately 12°C thought to have been forced by changes in atmospheric greenhouse gas composition (Hansen et al., 2008; Beerling & Royer, 2011). As suggested by Zachos et al. (1993), the transient climatic episode in this warm period, the PETM, at ~55.8 Ma and 52 Ma, may at face value seem an odd choice, given that it represents a period of extreme warmth. The Cenozoic warmth ~52 Ma corresponds with maximum reconstructed CO_2, and the rapid initiation of Antarctic glaciation to the Eocene-Oligocene boundary during ~34 Ma follows sharp fall in CO_2 (Royer, 2006). Crowley and Berner, (2001) and Royer et al. (2004&2006) reported clearest example in the Phanerozoic of a long-term positive coupling between CO_2 and temperature during the late Eocene to present day. The major changes in the Cenozoic earth's climate from a relatively warm and equable climate in the Paleocene to cold conditions with nearly freezing temperatures at the poles in the Pliocene, an evidence from various proxy record (Kennet & Barker, 1990; Lear et al., 2003; Zachos et al., 2001; Singh & Gupta, 2005), and like e.g. $\delta^{18}O$ for palaeotemperature (Zachos et al. 2001; Royer et al. 2004), $\delta^{11}B$ for palaeo-pH and palaeo-CO_2 (Pearson & Palmer, 2000), Sr/Ca from benthic foraminiferal calcite for weathering fluxes (Lear et al., 2003), etc. The late Oligocene and early Miocene an interval firmly linked with global warming and/or decreased ice mass in Antarctica (low $\delta^{18}O$ values; Miller et al., 1999 & 2004; Zachos et al., 2001; Barker & Thomas, 2004). The other section will narrate the vagaries of monsoon during last million years (glacial-interglacial cycles) and during last glacial maximum (LGM). A review of estimates of palaeo-atmospheric CO_2 levels from geochemical models, palaeosols, algae and foraminifera, plant stomata and boron isotopes concluded that there is no evidence that concentrations were ever more than 7500 ppm or less than 100 ppm during the past 300 m.y. (Crowley & Berner, 2001). Tripati et al. (2005) provide a strong, although indirect, case for three short lived glaciations at 42 – 41 Ma, 39 – 38 Ma, and 36.5 – 36 Ma based on independent reconstructions of sea level, temperature, and calcite compensation depth. The CO_2 record indicates moderately high CO_2 levels (1200 ppm) at 44 Ma, dropping to low levels (<500 ppm) just before the onset of the first cool event at 42 Ma (Fig. 3). This pattern of CO_2 (<1000 ppm during cool events and >1000 ppm elsewhere) generally persists across the remaining two events. Pearson and Palmer (2000) reported during Cenozoic pCO_2, indicating that this termination occurred at a time when green house gas levels were decreased or already low (Fig.2). ~25 Ma onwards the pCO_2 shows continuously decreasing trend due to the tectonic events such as mountain building or oceanic gateway reconfigurations, which can alter atmospheric circulation and water vapour transport, may have had a dominant role in triggering large scale shifts in climate as well (Kuztbach et al., 1993; Mikolajewicz et al., 1993; Driscoll & Haug, 1998, Zachol et al., 2001). Neftel et al. (1988) have been reported the atmospheric CO_2 concentration about 30% less than Holocene pre-industrial value during the last glaciation, although this change is thought to originate from ocean (Broecker & Pang, 1987; Archer & Maier-Reimer, 1994). The Cenozoic that represent examples of unique climate states of the Earth: the warm Eocene greenhouse world, which is associated with elevated atmospheric CO_2; the Miocene, when globally warm temperatures persisted but without the aid of high atmospheric CO_2 concentrations; and the Pliocene, a warm period that gradually cooled to become the Pleistocene ice ages (Lyle et al., 2008).

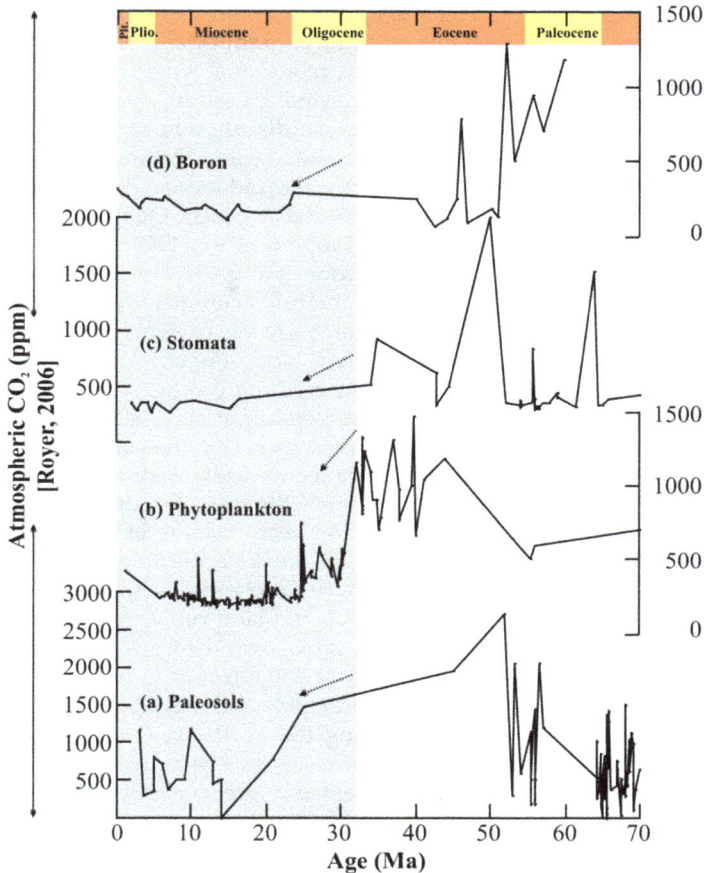

Fig. 3. Earth's atmospheric CO_2 history through the Cenozoic period. Various proxy generally track the estimates of atmospheric CO_2 (given separate panels a-d) reconstructed from the terrestrial and marine proxies following recent revisions (Royer, 2006). The gray zone shows from ~32 Ma to Holocene period, at 1000 and 500 ppm CO_2 represents the proposed CO_2 thresholds for, respectively, the initiation of globally cool events and full glacial (Royel, 2006). Early results provide increasing atmospheric CO_2 reconstructions.

5. Indian monsoon system and Cenozoic

The name "Monsoon" is derived from the Arabic word "mausim" meaning "season". Seasonal changes in the wind direction dominate the tropical climate of the Indian monsoon, bringing significant changes in ocean chemistry, biota and biogenic sedimentation budget. The seasonal reversal in the northern Indian Ocean wind system, called the Indian monsoon, is an important component of the global climate system affecting the weather and climate of the Asian and African regions between 30°N and 20°S latitudes (Fig. 1). The Indian monsoon is of great socio-economic importance, affecting the lives and livelihood of over 60 percentages of the world's population. In the Northern Hemisphere, during the summer monsoon, winds are southwesterly over the Arabian Sea and Bay of Bengal, whereas, during winter the monsoonal winds are directed away from the Asian continent, causing northeasterly wind transport over the area (Schott & McCreary, 2001). Several factors like the seasonal distribution of solar insolation, global ice volume, sea surface temperature, albedo, Himalayan-Tibetan heat budget, El Nino-Southern Oscillation (ENSO) and variability in the Indonesian Throughflow (ITF) may influence the monsoonal strength (Prell, 1984; Kutzbach, 1981; Quade et al., 1989; Clemens & Prell, 1990; Raymo & Ruddiman, 1992; Harrison et al., 1993; Kutzbach et al., 1993; Molnar et al., 1993; Raymo, 1994; Li, 1996; Gordon et al., 1997; An et al., 2001; Gupta et al., 2003, 2004). The monthly mean wind field for July (summer monsoon) are shown in Figure 1a., which induces ocean upwelling due to process of Ekman transport. During the summer monsoon season, the upwelling regions along the coasts of Somalia and Oman along with the southern coast of India experiences high productivity during the summer monsoon season (Banse, 1987; Nair et al., 1989; Naidu et al., 1992), making Arabian Sea one of the most productive location in the world (Fig. 1b). The summer monsoon season also accounts for 70-80% of the flux of organic matter to the sediments (Nair et al., 1989) and the southwest monsoon current drives upwelling and productivity changes in the Maldives region and other regions over Indian Ocean (Schulte et al., 1999). The summer monsoon anti-cyclonic gyre also results in weak upwelling along the coast of southwestern India and supports increased productivity during the summer season (Naidu et al., 1992). However local precipitation and land runoff results in a stratification and development of low-salinity layer (5-10m thick) over the up-welled waters (Schulte et al., 1999). Thus, inspite of a shallow thermocline the effect of upwelling is small. Due to the inflow of low salinity surface waters from the Bay of Bengal (Schulte et al., 1999) there is no increase in primary productivity during the northeast monsoon though there is deepening of the mixed layer during this time. Thus the upwelling and productivity changes in the Maldives region in the present day ocean could be related to summer monsoon variability. A spatial pattern documenting wide range of paleoceanographic and paleoclimatologic responses to the green house condition can be recorded from the sedimentary deposit from the Indian Ocean region. Monsoonal variability on different time scales ranging from annual cycles to long-term trends of millions of years can be well documented from these sedimentary records. A study on the records of the concentration and flux of biogenic components, such as calcium carbonate, organic carbon, and the composition of benthic and planktonic foraminiferal assemblages gives rise to indications of monsoonal variability as well as upwelling intensity (Prell et al., 1992).

5.1 The Indian summer monsoon (ISM) and climate change using multi proxy data

The ISM commenced about ~20 Ma ago because of uplift of the Tibetan Plateau beyond a critical height (Harrison et al., 1992; Prell & Kutzbach, 1992; Molnar et al., 1993). During summer, heating of the Tibetan plateau creates low atmospheric pressure, which acts as a powerful pump for moist air from the oceans to travel large distance, resulting in heavy rain (Quade et al., 1995). Reverse circulation of wind occurs during the winter; the radiative cooling of Tibetan Plateau causes flow of cold dry continental air towards the Indian Ocean. It has been argued that uplift of the Himalayas and Tibetan Plateau strengthen the land-sea thermal contrast (Hahn & Manabe, 1975) that led to the onset or major intensification of the Indian

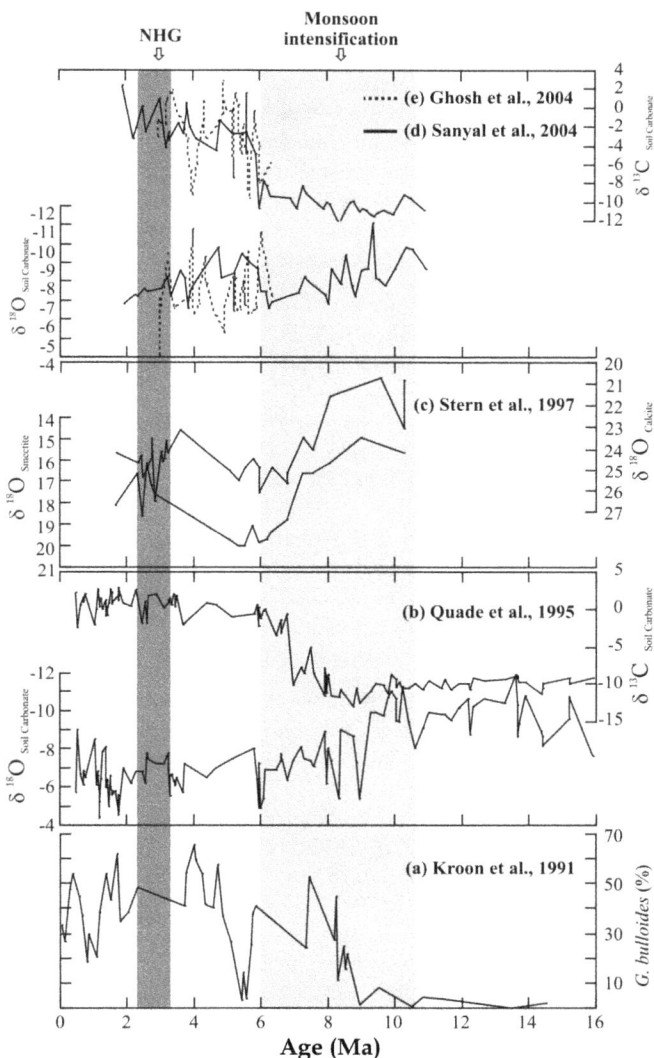

Fig. 4. (a) relative abundances of *G. bulloides* in time (Kroon et al., 1991); (b) The $\delta^{18}O$ and $\delta^{13}C$ of the soil carbonate from Siwalik Group fluvial sediments in the northern Pakistan (Quade et al., 1995); (c) $\delta^{18}O$ values of Calcite nodules and Smectite (Stren et al., 1997); (d) $\delta^{18}O$ and $\delta^{13}C$ of the soil carbonate from Siwalik basin (Sanyal et al., 2004); (e) $\delta^{18}O$ and $\delta^{13}C$ of the soil carbonate from Upper Siwalik (Ghosh et al., 2004). Grey zone shows a major excursion of monsoon intensification. Carbon isotope ratio of soil carbonate and associated organic matter indicates that vegetation was entirely of C_3 type from ~11 to 6 Ma. Post time is marked by appearance and expansion of C_4 grass. Dark grey zone (at ~3.2-2.3 Ma) indicates beginning of the major intensification of the Northern Hemisphere glaciation (NHG) (Zachos et al., 2001)

monsoon marked by the summer monsoon with heavy rainfall over the Indian subcontinent (Hastenrath, 1991) and cold dry winter monsoon. According to this hypothesis, changes in the elevation of the Himalayan-Tibetan region and its snow cover have modulated the development of the Indian or south Asian monsoons since the middle to late Miocene (Molnar et al., 1993; Clift et al., 2002). Many proxies indicate that the intensification of monsoonal winds led to increased upwelling over the Arabian Sea (Kroon et al., 1991) and eastern Indian Ocean (Singh & Gupta, 2004), a shift from C_3 to C_4 type vegetation on land (Quade et al., 1989), and increased terrigenous flux to the Indian Ocean as a result of increased weathering and erosion in the uplifted mountainous region (Prell & Kutzbach, 1992).

The high weathering rates increased nutrient flux (including phosphorus) to the oceans, increasing oceanic productivity (Filipelli, 1997). However, the timing of these events (like the uplift of the Himalayan-Tibetan Plateau, changes in the monsoon and changes in vegetation in south Asia) is not well constrained (Gupta et al., 2004). For instance, the dates for the elevation (~ 4 km) of the Tibetan Plateau, required to drive monsoon, vary from 35 to 8 Ma (Gupta et al., 2004). The long-term changes in the monsoon appear to have been linked with the uplift of the Himalayas and Tibetan Plateau (Rea, 1992; Prell et al., 1992; Molnar et al., 1993; Prell & Kutzbach, 1997; Filipelli, 1997) and Northern Hemisphere glaciations (Gupta & Thomas, 2003; Singh & Gupta, 2005). On the other hand, the short-term changes in the monsoon have been linked to the changes in the North Atlantic and Eurasian snow cover (Schulz et al., 1998; Anderson et al., 2002; Gupta et al., 2003), orbital cycle of eccentricity, tilt and precession (Prell & Kutzbach, 1987; Gupta et al., 2001) solar influence (Neff et al., 2001; Fleitmann et al., 2003; Gupta et al., 2005).

The Indian monsoon system is believed to have originated in the late Oligocene or early Miocene, and appears to have a major shift in its intensity at ~8.5 Ma. The ISM substantially increased its strength after 10-8 Ma (Quade et al., 1989; Kroon et al., 1991; An et al., 2001; Gupta et al., 2004). During middle Miocene (~15 Ma) the productivity record showed marked increment, significantly enhancing the productivity in all the oceans and reached maximum 10-8 Ma (Dickens & Owen, 1999; Hermoyian & Owen, 2001). The high productivity correspond to a "biogenic bloom" that began ~15-13 Ma and peaked 10-8 Ma (Pisias et al., 1995; Dickens & Owen, 1999; Hermoyian & Owen, 2001). Various proxy records have been interpreted as indicating that the monsoons started or strongly intensified between ~10 and 8 Ma, as a response to Himalayan-Tibetan uplift to at least about half of its present elevation (Prell & Kutzbach, 1992; Rea, 1992). The Indian Ocean high productivity event occurred (~10-8 Ma) at the end of a phase of build-up of the East Antarctic ice sheet and possibly the beginning of the formation of the west Antarctic ice sheet (Zachos et al., 2001, Barker & Thomas, 2004), and as well as in the global compilation of deep-sea oxygen isotope records (Zachos et al., 2001; Fig. 2). The biological, sedimentological and geochemical responses to this late Miocene event have been observed across the Indian, Atlantic and Pacific Oceans (Kroon et al., 1991; Dickens & Owen, 1999; Hermoyian & Owen, 2001; Gupta et al., 2004). During the late Miocene, the increased glaciation on Antarctica may have strengthened wind system, causing widespread open-ocean as well as coastal upwelling over a large part of the Atlantic, Indian and Pacific Ocean. This increased upwelling could have triggered the widespread biological productivity during the late Miocene (Gupta et al., 2004). Harrison et al. (1992) suggested rapid uplift around ~8 Ma of Tibetan Plateau, may have acted as an effective orographic barrier inducing depleted $\delta^{18}O$

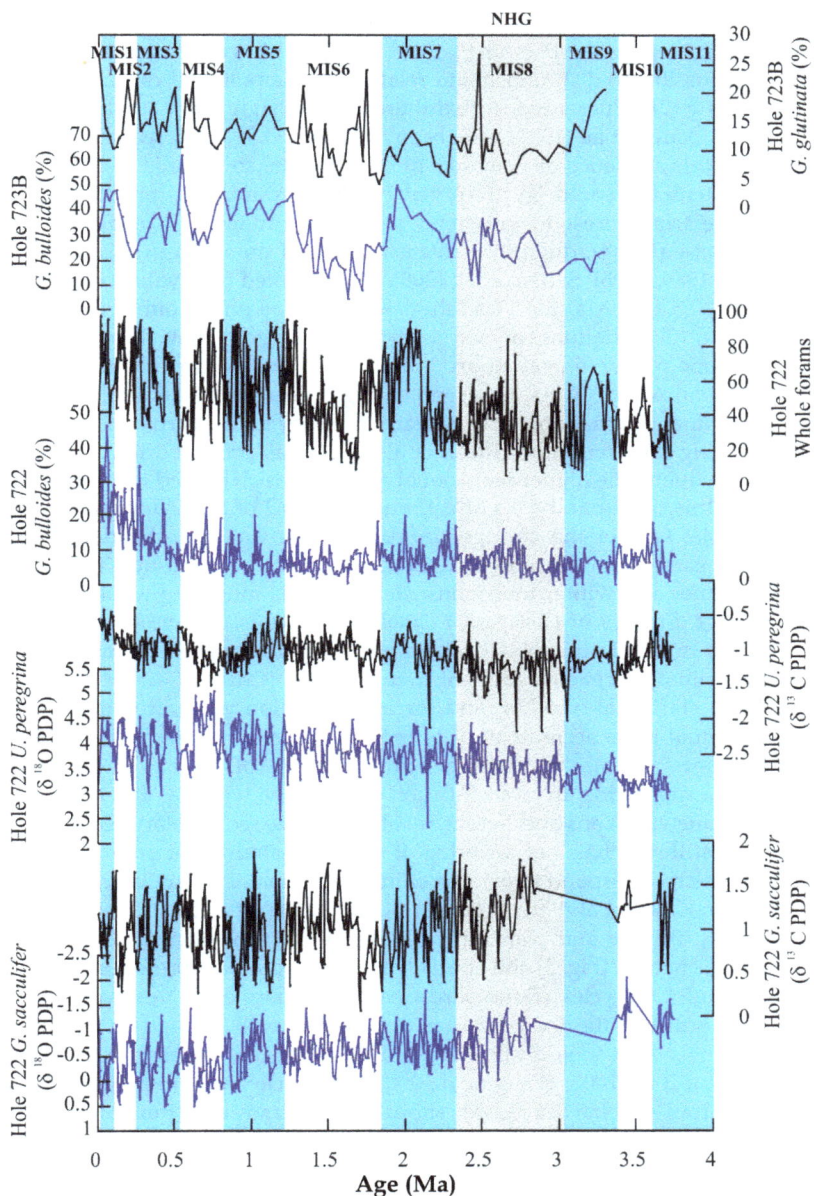

Fig. 5. Visual correlation between marine proxy records of the Indian summer monsoon during Plio-Pleistocene time at Arabian Sea using benthic and planktonic foraminifera with isotopes values (Clemens et al., 1996). Dark grey zone (at ~3.2-2.3 Ma) indicates beginning of the major intensification of the Northern Hemisphere glaciation (NHG) (Zachos et al., 2001; Clemens et al., 1991, 2003) and Blue color bars indicates Marine Isotope Stages (MIS 1-11)

value for the precipitation originating from the central Asia, while higher $\delta^{18}O$ values for the precipitation happening on the Potwar Plateau. The $^{18}O/^{16}O$ value increase observed at 8.5 and 6.5 Ma is impossible to distinguish and relate with the potential causal mechanism, but these clay mineral $\delta^{18}O$ values support that there was the significant climate changes (Stern et al, 1997; Fig. 4). Qaude et al. (1989) have been reported ~3.5 $°/_{oo}$ increase in $\delta^{18}O$ in the soil formed calcite in this sequence of paleosols at ~8.5-6.5 Ma, so there is a significant isotope change which potentially could be preserved by the clay minerals and there is dramatic vegetation change from forest to grassland slightly postdating (7.7-6.5 Ma) the oxygen isotope ratio increase (Fig. 4), due to the intensification of monsoon in this region at 8.5-6.5 Ma (Quade et al., 1989, 1995). Sanyal et al. (2004) documented $\delta^{13}C$ values of soil carbonates show that, from 10.5 to 6 Ma, the vegetation was C_3 type and around ~6 Ma C_4 grasses dominated. The $\delta^{18}O$ variations of soil carbonates suggest that the monsoon system intensified, with one probable peak at around 10.5 Ma and a clear intensification at 6 Ma, with peak at 5.5 Ma. After 5.5 Ma, monsoon strength decreased and attained the modern-day values with minor fluctuations, which is supported by marine proxy of upwelling in the Arabian Sea. During the same time Ghosh et al. (2004) has proposed that greater moisture availability ~4 Ma due to the higher monsoonal activity, which caused an abrupt transient of C_3 dominance in this region and $\delta^{18}O$ of carbonate and δD of clay support depleting trend observed at ~4 Ma. One of the strongest arguments for an intensification of the Indian monsoon ~10 Ma came from deep-ocean drilling in the Arabian Sea by Kroon et al. (1991). During both summer and winter monsoons, steady winds induce upwelling of cold, deep, nutrient-rich water. Studies of *Globigerina bulloides* from this region shows that thrives in cold water where nutrients upwell, becomes abundant during monsoons, particularly in summer, and dominates organic sediment accumulation (Curry et al. 1992, Prell & Curry 1981). *G. bulloides* started at ~14 Ma, and for a few million years, it comprised only few percent of the annual mass accumulation of organic sediment in the western Arabian Sea. Then beginning near ~10 Ma, it suddenly comprised tens of percent of organic sediment deposition (Fig. 4), suggesting an intensification of monsoon winds at that time (Kroon et al., 1991). The $\delta^{18}O$ value variations at different Siwalik sections, sedimentary record from foothill of Himalya suggest three phases of lowering of monsoon strength at around ~10.5 Ma, ~5.5 Ma and ~3 Ma, which correspond to periods of intensification of the Indian summer monsoon (Fig. 4a–d based on Kroon et al. 1991; Quade et al., 1995; Stern et al. 1997; Sanyal et al. 2004 and Ghosh et al., 2004). Benthic and planktonic foraminiferal ratios were used to estimate Plio-Pleistocene climate changes (Fig. 5) and specially there is prominent presence of MIS 3 (Marine Isotope Stages) and DO cycles (Dansgaard-Oeschger); marine C^{14} reservoir ages and the incremental time scale of the ice cores are not known well enough to resolve potential difference in timing which would allow inferences about the mechanism of global energy transfer (Sarnethein, 2001, 2002; Wang et al., 2005). DO cycle are not restricted to MIS 3 and monsoon proxies from Arabian Sea records shows that the transition from interglacial MIS 5 to glacial MIS 4 occurred in discrete steps, characterized by a number of warming rebounds. This record, chronologically constrained by the well dated sediments from Arabian Sea (Fig. 5). Clemens et al. (1991, 2003) observed together with *G. bulloides* other various proxy records of summer monsoon strength at 3.5 Ma. They suggested that of strong monsoon has systematically drifted over the past 2.6 Ma (refer Fig 5). During the initiation and growth of NGH ice sheets, the phase of strong monsoon moves away from the phase maximum ice-volume, systematically shifting by ~83 and ~124 k.y., at the precession and obliquity bands, respectively (Clemens et al., 1991, 2003).

5.2 Quaternary climate change

During the Quaternary time the Indian monsoon system also underwent parallel changes like glacial and interglacial mode (Schuldz et al., 1998; Leuschner & Sirocko, 2003). The cold spell intervals in the North Atlantic have been found to be associated with intervals of weak summer monsoon (Schuldz et al., 1998), the winter monsoon strengthened during the same time (Fontunge & Duplessy, 1986). The summer monsoon oscillated with millennial scale variability (the Dansgaard-Oescher-DO) and Henrich events concentrated at periodicities of ~1100, 1450 and 1750 years during the last glacial cycles (Naidu & Malmgren, 1995; Siriko et al., 1996). The pattern is almost similar to that of changes recorded from the Greenland ice cores (Schulz et al., 1998; Leuschner & Sirocko, 2003; Fig. 6). The atmospheric CO_2, ice core data (GSIP2) with Asian speleothems and Arabian Sea cores uniformly indicate the monsoon was weaker during cold intervals such as Heinrich event 1 (H1) and the Younger Dryas (YD), and stronger during the warming period observed during the Bølling–Allerød (B–A) period (Fig. 6; Schulz et al., 1998; Altabet et al., 2002; Ivanochko et al., 2005; Sinha et al., 2005; Wang et al., 2001; Dykoski et al., 2005; Sirocko et al., 1996). The deglaciation affecting Asian monsoon intensity as recorded in the speoleothems samples analysed from the Chinese and Indian sites (see Fig. 6; Sinha et al., 2005; Dykoski et al., 2005). These findings suggest a strong solar influence on the monsoon during the Holocene (Dykoski et al., 2005; Fleitmann et al., 2003; Wang et al., 2005; Neff et al., 2001; Fig. 7). The speleothem record from China and India region shows similar trend from ~16000–11000 calendar years before present (cal yr B.P.) (Fig. 6). Oxygen isotopes anomaly at D4 are also thought to be driven by an amount effect and are interpreted to reflect changes in the strength of the Asian monsoon (Dykoski et al., 2005). In general, these all speleothem time series are remarkably similar (Fig. 6 & 7). Shakun et al. (2007) suggested these climate events affected a large area of the monsoon region in the same way and to the same degree because the magnitudes of the transitions into the Bølling, the YD, and the Holocene are nearly identical in the speleothem records from these caves. Even more interesting, the structures of these climate changes are similar. In particular, the transitions into the Bølling period and the YD in all these records, as well as the Timta Cave record from northern India (Sinha et al., 2005), are remarkably gradual and take place over several centuries (Fig. 6).

The characteristic feature of ISM during Holocene period is presentence presence oscillating intensity with frequency of occurrences of events coinciding with millennial time scale. Goodbred & Kuel (2000) reported the ISM was stronger in the early Holocene, which is evident from the huge sediments deposits at Ganges-Brahmaputra, the rapid speleothem growth (see Fig. 1) eg. Hoti Cave (Neff et al., 2001) and Qunf Cave at Oman margin (Fleitmann et al., 2003), Socotra Island Yemen at northwestern Indian Ocean (Shakun, et al. 2007), Timta Cave at western Himalaya (Sinha et al., 2005), Hulu Cave at eastern China (Wang et al., 2004), Dongge Cave at southern China (Wang et al., 2005), Dandak Cave at central India (Berkelhammer, et al. 2010) and *G. bulloides* census data from Arabian Sea (Gupta et al., 2004). Northern Hemisphere temperatures peaked at ~10400 and 5500 cal yr B.P., and during which time Asian and African monsoon reached their maximum so called "Holocene Climatic Optimum". Alley et al. (1997) suggested the early Holocene monsoon maximum was interrupted by an abrupt cooling peak ~8200 cal yr B.P., the summer monsoon over the Indian subcontinent and tropical Africa weakened during same time (Gasse, 2000). After ~ 5500 cal yr B.P., consecutive shifts towards drier condition in northern Africa and Asian was noted (Overpec al., 1996; Gasse, 2000), which led to termination of

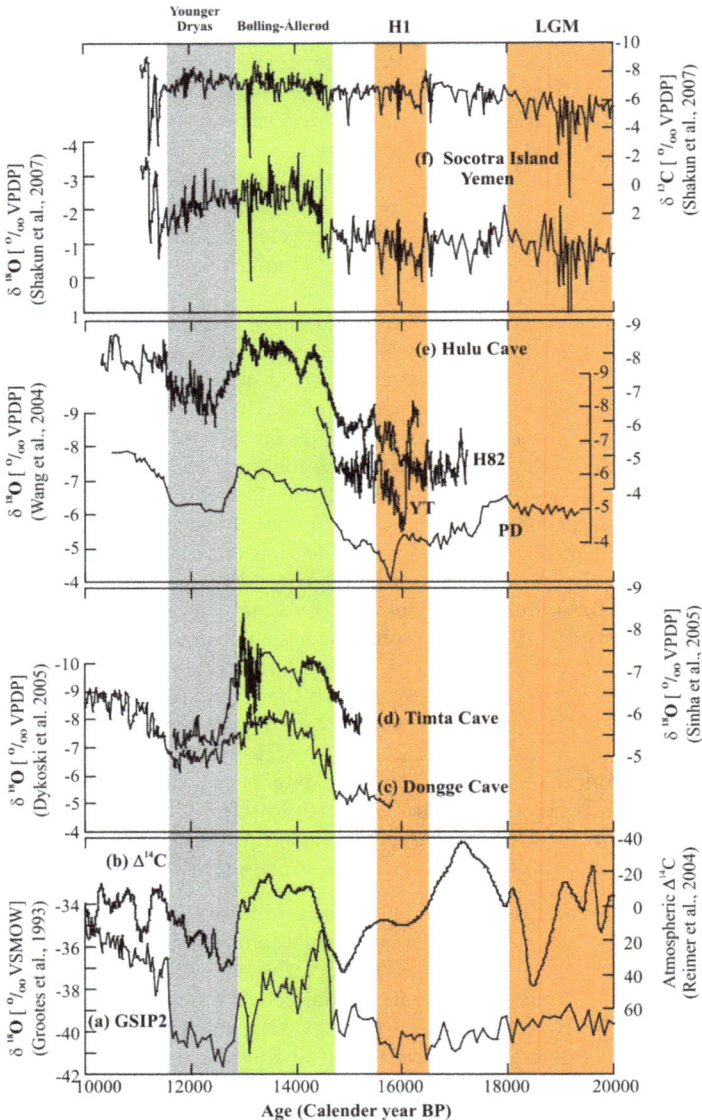

Fig. 6. Visual correlation between various proxy records of the Indian summer monsoon.
a) GSIP2 δ18O record (Grootes et al., 1993); (b) Atmospheric Δ14C record (Remier et al., 2004);
(c) Oxygen isotope record from Dongge Cave at southern China (Dykoski et al., 2006); (d)
Timta Cave δ18O at Western Himalaya (Sinha et al., 2005); (e) δ18O record from Hulu Cave
at eastern China (Wang et al., 2004) and (f) δ18O and δ13C record Socotra Island Yemen at
Northwestern Indian Ocean (Shakun, et al. 2007). Grey band indicate the timing and
duration of the Younger Dryas (YD) and green bars indicate Bølling–Allerød (B–A). Orange
color bars number H1 indicates Henrich events with LGM (Last glacial maximum).

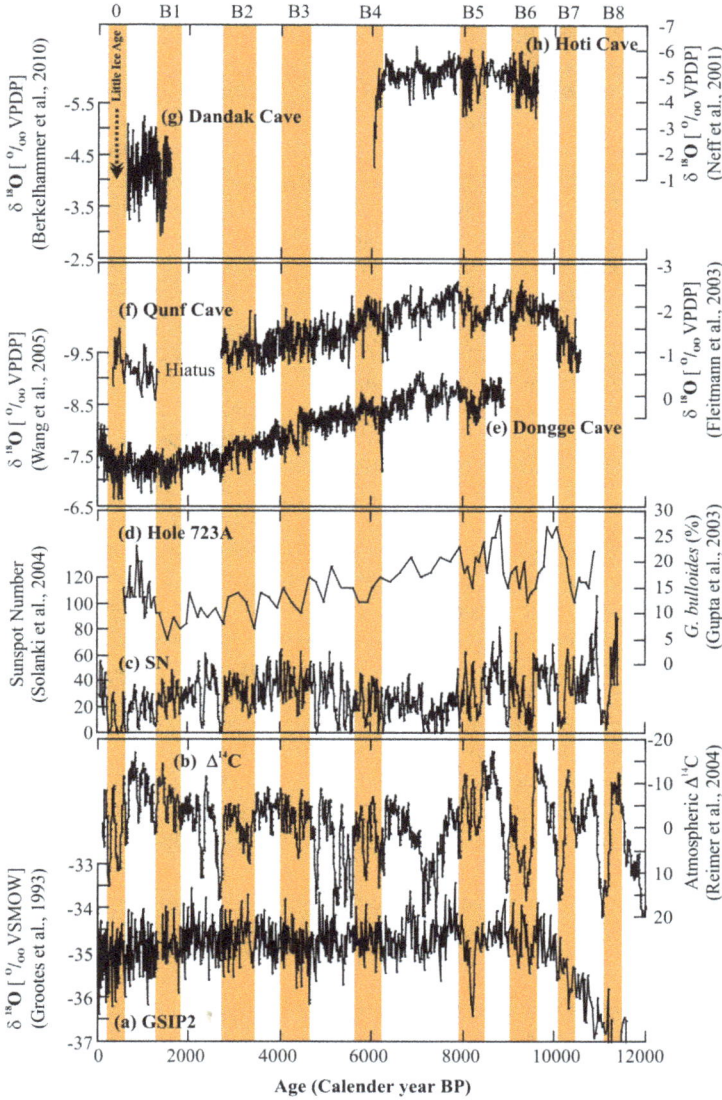

Fig. 7. Correlation between various proxy records of the Indian summer monsoon. (a) GSIP2 δ18O record (Grootes et al., 1993); (b) Atmospheric Δ14C record (Reimer et al., 2004); Sunspot number (Solanki et al., 2004); (d) *G. bulloides* census data from Arabian Sea (Gupta et al., 2004); (e) δ18O data from Dongge Cave at southern China (Wang et al., 2005); (f) δ18O record from Qunf Cave at Oman margin (Fleitmann et al., 2003); (g) , Dandak Cave δ18O at central India (Berkelhammer, et al. 2010) and δ18O Hoti Cave (Neff et al., 2001). Numbers from B1 to B8 indicate each of the eight Bond events and number 0 indicates Little Ice Age event (Bond et al., 2001)

several ancient civilization existed in this regions. During ~4200 cal yr B.P., persistent record of drought due to the aridification led to societal collapse both in the Egyptian and Mesopotamian civilization (Weiss et al., 1993; Cullen et al., 2000). The precipitation diminishes during the arid phase over Indian region as well (Sharma et al., 2004), and the Indus valley civilization transformed from an organized urban phase of smaller settlements migrated southward in search of water (Allchin & Allchin, 1997; Staubwasser et al., 2003; Gupta et al., 2004). The weakest summer monsoon occurred in the late Holocene around ~2500 to 1500 cal yr B.P. (Anderson et al., 2002; Gupta et al., 2003; Fig. 7). Gupta et al. (2003) have been suggested the summer monsoon intensified during the Medieval Warm Priod (AD. 900-1400) whereas during the most recent climatic event, the little ice age (AD. 1450-1850), there was a drastic reduction in the intensity of the ISM (Fig. 7; Gupta et al., 2003). The ISM is controlled by the Eurasian snow cover (Bamzai & Shukla, 1999) too and the amplitude and period of ENSO (Krishna Kumar et al., 1999). ISM variability on millennial scale may be attributed to the solar forcing (Neff et al., 2001; Fleitmen et al., 2003) and glacial and interglacial boundary condition (Burns et al., 2001). The $\delta^{18}O$ record from the stalagmite, which serves as a proxy for variations in the tropical circulation and monsoon rainfall, allows us to make a direct comparison of the $\delta^{18}O$ record with the $\Delta^{14}C$ record from tree rings (Fig. 7; Stuiver et al., 1998; Neff et al., 2001), which largely reflects changes in solar activity (Stuiver & Braziunas,, 1993; Beer et al., 2000). Bond et al. (2001) suggested that solar variability could be an affecting climate variation during the Holocene on the basis of analysis of ^{14}C records from the tree rings. The production rate of comogenic nuclides (^{14}C and ^{10}Be) that reflect changes in solar activity appear to closely follow the Bond Cycles (see Fig. 7). van Geel et al. (1999) has pointed out that the exact process responsible for the linking global climate through solar forcing is poorly understood, variations in ultra violet radiation and cosmic ray flux may trigger abrupt climate changes by the altering the heat budget of the stratosphere and changing the atmospheres optical parameters and radiation balance (Kodera, 2004). The summer monsoon strengthened has been related to changes solar insolation during Milankovitch cycles (Clemans et al., 1991), but centennial and decadal variations in the $\Delta^{14}C$ record are controlled by changes in solar activities or sun spot numbers, (see Fig. 7; Wang et al., 1999; Neff et al., 2001; Gupta et al., 2005). During the Holocene, the relation between intervals of low sunspot activity and low intensity of the summer monsoon were observed in marine records (Fig. 7; Gupta et al., 2005).

6. Conclusions

The past ~65 million years of geological history documented in the continental and marine sedimentary record allowed understanding the evolution of climate change in a longer time scales. The effect of green house gas concentration on global climate and moisture circulation can be explored through careful investigation of available proxy records. The impact of other factors like topography and albedo change can be addressed knowing the magnitude of upliftment rate of mountain region. The simultaneous analyses of proxy records from both land and marine location provided conclusive signature of climate change through time. The $\delta^{18}O$ of foraminiferal calcite from sea sediments provided independent estimate of time for commencement of Antarctic and Arctic glaciations (Zachos et al., 2001). The increasing and decreasing trends suggested based on the $\delta^{18}O$ record demarcate periods of global warming and cooling, responsible for growth and decay of ice sheet. The elevation (~ 4 km) of the Tibetan Plateau, required to drive the monsoon, varied significantly between

35 to 8 Ma. The combination of factors like pCO_2 concentration in the atmosphere and rate of uplift of Tibetan plateau were found responsible for modulating the intensity of Indian monsoon. The published data from land and oceanic region indicate a major changes land vegetation and oceanic productivity during period of intense monsoon. The high southern latitude cooling and increased volume of ice act as important factors responsible for lowering the strength of monsoonal circulation. The signature for strengthening of the upwelling, presumably from strengthening of seasonal winds over the Indian Ocean was noted as a proxy for monsoon intensification. The signature of increasing aridity and warming in northern Pakistan perhaps increases both in temperature and seasonal precipitation in the Indo-Gangetic plain just south of the Himalayas in Nepal and in the Himalayas are recorded from soil carbonate preserved in the sedimentary archives. Together with ice core record, speleothems occurrences from Asian region act as a recorder of Holocene climate and monsoonal variability. The role of sun, green house gases and tectonic adjustment of continental lithosphere played significant role controlling the monsoon over Indian region. Understanding the sensitivity of all these factors on rainfall or precipitation is yet to be understood.

7. Acknowledgment

The authors profusely thank MoES for fellowship (Scheme code: MESO-006) to Dr. KM.

8. References

Allchin, B. & Allchin, R. (1997). Origins of a Civilization. The Prehistory and Early Archaeology of South Asia. New Dehli: Penguin Books India

Alley, R.B., Mayewski, P.A., Sowers, T., Stuiver, M., Taylor, K.C. & Clark, P.U. (1997). Holocene climatic instability: a prominent, widespread event 8200 years ago. Geology, Vol. 25, pp. 483-486.

Altabet, M.A., Higginson, M.J. & Murray, D.W. (2002). The effect of millennial-scale changes in Arabian Sea denitrification on atmospheric CO_2. Nature, Vol. 415, pp. 159-162.

An, Z.S., Kutzbach, J., Prell, W.L. & Porter, S.C. (2001). Evolution of Asian monsoons and phased uplift of the Himalayan-Tibetan Plateau since late Miocene times: Nature, Vol. 411, pp. 62–66.

Anderson, D.M. & Prell, W.L. (1993). A 300 kyr record of upwelling off Oman during the late Quaternary: evidence of the Asian southwest monsoon. Paleoceanography, Vol. 8, pp. 193-208.

Anderson, D.M., Overpeck, J.T., & Gupta, A.K. (2002). Increase in the Asian SW Monsoon during the Past Four Centuries. Science, Vol. 297, pp. 596-599.

Archer, D. & Maier-Reimer, E. (1994). Effect of deep-sea sedimentary calcite preservation on atmospheric CO_2 concentration, Nature, Vol. 367, pp. 260-264.

Arrhenius, S. (1896). On the influence of carbonic acid in the air upon the temperature of the ground. Philosophical Magazine Series, Vol. 5, No. 4 (251), pp. 237–276.

Auras-Schudnagies, A., Kroon, D., Ganssen, G.M., Hemleben, C. & van Hinte, J.E., (1989). Biogeographic evidence from planktic foraminifers and pteropods for Red Sea anti-monsoonal surface currents. Deep-Sea Research., Vol. 10, pp. 1515-1533.

Ballantyne, C.K. (1994). Gibbsitic soils on former nunataks: implications for ice sheet reconstruction. Journal of Quaternary Science, Vol. 9, No. 1, pp. 73-80.

Ballantyne, C.K., McCarroll, D. & Stone, J.O. (2006). Vertical dimensions and age of the Wicklow Mountains ice dome, Eastern Ireland, and implications for the extent of the last Irish Ice Sheet. Quaternary Science Reviews, Vol. 25, No. 17-18, pp. 2048-2058.

Bamzai, A.S., & Shukla, J. (1999). Relation between Eurasian snow cover, snow depth, and the Indian summer monsoon: An observational study, Journal of Climate, Vol. 12, No. 10, pp. 3117-3132.

Banse, K. (1987). Seasonality of phytoplankton chlorophyll in the central and northern Arabian Sea. Deep Sea Research, Vol. 34, pp. 713-723.

Barker, P.F. & Thomas, E. (2004). Origin, signature and palaeoclimatic influence of the Antarctic Circumpolar Current: Earth-Science Reviews, Vol. 66, pp. 143–162

Be, A.W.H. & Hutson, W.H. (1977). Ecology of planktonic foraminifera and biogeographic patterns of life and fossil assemblages in the Indian Ocean. Micropaleontology, Vol. 23, pp. 369-414.

Beck, A., Sinha, A., Burbank, D.W., Seacombe, W. J. & Khan, S. (1998). In Late Paleocene-Early Eocene Climatic and Biotic Events in the Marine and Terrestrial Records (eds Aubry, M.P., Lucas, S.G. & Berggren, W. A.) 103-117 (Columbia Univ. Press, New York).

Beer, J., Mende, W. & Stellmacher, R. (2000). The role of the sun in climate forcing. Quaternary Science Reviews, Vol. 19, pp. 403–415.

Beerling, D.J., & Royer, L.D. (2011). Convergent Cenozoic CO_2 history. Nature Geoscience, Vol. 4, No. 7., pp. 418-420. doi:10.1038/ngeo1186.

Berger, W.H. & Vincent, E. (1986) Deep-sea carbonates: reading the carbon-isotope signal. Geologische Rundschau Vol. 75, pp. 249-269.

Berkelhammer, M., Sinha, A., Mudelsee, M., Cheng, H., Edwards, R.L. Cannariato, K.G. (2010). Persistent multidecadal power of the Indian Summer Monsoon Earth and Planetary Science Letters, Vol. 290, No. 1-2, pp. 166-172,

Berner, R.A., Lasaga, A.C. & Garrels, R.M. (1993). The carbonate-silicate geochemical cycle and its effect on atmospheric carbon dioxide over the past 100 million years. Am. J. Sci. Vol. 283, pp. 641-683.

Bond, G., Kromer, B., Beer, J., Muscheler, R., Evans, M.N., Showers, W., Hoffmann, S., Lotti-Bond, R., Hajdas, I. & Bonani, G. (2001). Persistent solar influence on North Atlantic climate during the Holocene. Science, Vol. 294, pp. 2130-2136.

Brady, P.V. (1991). The effect of silicate weathering on global temperature and atmospheric CO_2. J. Geophys. Res. Vol. 96, pp. 18101-18106.

Broecker, W.S. & Pang, T.H. (1987). The role of $CaCO_3$ compensation in the glacial to interglacial atmospheric CO_2 change. Global Biogeochemical Cycles, Vol. 1 pp. 15-29

Buggisch, W. (1991). The global Frasnian–Famennian Kellwasser Event. Geologische Rundschau, Vol. 80, No. 1, pp. 49–72.

Burns, S.J., Fleitmann, D., Matter, A., Neff, U. & Mangini, A. (2001). Speleothem evidence from Oman for continental pluvial events during interglacial periods, Geology, Vol. 29, pp. 623 - 626.

Cerling, T.E. (1991). Carbon dioxide in the atmosphere: evidence from Cenozoic and Mesozoic paleosols. American Journal of Science, Vol. 291, pp. 377–400.

Cerling, T.E. Quade, J. Wang, Y. & Bowman., J.R. (1989). Carbon isotopes in soils and palaeosols as ecology and paleoecology indicators. Nature, Vol. 341, pp.138–139.

Chamberlin, T.C. (1899). An attempt to frame a working hypothesis of the cause of glacial periods on an atmospheric basis. Journal of Geology, Vol. 7, pp. 545–584, 667–685, 751–787.

Chamley, H. (1989). Clay Sedimentology. Springer, Berlin.

Chamley, H. (1998). Clay mineral sedimentation in the ocean. In: Paquet, H. & Clauer, N. (eds) Soils and Sediments, Mineralogy and Geochemistry. Springer, Heidelberg, pp. 269–302.

Clemens, S.C., Murray, D.W. & Prell. W.L. (1996). Nonstationary phase of the Plio-Pleisotcene Asian Monsoon. Science, Vol. 274, pp. 943-948.

Clemens, S., Prell, W., Murray, D., Schimmield, G. & Weedon, G. (1991). Forcing mechanisms of the Indian Ocean monsoon. Nature, Vol. 353, pp. 720-725.

Clemens, S.C. & Prell, W.L. (1990). Late Pleistocene variability of Arabian Sea summer monsoon winds and continental aridity: Eolian records from the lithogenic components of deep sea sediments. Paleoceanography, Vol. 5, No. 2, pp. 109-145.

Clift, P.D., Lee, J.I., Clark, M.K. & Blusztajn, J.S. (2002). Erosional response of South China to arc rifting and monsoonal strengthening: a record from the South China Sea. Marine Geology, Vol. 184, pp. 206–266.

Crowley, T.J. & Berner, R.A. (2001). CO_2 and climate change. Science, Vol. 292, pp. 870–872.

Cullen, J.L. & Prell, W.L. (1984). Planktonic Foraminifera of the northern Indian Ocean: distribution and preservation in surface sediments. Marine Micropaleontology, Vol. 9, pp. 1–52.

Cullen, H.M., deMenocal, P.B., Hemming, S., Hemming, G., Brown, F.H., Guilderson, T. & Sirocko, F. (2000). Climate change and the collapse of the Akkadian empire: Evidence from the deep sea, Geology, Vol. 28, pp. 379-382.

Curry, W.B., Ostermann, D.R., Guptha, M.V.S. & Ittekkot, V. (1992). Foraminiferal production and monsoonal upwelling in the Arabian Sea: evidence from sediment traps. Upwelling Systems: Evolution since the Miocene. Geological Society Special Publication, Vol. 64, pp. 93-106.

Demicco, R.V., Lowenstein, T.K. & Hardie, L.A. (2003). Atmospheric $pCO_{(2)}$ since 60 Ma from records of seawater pH, calcium, and primary carbonate mineralogy. Geology, Vol. 31, No. 9, pp. 793–796.

Dickens, G.R. & Owen, R.M. (1999). The latest Miocene– early Pliocene biogenic bloom: a revised Indian Ocean perspective. Marine Geology, Vol. 161, pp. 75–91.

Driscoll, N.W. & Haug, G.H. (1998). A short circuit in thermohaline circulation: A cause for Northern Hemisphere glaciation? Science, Vol. 282, pp. 436-438.

Dykoski, C.A., Edwards, R.L., Cheng, H., Yuan, D., Cai, Y., Zhang, M., Lin, Y., Qing, J., An, Z. & Revenaugh, J. (2005). A high-resolution absolute-dated Holocene and deglacial Asian monsoon record from Dongge Cave, China. Earth Planetary Science Letters, Vol. 233, pp. 71–86.

Ehrmann, W., Setti, M. & Marinoni, L. (2005). Clay minerals in Cenozoic sediments off Cape Roberts (McMurdo Sound, Antarctica) reveal palaeoclimatic history.

Filipelli, G.M. (1997). Intensification of the Asian monsoon and a chemical weathering event in the late Miocene-early Pliocene: implications for late Neogene climate change. Geology, Vol. 25, pp. 27-30.

Fleitmann, D., Burns, S.J., Mudelsee, M., Neff, U., Kramers, J., Mangini, A. & Matter, A. (2003). Holocene forcing of the Indian monsoon recorded in a stalagmite from Southern Oman. Science, Vol. 300, pp. 1737-1739.

Fletcher, B.J., Beerling, D.J., Royer, D.L. & Brentnall, S.J. (2005). Fossil brophytes as recorders of ancient CO_2 levels: Experimental evidence and a Cretaceous case study. Global Biogeochemical Cycle, Vol. 19, GB3012. doi:10.1029/2005GB002495.

Frakes, L.A., Francis, J.E. & Syktus, J.I. (1992). Climate modes of the Phanerozoic – The history of the Earth's climate over the past 600 million years. Cambridge University Press, Cambridge.

Gasse, F. (2000). Hydrological changes in the African tropics since the Last Glacial Maximium. Quaternary Science Reviews, Vol. 19, pp. 189-211

Garzione, C.N., Dettman, D.L. & Horton, B.K. (2004). Carbonate oxygen isotope paleoaltimetry: evaluating the effect of diagenesis on paleoelevation estimates for the Tibetan plateau. Palaeogeography, Palaeoclimatology, Palaeoecology, Vol. 212, pp. 119–140

Ghosh, P., Ghosh, P. & Bhattacharaya, S.K. 2001. levels in the late Palaeozoic and Mesozoic atmosphere from the $\delta^{13}C$ values of the pedogenic carbonate and the organic matter, Palaeogeography, Palaeoclimatology, Palaeoecology, Vol. 170, pp. 285-296.

Ghosh, P., Padia, J.T. & Rakesh Mohindra. (2004). Stable isotopic studies of palaeosol sediment from Upper Siwalik of Himachal Himalaya: evidence for high monsoonal intensity during late Miocene?. Palaeogeography, Palaeoclimatology, Palaeoecology, Vol. 206, pp. 103-114.

Goodbred, S.L. & Kuehl, S.A. (2000). Enormous Ganges-Brahmaputra sediment discharge during strengthened early Holocene monsoon. Geology, Vol. 28, pp. 1083–1086.

Gordon, A.L., Ma, S. , Olson, D.B., Hacker, P., Ffield, A., Talley, L.D., Wilson, D. & Baringer, M. (1997). Advection and Diffusion of Indonesian Throughflow within the Indian Ocean South Equatorial Current, Geophysical Research Letters, Vol. 24, No. 21, pp. 2573-2576.

Grootes, P.M., Stuiver M., White J. W.C., Johnsen S. & Jouzel, J. (1993). Comparison of Oxygen Isotope Records from the GISP2 and GRIP Greenland Ice Cores. Nature, Vol. 366, pp. 552-554.

Gupta, A.K., Anderson, D.M. & Overpeck, J.T. (2003). Abrupt changes in the Asian southwest monsoon during the Holocene and their links to the North Atlantic Ocean. Nature, Vol. 421, pp. 354-357.

Gupta, A.K. & Srinivasan, M.S. (1992). Species diversity of Neogene deep sea benthic foraminifera from northern Indian Ocean sites 214 and 216A. In: Takayanagi, Y., Saito, T. (Eds.), Studies in Benthic Foraminifera, Benthos'90, Tokai University press, Shizuoka, Japan, pp. 249-254.

Gupta, A.K. & Thomas, E. (1999). Latest Miocene-Pleistocene productivity and deep-sea ventilation in the northwestern Indian Ocean (DSDP Site 219). Paleoceanography, Vol. 14, pp. 62-73.

Gupta, A.K. & Thomas, E. (2003). Initiation of Northern Hemisphere glaciation and strengthening of the northeast Indian monsoon: Ocean Drilling Program Site 758, eastern equatorial Indian Ocean. Geology, Vol. 31, pp. 47-50.

Gupta, A.K., Das, M. & Anderson, D.M. (2005). Solar influence on the Indian summer monsoon during the Holocene. Geophysical Research Letters, Vol. 32, L17703, pp. 1-4.

Gupta, A.K., Joseph, S. & Thomas E. (2001). Species diversity of Miocene deep-sea benthic foraminifera and water mass stratification in the northeastern Indian Ocean. Micropaleontology, Vol. 47, pp. 111-124.

Gupta, A.K., Singh, R.K., Joseph, S. & Thomas, E. (2004). Indian Ocean high-productivity event (10-8 Ma): linked to global cooling or to the initiation of the Indian monsoons? Geology, Vol. 32, pp. 753-756

Hahn D.G. & Manabe, S. (1975). The role of mountain in the south Asian monsoon circulation. Jouranl of Atmospheric Science, Vol. 32, pp. 1515-1541.

Hansen, J., Sato, M., Kharecha, P., Beerling, D., Berner, R., Masson-Delmotte, V., Pagani, M., Raymo, M., Royer, D.L. & Zachos, J.C. (2008), Target atmospheric CO_2 : where should humanity aim?, Open Atmospheric Science Journal, Vol. 2, pp. 217–231.

Harrison, T.M., Copeland, P., Hall, S.A., Quade, J., Burner, S., Ojha, T.P. & Kidd, W.S.F. (1993). Isotopic preservation of Himalayan/Tibetan uplift, denudation, and climatic histories in two molasse doposits. Journal of Geology, Vol. 101, pp. 155–177

Harrison, T.M., Copeland, P., Kidd, W.S.F. & Yin, A. (1992). Rising Tibet. Science, Vol. 255, pp. 1663–1670

Hastenrath, S. (1991), Climate dynamics of the tropics: Dordrecht, Netherlands, Kluwer Academic Publishers, pp. 488.

Hay, W.W., Soeding, E., de Conto, R.M. & Wold, C.N. (2002). The late Cenozoic uplift–climate change paradox: International Journal of Earth Sciences (Geologische Rundschau), Vol. 91, pp. 746–774.

Henderson, G.M. (2002). New oceanic proxies for paleoclimate. Earth and Planetary Science Letters, Vol. 203, No. 1, pp. 1–13.

Hermoyian, C.S. & Owen, R.M. (2001). Late Miocene–early Pliocene biogenic bloom: evidence from low-productivity regions of the Indian and Atlantic Oceans. Paleoceanography, Vol. 16, pp. 95– 100.

Hoang, L.V., Clift, P.D., Schwab, A.M., Huuse, M., Nguyen, D.A., & Zhen,. S. (2010). Monsoon Evolution and Tectonics–Climate Linkage in Asia. Geological Society, London, Special Publications, 342, 219–244.

Hoorn, C., Ohja, T. & Quade, J. (2000). Palynological evidence for vegetation development and climatic change in the sub-Himalayan zone (Neogene, Central Nepal). Palaeogeography, Palaeoclimatology, Palaeoecology, Vol. 163, pp. 133 161.

Imbrie, J. & Newell, N. (1964). Approaches to Paleoecology. John Wiley and Sons Inc., NewYork.

Ivanochko, T.S., Ganeshram, R.S., Brummer, G.J.A., Ganssen, G., Jung, S.J.A., Moreton, S.G. & Kroon, D. (2005). Variations in tropical convection as an amplifier of global climate change at the millennial scale. Earth Planetary Science Letters, Vol. 235, pp. 302-314.

Kasemann, S.A., Hawkesworth, C.J., Prave, A.R., Fallick, A.E. & Pearson, P.N. (2005). Boron and calcium isotope composition in Neoproterozoic carbonate rocks from Namibia: evidence for extreme environmental change. Earth and Planetary Science Letters, Vol. 231, No. 1-2, pp. 73–86.

Kennett, J.P. & Barker, P.F. (1990). Latest Cretaceous to Cenozoic climate and oceanographic developments in the Weddell Sea, Antarctica: an Ocean-Drilling Perspective. Proc. Ocean Drill. Prog., Sci. Results, Vol. 113, pp. 937–960.

Kennett, J.P. & Srinivasan, M.S. (1983). Neogene Planktonic Foraminifera: A Phylogenetic Atlas: Stroudsburg, P.A. (Hutchinson Ross).

Kennett, J.P., Elmstrom, K. & Penrose, N.L. (1985). The last deglaciation in Orca Basin, Gulf of Mexico: High-resolution planktonic foraminiferal changes. Vol. 50, pp. 189-216.

Kerrick, D. M. & Caldeira, K. (1998). Metamorphic degassing from orogenic belts. Chem. Geol. 145, 213- 232.

King, A.L. & Howard, W.R. (2005). $\delta^{18}O$ seasonality of planktonic foraminifera from Southern Ocean sediment traps: Latitudinal gradients and implications for paleoclimate reconstructions. Marine Micropaleontology, Vol. 56, pp. 1-24.

Kitoh, A. (2004). Effects of mountain uplift on East Asian summer climate investigated by a coupled atmosphere ocean GCM. Journal of Climatology, Vol. 17, pp. 783–802.

Kodera, K. (2004), Solar influence on the Indian Ocean monsoon through dynamical processes, Geophys. Res. Lett., Vol. 31, L24209, doi:10.1029/ 2004GL020928.

Krishna Kumar, K., Rajogopalan, B. & Cane, M.A. (1999). On the weakening relationship between the Indian monsoon and ENSO. Science, Vol. 284, pp. 2156–2159.

Kroon, D., Steens, T.N.F. & Troelstra, S.R. (1991). Onset of monsoonal related upwelling in the western Arabian Sea as revealed by planktonic foraminifers. In: Prell, W.L., Niitsuma, N. et al., Proceedings of the ocean drilling program, scientific results, Ocean Drilling Program, College Station, TX, Vol. 117, pp. 257-264.

Kutzbach, J.E. (1981). Monsoon climate of the early Holocene: Climate experiment with the earth's orbital parameters from 9000 years ago. Science, Vol. 214, pp. 59-61.

Kutzbach, J.E. & Gallimore, R.G. (1989). Pangaean climates: megamonsoons of the megacontinent. Journal of Geophysical Research-Atmospheres, Vol. 94 (D3), pp. 3341–3357.

Kutzbach, J.E., Guetter, P.J., Behling, P.J & Selin, R. (1993). Simulated climatic changes: results of the COHMAP climate-model experiments. Chapter 4 In "Global Climates since the Last Glacial Maximum" (H.E. Wright, Jr, J.E. Kutzbach, T. Webb III, W.F. Ruddiman, F.A. Street-Perrott, & P.J. Bartlein, eds.), University of Minnesota Press, Minneapolis, MN, pp. 24-93.

Lacis, A.A, G.A. Schmidt, D. Rind, & R.A. Ruedy. (2010): Atmospheric CO_2: Principal control knob governing Earth's temperature.Science, Vol. 330, pp. 356-359, doi:10.1126/science.1190653.

Lear, C.H., Elderfield, H. & Wilson, P.A. (2003). A Cenozoic seawater Sr/Ca record from benthic foraminiferal calcite and its application in determining global weathering fluxes. Earth and Planetary Science Letters, Vol. 208 (1-2), pp. 69–84.

Leuschner, D.C. & Sirocko, F. (2003). Orbital insolation forcing of the Indian Monsoon - a motor for global climate changes? Palaeogeography, Palaeoclimatology, Palaeoecology, Vol. 197, pp. 1-2, 83-95.

Li, T. (1996). The process and mechanism of rise of the Qinghai–Tibet Plateau. Tectonophysics, Vol. 260, pp. 45–53.

Lu, H.Y., Wang, X.Y., An, Z.S., Miao, X.D., Zhu, R.X., Ma, H.Z., Li, Z., Tan, H.B., & Wang, X. Y. (2004). Geomorphologic evidence of phased uplift of the northeastern Qinghai-Tibet Plateau since 14 million years ago, Sci. China Ser. D, 47, 822–833.

Lyell, C. 1830. Principles of Geology.

MacLeod, K.G., Huber, B.T. & Isaza-Londoño, C. (2005). North Atlantic warming during global cooling at the end of the Cretaceous. Geology, Vol. 33, pp. 437–440.

McElwain, J.C. & Chaloner, W.G. (1995). Stomatal density and index of fossil plants track atmospheric carbon dioxide in the Palaeozoic. Ann. Bot. Vol. 76, pp. 389–395.

McGowran, B. (1989). Silica burp in the Eocene ocean. Geology 17, 857-860.

Mikolajewicz, U., Maierreimer, E., Crowley, T.J. & Kim, K.Y. (1993). Effect of Drake and Panamanian gateways on the circulation of an ocean model. Paleoceanography, Vol. 8, No. 4, pp. 409-426.

Miller, K.G., Barrera, E., Olsson, R.K., Sugarman, P.J. & Savin, S.M. (1999). Does ice drive early Maastrichtian eustasy? Geology, Vol. 27, pp. 783–786.

Miller, K.G., Sugarman, P.J., Browning, J.V., Kominz, M.A., Olsson, R.K., Feigenson, M.D. & Herna´ndez, J.C. (2004). Upper Cretaceous sequences and sea level history, New Jersey Coastal Plain. Geol. Soc. Am. Bull. Vol. 116, pp. 368–393

Molnar, P., England, P., & Martiod, J. (1993). Mantle dynamics, uplift of the Tibetan Plateau and the Indian monsoon development: Reviews of Geophysics, Vol. 34, pp. 357–396.

Naidu, P.D. & Malmagren, B.A. (1995). Do benthic foraminifer records represent a productivity index in oxygen minimum zone areas? An evaluation from the Oman Margin, Arabian Sea. Marine Micropaleontology, Vol. 26, pp. 49-55.

Naidu, P.D., Prakash Babu, C., & Rao, C.M. (1992). The upwelling record in the sediments of the western continental margin of India. Deep-Sea Research, Vol. 39, No. 3/4, pp. 715-723.

Nair, R.R., Ittekot, V., Manganini, S.J., Ramaswamy, V., Haake, B., Degens, E.T., Desai, B.N. & Honjo, S. (1989). Increased particle flux to the deep ocean related to monsoons. Nature, Vol. 338, pp. 749-751.

Neff, U., Burns, S.J., Mangini, A., Mudelsee, M., Fleitmann, D. & Matter, A. (2001). Strong coherence between solar variability and the monsoon in Oman between 9 and 6 kyr ago. Nature, Vol. 411, pp. 290-293.

Neftel, A., Oeschger., H, Staffelbach, T. & Stauffer, B. (1988) CO_2 record in the Byrd ice core 50,000-5,000 years BP. Nature, Vol. 331, pp. 609-611

Nigrini, C. (1991). Composition and biostratigraphy of radiolarian assemblages from an area of upwelling (northwestern Arabian Sea, Leg 117) . In Prell, W.L., Niitsuma, N., et al., Proceedings of the Ocean Drilling Program, Scientific results,: College Station, Texas, Ocean Drilling Program, Vol. 117, pp. 89–126.

Overpeck J.T., Anderson, D., Trumbore, S. & Prell, W. (1996). The southwest Indian monsoon over the last 18,000 years. Climate Dynamics, Vol. 12, pp. 213-225.

Owen, R.M. & Rea, D.K. (1985). Sea floor hydrothermal activity links climate to tectonics-the Eocene carbon dioxide greenhouse. Science Vol. 227, pp. 166-169.

Pagani, M. (2002). The alkenone-CO_2 proxy and ancient atmospheric carbon dioxide. Philosophical Transactions of the Royal Society of London Series A-Mathematical Physical and Engineering Sciences, Vol. 360, No. 1793, pp. 609–632.

Pagani, M., Freeman, K.H. & Arthur, M.A. (1999). Late Miocene atmospheric CO_2 concentrations and the expansion of C_4 grasses. Science, Vol. 285, pp. 876–879.

Pagani, M., Lemarchand, D., Spivack, A. & Gaillardet, J. (2005a). A critical evaluation of the boron isotope-pH proxy: the accuracy of ancient ocean pH estimates. Geochimica et Cosmochimica Acta, Vol. 69, pp. 953–961.

Pagani, M., Zachos, J.C., Freeman, K.H., Tipple, B. & Bohaty, S. (2005b). Marked decline in atmospheric carbon dioxide concentrations during the Paleogene. Science, Vol. 309, No. 5734, pp. 600–603.

Pearson, P.N. & Palmer, M.R. (2000). Atmospheric carbon dioxide concentrations over the past 60 million years. Nature, Vol. 406, No. 6797, pp. 695–699.

Pearson, P.N. & Palmer, M.R. (2002). The boron isotope approach to paleo-pCO_2 estimation. Geochimica et Cosmochimica Acta, Vol. 66, No. 15A, pp. A586–A586.

Pearson, P.N., Palmer, M.R. (1999). Middle Eocene seawater pH and atmospheric carbon dioxide concentrations. Science, Vol. 284, pp. 1824–1826.

Peeters, F.J.C., Brummer G.J.A. & Ganssen, G. (2002). The effect of upwelling on the distribution and stable isotope composition (planktic foraminifera) in modern surface waters of the NW Arabian Sea. Global and Planetary Change, Vol. 34, pp. 269-291.

Petit, J.R., Jouzel, J., Raynaud, D., Barkov, N.I., Barnola, J.-M., Basile, I., Bender, M., Chappellaz, J., Davis, M., Delaygue, G., Delmotte, M., Kotlyakov, V.M., Legrand, M., Lipenkov, V.Y., Lorius, C., Pe´pin, L., Ritz, C., Saltzman, E. & Stievenard, M. (1999). Climate and atmospheric history of the past 420,000 years from the Vostok ice core, Antarctica. Nature, Vol. 399, pp. 429–436.

Phadtare, N.R., Kumar, R. & Ghosh, S.K. (1994). Stratigraphic palynology, floristic succession and the Tatrot/Pinjor boundary in upper Siwalik sediments of Haripur Khol area, district Sirmaur (H.P), India. Himalayan Geology, Vol. 15, pp. 69–82.

Pisias, N.G., Mayer, L.A., & Mix, A.C. (1995), Paleoceanography of the eastern equatorial Pacific during the Neogene: Synthesis of Leg 138 drilling results, in Pisias, N.G., Mayer, L.A., et al., Proceedings of the Ocean Drilling Program, Scientific results,: College Station, Texas, Ocean Drilling Program, Vol. 138, pp. 5–21.

Plaut, G., Ghil, M. & Vautard, R. (1995). Interannual and interdecadal variability in 335 years of Central England Temperatures. Science, Vol. 268, No. 5211, pp. 710–713.

Prell, W.L. & Curry, W.B. (1981). Faunal & isotopic indices of monsoonal upwelling: Western Arabian Sea. Oceanologica acta, Vol. 4, pp. 91-98

Prell, W.L. & Kutzbach, J.E. (1987). Monsoon variability over the past 150,000 years. Journal of Geophysical Research, Vol. 92, pp. 8411-8425.

Prell, W.L. & Kutzbach, J.E. (1997). The Impact of Tibet-Himalayan Elevation on the Sensitivity of the Monsoon Climate System to Changes in Solar Radiation, in Tectonic Uplift and Climate Change, Plenum Press, New York, pp. 171-201.

Prell, W.L. (1984). Variation of monsoonal upwelling: A response to changing solar radiation. In: Hansen, J.E. & Takahashi, T. (eds): Climate processes and climate sensitivity, Washington, D.C. (AGU). Geophysical Monograph Series, Vol. 29, pp. 48-57.

Prell, W.L., & Kutzbach, J.E. (1992), Sensitivity of the Indian monsoon to forcing parameters and implications for its evolution: Nature, Vol. 360, pp. 647–651.

Prell, W.L., Murray, D.W., Clemens, S. & Anderson, D.M. (1992). Evolution and variability of the Indian Ocean Summer Monsoon Evidence from western Arabian Sea Drilling Program. Geophysical Monograph, AGU, Washington, D.C., Vol. 70, pp. 447-469.

Quade, J., Cater, M. L. J., Ojha, P. T., Adam, J. & Harrison, M. T. (1995). Late Miocene environmental change in Nepal and the northern Indian subcontinent: stable isotopic evidence from paleosols. Geological Society of America Bulletin, Vol. 107, pp. 1381–1397.

Quade, J., Cerling, T.E. & Bowman, J.R. (1989). Development of Asian Monsoon revealed by marked ecological shift during the latest Miocene in northern Pakistan. Nature, Vol. 342, pp. 163– 166.

Raymo, M.E. & Ruddiman, W.F. (1992). Tectonic forcing of late Cenozoic climate. Nature, Vol. 359, pp. 117-122.

Raymo, M.E. (1994). The Himalayas, organic carbon burial, and climate in the Miocene. Paleoceanography, Vol. 9, pp. 399-404.

Rea, D.K. (1992). Delivery of Himalayan sediment to the northern Indian Ocean and its relation to global climate, sea level, uplift, and seawater strontium. In Duncan, R.A., Rea, D.K., Kidd, R.B., von Rad, U., and Weissel, J.K. (Eds.), Synthesis of Results from Scientific Drilling in the Indian Ocean. Am. Geophys. Union, Washington, D.C. Geophysical Monograph, Vol. 70, pp. 387–402.

Reimer, J., Baillie, M.G.L., Bard, E., Bayliss, A., Beck, J.W., Bertrand, C., Blackwell, P.G., Buck, C.E., Burr, G., Cutler, K.B., Damon, P.E., Edwards, R.L., Fairbanks, R.G., Friedrich, M., Guilderson, T.P., Hughen, K.A., Kromer, B., McCormac, F.G., Manning, S.W., Ramsey, C.B., Reimer, R.W., Remmele, S., Southon, J.R., Stuiver, M., Talamo, S., Taylor, F.W., van der Plicht, J. & Weyhenmeyer, C.E. (2004). Residual delta ^{14}C around 2000 year moving average of IntCal04. Radiocarbon, Vol. 46, pp. 1029-1058.

Retallack, G.J. (1995). Paleosols of the Siwalik Group as a 15 Ma record of South Asian paleoclimate. In: Wadia, S., Korisettar, R. & Kale, V.S. (eds) Quaternary Environments and Geoarchaeology of India: Essays in Honour of S.N. Rajaguru. Memoir of the Geological Survey of India, Vol. 32, pp. 36–51

Retallack, G.J. (2001). Soils of the past: an introduction to paleopedology. Blackwell Science, Oxford.

Royer, D.L. (2006) . CO_2-forced climate thresholds during the Phanerozoic. Geochimica et Cosmochimica Acta, Vol. 70, pp. 5665–5675

Royer, D.L., Berner, R.A. & Beerling, D.J. (2001). Phanerozoic atmospheric CO_2 change: evaluating geochemical and paleobiological approaches. Earth-Science Reviews, Vol. 54, No. 4, pp. 349–392.

Royer, D.L., Berner, R.A., Montañez, I.P., Tabor, N.J. & Beerling, D.J. (2004). CO_2 as a primary driver of Phanerozoic climate. GSA Today, Vol. 14, No. 3, pp. 4–10.

Ruddiman, W.F., Kutzbach, J.E. (1989). Forcing of late Cenozoic northern hemisphere climate by plateau uplift in southern Asia and the American West. Journal of Geophysical Research, Vol. 94, pp. 18409–18427.

Sangode, S.J. & Bloemendal, J. (2004). Pedogenic transformation of magnetic minerals in Pliocene– Pleistocene palaeosols of the Siwalik Group, NW Himalaya, India. Palaeogeography, Palaeoclimatology, Palaeoecology, Vol. 212, pp. 95–118.

Sanyal, A., Hemming, N.G., Hanson, G.N. & Broecker, W.S. (1995). Evidence for a higher pH in the glacial ocean from boron isotopes in foraminifera. Nature, Vol. 372 (6511), pp. 234– 236, 10.1038/373234a0.

Sanyal, P., Bhattacharya, S.K., Kumar, R., Ghosh, S.K. & Sangode, S.J. (2004). Mio -Pliocene monsoonal record from Himalayan Foreland basin (Indian Siwalik) and its relation to the vegetational change. Palaeogeography, Palaeoclimatology, Palaeoecology, Vol. 205, pp. 23–41.

Schott, F.A. & McCreary, J.P. Jr. (2001). The monsoon circulation of the Indian Ocean Progress in Oceanography, Vol. 51, pp. 1-123

Schouten, S., Van Kaam-Peters, H.M.E., Rijpstra, W.I.C., Schoell, M. & Damste, J.S.S. (2000). Effects of an oceanic anoxic event on the stable carbon isotopic composition of Early Toarcian carbon. American Journal of Science, Vol. 300, No. 1, pp. 1–22.

Schulz, H., von Rad, U. & Erlenkeuser, H. (1998). Correlation between Arabian Sea and Greenland climate oscillations of the past 110, 000 years. Nature, Vol. 393, pp. 54-57.

Sellwood, B.W., Price, G.D., Shackleton, N.J. & Francis, J.E. (1993). Sedimentary facies as indicators of Mesozoic palaeoclimate [and discussion]. Philosophical Transactions of the Royal Society of London Series B-Biological Sciences, Vol. 341 (1297), pp. 225-233.

Sen Gupta, B.K. & Machain-Castillo, M.L. (1993). Benthic foraminifera in oxygen-poor habitats. Marine Micropaleontology, Vol. 20, pp. 183-201.

Shakun, J.D., Burns, S. J., Fleitmann, D., Kramers, J., Matter, A., & Al-Subary, A. (2007). A high-resolution, absolute-dated deglacial speleothem record of Indian Ocean climate from Socotra Island, Yemen. Earth and Planetary Science Letters, Vol. 259, pp. 442-456.

Siegenthaler, U., Stocker, T.F., Monnin, E., Luthi, D., Schwander, J., Stauffer, B., Raynaud, D., Barnola, J.M., Fischer, H., Masson-Delmotte, V. & Jouzel, J. (2005). Stable carbon cycle-climate relationship during the late Pleistocene. Science, Vol. 310, No. 5752, pp. 1313– 1317.

Singh, R.K. & Gupta, A.K. (2004). Late Oligocene-Miocene paleoceanographic evolution of the southeastern Indian Ocean: Evidence from deep-sea benthic foraminifera (ODP Site 757). Marine Micropaleontology, Vol. 51, pp. 153-170.

Singh, R.K. & Gupta, A.K. (2005). Systematic decline in benthic foraminiferal species diversity linked to productivity increases over the last 26 Ma in the Indian Ocean. Journal of Foraminiferal Research, Vol. 35, pp. 219-227.

Sinha, A., Cannariato, K.G., Stott, L.D., Li, H.C., You, C.F., Cheng, H., Edwards, R.L. & Singh, I.B. (2005). Variability of southwest Indian summer monsoon precipitation during the Bølling–Allerød. Geology, Vol. 33, pp. 813–816.

Sirocko, F., Garbe-Schönberg, D., McIntyre, A. & Molfino, B. (1996). Teleconnections between the subtropical monsoons and high-latitude climates during the last deglaciation. Science, Vol. 272, pp. 526-529.

Solanki, S. K., Usoskin, I.G., Kromer, B., Schüssler, M., & Beer, J. (2004), Unusual activity of the sun during recent decades compared to the previous 11,000 years, Nature, Vol. 431, pp. 1084–1087.

Solomon et al. & IPCC (2007). Climate Change: The Physical Science Basis. Contribution of Working Group I to the Fourth Assessment Report of the Intergovernmental Panel on Climate Change [Solomon, S., D. Qin, M. Manning, Z. Chen, M. Marquis, K.B. Averyt, M.Tignor & H.L. Miller (eds.)]. Cambridge University Press, Cambridge, United Kingdom and New York, NY, USA.

Staubwasser, M., Sirocko, F., Grootes, P.M. & Segl, M. (2003). Climate change at the 4.2 ka BP termination of the Indus valley civilization and Holocene south Asian monsoon variability, Geophysical Research Letters, Vol. 30: L016822

Stern, A.L., Chamberlain, R.C., Reylonds, C.R. & Johnson, D.J. (1997). Oxygen isotope evidence of climate change from pedogenic clay minerals in the Himalayan molasse. Geochimica et Cosmochimica Acta, Vol. 61, pp. 731–744.

Strauss, H. & Peters-Kottig, W. (2003). The Paleozoic to Mesozoic carbon cycle revisited: the carbon isotopic composition of terrestrial organic matter. Geochemistry Geophysics Geosystems, Vol. 4, No. 10, pp. 1–15, 1083, 0.1029/2003GC000555.

Stuiver, M. & Braziunas, T.F. (1993). Sun, ocean climate and atmospheric $^{14}CO_2$: an evaluation of causal and spectral relationships. Holocene, Vol. 3, pp. 289-305.

Stuiver, M. et al. (1998). INTCAL98 Radiocarbon age calibration, 24,000±0 cal BP. Radiocarbon, Vol. 40, pp. 1041-1083.

Thiede, R.C., Bookhagen, B., Arrowsmith, J.R., Sobel, E. R. & Strecker, M.R. (2004). Climatic control on rapid exhumation along the Southern Himalayan Front. Earth and Planetary Science Letters, Vol. 222, pp. 791–806.

Thiry, M. (2000). Palaeoclimatic interpretation of clay minerals in marine deposits: an outlook from the continental origin. Earth-Science Reviews, Vol. 49, No. 1-4, pp. 201–221.

Thomas, E., Booth, L., Maslin, M. & Shackleton, N.J. (1995). Northeastern Atlantic benthic foraminifera during the last 45,000 years: productivity changes as seen from the bottom up. Paleoceanography, Vol. 10, pp. 545-562.

Thomas, J. V., Prakash, B. & Mahindra, R. (2002). Lithofacies and paleosol analysis of the Middle and Upper Siwalik Groups (Plio-Pleistocene), Haripur- Kolar section, Himachal Pradesh, India. Sedimentary Geology, Vol. 150, pp. 343–366.

Thoning, K.W., Tans, P.P. & Komhyr, W.D. (1989). Atmospheric carbon dioxide at Mauna Loa Observatory 2. Analysis of the NOAA GMCC data, 1974-1985, Journal of Geophysics Research, Vol. 94, pp. 8549-8565,

Thunell, R. & Sautter, L.R. (1992). Planktonic foraminiferal faunal and stable isotopic indices of upwelling. A sediment trap study in the San Pedro Basin, Southern California Bight. In: Summerhays, C.P., et al. (Ed.), Upwelling Systems. Evolution since the Early Miocene. Geological Society Special Publication, Vol. 64, pp. 77-91.

Tripati, A., Backman, J., Elderfield, H. & Ferretti, P. (2005). Eocene bipolar glaciation associated with global carbon cycle changes. Nature, Vol. 436, No. 7049, pp. 341–346.

van der Burgh, J., Visscher, H., Dilcher, D.L. & Ku¨rschner, W.M. (1993). Paleoatmospheric signatures in Neogene fossil leaves. Science, Vol. 260, pp. 1788–1790.

van Geel, B., Raspopov, O.M., Renssen, H., van der Plicht, J., Dergachev, V.A. & Meijer, H.A.J. (1999). The role of solar forcing upon climate change. Quaternary Science Reviews, Vol. 18, pp. 331-338.

Vaughan, A.P.M. (2007). Climate and Geology – a Phanerozoic perspective. In: Williams, M.; Haywood, A.M.; Gregory, F.J.;Schmidt, D.N., (eds.) Deep-time perspectives on climate change: marrying the signal from computer models and biological proxies. London, Geological Society of London, pp. 5-59. (Micropalaeontological Society Special Publications)

Wang, L., Sarnthein, M., Erlenkeuser, H., Grimalt, J., Grootes, P., Heilig, S., Ivanova, E., Kienast, M., Pelejero, C. & Pflaumann, U. (1999). East Asian monsoon climate during the late Pleistocene: high-resolution sediment records from the South China Sea. Marine Geology, Vol. 156, pp. 245–284.

Wang, P.X., Tian, J., Cheng, X.R., Liu, C.L. & Xu, J. (2004). Major Pleistocene stages in a carbon perspective: The South China Sea record and its global comparison. Paleoceanography, Vol. 19, No. 4.

Wang, Y., Cheng, H., Edwards, R.L., He, Y., Kong, X., An, Z., Wu, J., Kelly, M.J., Dykoski, C.A. & Li, X. (2005). The Holocene Asian monsoon, links to solar changes and North Atlantic climate. Science, Vol. 308, pp. 854-857.

Wang, Y.J., Cheng, H., Edwards, R.L., An, Z.S., Wu, J.Y., Shen, C.-C., & Dorale, J.A. (2001). A high-resolution absolute-dated late Pleistocene monsoon record from Hulu Cave, China: Science, Vol. 294, pp. 2345–2348, doi: 10.1126/science.1064618.

Wefer, G., Berger, W.H., Bijma, J. & Fischer, G. (1999). Clues to ocean history: a brief overview of proxies. In Fischer, G. & Wefer, G. (Eds.) Uses of Proxies in Paleoceanography: Examples from the South Atlantic. Springer-Verlag Berlin Heidelberg, pp. 1-68.

Wells, P., Wells, G., Cali, J. & Chivas, A. (1994). Response of deep-sea benthic foraminifera to late Quaternary climate changes, southeast Indian Ocean, offshore Western Australia. Marine Micropaleontology, Vol. 23, No. 3, pp. 185-229.

Weiss, H., Courty, M.A., Wetterstrom, W., Guichard F., Senior L., Meadow R. & Curnow A. (1993). The Genesis and Collapse of Third Millennium North Mesopotamian Civilization: Science, Vol. 261, pp. 995-1004.

Worsley, T.R., Moore, T.L., Fraticelli, C.M. & Scotese, C.R. (1994). Phanerozoic CO_2 levels and global temperatures inferred from changing paleogeography. Geol. Soc. Am. (Special Paper) 288, 57-73.

Wright, H.T., Jr., Kutzbach, J.E., Webb III, T., Ruddiman, W.E.F., Street-Perrott, F.A., Bartlein, P.J. (Eds.). (1993). Global Climates Since the Last Glacial Maximum. University of Minnesota Press, Milleapolis, pp. 569.

Yapp, C.J., Poths, H., (1992). Ancient atmospheric CO_2 pressures inferred from natural goethites. Nature, Vol. 355, pp. 342–344.

Yemane, K. & Kelts, K. (1996). Isotope geochemistry of Upper Permian early diagenetic calcite concretions: Implications for Late Permian waters and surface temperatures in continental Gondwana. Palaeogeography Palaeoclimatology Palaeoecology, Vol. 125, No. 1- 4, pp. 51–73.

Zachos, J., Pagani, M., Sloan, L., Thomas, E. & Billups, K. (2001). Trends, rhythms, and aberrations in global climate 65 Ma to present. Science, Vol. 292, No. 5517, pp. 686–693

Zachos, J.C., Lohmann, K.C., Walker, J.C.G. & Wise, S.W. (1993). Abrupt climate change and transient climates during the Paleogene: a marine perspective. Geology, Vol. 101, No. 2, pp. 191–213.

Zheng, H., Powell, C.M., Rea, D.K., Wang, J. & Wang, P. (2004). Late Miocene and mid-Pliocene enhancement of the east Asian monsoon as viewed from the land and sea. Global and Planetary Change, Vol. 41, pp. 147-155.

10

Paleotempestology: Reconstructing Atlantic Tropical Cyclone Tracks in the Pre-HURDAT Era

Jill S. M. Coleman and Steven A. LaVoie
Ball State University
USA

1. Introduction

The study of past tropical cyclone activity by means of geological proxies and/or historical documentary records is known as paleotempestology. This scientific discipline has become prominent over the course of the last decade partially in response to the recent increase in tropical cyclone count and intensity in the North Atlantic basin witnessed since 1995. The field has also developed due to the socioeconomic impacts of tropical cyclones particularly along vulnerable coastal regions. During the twenty-five years prior to the start of the most recent increase in hurricane activity, major (Category 3, 4, or 5) hurricanes were less frequent than in previous decades. Yet, property losses from the hurricanes that did make landfall in the United States increased during this period due to development in damage prone areas (NOAA Paleoclimatology Program, 2000). Many researchers hence stress the importance of identifying historical tropical cyclones to understand long term trends in tropical cyclone climatology and to determine the influence of anthropogenic global warming on tropical cyclone activity and intensity. The North Atlantic Hurricane Database (HURDAT) has been one of the authoritative sources for examining North Atlantic tropical cyclone activity trends since 1850. However, some of the deadliest known hurricanes and potentially most active seasons in the North Atlantic basin occurred prior to the beginning of the HURDAT record (Table 1), including the Great Hurricane of 1780 that killed an estimated 22,000 people and was one of eight known tropical cyclones during that season. This chapter will provide a brief overview of some paleotempestology techniques and illustrate a methodology for identifying and reconstructing historical North Atlantic tropical cyclone tracks in the pre-HURDAT era employing a Geographic Information System (GIS) and utilizing readily accessible archival data.

2. Paleotempestology methods

2.1 Geological proxies

Some paleotempestological studies utilize evidence of tropical cyclones found in the physical landscape, such as sedimentary records, tree rings, and other geological proxies. When an ocean storm or tsunami produces a surge, coastal sediments are brought inland up to several miles away from the ocean. These sediments are preserved in lakes and marshes located near the ocean and are collected by scientists to estimate the dates of significant surge events generated from tropical cyclones, strong winter cyclones, or tsunamis. North

Atlantic sediment-based overwash reconstructions are limited yet extend geographically throughout major tropical cyclone impact zones within the basin, such as along the U.S. eastern and Gulf of Mexico coasts and the Caribbean. In a Rhode Island sedimentary record analysis, Donnelly et al. (2001) determined at least seven major hurricanes made landfall in New England in the last 700 years, with three occuring prior to the start of HURDAT (1635, 1638, and 1815) and two before European settlement. Basin-wide sediment analysis reconstruction over the past 1,500 years have shown Atlantic tropical cyclone activity to have peaked during medieval times (circa AD 1000) and decreased in the modern era (Mann et al. 2009), particularly during the 1970s and 1980s (Nyberg, et al., 2007). In the Gulf of Mexico, Belize sedimentary records of the past 5000 years show major hurricanes may have struck central America an average of once every decade with tropical cyclone activity being especially active about 2,500 to 4,500 years before present (McCloskey and Keller, 2009). Liu (2007) and Liu & Fearn (2009) note a similar period of hyper-hurricane activity along the central-eastern U.S. Gulf coast between 3,800 and 1,000 years ago based on the multiple sand layers found in sediment cores taken between Lousiana and western Florida.

Rank	Year	Name	Deaths	Rank	Year	Name	Deaths
1	1780	The Great Hurricane	22,000+	16	1831	Great Barbados Hur.	2,500
2	1997	Mitch	19,325	17	1931	Unnamed	2,500
3	1900	Galveston Hurricane	12,000	18	1935	Unnamed	2,150+
4	1974	Fifi	10,000	19	1979	David	2,068+
5	1930	Hurricane San Zeon	8,000	20	1781	Unnamed	2,000+
6	1963	Flora	8,000	21	1893	Sea Islands Hurricane	2,000+
7	1776	Pointe a Pitre Bay Hur.	6,000+	22	1780	Solano's Hurricane	2,000
8	1775	Newfoundland Hurricane	4,000	23	1870	Hurricane San Marcos	2,000
9	1899	Hurricane San Ciriaco	3,433+	24	1893	Chenier Caminada Hurr.	2,000
10	1928	Okeechobee Hurricane	3,411+	25	1666	Unnamed	2,000
11	1932	Unnamed	3,107+	26	2005	Katrina	1,836
12	2004	Jeanne	3,025	27	2005	Stan	1,628
13	1813	Unnamed	3,000+	28	1767	Unnamed	1,600
14	1934	Unnamed	3,000+	29	1909	Unnamed	1,500
15	1791	Hurricane Las Puentas	3,000	30	1644	Unnamed	1,500

Table 1. The 30 deadliest recorded Atlantic hurricanes. Hurricanes that occurred prior to 1851 (the start of HURDAT) are highlighted. (Source: National Hurricane Center).

Tree ring analysis or dendrochronology is a relatively new tool for identification of pre-HURDAT tropical cyclones. Tree-ring cellulose store information on the oxygen isotopic composition (O^{18}) of source water around the time of growth, particularly in tree species with shallow root systems. Lower oxygen isotope water environments are common after extremely intense rainfall periods, such as after a tropical cyclone passage, and diminshed O^{18} levels are recorded in the cellulose in the weeks following an event. Miller et al. (2006) present a 220-year record of oxygen isotope values in longleaf pine tree rings in which they link anomalously low oxygen isotope values in the latewood portion of the ring (i.e., summer-early autumn growth) with tropical cyclones that impacted the southeastern United States. For example, tree ring samples from Valdosta, Georgia showed evidence of a 1780 hurricane, most likley Solano's Hurricane, that tracked from eastern Gulf of Mexico, made landfall in northwestern Florida, and then moved into Georgia before heading into the Atlantic.

2.2 Historical documents

Another aspect of paleotempestological research focuses on the examination of historical documents to reconstruct the tracks of tropical cyclones. When a strong tropical storm or hurricane makes landfall in a populated area, its impacts are typically recorded and can range from a qualitative description of the overall damage to detailed hourly observations pre- and post-impact. Most landfall accounts are in the form of newspapers, some of which simply reprint a story featured in another newspaper in the afflicted area and are reported often months after an event. For example, the impacts of an August 1788 Caribbean hurricane were not reported in London newspapers until more than two months after the event (Fig. 1). This introduces a potential error source on tropical cyclone landfall dates and impact times from newspapers and reports may need to be cross-referenced using government records and shipping logs. When tropical cyclones are over the open ocean, the logs of various merchant, transport, and naval ships are often our only information source in the pre-HURDAT era. The annual frequency of ship reports in the Atlantic varied according to the amount of trade between the Americas and the Old World and the level of conflict, if any. Based on ship track densities, Vecchi & Knutson (2011) suggest pre-satellite era Atlantic hurricane counts may be exceedingly underestimated and that increased hurricane activity in the late 20th century is a function of technological detection and not anthropogenic warming.

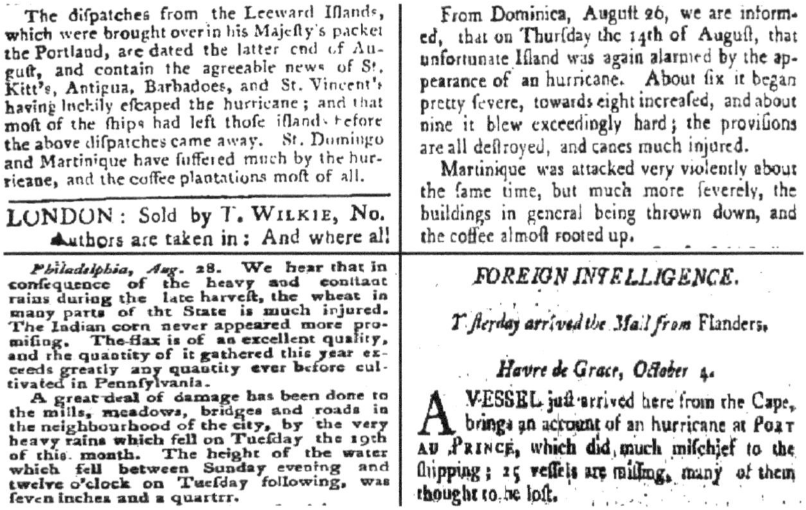

The diſpatches from the Leeward Iſlands, which were brought over in his Majeſty's packet the Portland, are dated the latter end of Auguſt, and contain the agreeable news of St. Kitt's, Antigua, Barbadoes, and St. Vincent's having luckily eſcaped the hurricane; and that moſt of the ſhips had left thoſe iſlands before the above diſpatches came away. St. Domingo and Martinique have ſuffered much by the hurricane, and the coffee plantations moſt of all.

LONDON: Sold by T. WILKIE, No.
Authors are taken in: And where all

From Dominica, Auguſt 26, we are informed, that on Thurſday the 14th of Auguſt, that unfortunate Iſland was again alarmed by the appearance of an hurricane. About ſix it began pretty ſevere, towards eight increaſed, and about nine it blew exceedingly hard; the proviſions are all deſtroyed, and canes much injured.

Martinique was attacked very violently about the ſame time, but much more ſeverely, the buildings in general being thrown down, and the coffee almoſt rooted up.

Philadelphia, Aug. 28. We hear that in conſequence of the heavy and conſtant rains during the late harveſt, the wheat in many parts of the State is much injured. The Indian corn never appeared more promiſing. The flax is of an excellent quality, and the quantity of it gathered this year exceeds greatly any quantity ever before cultivated in Pennſylvania.

A great deal of damage has been done to the mills, meadows, bridges and roads in the neighbourhood of the city, by the very heavy rains which fell on Tueſday the 19th of this month. The height of the water which fell between Sunday evening and twelve o'clock on Tueſday following, was ſeven inches and a quarter.

FOREIGN INTELLIGENCE.

Yeſterday arrived the Mail from Flanders.

Havre de Grace, October 4.

A VESSEL juſt arrived here from the Cape, brings an account of an hurricane at PORT AU PRINCE, which did much miſchief to the ſhipping; 25 veſſels are miſſing, many of them thought to be loſt.

Fig. 1. A collection of 1788 London newspaper articles summarizing the impacts of an August Caribbean hurricane. From upper left to lower right: Oct. 23rd *Chronicle*; Oct. 18th *Bristol Journal*; Nov. 14th *Public Advertiser*; and Oct. 7th *Times* (Heritage Archives, Inc., Available at: www.newspaperarchive.com)

Historical data from the late 18th through the 19th centuries are frequently employed in individual Atlantic and Gulf of Mexico hurricane track reconstructions. Early station observations, marine log books, colonial government reports and *London Gazette* newspaper articles were used to reconstruct the track of a 1680 tropical cyclone that formed near the

Cape Verde Islands, impacted the Caribbean Lesser Antilles, and struck the British Isles as a powerful extratropical cyclone (Wheeler et al., 2009). During the onset of the American Revolution, the "Independence Hurricane" made landfall in North Carolina on September 2nd, 1775 and then was thought to have tracked northward into Newfoundland on September 12th, leaving approximately 4,000 dead. Historic document analysis showed two separate cyclones rather than a single hurricane track a more likely scenario (Rapport & Ruffman, 1999; Williams, 2008).

Hurricane track reconstructions are also useful for comparing with the modern record. Mock et. al. (2010) reconstructed the Great Louisiana Hurricane of 1812 track using largely qualitative descriptions of prevailing wind direction and strength, precipitation intensity, and damage reports from merchant and naval ship logs, personal diaries, and newspaper accounts. The resultant track highlighted a "worst case scenario" for New Orleans and mimicked aspects of 2005 Hurricane Katrina. The landfall of October 2005 Hurricane Vince in Spain was widely reported as the first European tropical cyclone; however, historical archives have suggested a storm with a similar genesis and track occurring in October 1842 (Vaquero et al., 2008).

2.3 Chronologies

Historians have used geological proxies, historical documents and other archival data to construct comprehensive chronologies of Atlantic tropical cyclones. A chronology is a sequential listing or timeline of tropical cyclone events by date and locations impacted. The foundation for many tropical cyclone chronologies written over the last century stems from a paper written by Andrés Poey (1855) that synthesized data from shipping logs and newspaper accounts to catalog 400 cyclones in the North Atlantic and Caribbean basins between 1493 and 1855. The Poey chronology contains several serious flaws, such as the inclusion cyclones of non-tropical origins and structure and multiple entries for a single cyclone; these unintentional errors persist in other uncorrected Poey-based chronologies, such as in Tannehill (1938).

Perhaps the most widely used chronology referenced, the Tannehill (1938) chronology was the first considerable update of the original Poey listing. In addition to increasing the tropical cyclone event timeline by over 80 years, Tannehill added qualitative information on hurricane events and removed many clearly non-tropical cyclones, particularly those occurring outside the traditional bounds of the Atlantic hurricane season (i.e., December-May). Tannehill also produced several hurricane track reconstructions (e.g., Fig. 2), a basis for the modern "best-track" data approach of the HURDAT database. However, the Tannehill chronology does not combine entries from a single cyclone that impacted multiple locations, thereby inflating the number of different tropical cyclones. Subsequent chronologies in the latter half of the 20th century have strived to amend these inaccuracies.

Significant improvements and renewed interest in tropical cyclone chronologies occurred in the 1960s, coinciding with technological innovations in weather monitoring (i.e., satellite and radar) and several active hurricane seasons during the previous decade. *Atlantic Hurricanes* (Dunn & Miller, 1960) was one of the first exclusively American chronologies, categorizing tropical cyclones by six coastal impact zones for a 400-year period. *Early American Hurricanes, 1492-1870* (Ludlum, 1963) and *Hurricanes of the Caribbean and Adjacent Regions, 1492-1800* (Millás, 1968) reexamined earlier chronologies, most notably Poey (1855) and Tannehill (1938), on a case-by-case basis and provided additional evidence supporting the decision to accept or reject a particular entry as a valid tropical cyclone. In particular, Ludlum (1963) examines historical documents from a meteorological perspective, estimating the location and movement of some of the more prominent tropical cyclones of the early American period.

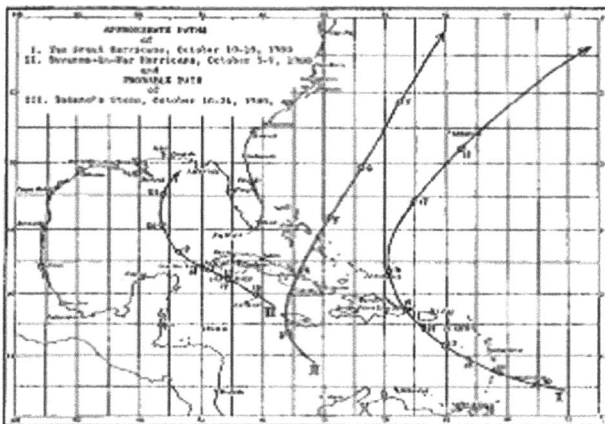

Fig. 2. The reconstructed tracks of three October 1780 hurricanes (Figure 70,Tannehill (1938))

Two notable modern chronologies incorporate, expand and correct earlier work and utilize newly available data sources and technological methods. Bossak & Elsner (2003) created the Historical Hurricane Information Tool (HHIT), a geographic information system (GIS) database that provides not only track data but supporting evidence for tropical cyclones of the early 19th century. The HHIT is publicly available database that uses historical archives (including previous chronologies, particularly Ludlum (1963)), knowledge of the behavior and structure of tropical cyclones, and GIS to document the impacts of 91 individual tropical cyclones that impacted the American coastline between 1800 and 1850. Modern technology has enabled easier accessibility of historical documents through online digital archives and language translation programs. Utilizing data sources not previously available, Chenoweth (2006) revises earlier chronologies, particularly that of Poey (1855) and Millás (1968), to create a comprehensive listing of historical Atlantic tropical cyclones with multiple supporting data sources. The data and methodologies defined in the Bossak & Elsner and Chenoweth studies serve as the foundation for the Atlantic tropical cyclone track reconstructions presented here.

3. Atlantic tropical cyclone track reconstruction

Prior to the start of HURDAT in 1851, a myriad of sources are needed to reconstruct hurricane frequencies and tracks, such as newspaper accounts, shipping logs, diaries, descriptive summaries, and chronologies. Taking these mostly qualitative, and occasionally conflicting or inaccurate, accounts of tropical cyclones and extracting information to develop a series of geographic coordinates is a significant challenge of historical hurricane database construction. The hurricane track reconstructions shown here are a result of identifying potential sources of quantitative and qualitative data, analyzing the historical archives, creating a tabular database for further data analysis, and finally creating the spatial dataset, including ArcGIS shapefiles (as outlined in LaVoie, 2011). This method advances other chronological or GIS based tropical cyclone reconstructions studies (e.g., Chenoweth, 2006; Bossak & Elsner, 2003) by utilizing multiple primary and secondary sources to determine the geographical locations impacted by a hurricane and presenting the resultant tracks in a visual medium. In addition to example track reconstructions, the long-term trends in tropical cyclone frequency is examined.

3.1 Data

3.1.1 Data sources: HURDAT and pre-HURDAT

The North Atlantic Hurricane Database (HURDAT) is considered one of the most comprehensive and authoritative databases for Atlantic hurricane track and landfall information currently available. Initially created in the 1960s to provide tropical cyclone forecasting guidance, the original database contained six-hourly "best track" and intensity information for all known tropical cyclones starting in 1886 (Jarvinen et. al, 1984). During the past decade, HURDAT has been going through an extensive reanalysis led by the Hurricane Research Division (United States) and Christopher Landsea to extend and correct the Atlantic hurricane database of individual cyclone tracks and intensities as well as U.S. landfalling storms. For example, 1992 Hurricane Andrew has since been upgraded from a Category 4 to Category 5 hurricane at Florida landfall and in August 2011 alone, about 20 different years have had alterations to the track and/or intensity of some tropical cyclones (AOML, accessed 2011). The current HURDAT extends from 1851 to present and is used to illustrate similarities in pre-HURDAT historical hurricane tracks with the modern era.

In order to conduct a temporal study of tropical cyclone activity prior to the start of HURDAT, data must first be extracted from historical archives containing mainly descriptive information and nominal level data. Quantitative measurements such as temperature and barometric pressure are very sparse in the pre-HURDAT era and measurements were limited to landfalls near larger population centers and to shipping lanes. Table 2 lists the data sources used in the pre-HURDAT analysis according to author, period of record, number of relevant tropical cyclone entries (i.e., confirmed tropical cyclones occurring during the analysis period from 1751-1850), and regional focus. These sources include many of the chronologies previously discussed (see Section 2.3) as well as several others with a more regional U.S. focus. Although some of these references are contemporaries of one another, each source provides a different type of information (e.g., quantitative data, qualitative descriptions, landfall impacts, track data, etc.).

In addition to the sources listed in Table 3.2, newspapers and shipping reports are used. In populated cities of the around the North Atlantic, newspapers would summarize accounts of hurricanes, both firsthand from their city as well as stories from sailors visiting their ports. Historical newspapers are acquired from those publicly available through the Heritage Archives, Inc. website (www.newspaperarchive.com). Shipping logs provide one of the few available sources of both wind speed and direction. These ship reports can help locate the center of circulation and are especially valuable over the open ocean. Unless specifically identified in one of the chronologies, ship reports are derived from the International Comprehensive Ocean-Atmosphere Data Set (ICOADS).

Source Name	Record	Relevant Entries	Region
Poey (1855)	1492-1855	296	North Atlantic
Perley (1891)	1635-1890	8	New England
Tannehill (1938)	1492-1855	327	North Atlantic
Dunn & Miller (1960)	1559-1958	101	United States
Ludlum (1963)	1492-1870	76	United States
Millás (1968)	1492-1800	136	Caribbean
Barnes (2001)	1546-2006	35	Florida
Sandrik & Landsea (2003)	1565-1899	44	Georgia, NE Florida
Bossak & Elsner (2003)	1800-1850	91	United States
Chenoweth (2006)	1700-1855	308	North Atlantic
Barnes (2007)	1524-1999	15	North Carolina
Hairr (2008)	1775-1999	7	North Carolina

Table 2. Pre-HURDAT analysis data sources arranged by publication date. The number of relevant entries refers to tropical cyclone events occurring between 1750 and 1850 in the North Atlantic (including the Caribbean and Gulf of Mexico).

Despite the extensive range of resources for historical hurricane track reconstruction, each source has particular caveats that must be considered when extracting data from them. HURDAT is undergoing incessant revisions and it is likely that an average of three tropical cyclones are not accounted for in the database each year prior to 1966, the commencement of continuous satellite monitoring of the North Atlantic basin (Landsea, 2007). The HHIT database provides users with track information for recorded cyclones that impacted only the United States between 1800 and 1850 (versus the entire ocean basin) and does not account for potential error sources in the early chronologies used as the foundation for the track reconstructions. As noted in Section 2.3, the primary difficulty with utilizing chronologies is that most authors use previous chronologies as a foundation for their studies and as a result, some incorrect data is transferred into the updated listing. The disadvantage of many newspaper accounts is that many of them contain secondhand reproductions from other newspapers or sources which can result in a miscommunication of details. While shipping logs are our most likely data source for hurricane tracks at sea, anemometers were generally unavailable and wind speed was estimated using the Beaufort scale (i.e., the relative wave height) or qualitatively described as "hurricane force" winds. Each of these limitations is addressed on a case-by-case basis whereby a tropical cyclone is confirmed using multiple sources (as opposed to a single chronology) before being incorporated into the master database list.

3.1.2 Methodology

A master database of North Atlantic tropical cyclones from 1751-1850 was created using quantitative and qualitative information from the aforementioned data sources. Each tropical cyclone event was coded based on the number of data sources and information origin (e.g., chronology, newspaper, etc.). This included linking the HHIT database track positions for tropical occurring between 1800 and 1850. Each entry was then subsequently cross-referenced for spatial and temporal consistency among the data sources. A tropical cyclone from the 1788 Atlantic hurricane season is used to illustrate this process.

In mid-August 1788, the third known tropical cyclone of that season impacted the Caribbean and United States. Multiple data sources document the existence of an Atlantic tropical cyclone during this period, but disagree on whether or not the impacts in the Caribbean and United States are from a single tropical cyclone or are two separate events (Table 3). Locally known as "Hurricane San Rouge" in the Caribbean and the "Western New England Hurricane" in the northeastern United States, the Millás (1968), Chenoweth (2006), Poey (1855) and Tannehill (1938) chronologies list multiple events for the third (or fourth) 1788 tropical cyclone. Chenoweth (2006) lists this hurricane as two individual cyclones; Hurricane San Rouge dissipating after impacting Haiti on August 16th and a different tropical storm (not a hurricane) impacting the United States a few days later. In contrast, Ludlum (1963) identifies only a single tropical cyclone impacting both regions between August 14th and August 19th. Additional supporting evidence from newspapers, ship logs and overwash sediments in New Jersey (Donnelly et al., 2004) suggest that these cyclones should be classified as a single hurricane that made landfall in the Caribbean and the northeastern United States. Consequently, all entries related to a tropical cyclone impacting the Caribbean around August 14th and the northeastern United States by August 19th are merged into single event and are used to construct the tropical cyclone track positions.

After creation of the master database, six-hourly geographic positions for each storm are estimated. Six-hour intervals were chosen to coincide with the HURDAT format. Pre-HURDAT tropical cyclones often lack the quantitative data of modern-era storms; however the storm track and intensity can be estimated from reports detailing the damage extent, landfall time, wind direction shifts, and relative storm location observations. In addition, the observed tracks of similar tracking tropical cyclones can be used to estimate the likely cyclone trajectory in areas where data is sparse. The coordinates are then plotted using mapping software applications (e.g., Google Earth) or a geographic information system. The ESRI ArcGIS system enables the transfer and display of multiple sets of coordinates and manipulation of different storm attributes, including linking qualitative impact information.

Using the information presented in Table 3, the estimated track of the third 1788 Atlantic tropical cyclone is shown in Figures 3 and 4. The data indicates the hurricane center struck the island of Martinique shortly after midnight on August 15th and moved directly west since the islands of Antigua to the north and Trinidad to the south did not report any impacts from the hurricane. After damaging primarily the southwestern portion of Puerto Rico, the cyclone crossed the island of Hispaniola from southeast to northwest on August 16th and 17th before moving rapidly north towards New England. Based on Perley (1891) and contemporary newspaper accounts, the hurricane made landfall in southern New Jersey on August 19th and quickly transitioned into an extratropical storm, a typical pattern of poleward moving storms though the northeastern United States (Fig. 4).

The final master database for pre-HURDAT Atlantic tropical cyclones contained 408 entries for the 1751-1850 period. Each entry on average appeared in three or more primary sources. While sufficient information is not available to reconstruct tracks for all 408 entries with reasonable confidence, the process outlined here shows the potential for extending the HURDAT record. The reconstructed tracks for two example pre-HURDAT Atlantic tropical cyclones are presented in the following section. One example illustrates track reconstruction using archival documents and the other shows how the atmospheric environment during the hurricane life cycle can be approximated in cases when meteorological observations are sparse or non-existent.

Source	Data
Chenoweth (2006)	August 14-18th hurricane impacting central Leeward Is. and Haiti August 17-19th tropical storm impacting southeastern Pennsylvania and New England
Millás (1968)	Tropical cyclone near Martinique, Dominica and Guadeloupe on August 14-15th "Hurricane San Rouge" south and near Puerto Rico and Hispaniola August 16-17th
Ludlum (1963)	Tropical cyclone impacting Martinique, Puerto Rico, Pennsylvania (PA), New Jersey (NJ), New York (NY), Connecticut (CT), Massachusetts (MA), Vermont (VT), New Hampshire (NH), and Maine (ME) August 14-19th
Dunn and Miller (1960)	Minimal hurricane with six deaths in NH, MA, and CT Much hurricane damage in Eastern New York
Tannehill (1938)	Tropical cyclone near Martinique August 14-15th Tropical cyclone near Puerto Rico and Hispaniola August 16-17th Tropical cyclone impacting the United States August 19th
Poey (1855)	Tropical cyclone near Martinique, Puerto Rico and Hispaniola August 14-16th Tropical cyclone impacting the United States August 19th
Perley (1891)	During the afternoon of August 19th, a gale impacted portions of western New England and eastern NY. The event lasted from around noon until approximately 4pm. There were reports of damage in southwestern CT, western MA, southern VT, and southwestern NH. Winds during the event were variable.
Ship Log Summary	Available observations on August 14th suggest that a large area of high pressure dominated the eastern Atlantic. A ship located between St. Kitts and Antigua reports a NE wind at 20mph, the outer edge of the hurricane. The same ship, located north of the Lesser Antilles on the 16th, experiences a strong pressure gradient between the hurricane and the Bermuda-Azores high. A ship moving just off the east Florida coast reports northerly winds as the hurricane moves well to the east. A ship located to the east of Boston reports strong winds albeit below tropical storm force, from the south on August 19th.
Newspapers	From Millás (1968): Guadeloupe and Martinique suffered from a gale the night of August 14th. Dominica saw much damage to sugar canes and provisions but damage was much worse in Martinique. St. Lucia did not suffer much and the hurricane did not impact Antigua which was suffering from a drought. The hurricane passed south of Puerto Rico and crossed Hispaniola from southeast to northwest on August 16th and 17th. A Spanish ship sunk in the western Bahamas. From various London Newspapers: The islands of St. Kitts, Antigua, Barbados, and St. Vincent escaped the hurricane while Hispaniola and Martinique suffered much. 25 vessels are missing at Port au Prince. The hurricane started at Dominica at 6pm (on August 14th) and increased in violence throughout the night. The hurricane was more severe at Martinique. Much damage was done to crops in Pennsylvania (on the 19th).

Source	Data
Final Result	Tropical Cyclone 1788-3: Hurricane San Rouge/Western New England Hurricane August 14th: Landfall in Martinique moving West August 15th: Northeastern Caribbean Sea August 16th: Approaching Puerto Rico August 17th: Impacting Puerto Rico and Hispaniola August 18th: Moving rapidly towards the North-Northeast August 19th: US landfall; moving North rapidly

Table 3. Listing of available sources for the third 1788 tropical cyclone and the resulting entry used for track reconstruction.

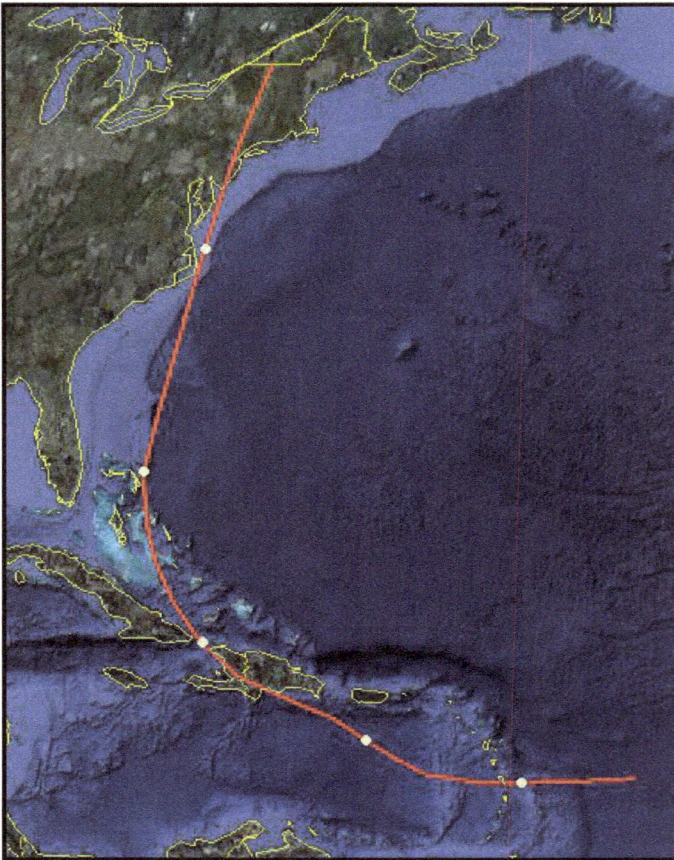

Fig. 3. Estimated track of the 1788 Hurricane San Rouge/Western New England Hurricane. White dots are 0Z positions from August 15th through August 19th.

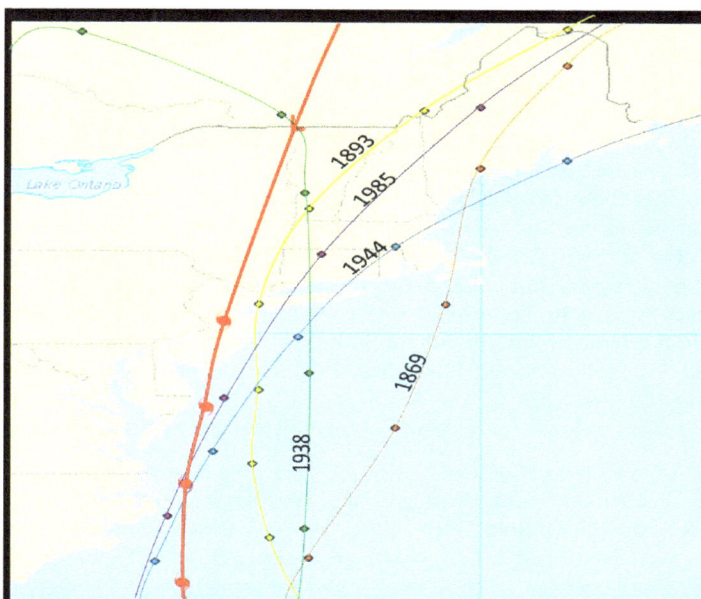

Fig. 4. The estimated United States impact track of 1788 Hurricane San Rouge/Western New England Hurricane and five other notable New England hurricanes with similar trajectories.

4. Pre-HURDAT data analysis: Individual track reconstructions

4.1 Analysis of the Great Hurricane of 1780

The deadliest known Atlantic hurricane on record occurred in October 1780, impacting primarily the eastern Caribbean and resulting in an estimated 22,000 deaths (Table 1). Known simply as the "Great Hurricane", the tropical cyclone is a prominent entry in nearly all archival data sources, including the Ludlum (1963) chronology that does not focus on areas outside the United States. Although most sources depict a track beginning in the Lesser Antilles and culminating in the vicinity of Bermuda, there are several key differences.

Figures 5 and 6 depict the "best track" of the Great Hurricane of 1780 (solid red line) analyzed using the master database sources (see Section 3.1) compared with five alternative tracks produced by previous authors (Norcross, 2007; Millás (1968); Ludlum (1963); Tannehill (1938); and Reid (1838)). Each storm description indicates a hurricane formed far southeast of the Lesser Antilles and made landfall (or very near) Barbados at nearly peak intensity before heading northwestward within the vicinity of St. Lucia and Martinique. Ship reports indicate that the hurricane moved very slowly in the eastern Caribbean before passing over or very near the small island of Mona located between Puerto Rico and Hispaniola. The storm then recurved northeastward around 70° W longitude and passed some distance to the southeast of Bermuda before moving into the open Atlantic. The track variations among different authors may be a result of different interpretations of wind information and reports of damage on land. In particular, the lack of damage reports from certain locations is an important component to understanding the actual track of this hurricane.

The Great Hurricane of 1780 appears to have taken a track that is more unusual than depicted by previous historians (Fig. 5 and 6). Sources agree the hurricane took an unusually long time to traverse the eastern Caribbean, taking nearly a week to go from Barbados to eastern Hispaniola. Local weather reports throughout the Caribbean indicate the rate of storm motion was not constant and the hurricane may have stalled and/or changed direction after passing into the Caribbean Sea. Both Reid (1838) and Tannehill (1938) assumed that the hurricane did not deviate significantly from a northwesterly track through the Caribbean but this fails to explain the high death toll (>> 1,000) recorded from St. Eustatius in the northern Lesser Antilles. In contrast, Ludlum (1963) and Norcross (2007) believe that the hurricane did pass close to the northern Lesser Antilles, with a track suggesting direct impacts to Antigua, St. Croix, and Montserrat; however, Antigua, located only 75 miles (120 kilometers) to the east of St. Eustatius, received only minor damages and Montserrat and St. Croix reported none. Millás (1968) compensated for the disparities in damage reports but the track of the hurricane still does not pass close enough to St. Eustatius and St. Kitts to cause the level of destruction reported on these islands.

The "best track" of the Great Hurricane proposed here (Fig. 5 and 6) modifies the previously discussed track positions by incorporating qualitative information from newspaper accounts and several key ship observations from military vessels throughout the region. The log of the HMS Albemarle, stationed off the southwestern coast of Barbados, indicates that the hurricane passed just north of its location where northeasterly winds eventually shifted to the west and then south. The HMS Alcmene, located off the southwestern coast of Martinique, reported a gradual change in the wind direction as the eye of the hurricane

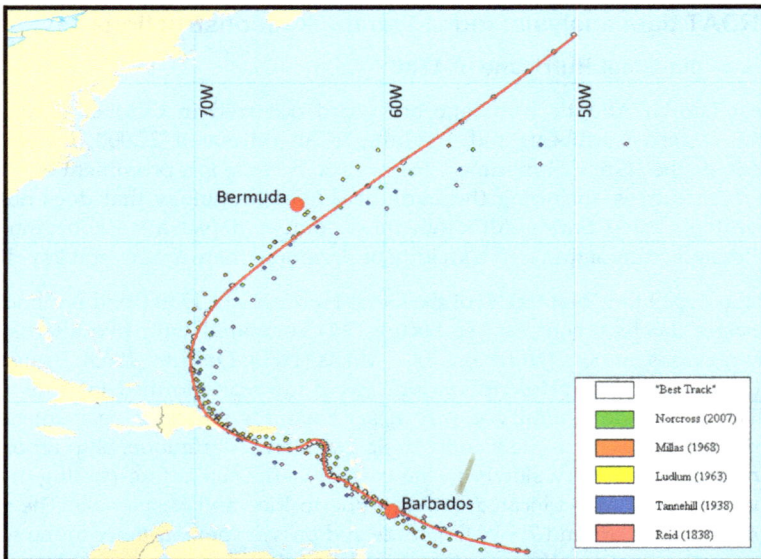

Fig. 5. The "best track" of the Great Hurricane of 1780 (solid red line) compared with five alternative tracks suggested by previous authors (Norcross, 2007; Millás (1968); Ludlum (1963); Tannehill (1938); and Reid (1838)).

passed very close to that island. Finally, the HMS Star located near Antigua reported four days of winds out of the east before shifting to the southeast on the 14th of October. Land reports show the most severe storm damage came from Barbados, St. Lucia, St. Vincent, Martinique, southern Guadeloupe, St. Kitts, St. Eustatius, southwestern Puerto Rico, and southeastern St. Domingo (Hispaniola). This information suggests that after the hurricane passed Martinique and Dominica, the cyclone turned north towards St. Kitts before abruptly shifting west/southwest and eventually northwest across extreme southeastern Hispaniola (Fig. 6). While this is an atypical track for any tropical cyclone in the North Atlantic, similar behavior has been observed in other hurricanes (e.g., Hurricane Marilyn in 1995).

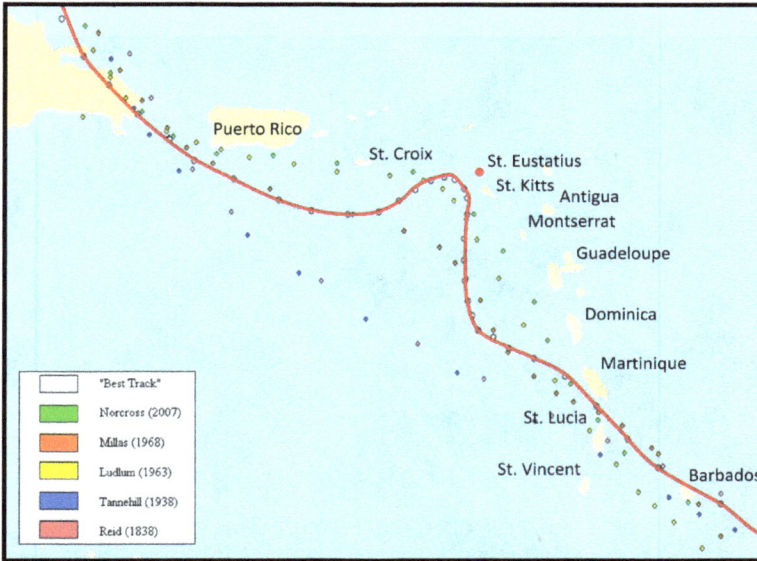

Fig. 6. Same as Figure 5 except enlarged and focused on the Eastern Caribbean.

4.2 The 1821 New England hurricane

The 1821 New England hurricane, also known as the "Norfolk and Long Island Hurricane", was the first tropical cyclone of that season. First detected north of Puerto Rico on September 1st, this fast moving and powerful hurricane passed north of the Bahamas, turned towards the north, and made the first of several landfalls about 20 miles east of Moorehead City, North Carolina late on September 2nd. The center then stayed within 30 miles of the East Coast of the United States. This 1821 hurricane was the only major hurricane to directly impact the modern limits of New York City, making landfall on September 3rd. Following a brief passage over Long Island, the hurricane made its final landfall in Connecticut on September 4th and passed through New England and into Canada.

4.2.1 An analysis of the 1821 Hurricane using 20th century data

While meteorological observations during the pre-HURDAT era are scarce, the analogs of modern tropical cyclones can be used to visualize the general synoptic weather pattern that prevailed when a hurricane made landfall. The track of the 1821 New England hurricane (and hurricanes in the 1820s and 1830s in general) followed the hurricane track patterns of many storms of the 1950s and 1960s that featured similar number of landfalls along the eastern United States. In particular, the tracks of Hurricanes Carol (1954), Edna (1954), Donna (1960) and Gerda (1969) display a comparable U.S. landfall trajectory, stretching from Cape Hatteras through New England (Fig. 7). Hurricanes Carol, Edna and Donna are used to create a composite of the overall synoptic environment around the time of landfall, thus serving as a proxy for meteorological conditions during the 1821 New England Hurricane; Hurricane Gerda was excluded from the final composite analysis since the hurricane center did not cross into New England.

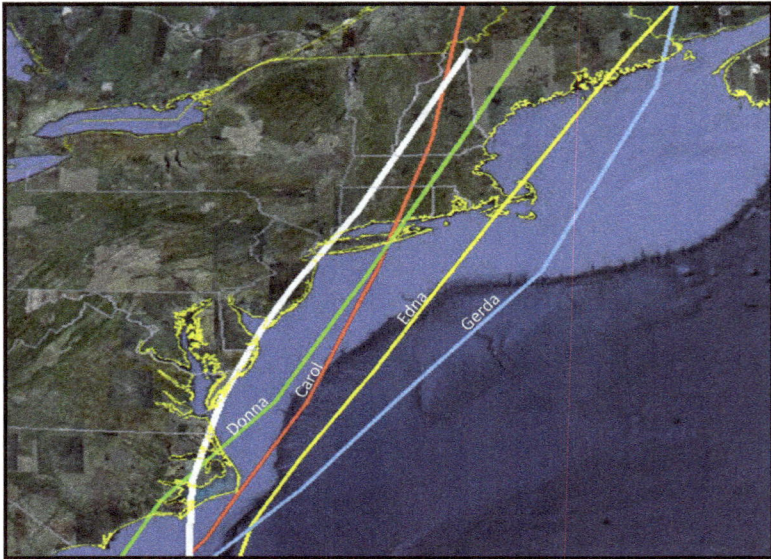

Fig. 7. Reconstructed track of the 1821 New England Hurricane (thick white line) and similar track hurricanes used for the synoptic environment composite analysis (Hurricanes Carol (1954); Edna (1954); Donna (1960); and Gerda (1969)).

Using the three selected 20th century tropical cyclones, composites of the 500-hPa geopotential heights (in meters) and mean sea level pressure (SLP) (in hPa) were generated for five time periods: two days, one day, and twelve hours prior to landfall; during landfall; and twelve hours after landfall. The composites were generated using six-hourly data from the NCEP/NCAR Reanalysis dataset (Kalnay, et. al, 1996) and images generated from the NOAA/ESRL Physical Sciences Division website. The composite analysis highlights the rapid alteration of the synoptic environment surrounding the time of landfall (Fig. 8a-e).

Two days prior to a hurricane making landfall in New England (Fig. 8a), the Bermuda-Azores High dominates the eastern and central Atlantic, centered around 45° W. In the mid-latitudes, a generally zonal pattern quickly becomes increasingly meridional as the hurricane moves up the eastern United States coast (Fig. 8b-d). The ridge over the central

Atlantic strengthens in response to a deepening trough over the eastern Great Lakes region. By twelve hours prior to landfall (Fig. 8c), a strong geopotential height gradient is developed and the hurricane follows this southwest to northeast orientated gradient which continues to intensify even after the hurricane has made landfall.

The composite presented in Figure 8 represents the likely synoptic environment for the 1821 New England Hurricane that followed a similar trajectory to its mid-20th century counterparts. During the time the hurricane was rapidly approaching New York City from the southwest (September 3rd at 22Z), wind direction observations and weather conditions from ICOADS ship reports indicate a hurricane along the northeast U.S. coast and also suggest the presence of a geographically extensive high pressure system in the north-central Atlantic (Fig. 9). The wind behavior of the ten meteorological observations in the northeastern United States clearly indicates the counterclockwise rotation around the hurricane center (Fig. 10). These observations confirm the results of the composite analysis.

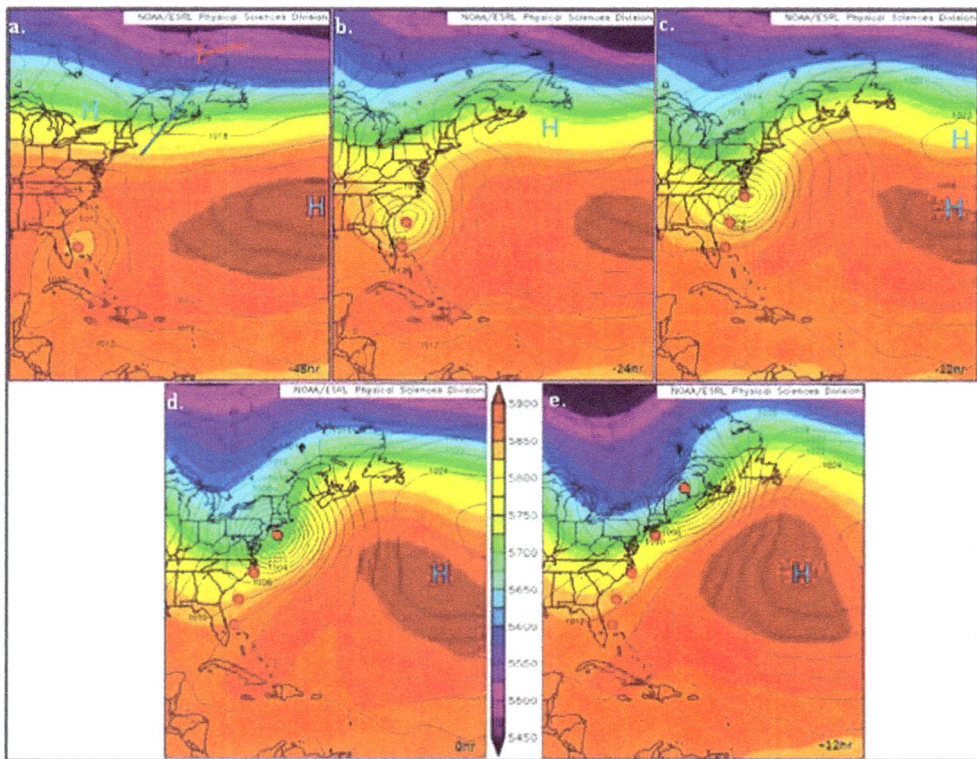

Fig. 8. Composite time series for Hurricanes Carol (August 31, 1954 at 12Z), Edna (September 11, 1954 at 18Z), and Donna (September 12, 1960 at 18Z). Solid contours are SLP (2 hPa intervals) and shaded contours are 500-hPa geopotential heights (50 m intervals). Time series from top left to bottom right are: (a) 48 hours, (b) 24 hours, and (c) 12 hours prior to landfall; (d) during landfall; and (e) 12 hours after landfall.

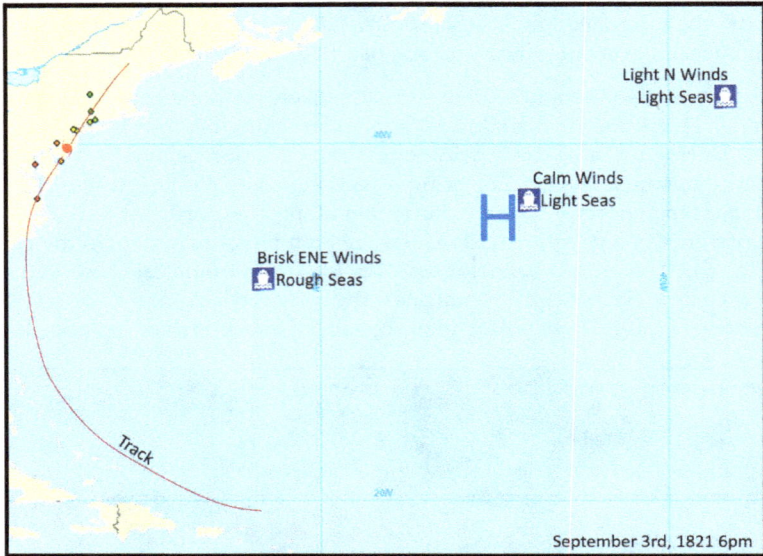

Fig. 9. The estimated 1821 New England Hurricane track and the locations of the available meteorological and ship report observations for September 3rd at approximately 22Z.

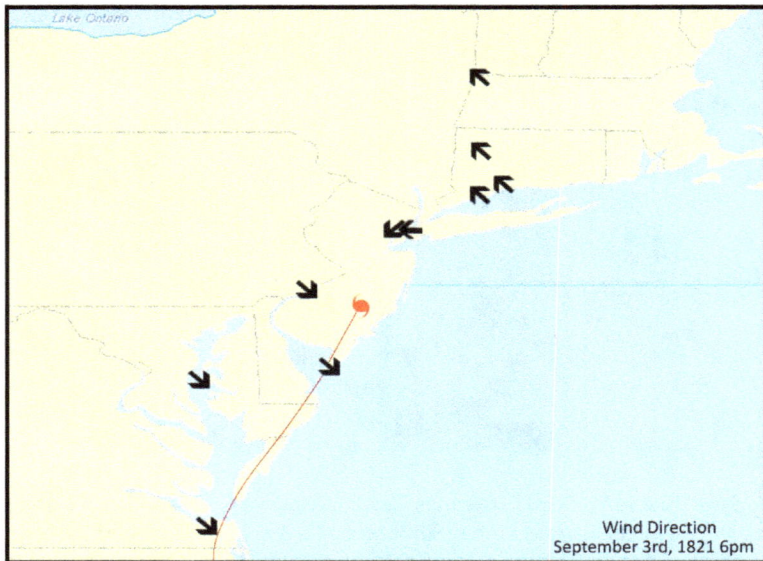

Fig. 10. The approximate location of the 1821 New England Hurricane (New Jersey) and wind direction at ten land locations taken around 22Z on September 3rd.

5. Pre-HURDAT era analysis: Long-term trends

In the past decade much attention has been given to the potential relationship between climate variability and change and relative tropical cyclone activity over time. The debate centers on whether or not the dramatic increase in the number of tropical cyclones observed since 1995 is due to natural variability in conjunction with our increasing ability to detect tropical cyclones or simply due to anthropogenic effects (e.g., Mann and Emanuel, 2006; Knutson et al., 2010; Holland and Webster, 2007). Many studies analyzing this issue analyze only post-satellite era data, which may not accurately reflect long-term trends or cycles in tropical cyclone activity.

Tropical cyclone and subtropical counts for the period 1851-2010 were derived from the official HURDAT "best track" database (AOML, accessed 2011) and combined with the pre-HURDAT era (1751-1850) count estimates derived from the master database. Based on the 1751-2010 time frame a gradual increase in the Atlantic tropical cyclone counts is apparent, with the largest increase beginning in the late 1990s that continues into the 21st century (Fig. 11). The most named tropical cyclones observed in the Atlantic basin occurred in 2005 (n = 28); however, several other peak years are apparent. The hurricane seasons that show the most robust signal are 2005, 1969 (n = 18), 1933 (n = 21), 1887 (n = 19), and 1837 (n = 12) with an average 42 years between these peak events. The record 2005 year (and the higher tropical cyclone activity in the past two decades in general) may not be an unusual occurrence, especially if pre-satellite era counts are adjusted for non-landfalling tropical cyclones that may have gone undetected.

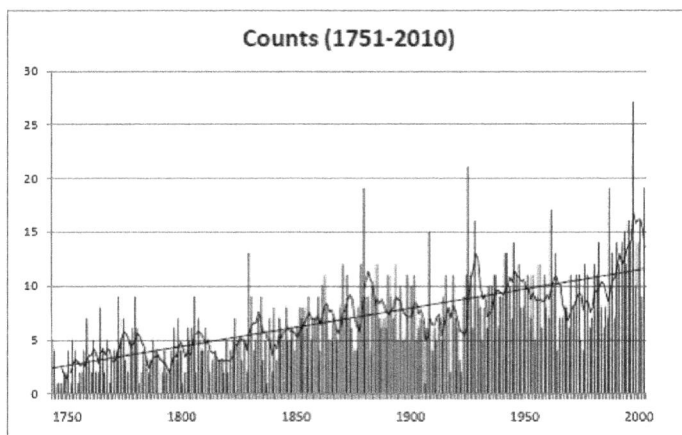

Fig. 11. Tropical cyclone frequencies by year for the period 1751-2010 with trend line and five-year moving average.

Periods of "hyperactivity" are often followed by a phase of subdued tropical cyclone activity (Fig. 12). During the 20th century two periods of below normal Atlantic tropical cyclone counts (1900 to 1931 and 1972 to 1995) were followed by sudden extended increases (1932-1971 and 1995 to present). The pre-HURDAT era shows a similar cyclical, multi-decadal pattern in tropical cyclone frequency, albeit shorter in duration than in the past century. Guan and Nigam (2009) note four leading modes of Atlantic sea surface temperature (SST) variability in the North Atlantic (the Atlantic Multidecadal Oscillation; a two-mode, El Niño-like pattern in the tropical eastern Atlantic; and an SST tripole resembling the North Atlantic Oscillation) that can produce warm and cold Atlantic temperature phases with seasonal to multi-decadal time scale. These modes may be important for explaining the cyclical and linear trend in the number of Atlantic tropical cyclones, but not necessarily any changes in the intensity (Briggs, 2008).

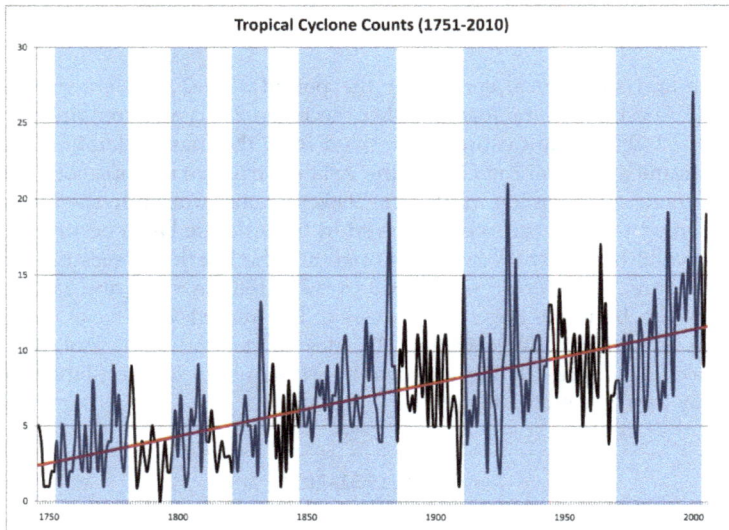

Fig. 12. Tropical cyclone frequencies by year for the period 1751-2010 with trend line and five-year moving average with period of increasing tropical cyclone activity highlighted.

6. Conclusion

In addition to a broad overview of paletempestology methods, a simplistic methodology for taking mainly qualitative archival information and reconstructing the tracks of historical tropical cyclones is presented. This paper shows analysis of North Atlantic hurricane seasons prior to the start of the HURDAT database (before 1851) is feasible using readily available public data. Furthermore, historical hurricane tracks need to be continually analyzed if new data sources become available, as shown in the case of the Great Hurricane of 1780. A comparison of landfall patterns associated with similar landfalling east coast hurricanes in the pre-HURDAT period and the modern record highlights the potential for using historical hurricane track patterns for real-time hurricane track forecasting.

By extending the North Atlantic tropical cyclone record back to 1751 using basic historical documents alone, the frequency of tropical cyclones in the Atlantic appears to be increasing. A cyclical pattern in activity also emerges, alternating by periods of 10-20 years of active and non-active seasons. Further research needs to be done to confirm whether or not these trends are a product of anthropogenic induced climate change or are a natural response to shifts in long-term atmospheric-oceanic patterns, such as the Atlantic Multidecadal Oscillation (AMO).

7. References

Atlantic Oceanographic and Meteorological Laboratory (AOML), Hurricane Research Division. (Accessed 2011). North Atlantic Hurricane Database (HURDAT) and Reanalysis Project, Available from http://www.aoml.noaa.gov/hrd/hurdat.

Barnes, J., 2001: *North Carolina's Hurricane History*. 3rd ed. Univ. of Barnes, J. (2007) *Florida's Hurricane History*. University of North Carolina, 407 pp.

Barnes, J. (2001). *North Carolina's Hurricane History*. 3rd ed. University of North Carolina, 319 pp.

Briggs, W. M. (2008). On the Changes in the Number and Intensity of North Atlantic Tropical Cyclones. *Journal of Climate*, Vol. 21, pp. 1387-1402.

Bossak, B.; & Elsner, J. (2003). Use of GIS in Plotting Early 19th Century Hurricane Information, *Historical Hurricane Information Tool*, Available from http://mailer.fsu.edu/~jelsner/HHITProject/FullText.pdf

Chenoweth, M. (2006). Reassessment of Historical Atlantic Basin Tropical Cyclone Activity,1700-1855. *Climatic Change*, Vol. 74, DOI: 10.1007/s10584-005-9005-2.

Donnelly, J.; Butler, J.; Roll, S.; Wengren, M. & Webb III, T. (2004). A Backbarrier Overwash Record of Intense Storms from Brigantine, New Jersey, *Marine Geology*, Vol. 210, pp. 107–121.

Dunn, G., & Miller, B. (1960). *Atlantic Hurricanes*. Louisiana State University, 326 pp.

Guan, B., & Nigam, S. (2009). Analysis of Atlantic SST Variability Factoring Interbasin Links and the Secular Trend: Clarified Structure of the Atlantic Multidecadal Oscillation. *Journal of Climate*, Vol. 22, pp. 4228–4240.

Hairr, J. (2008). *The Great Hurricanes of North Carolina*. History Press, 160 pp. Heritage Archives, Inc. *Newspaper Archives. Com*, Available from http://www.newspaperarchive.com

Holland, G.; & Webster, P. (2007). Heightened Tropical Cyclone Activity in the North Atlantic: Natural Variability or Climate Trend? *Philosophical Transactions of the Royal Society*, Volume 365, pp. 2695–2716.

International Comprehensive Ocean-Atmosphere Data Set (ICOADS). (Accessed 2011), Release 2.5., Available from http://icoads.noaa.gov.

Jarvinen, B.; Neumann, C.; & Davis, M. (1984). A Tropical Cyclone Data Tape for the North Atlantic Basin, 1886–1983: Contents, Limitations and Uses, Technological Memo. NWS NHC-22, 21 pp.

Kalnay, E. & Coauthors. (1996). The NCEP/NCAR Reanalysis 40-year Project, *Bulletin of the American Meteorological Society*, Volume 77, pp. 437-471.

Knutson, T.; Landsea, C.; & Emanuel, K. (2010). Tropical Cyclones and Climate Change, *A Review in Global Perspectives on Tropical Cyclones: From Science to Mitigation*, World Scientific Publishing Company, pp. 243-284.

Landsea, C. (2007). Counting Atlantic Tropical Cyclones Back to 1900. *EOS Transactions, American Geophysical Union*, Vol. 88, pp. 197-202.

LaVoie, S. (2011). *An Analysis of Hurricane Seasons in the Pre-HURDAT Era (1751-1850)*, Masters Thesis, Ball State University, 100 pp.

Liu, K. (2007). Uncovering Prehistoric Hurricane Activity, *American Scientist*, Vol. 95, pp.126-133.

Liu, K.; & Fearn, M. (2000). Reconstruction of Prehistoric Landfall Frequencies of Catastrophic Hurricanes in Northwestern Florida from Lake Sediment Records, *Quaternary Research*, Vol. 54, pp. 238–245.

Ludlum, D. (1963). *Early American Hurricanes, 1492-1870*. American Meteorological Society, 342 pp.

Mann, M.; & Emanuel, K. (2006). Atlantic Hurricane Trends Linked to Climate Change. *EOS Transactions, American Geophysical Union*, Volume 87, pp. 233–241.

Mann, M. E.; Woodruff, J. D.; Donnelly, J. P.; & Zhang, Z. (2009). Atlantic Hurricanes andClimate Over the Past 1,500 Years, *Nature*, Vol. 460, pp. 880-883.

McCloskey, T; & Keller, G. (2009). 5000-Year Sedimentary Record of Hurricane Strikes onthe Central Coast of Belize, *Quaternary International*, Vol. 195, pp. 53–68.

Millás, J., (1968). *Hurricanes of the Caribbean and Adjacent Regions, 1492-1800*. Academy of the Arts and Sciences of the Americas, 328 pp.

Miller, D.; Mora, C.; Grissino-Mayer, H.; Mock, C.; Uhle, M. & Sharp, Z. (2006). Tree-ringIsotope Records of Tropical Cyclone Activity, *PNAS*, Vol. 103, pp. 14294-14297.

Mock, C. J.; Chenoweth, M.; Altamirano, I.; Rodgers, M. D.; & García-Herrera, R. (2010). The Great Lousiana Hurricane of 1812, *Bulletin of the American Meteorological Society*, Vol. 91, pp. 1653-1663.

NOAA/ESRL Physical Sciences Division. Atmospheric Variables Plotting Page. Boulder Colorado, Available from http://www.esrl.noaa.gov/psd

NOAA Paleoclimatology Program. (2000). Paleotempestology Resource Center, Available from http://www.ncdc.noaa.gov/paleo/hurricane

Norcross, B. (2007). *Hurricane Almanac: The Essential Guide to Storms Past, Present, and Future*,St. Martin Griffin, 336 pp.

Nyberg, J.; Malmgren, B.A. ;Winter, A.; Jury, M. R.; Kilbourne, K. H. & Quinn, T. M. (2007).Low Atlantic Hurricane Activity in the 1970s and 1980s Compared to the Past 270 Years, *Nature*, Vol. 447, pp. 698–701.

Perley, S. (1891). *Historic Storms of New England*. Commonwealth Editions, 320 pp.

Poey, A. (1855). A Chronological Table Comprising 400 Cyclonic Hurricanes which have Occurred in the West Indies and in the North Atlantic within 362 years, from 1493-1855. *The Journal of the Royal Geographical Society*, Vol. 25, pp. 291-328.

Rappaport, E.; & Ruffman, A. (1999). The Catastrophic 1775 Hurricane(s): the Search for Data and Understanding, *Preprints, 23rd Conference on Hurricanes and Tropical Meteorology*, Dallas, TX. American Meteorological Society, pp. 787-790.

Reid, W. (1838). *An Attempt to Develop the Law of Storms by Means of Facts, Arranged According to Place and Time*. J. Weale, 463 pp.

Sandrik, A.; & Landsea, C. (2003). Chronological Listing of Tropical Cyclones Affecting North Florida and Coastal Georgia 1565-1899 NOAA. Hurricane Research Division. [Available online at http://www.aoml.noaa.gov/hrd/Landsea].

Tannehill, I. (1938). *Hurricanes*. Princeton University Press, 504 pp.

Vaquero, J. M.; García-Herrera, R.; Wheeler, D.; M. Chenoweth; & Mock, C. J. (2008). A Historical Analog of 2005 Hurricane Vince, *Bulletin of the American Meteorological Society*, Vol. 89, pp. 191-201.

Vecchi, G. A.; & Knutson, T. R. (2011). Estimating Annual Numbers of Atlantic Hurricanes Missing from the HURDAT Database (1878-1965) Using Ship Track Density, *Journal of Climate*, Vol. 24, pp. 1736-1746.

Wheeler, D.; Garcia-Herrera, R.; Vaquero, J. M.; Chenoweth, M.; & Mock, C. J. (2009). Reconstructing the Trajectory of the August 1680 Hurricane from Contemporary Records, *Bulletin of the American Meteorological Society*, Vol. 90, pp. 971-978.

Williams, T. (2008). *Hurricane of Independence: the Untold Story of the Deadly Storm at the Deciding Moment of the American Revolution*. Sourcebooks, 320 pp.

11

Climate Change:
Is It More Predictable Than We Think?

Rafail V. Abramov
Dept. of Mathematics, Statistics and Computer Science, University of Illinois at Chicago
USA

1. Introduction

The global climate system ties together many physical variables, such as flow velocity, density, pressure, temperature, to name a few. The core equations of the climate system are the primitive evolution equations of the atmosphere and ocean (Lions et al., 1992a;b; 1993a;b; 1995; Majda, 2003), which directly involve the flow velocity (or, alternatively, streamfunction and vorticity), density and pressure. To incorporate the effects of other relevant physical processes which supply the energy to or draw it from the motion of the flow, the primitive equations are coupled to other physical processes through temperature, water vapor, ocean surface pressure, and other variables. The coupling terms often preserve energy balance, that is, at any moment, the sum of energy transfer rates between all coupled processes is zero.

The main difficulty in the study of the behavior of primitive equations lies in the nonlinearity of the dynamics of velocity or streamfunction-vorticity in the advection term. The nonlinearity of the primitive equations is also the main source of chaos and lack of predictability for long times in the weather and climate prediction. As has first been recognized by Lorenz (1963), even a simple three-variable nonlinear dynamical system (the so-called Lorenz attractor), based on the idealized convection cell with cooling at the top and heating at the bottom, exhibits extreme sensitivity to initial conditions. Nowadays, the Lorenz attractor is considered a canonical textbook example of chaos in a nonlinear dynamical system, with many illustrations depicting two nearly identical initial conditions evolving into two unrelated trajectories after a short period of time. In more complex dynamical systems with advection terms, nonlinear chaos develops in much more sophisticated fashion, making long-term forecasts difficult and uncertain.

Despite nonlinearity and chaos in the dynamics of the atmosphere and oceans, it has long been recognized by scientists that that the observed motion of the flow can be decomposed into a multitude of different spatio-temporal scales, ranging from thousands of kilometers and many years to few hundred meters and several minutes. The slowest-varying modes, constituting low frequency variability (LFV), involve large scale spatial patterns, usually called "oscillations". The examples of these patterns are the well-known El Niño Southern Oscillation (ENSO), Arctic Oscillation (AO), Antarctic Oscillation (AAO), and North Atlantic Oscillation (NAO). The rest of the spatio-temporal scales constitute much faster fluctuations, superimposed with the slowly varying LFV modes. It is believed by a number of scientists that the mutual combination of the states of LFV modes plays a major role in the present planetary climate (Crowley, 2000; Delworth & Knutson, 2000), and, therefore, the projection of behavior of LFV modes into the future is one of the key considerations for the climate change prediction.

One of the interesting questions about the dynamics of low frequency variability is the effect of its coupling with the faster small scale processes on nonlinear chaotic behavior of LFV modes. Does short time scale chaos and rapid mixing make the long time behavior of LFV modes more chaotic, or otherwise? The nonlinear variables of the low frequency variability, as well as small scale fast processes, are the velocity/streamfunction fields, due to nonlinear advection terms of the primitive equations. In these variables, the energy emerges as a positive-definite quadratic form, and, therefore, the direct coupling between velocity/streamfunction fields of LFV and fast dynamics should preserve such a form to reflect the energy balance during its transfer between the large scale slow motion and fast small scale processes. In Section 2 of this chapter, we present the results of the recent study (Abramov, 2011c) how the fast small scale variables affect chaotic properties of the slow variables through linear energy-preserving coupling in a general nonlinear two-scale model with quadratic energy. White linear coupling is the simplest form of coupling between slow and fast variables, it is nonetheless common in interactions between velocity/streamfunction variables (see, for example, the model of mean flow – small scale interactions via topographic stress in Grote et al. (1999), where the total energy conservation in the coupling between the zonal mean flow and small scale fluctuations is a key requirement). For the simple two-scale Lorenz 96 model (Abramov, 2010; Fatkullin & Vanden-Eijnden, 2004; Lorenz, 1996; Lorenz & Emanuel, 1998), which mimics certain large-scale features of the atmosphere such as the Rossby waves, we show through numerical simulations that chaos at slow variables can be suppressed by the rapidly mixing fast variables, even to the point when the behavior of the slow variables becomes completely predictable.

Another interesting question arises immediately from the first one. Often, it is not possible to model a large multiscale dynamical system through a direct numerical simulation, due to both the exceedingly large number of fast variables, and the need to use a small time discretization step in the numerical scheme to resolve the motion of the fast variables. In this case, one solution is to make a suitable closure for the slow variables only, using the averaging formalism for the fast variables (Papanicolaou, 1977; Vanden-Eijnden, 2003; Volosov, 1962). However, there is a technical difficulty associated with the averaging formalism: at every given state of the slow variables, one has to know the statistics of the motion of the isolated fast variables with given slow state treated as a parameter. Generally, the statistics of a nonlinear fast dynamics are not known explicitly, and, therefore, either a statistical numerical simulation with the fast variables has to be performed at a each time step of the slow variables, or a suitable approximation for relevant fast statistics has to be constructed. Sometimes, the impact of fast unresolved variables in LFV dynamics is modeled through a constant reference forcing (Abramov & Majda, 2009; Franzke, 2002; Selten, 1995), which does not reflect the response of the statistics of the fast variables to changes in the state of the slow variables. However, the results in Section 2 clearly indicate that chaos at slow variables can be suppressed by the interactions with fast variables. Thus, the question is: how to capture the suppression of chaos at slow variables in a closed model for slow variables only, through a suitable but simple approximation? In Section 3 we demonstrate that an additional linear correction to the reduced equations for the slow dynamics is sufficient to reproduce major statistics of the slow variables in a fully coupled model. This correction emerges from the approximation of the statistical response of the fast variables to changes in slow variables, based on the linear fluctuation-dissipation theorem (Majda et al., 2005; Risken, 1989). Section 3 is based on Abramov (2011b).

2. Suppression of chaos at slow variables via linear energy-preserving coupling

Dynamical systems, where the evolution of variables is separated between two or more different time scales, are common in the atmospheric/ocean science (Buizza et al., 1999; Franzke et al., 2005; Hasselmann, 1976; Palmer, 2001). The structure of these systems is typically characterized by the existence of a special subset of slow variables, which evolve on a much longer time scale than the rest of the variables. In particular, one can think of the low-frequency variability models in the atmospheric science, where the slow variables, usually the large scale empirical orthogonal functions describing the large-scale slowly-varying patterns in the atmosphere (such as the Arctic or North Atlantic oscillations, for example), are coupled with small-scale fast processes, which are often very chaotic, turbulent and unpredictable (with respect to the slow time scale, that is). One of the key questions about the behavior of multiscale dynamics is the effect of the rapidly mixing turbulent fast dynamics on the chaotic properties of the slow variables.

In this section we present the results of Abramov (2011c), where the chaotic behavior of slow variables is studied by applying the averaging formalism to the dynamics of the linearized model for the slow variables in a two-scale dynamical system with linear energy-preserving coupling. In particular, we consider a two-scale system of autonomous ordinary differential equations of the form

$$\frac{d\boldsymbol{x}}{dt} = \boldsymbol{f}(\boldsymbol{x}) + \lambda_y \boldsymbol{L} \boldsymbol{y}, \qquad \frac{d\boldsymbol{y}}{dt} = \boldsymbol{g}(\boldsymbol{x}) - \lambda_x \boldsymbol{L}^T \boldsymbol{x}, \tag{1}$$

where $\boldsymbol{x} = \boldsymbol{x}(t) \in \mathbb{R}^{N_x}$ are the slow variables, $\boldsymbol{y} = \boldsymbol{y}(t) \in \mathbb{R}^{N_y}$ are the fast variables, with $N_y \gg N_x$, $\boldsymbol{f}$ and $\boldsymbol{g}$ are N_x and N_y vector-valued nonlinear functions of $\boldsymbol{x}$ and $\boldsymbol{y}$, respectively, $\boldsymbol{L}$ is a constant $N_x \times N_y$ matrix, and $\lambda_x, \lambda_y > 0$ are the coupling parameters. Here, for simplicity, we assume that $\boldsymbol{L}$ has the full rank, that is, N_x. It can be shown directly that the coupling terms in (1) preserve the quadratic energy of the form

$$E = \lambda_x E_x + \lambda_y E_y, \qquad E_x = \frac{1}{2}\|\boldsymbol{x}\|^2, \qquad E_y = \frac{1}{2}\|\boldsymbol{y}\|^2. \tag{2}$$

For simplicity of presentation, here we assume that the total energy is a weighted sum of squares of the components of $\boldsymbol{x}$ and $\boldsymbol{y}$; the more general case with energy being an arbitrary positive-definite quadratic form is discussed in Abramov (2011c). Here note that $\boldsymbol{f}(\boldsymbol{x})$ and $\boldsymbol{g}(\boldsymbol{y})$ are not required to preserve the energy, as they might contain forcing and dissipation, which frequently happens in atmosphere/ocean dynamics. For the two-scale system in (1), we look at the averaged dynamics for $\boldsymbol{x}$ alone (Abramov, 2010; Papanicolaou, 1977; Vanden-Eijnden, 2003; Volosov, 1962). The averaging formalism produces the closed system for $\boldsymbol{x}$ in the form

$$\frac{d\boldsymbol{x}}{dt} = \boldsymbol{f}(\boldsymbol{x}) + \lambda_y \boldsymbol{L} \bar{\boldsymbol{z}}(\boldsymbol{x}), \tag{3}$$

where $\bar{\boldsymbol{z}}(\boldsymbol{x})$ is the mean state of the fast variables with $\boldsymbol{x}$ given as a constant parameter:

$$\frac{d\boldsymbol{z}}{dt} = \boldsymbol{g}(\boldsymbol{z}) - \lambda_x \boldsymbol{L}^T \boldsymbol{x}, \qquad \bar{\boldsymbol{z}}(\boldsymbol{x}) = \lim_{r \to \infty} \frac{1}{r} \int_0^r \boldsymbol{z}(t)\, dt. \tag{4}$$

To observe the chaotic properties of (3), and, in particular, the sensitivity to initial conditions, one has to look at the linearized dynamics of (3), given by

$$\frac{dv}{dt} = \left[\frac{\partial f}{\partial x}(x(t)) + \lambda_y L \frac{\partial}{\partial x} \bar{z}(x(t)) \right] v, \tag{5}$$

where $x(t)$ is the solution of (3). It is not difficult to see that, for two nearby initial conditions x_0 and x_0^* of (3) with $v_0 = x_0^* - x_0$, the solution $v(t)$ of (5) is an approximation of the difference of the trajectories $(x^*(t) - x(t))$, as long as this difference remains small. Then, the sensitivity of (3) to initial conditions is given by the rate of growth of $\|v(t)\|$ in time. For the rate of growth, one can write

$$\frac{1}{2}\frac{d}{dt}\|v\|^2 = v^T \frac{\partial f}{\partial x}(x(t))v + \lambda_y v^T \left[L \frac{\partial}{\partial x} \bar{z}(x(t)) \right] v. \tag{6}$$

Above, observe that the first term in the right-hand side comes from the uncoupled part of (3), while the second term represents the effect of coupling of the slow variables x to the fast variables y. At this point, consider (4), perturbed by a small constant forcing δg,

$$\frac{dz}{dt} = g(z) - \lambda_x L^T x + \delta g, \tag{7}$$

with the perturbed mean state $\bar{z} + \delta\bar{z}$, and denote

$$R(x) = \frac{\delta\bar{z}}{\delta g}. \tag{8}$$

Then, assuming that the small constant perturbation δg comes from a small change in x, that is,

$$\delta g = -\lambda_x L^T \delta x, \tag{9}$$

for the derivative of $\bar{z}(x)$ one obtains, by the chain rule,

$$\frac{\partial}{\partial x}\bar{z}(x) = \frac{\delta\bar{z}}{\delta g}\frac{\partial\delta g}{\partial\delta x} = -\lambda_x R(x) L^T. \tag{10}$$

Now, taking into account (10), for (6) we obtain

$$\frac{1}{2}\frac{d}{dt}\|v\|^2 = v^T \frac{\partial f}{\partial x}(x(t))v - \lambda_x \lambda_y v^T L R(x(t)) L^T v. \tag{11}$$

At this point, one can see that if $R(x)$ is positive-definite, then the second term in the right-hand side of (11) becomes negative for arbitrary nonzero v, which means that the coupling to the fast variables reduces the rate of growth of $\|v\|$ and decreases chaos in (3), as well as its sensitivity to changes in initial conditions.

When is $R(x)$ positive-definite? Generally, since

$$\delta\bar{z} = R(x)\delta g, \tag{12}$$

where δg is a small constant perturbation in (7), and $\delta\bar{z}$ is the response of the mean state of (7) to δg, the positive-definiteness of $R(x)$ means that

$$\delta\bar{z}^T \delta g = \delta g^T R(x)\delta g > 0 \text{ for all sufficiently small } \delta g, \tag{13}$$

that is, the response of the mean state $\delta\bar{z}$ does not develop against the perturbation δg, as long as δg is sufficiently small. It is not difficult to show that the following identity holds whenever δg vanishes:

$$\frac{\mathrm{d}}{\mathrm{d}t}\delta\bar{z} = -\delta g, \tag{14}$$

that is, at the moment the small constant perturbation δg vanishes from (7), the time derivative of $\delta\bar{z}$ equals δg with the opposite sign. Multiplying the above relation by $\delta\bar{z}^T$ on both sides, we obtain

$$\delta\bar{z}^T\frac{\mathrm{d}}{\mathrm{d}t}\delta\bar{z} = \frac{1}{2}\frac{\mathrm{d}}{\mathrm{d}t}\|\delta\bar{z}\|^2 = -\delta\bar{z}^T\delta g < 0, \tag{15}$$

that is, any sufficiently small perturbation of the mean state $\delta\bar{z}$ decreases in time at the moment when the external perturbation δg is removed (stability of mean state under perturbations). An example of such dynamics is the Ornstein-Uhlenbeck process (Uhlenbeck & Ornstein, 1930):

$$\frac{\mathrm{d}z}{\mathrm{d}t} = -\Gamma z + g + \sigma\frac{\mathrm{d}W_t}{\mathrm{d}t} - \lambda_x L^T x, \tag{16}$$

where Γ is a constant positive-definite matrix, g is a constant vector, σ is a constant matrix, and W_t is a Wiener process. Indeed, applying statistical averages on both sides of (16) yields

$$\bar{z} = \Gamma^{-1}(g - \lambda_x L^T x), \tag{17}$$

and, therefore,

$$R = \Gamma^{-1}, \tag{18}$$

which ascertains the positive-definiteness of R. In the case of general nonlinear dynamics in (4), it is shown in Abramov (2011c) that, if the statistical distribution of the solution of (4) can be approximated by the Gaussian probability density with mean state $\bar{z}$ and covariance matrix Σ, then R can be approximated by the integral of the time autocorrelation matrix:

$$R = \left[\int_0^\infty C(s)\,\mathrm{d}s\right]\Sigma^{-1}, \qquad C(s) = \lim_{r\to\infty}\frac{1}{r}\int_0^r z(t+s)(z(t) - \bar{z})^T\,\mathrm{d}t. \tag{19}$$

This is the quasi-Gaussian approximation of $R(x)$ (Abramov, 2009; 2010; 2011a; Abramov & Majda, 2007; 2008; 2009; 2011; Majda et al., 2005). In this case, it is argued in Abramov (2011c) that the positive-definiteness of R must be associated with the situation where the typical Poincaré recurrence time of nonlinear motion around the mean state in (4) (which can be viewed as an advective time scale) is not much shorter than the turbulent mixing autocorrelation time. The reason is that, since $C(0)\Sigma^{-1}$ is the identity matrix, then there always exists $a^* > 0$ such that

$$R_a = \left[\int_0^a C(s)\,\mathrm{d}s\right]\Sigma^{-1} \text{ is positive definite for all } a, \quad 0 < a \leq a^*. \tag{20}$$

The positive-definiteness of R_a for larger a can probably be violated by the domination of the rotation part in $C(s)$ for larger s, which evolves on the advective time scale of (4). However, this effect can be prevented by a sufficiently rapid decay of $\|C(s)\|$ for large s, which is governed by the turbulent mixing autocorrelation time. Thus, in general, one can expect the positive-definiteness of R to appear in the situations where the turbulent mixing autocorrelation time scale is not much longer than the advective time scale.

Below we demonstrate through the series of numerical experiments with the two-scale Lorenz 96 model that increasing turbulent mixing at the fast variables promotes positive-definiteness of R, as well as decreases chaos at the slow variables.

2.1 The two-scale Lorenz 96 model

The Lorenz 96 model was suggested by Lorenz (1996) as a simple two-scale system which mimics certain large scale features of the atmospheric dynamics, such as Rossby waves. Abramov (2011c) rescaled the Lorenz 96 model using energy rescaling similar to the one used by Majda et al. (2005) for the one-scale Lorenz 96 model. The rescaled model is given by

$$\dot{x}_i = x_{i-1}(x_{i+1} - x_{i-2}) + \frac{1}{\beta_x}\left(\bar{x}(x_{i+1} - x_{i-2}) - x_i\right) + \frac{F_x - \bar{x}}{\beta_x^2} - \frac{\lambda_y}{J}\sum_{j=1}^{J} y_{i,j},$$

$$\dot{y}_{i,j} = \frac{1}{\varepsilon}\left[y_{i,j+1}(y_{i,j-1} - y_{i,j+2}) + \frac{1}{\beta_y}\left(\bar{y}(y_{i,j-1} - y_{i,j+2}) - y_{i,j}\right) + \frac{F_y - \bar{y}}{\beta_y^2} + \lambda_x x_i\right],$$

$$(21)$$

where $1 \leq i \leq N_x, 1 \leq j \leq J$. Additionally, we write the fast limiting dynamics for (21) as

$$\dot{z}_{i,j} = z_{i,j+1}(z_{i,j-1} - z_{i,j+2}) + \frac{1}{\beta_y}\left(\bar{z}(z_{i,j-1} - z_{i,j+2}) - z_{i,j}\right) + \frac{F_y - \bar{z}}{\beta_y^2} + \lambda_x x_i, \qquad (22)$$

where x is given as an external parameter, as in (4). The following notations are adopted above:

- x is the set of the slow variables of size N_x. The following periodic boundary conditions hold for x: $x_{i+N_x} = x_i$;
- y is the set of the fast variables of size $N_y = N_x J$ where J is a positive integer. The following boundary conditions hold for y: $y_{i+N_x,j} = y_{i,j}$ and $y_{i,j+J} = y_{i+1,j}$;
- F_x and F_y are the constant forcing parameters;
- λ_x and λ_y are the coupling parameters;
- ε is the time scale separation parameter;
- $\bar{x}$ and $\bar{y}$ are the statistical mean states, and β_x and β_y are the statistical standard deviations, respectively, of the corresponding uncoupled dynamics

$$\frac{d}{dt}x_i = x_{i-1}(x_{i+1} - x_{i-2}) - x_i + F_x,$$

$$\frac{d}{dt}y_{i,j} = y_{i,j+1}(y_{i,j+1} - x_{i,j+2}) - y_{i,j} + F_y,$$

$$(23)$$

separately for slow and fast variables.

In the rescaled Lorenz 96 model (21), F_x and F_y regulate the chaos and mixing of the x and y variables, respectively. However, in the absence of linear rescaling through the mean states $\bar{x}, \bar{y}$, and standard deviations β_x, β_y, the mean state and mean energy would also be affected by the changes in forcing, which affects the mean and energy trends in coupling for the fixed coupling parameters. Thus, the rescaling of the two-scale Lorenz 96 model is needed to adjust the effect of coupling independently of forcing, such that for any F_x and F_y, the mean states and variances of all x_i and $y_{i,j}$ in (21) are approximately zero and one, respectively. At this point, we can observe that the coupling in the rescaled Lorenz 96 model in (21) preserves the energy

$$E = \lambda_x E_x + \frac{\varepsilon \lambda_y}{J} E_y, \qquad E_x = \frac{1}{2} \sum_{i=1}^{N_x} x_i^2, \qquad E_y = \frac{1}{2} \sum_{i=1}^{N_x} \sum_{j=1}^{J} y_{ij}^2, \qquad (24)$$

and the coupling matrix L is given by

$$(Ly)_i = - \sum_{j=1}^{J} y_{ij}, \qquad (L^T x)_{ij} = -x_i. \qquad (25)$$

2.2 Suppression of chaos at slow variables by increasing mixing at fast variables

Below, we show the computed statistics of the rescaled Lorenz 96 model (21) with the following parameters: $N_x = 10$, $N_y = 40$, $F_x = 6$, $F_y = 6, 8, 12, 16$ and 24, $\lambda_x = \lambda_y = 0.25$, $\varepsilon = 0.01$ (the time scale separation between x and y is 100 times). The slow forcing parameter $F_x = 6$ is chosen so that the slow dynamics are not too chaotic, mimicking the behavior of low-frequency variability in the atmosphere (it is known from the previous work, such as Abramov (2009; 2010; 2011a); Abramov & Majda (2003; 2007; 2008); Majda et al. (2005) that for $F = 6$ the dynamics of the uncoupled model in (23) are weakly chaotic). The coupling parameters λ_x and λ_y are set to 0.25 so that they are neither too small, nor too large, to ensure rich interaction between the slow and fast variables without linearizing the rescaled Lorenz 96 system too much. The time-scale separation parameter $\varepsilon = 0.01$ is, again, chosen so that it is neither too large, nor too small (the time scale separation by two orders of magnitude is consistent, for instance, with the separation between annual and diurnal cycles in the atmosphere).

In the rescaled Lorenz 96 model (21), it turns out that the values of F_x and F_y do not significantly affect the mean state and mean energy for both the slow variables x and fast variables y. To show this, in Table 1 we display the mean states and variances of both x and y for the rescaled Lorenz 96 model in (21). Observe that, despite different forcing regimes, the means and variances for both x and y are almost unchanged, the mean states being near zero while the variances being near one, as designed by the rescaling. Here note that while the rescaling was carried out for the corresponding uncoupled model (where it sets the mean state to zero and variance to one precisely), using the same rescaling parameters in the coupled model (21) still sets its means and variances near prescribed values zero and one, respectively (although not precisely).

At this point, we turn our attention to the chaotic behavior of the slow variables x and mixing behavior of the fast variables y for the same range of parameters. Here we observe the average divergence behavior in time between the short-time (half of the time unit) running averages $\langle x \rangle (t)$ of the slow time series $x(t)$, which are initially generated very closely to each other (for technical details of this simulation, see Abramov (2011c)). The short time-averaging window of half of the time unit for the running average $\langle x \rangle (t)$ ensures that the slow variables $x(t)$ do not change much during this window, while the fast time series $y(t)$ mix completely during the same short time averaging window. The results of this simulation, together with the the time autocorrelation functions for the decoupled fast variables for the same set of parameters with x set to its mean state, are shown in Figure 1. Remarkably, the chaos in the slow x-variables is consistently suppressed as the fast forcing F_y increases, as the unperturbed

$N_x = 10,\ N_y = 40,\ F_x = 6,\ \lambda_x = \lambda_y = 0.25,\ \varepsilon = 0.01$				
F_y	x-mean	x-var	y-mean	y-var
6	$9.64 \cdot 10^{-3}$	0.9451	$-2.38 \cdot 10^{-3}$	1.066
8	$2.817 \cdot 10^{-2}$	0.9514	$-1.466 \cdot 10^{-2}$	1.098
12	$2.05 \cdot 10^{-2}$	0.9336	$-2.719 \cdot 10^{-2}$	1.139
16	$-1.353 \cdot 10^{-2}$	0.9006	$-4.028 \cdot 10^{-2}$	1.153
24	$-6.972 \cdot 10^{-2}$	0.8434	$-6.075 \cdot 10^{-2}$	1.167

Table 1. The mean states and variances of the x and y variables for the rescaled Lorenz 96 model in (21) with the following parameters: $N_x = 10$, $N_y = 40$, $F_x = 6$, $F_y = 6, 8, 12, 16$ and 24, $\lambda_x = \lambda_y = 0.25$, $\varepsilon = 0.01$.

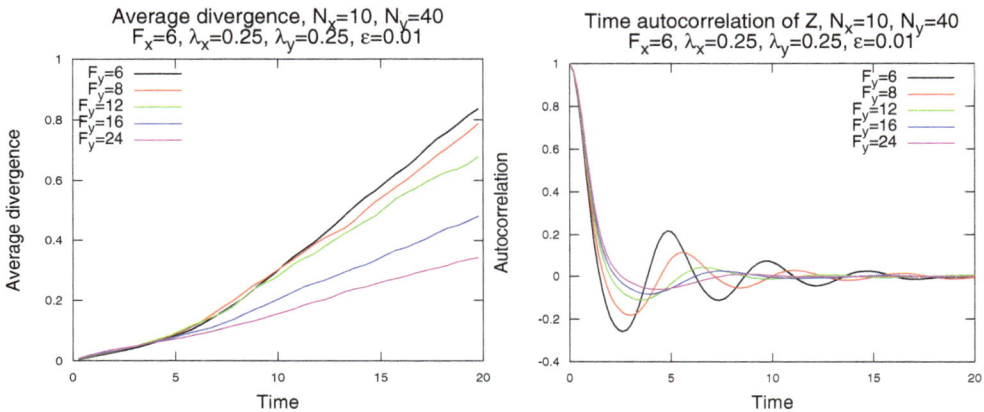

Fig. 1. Left: average divergence between perturbed and unperturbed running averages of the slow variables of (21). Right: the time autocorrelation functions of (22) with x_i fixed at its statistical mean state. The following parameters are used: $N_x = 10$, $N_y = 40$, $F_x = 6$, $F_y = 6, 8, 12, 16$ and 24, $\lambda_x = \lambda_y = 0.25$, $\varepsilon = 0.01$.

and perturbed slow running averages $\langle x \rangle(t)$ diverge from each other slower and slower in time. It cannot be caused by the changing statistical mean or variance of the slow or fast variables creating average counteracting forcing at the slow variables, as Table 1 clearly indicates that the mean states and variances of both the slow and fast variables do not change by a significant amount for different F_y. At the same time, observe that more rapid decay of the time autocorrelation functions for the fast variables is observed as chaos at slow variables is suppressed, supporting the theory developed above and in Abramov (2011c).

In addition, in Figure 2 we demonstrate viability of the quasi-Gaussian approximation to the response $R(x)$ of the mean state $\bar{z}$ in (4) to small constant external forcing, where x is set to the statistical mean state of the slow variables. Observe that as F_y increases and chaos at the slow variables is suppressed, the smallest eigenvalue of the symmetric part of R grows systematically, thus "increasing" positive-definiteness of R. Additionally, the probability density functions of the fast variables are shown in Figure 2 to demonstrate that the statistical distribution of the fast variables is close to Gaussian.

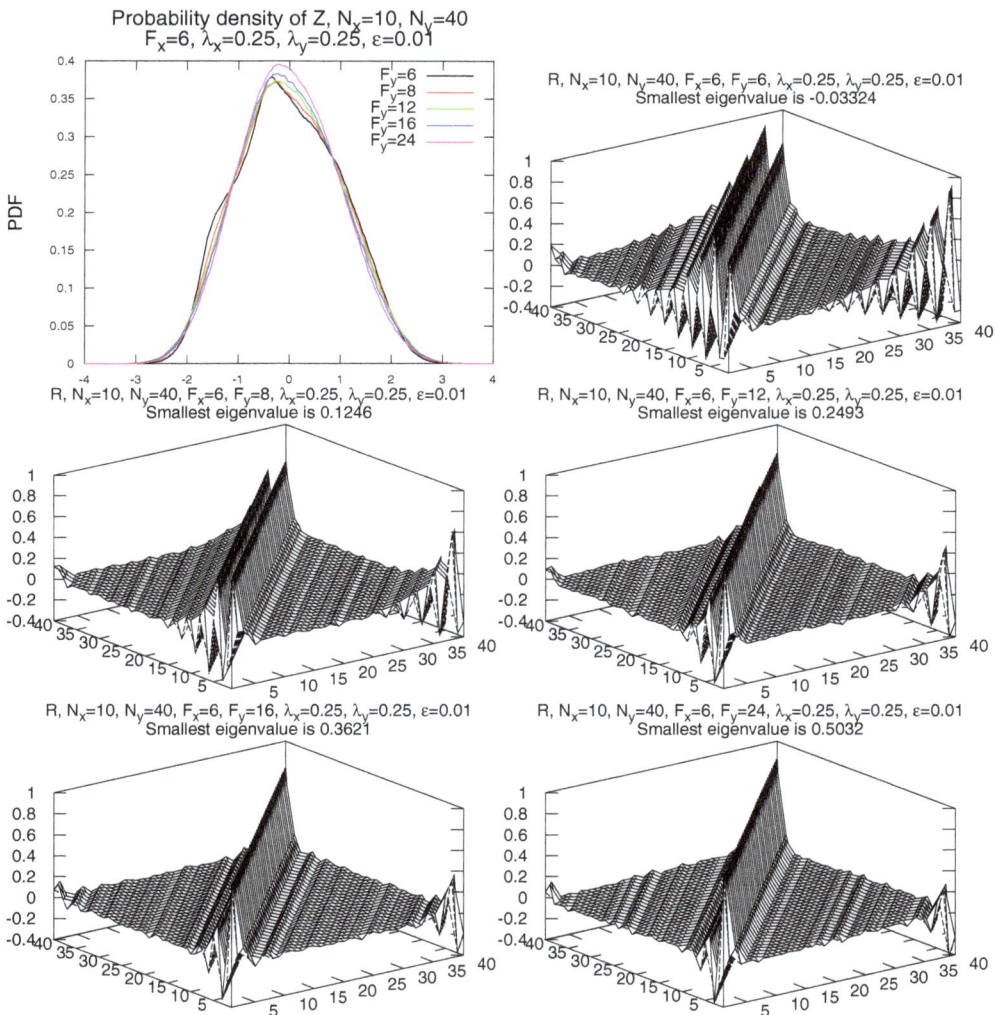

Fig. 2. Upper-left: probability density functions of (22) with x_i fixed at its statistical mean state. The rest: quasi-Gaussian approximations of averaged infinite-time linear response operators $\boldsymbol{R}(\bar{\boldsymbol{x}})$ for the rescaled Lorenz 96 model in (21) with the following parameters: $N_x = 10$, $N_y = 40$, $F_x = 6$, $F_y = 6, 8, 12, 16$ and 24, $\lambda_x = \lambda_y = 0.25$, $\varepsilon = 0.01$.

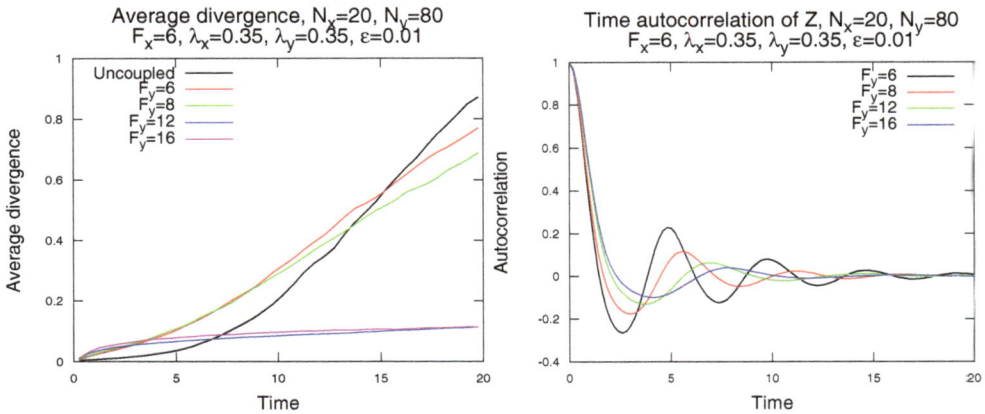

Fig. 3. Left: average divergence between perturbed and unperturbed running averages of the slow variables of (21). Right: the time autocorrelation functions of (22) with x_i fixed at its statistical mean state. The following parameters are used: $N_x = 20$, $N_y = 80$, $F_x = 6$, $F_y = 6, 8, 12$ and 16, $\lambda_x = \lambda_y = 0.35$, $\varepsilon = 0.01$, as well as for the uncoupled rescaled Lorenz 96 model with $N = 20$ and $F = 6$.

F_y	x-mean	x-var	y-mean	y-var
6	$6.216 \cdot 10^{-3}$	0.8878	$-9.982 \cdot 10^{-3}$	1.119
8	$2.34 \cdot 10^{-2}$	0.8728	$-2.791 \cdot 10^{-2}$	1.19
12	-0.1363	0.6927	$-5.553 \cdot 10^{-2}$	1.168
16	-0.1444	0.6703	$-8.669 \cdot 10^{-2}$	1.199

$N_x = 20$, $N_y = 80$, $F_x = 6$, $\lambda_x = \lambda_y = 0.35$, $\varepsilon = 0.01$

Table 2. The mean states and variances of the x and y variables for the rescaled Lorenz 96 model in (21) with the following parameters: $N_x = 20$, $N_y = 80$, $F_x = 6$, $F_y = 6, 8, 12$ and 16, $\lambda_x = \lambda_y = 0.35$, $\varepsilon = 0.01$.

Another key question in the atmosphere/ocean science is whether the uncoupled system, consisting of slow variables only, is more or less chaotic than its original version, coupled with the fast, often unresolved or underresolved variables. Indeed, often scientists work with uncoupled models consisting of slow variables only, where coupling terms were replaced with the estimates of the long-term averages of the corresponding fast variables, such as the T21 barotropic model with the realistic Earth topography (Abramov & Majda, 2009; Franzke, 2002; Selten, 1995), and study dynamical properties of the slow variables in the uncoupled models. The common sense in this case suggests that if the uncoupled slow model is chaotic, then, naturally, its original version coupled with fast rapidly mixing dynamics should be even more chaotic.

Remarkably, the common sense logic in this situation is deceiving. In fact, it turns out to be possible even to reach the transition from the chaotic to stable slow dynamics by increasing the turbulent mixing at the fast variables, while the uncoupled slow dynamics remain chaotic. Here we demonstrate such an example for the rescaled Lorenz 96 model in (21) with the following set of parameters: $N_x = 20$, $N_y = 80$, $F_x = 6$, $\lambda_x = \lambda_y = 0.35$, $\varepsilon = 0.01$, and compare it with the uncoupled rescaled Lorenz 96 model with the same parameters $N = 20$ and $F = 6$ for the slow variables. In Figure 3 we show the average divergence of a perturbed trajectory from an unperturbed one for this set of parameters. Observe that, while the slow dynamics for $F_y = 6, 8$ and for uncoupled dynamics with $F = 6$ are clearly chaotic, for $F_y = 12$ and greater values the abrupt transition occurs, where the difference between the perturbed and unperturbed x time series does not grow much beyond 10%. The time autocorrelations of the fast variables for the same set of parameters are also shown in Figure 3, while the statistical mean states and variances for and Table 2. Here, the chaos at slow variables is suppressed purely by the dynamical mechanism uncovered in this work; indeed, the mean state and variance of both the slow and fast variables do not change substantially enough to suppress chaos by creating a counteracting mean forcing term at the slow dynamics to suppress F_x (see Table 2 for confirmation), while the time autocorrelation functions for the fast variables in Figure 3 with x set to the mean state decay faster for larger values of F_y, indicating stronger mixing. The key observation here is that the behavior of the uncoupled model with just the slow variables in the same regime is deceiving – it is clearly chaotic, while the full two-scale model loses chaos at the slow variables in more turbulent regimes of the fast dynamics.

3. Capturing statistics of coupled dynamics via simple closure for slow variables

The time-space scale separation of atmospheric dynamics causes its direct numerical simulation to be computationally expensive, due both to the large number of the fast variables and necessity to choose a small discretization time step in order to resolve the fast components of dynamics. In the climate change prediction the situation is further complicated by the fact that climate is characterized by the long-term statistics of the slow LFV modes, which, under small changes of parameters (such as the solar radiation forcing, greenhouse gas content, etc) change over even longer time scale than the motion of the slow variables themselves. In this situation, where long-term statistics of the slow motion patterns need to be captured, the direct forward time integration of the most comprehensive global circulation models (GCM) is subject to enormous computational expense.

As a more computationally feasible alternative to direct forward time integration of the complete multiscale model, it has long been recognized that, if a closed simplified model for the slow variables alone is available, one could use this closed slow-variable model instead to simulate the statistics of the slow variables. Numerous closure schemes were developed for multiscale dynamical systems (Crommelin & Vanden-Eijnden, 2008; Fatkullin & Vanden-Eijnden, 2004; Majda et al., 1999; 2001; 2002; 2003), which are based on the averaging principle over the fast variables (Papanicolaou, 1977; Vanden-Eijnden, 2003; Volosov, 1962). Some of the methods (such as those in Majda et al. (1999; 2001; 2002; 2003)) replace the fast nonlinear dynamics with suitable stochastic processes (Wilks, 2005) or conditional Markov chains (Crommelin & Vanden-Eijnden, 2008), while others (Fatkullin & Vanden-Eijnden, 2004) provide direct closure by suitable tabulation and curve fitting. Majda et al. (2010) used the stochastic mode reduction method in a nonlinear stochastic model which mimicked the behavior of a GCM. However, it seems that all these approaches require either extensive computations to produce a closed model (for example, the methods in Crommelin & Vanden-Eijnden (2008) and Fatkullin & Vanden-Eijnden (2004) require multiple simulations of fast variables alone with different fixed states of slow variables), or somewhat *ad hoc* determination of closure coefficients by matching areas under the time correlation functions (Majda et al., 1999; 2001; 2002; 2003).

In this section we present a simple method of determining the closed model for slow variables alone, which requires only a single computation of appropriate statistics for the fast dynamics with a certain fixed state of the slow variables, developed in Abramov (2011b). The method is based on the first-order Taylor expansion of the averaged coupling term for the fast variables with respect to the slow variables, which was already computed in Section 2. We show through the computations with the appropriately rescaled two-scale Lorenz 96 96 model (21) that, with simple linear coupling in both slow and fast variables, this method produces quite comparable statistics to what is exhibited by the slow variables of the complete two-scale Lorenz 96 model. The main advantage of the method is its simplicity and easiness of implementation, partly due to the fact that the fast dynamics need not be explicitly known (that is, the fast dynamics can be provided as a "black-box" observations), and the parameters of the closed model for the slow variables are determined from the appropriate statistics of the fast variables for a given fixed state of the slow variables. Additionally, the method can be applied even when the statistics for both the slow and fast variables of the full multiscale model are not available due to computational expense.

For the purpose of this work, here we assume that the computation of (4) is practically feasible only for a single choice of the constant parameter $x = x^*$, where x^* is a suitable point, in the vicinity of which the motion occurs, such as the mean state of the original multiscale dynamics in (1), or a nearby state. A poor man's approach in this case is to compute the approximate average at a single point $x = x^*$, which is a zero order approximation:

$$\bar{z}(x) = \bar{z}(x^*) + O(\|x - x^*\|). \tag{26}$$

Here, one has to compute the time average $\bar{z}$, needed for the averaged dynamics in (3), only once, for the time series of (4) corresponding to $x = x^*$. However, as recently found in Abramov (2011c) and demonstrated above in Section 2, this approximation may fail to capture the chaotic properties of the slow variables in (1), because the coupling term in the averaged linearized dynamics (5) would not be reproduced. Here we propose the following first order correction:

$$\bar{z}(x) = \bar{z}(x^*) + \frac{\partial \bar{z}}{\partial x}(x^*)(x - x^*) + O(\|x - x^*\|^2) =$$
$$= \bar{z}(x^*) - \lambda_x R(x^*) L^T (x - x^*) + O(\|x - x^*\|^2), \tag{27}$$

where the derivative of $\bar{z}$ with respect to x is already computed in (10) through the linear response R of the statistical mean state to small constant external forcing. The first-order approximation above, applied to (3), leads to the following closed system for the slow variables alone:

$$\frac{dx}{dt} = f(x) + \lambda_y L \bar{z}^* - \lambda_x \lambda_y L R^* L^T (x - x^*), \tag{28}$$

where $\bar{z}^* = \bar{z}(x^*)$ is the time-average of the trajectory in (4) for $x = x^*$, while $R^* = R(x^*)$ is given by (19) for the same value of $x = x^*$.

Even with the linear coupling, the function $\bar{z}(x)$ (the dependence of the mean state of (4) on x) is not generally linear. Thus, the validity of the linear approximation in (28) depends on the influence (or lack thereof) of the nonlinearity of the function $\bar{z}(x)$. While rigorous estimates of the validity of the linear approximation in (28) can hardly be provided in general case, here, instead, we try to justify it by comparing the fast limiting system in (4) to the

Ornstein-Uhlenbeck process in (16). It is easy to see, by applying statistical averages on both sides, that the difference between the statistical mean states of (16) corresponding to x and x^* is

$$\bar{z}_{OU} - \bar{z}^*_{OU} = \mathbf{\Gamma}^{-1}L_x(x - x^*), \tag{29}$$

which is valid for $(x - x^*)$ of an arbitrary norm. At the same time, it was already shown in (18) (Section 2) that $R = \mathbf{\Gamma}^{-1}$, which means that, in the case of the Ornstein-Uhlenbeck process, the relation (27) is exact for an arbitrarily large perturbation $(x - x^*)$. Hence, if the nonlinear process in (4) behaves statistically similarly to the Ornstein-Uhlenbeck process in (16), the averaged system in (28) can be expected to behave statistically similarly to the slow part of (1). Below we numerically test the approximation for slow variables with linear coupling using the two-scale Lorenz 96 model, the same that was used in Section 2.

3.1 Direct numerical simulation

Here we present a numerical study of the proposed approximation for the slow dynamics, applied to the two-scale Lorenz 96 model in (21). We compare the statistical properties of the slow variables for the three following systems:

1. The complete two-scale Lorenz 96 system from (21);
2. The approximation for the slow dynamics alone from (28);
3. The poor man's version of (28) with the first-order correction term R^* set to zero (further referred to as the "zero-order" system).

The fixed parameter x^* for the computation of R^* was set to the long-term mean state $\bar{x}$ of (21) (in practical situations, a rough estimate could be used).

Due to translational invariance of the studied models, the statistics are invariant with respect to the index shift for the variables x_i. For diagnostics, we monitor the following long-term statistical quantities of x_i:

a. The probability density functions (PDF), computed by bin-counting. A PDF gives the most complete information about the one-point statistics of x_i, as it shows the statistical distribution of x_i in the phase space.
b. The time autocorrelation functions $\langle x_i(t)x_i(t + s)\rangle$, where the time average is over t, normalized by the variance $\langle x_i^2\rangle$ (so that it always starts with 1).

The success (or failure) of the proposed approximation of the slow dynamics depends on several factors. First, as the quasi-Gaussian linear response formula (19) is used for the computation of R^*, the precision will be affected by the non-Gaussianity of the fast dynamics. Second, it depends how linearly the mean state $\bar{z}$ for the fast variables depends on the slow variables x. Here we observe the limitations of the proposed approximation by studying a variety of dynamical regimes of the rescaled Lorenz 96 model in (21):

- $N_x = 20, J = 4$ (so that $N_y = 80$). Thus, the number of the fast variables is four times greater than the number of the slow variables.
- $\lambda_x = \lambda_y = 0.3, 0.4$. These values of coupling are chosen so that they are neither too weak, nor too strong (although 0.3 is weaker, and 0.4 is stronger). Recall that the standard deviations of both x_i and $y_{i,j}$ variables are approximately 1, and, thus, the contribution to the right-hand side from coupled variables is weaker than the self-contribution, but still of the same order.

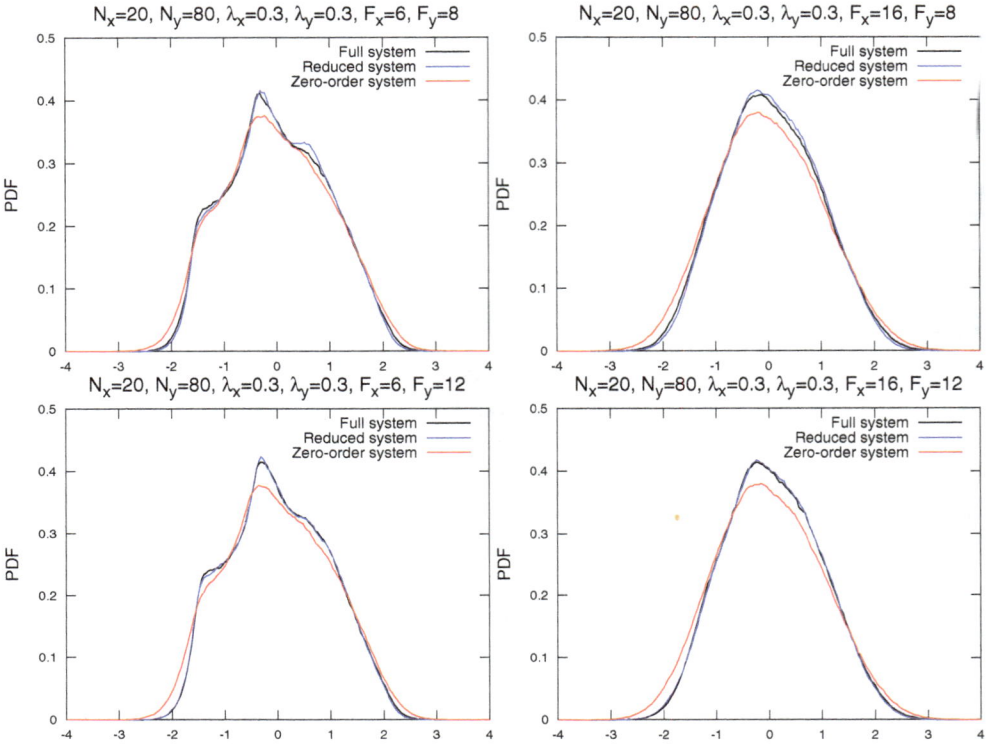

Fig. 4. Probability density functions of the slow variables. The following parameters are used: $N_x = 20$, $N_y = 80$, $F_x = 6, 16$, $F_y = 8, 12$, $\lambda_x = \lambda_y = 0.3$, $\varepsilon = 0.01$.

$\lambda_{x,y} = 0.3, F_y = 8$			$\lambda_{x,y} = 0.3, F_y = 12$		
	Red.	Z.O.		Red.	Z.O.
$F_x = 6$	$5.036 \cdot 10^{-3}$	$1.165 \cdot 10^{-2}$	$F_x = 6$	$2.581 \cdot 10^{-3}$	$1.576 \cdot 10^{-2}$
$F_x = 16$	$5.593 \cdot 10^{-3}$	$1.469 \cdot 10^{-2}$	$F_x = 16$	$2.71 \cdot 10^{-3}$	$1.818 \cdot 10^{-2}$
$\lambda_{x,y} = 0.4, F_y = 8$			$\lambda_{x,y} = 0.4, F_y = 12$		
	Red.	Z.O.		Red.	Z.O.
$F_x = 6$	0.1022	$8.857 \cdot 10^{-2}$	$F_x = 6$	$9.28 \cdot 10^{-2}$	0.1113
$F_x = 16$	$3.725 \cdot 10^{-3}$	$2.703 \cdot 10^{-2}$	$F_x = 16$	$5.885 \cdot 10^{-3}$	$3.209 \cdot 10^{-2}$

Table 3. L_2-errors between the PDFs of the slow variables of the full two-scale Lorenz 96 model and the two reduced models. Notations: "Red." stands for "Reduced" (that is, (28)), and "Z.O." stands for "Zero-order", the poor man's version of (28).

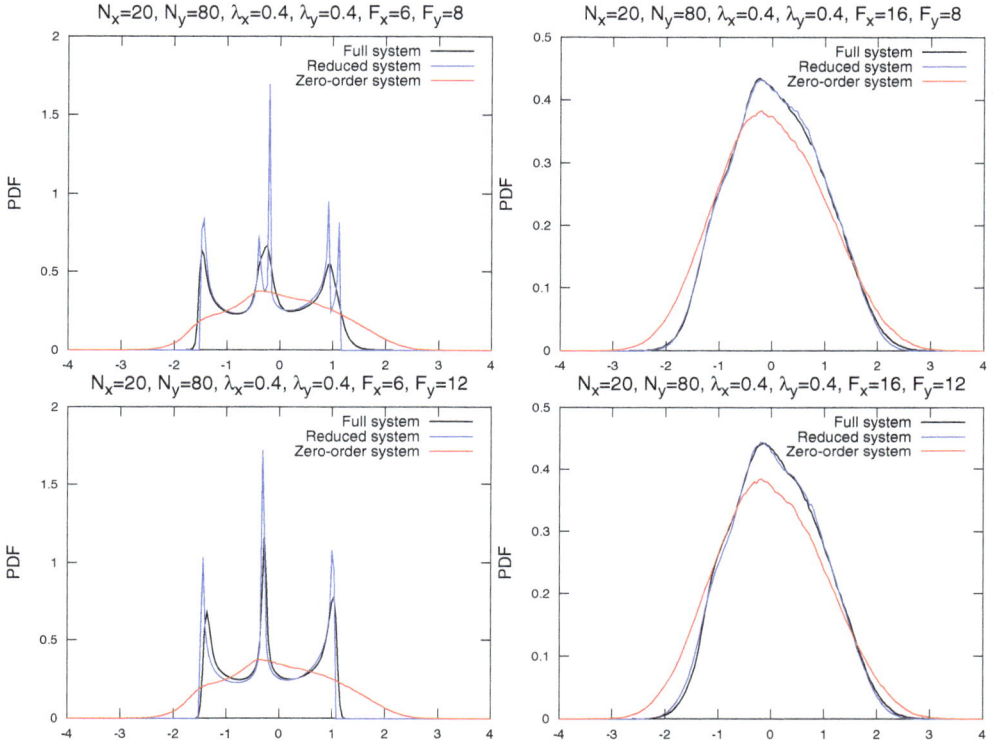

Fig. 5. Probability density functions of the slow variables. The following parameters are used: $N_x = 20$, $N_y = 80$, $F_x = 6, 16$, $F_y = 8, 12$, $\lambda_x = \lambda_y = 0.4$, $\varepsilon = 0.01$.

- $F_x = 6, 16$. The slow forcing F_x adjusts the chaos and mixing properties of the slow variables, and in this work it is set to a weakly chaotic regime $F_x = 6$, and strongly chaotic regime $F_x = 16$.

- $F_y = 8, 12$. The fast forcing adjusts the chaos and mixing properties of the fast variables. Here the value of F_y is chosen so that the fast variables are either moderately chaotic for $F_y = 8$, or more strongly chaotic for $F_y = 12$.

- $\varepsilon = 0.01$. The time scale separation of two orders of magnitude is consistent with typical real-world geophysical processes (for example, the annual and diurnal cycles of the Earth's atmosphere).

In Figures 4 and 5 we show the probability density functions of the slow dynamics for the full two-scale Lorenz 96 model, the reduced closed model for the slow variables alone in (28), and its poor man's zero order version without the linear correction term. In addition, in Table 3 we show the L_2-errors in PDFs between the full two-scale Lorenz 96 model and the two reduced models. Observe that for the more weakly coupled regimes with $\lambda_x = \lambda_y = 0.3$ all PDFs look similar, however, the reduced model with the correction term reproduces the PDFs much closer to those of the full two-scale Lorenz 96 model, than the zero-order model. In the more strongly coupled regime with $\lambda_x = \lambda_y = 0.4$ the situation tilts even more in favor of the reduced model with linear correction term in (28): observe that for the weakly chaotic

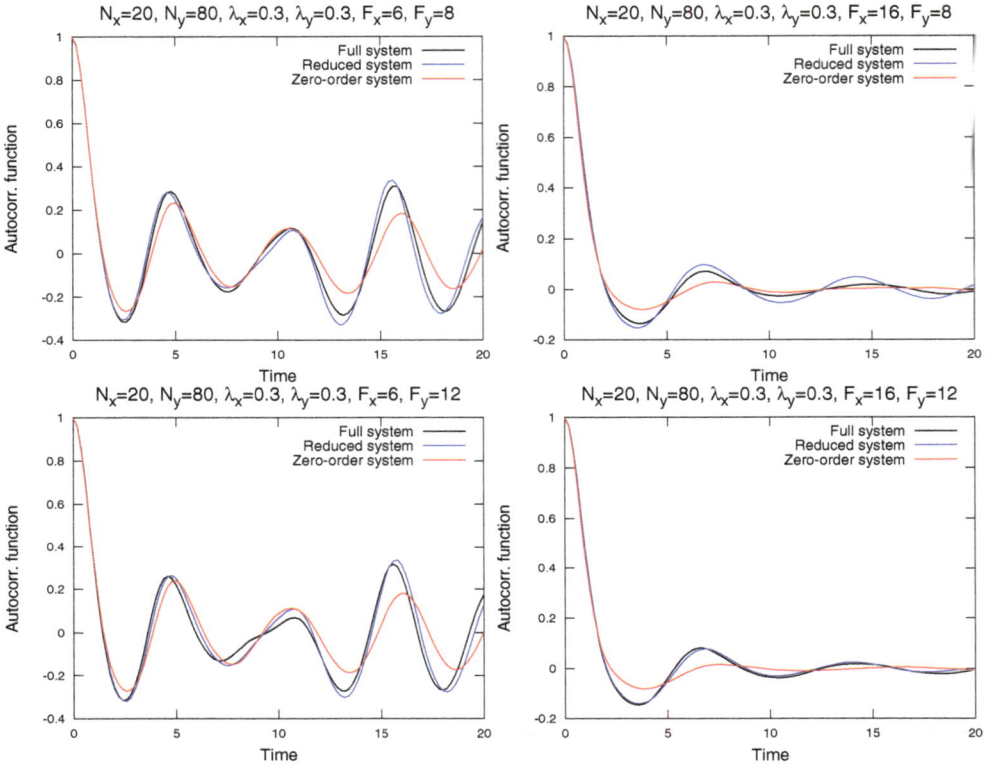

Fig. 6. Time autocorrelation functions of the slow variables. The following parameters are used: $N_x = 20$, $N_y = 80$, $F_x = 6, 16$, $F_y = 8, 12$, $\lambda_x = \lambda_y = 0.3$, $\varepsilon = 0.01$.

$\lambda_{x,y} = 0.3, F_y = 8$			$\lambda_{x,y} = 0.3, F_y = 12$		
	Av.	Z.O.		Av.	Z.O.
$F_x = 6$	$5.841 \cdot 10^{-2}$	0.1211	$F_x = 6$	$6.539 \cdot 10^{-2}$	0.1572
$F_x = 16$	$4.079 \cdot 10^{-2}$	$5.342 \cdot 10^{-2}$	$F_x = 16$	$1.559 \cdot 10^{-2}$	$7.396 \cdot 10^{-2}$
$\lambda_{x,y} = 0.4, F_y = 8$			$\lambda_{x,y} = 0.4, F_y = 12$		
	Av.	Z.O.		Av.	Z.O.
$F_x = 6$	$5.538 \cdot 10^{-2}$	0.3677	$F_x = 6$	0.2981	0.3986
$F_x = 16$	$8.534 \cdot 10^{-2}$	0.1355	$F_x = 16$	$4.835 \cdot 10^{-2}$	0.1482

Table 4. L_2-errors between the time autocorrelation functions of the slow variables of the full two-scale Lorenz 96 model and the two reduced models. Notations: "Red." stands for "Reduced" (that is, (28)), and "Z.O." stands for "Zero-order", the poor man's version of (28).

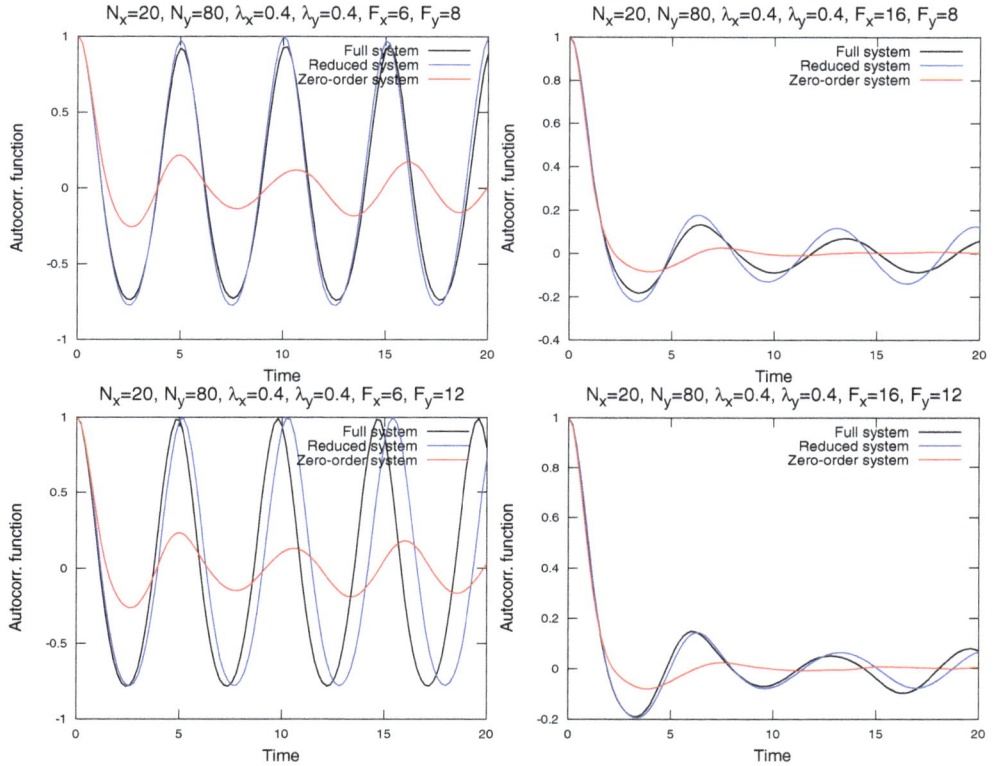

Fig. 7. Time autocorrelation functions of the slow variables. The following parameters are used: $N_x = 20$, $N_y = 80$, $F_x = 6, 16$, $F_y = 8, 12$, $\lambda_x = \lambda_y = 0.4$, $\varepsilon = 0.01$.

regime with $F_x = 6$ the PDFs of the full two-scale Lorenz 96 model have three sharp peaks, indicating strong non-Gaussianity. The reduced model in (28) reproduces these peaks, while its zero-order version fails. In Figures 6 and 7 we show the time autocorrelation functions of the slow dynamics for the full two-scale Lorenz 96 model, the reduced closed model for the slow variables alone in (28), and its poor man's zero order version without the linear correction term. Just as the PDFs, for the more weakly coupled regimes with $\lambda_x = \lambda_y = 0.3$ the time autocorrelation functions look similar, yet the reduced model with the correction term reproduces the time autocorrelation functions more precisely than the zero-order model. In the more strongly coupled regime with $\lambda_x = \lambda_y = 0.4$ the difference between the reduced model in (28) and its poor man's zero-order version is even more drastic: observe that for the weakly chaotic regime with $F_x = 6$ the time autocorrelation functions of the full two-scale Lorenz 96 model do not exhibit decay (indicating very weak mixing), and the reduced model in (28) reproduces the autocorrelation functions of the full two-scale Lorenz 96 model rather well, while its zero-order version fails. In addition, in Table 4 we show the L_2-errors in time autocorrelation functions (for the correlation time interval of 20 time units, as in Figures 6 and 7) between the full two-scale Lorenz 96 model and the two reduced models. Observe that, generally, the reduced system in (28) produces more precise results than its poor man's version without the correction term.

4. Conclusions

In this work we made an initial attempt to estimate and reproduce the chaotic properties of a nonlinear multiscale system with linear energy-preserving coupling which mimics major features of low frequency variability dynamics in real-world climate. In particular, we find that, due to the energy-preserving coupling, the sensitivity to initial conditions of the slow variables can be reduced by rapid mixing strong chaos of the fast variables. In addition, we develop a simple closure scheme for the slow variables alone, which captures major statistics of the full multiscale system. These two studies suggest that the predictability of the low frequency variability in multiscale climate dynamics could in practice be better than what is normally presumed today, and future projections of the low frequency variability modes could be captured by a much simpler reduced model for slow variables alone, which could potentially lead to improved climate change prediction. Below we list the major developments and observations of this work.

- A suitable theory of the effect of the fast rapidly mixing dynamics on the chaos at slow variables is developed by applying the averaging formalism to the linearized dynamics of the system. It is found that the linear energy-preserving coupling creates a systematic damping effect on the chaos at the slow scales when the fast dynamics is rapidly mixing.

- This effect is vividly demonstrated for the two-scale Lorenz 96 model, which is also appropriately rescaled so that the adjustments for the mixing regime at the fast variables do not affect the mean state and variance of both the slow and fast variables. In particular, it is shown through direct numerical simulations that the uncoupled slow dynamics may remain chaotic, while the full coupled system loses chaos and becomes completely predictable at the slow scales as the dynamics at the fast scales become more turbulent.

- With help of the observations above, we develop a simple method of constructing the closed reduced model for slow variables of a multiscale model with linear coupling, which requires only a single computation of the mean state and the time autocorrelation function for the fast dynamics with a fixed state of the slow variables. The method is based on the first-order Taylor expansion of the averaged coupling term for the fast variables with respect to the slow variables, which is computed using the linear fluctuation-dissipation theorem. We demonstrate through the computations with the same rescaled Lorenz 96 model that, with simple linear coupling in both slow and fast variables, the developed reduced model produces quite comparable statistics to what is exhibited by the complete two-scale Lorenz 96 model.

Given the above results, the question of improved predictability of low frequency variability and climate becomes more interesting. Indeed, if the coupling to the fast dynamics makes slow processes less chaotic, then the reduced models with constant parameterizations of interactions with fast dynamics could be more chaotic than the actual physical processes they describe. If one can create reduced climate models which capture the chaos suppression effect with adequate skill, the climate change projections can potentially become less uncertain. In the future, the author intends to develop such reduced climate models for a more realistic Earth-like setting, possibly in collaboration with climate scientists. It remains to be seen whether Earth's climate is more predictable than we tend to think.

5. Acknowledgments

The author is supported by the National Science Foundation CAREER grant DMS-0845760, and the Office of Naval Research grants N00014-09-0083 and 25-74200-F6607.

6. References

Abramov, R. (2009). Short-time linear response with reduced-rank tangent map, *Chin. Ann. Math.* 30B(5): 447–462.

Abramov, R. (2010). Approximate linear response for slow variables of deterministic or stochastic dynamics with time scale separation, *J. Comput. Phys.* 229(20): 7739–7746.

Abramov, R. (2011a). Improved linear response for stochastically driven systems, *Front. Math. China*, accepted.

Abramov, R. (2011b). A simple linear response closure approximation for slow dynamics of a multiscale system with linear coupling, *Mult. Mod. Simul.*, accepted.

Abramov, R. (2011c). Suppression of chaos at slow variables by rapidly mixing fast dynamics through linear energy-preserving coupling, *Comm. Math. Sci.*, accepted.

Abramov, R. & Majda, A. (2003). Quantifying uncertainty for non-Gaussian ensembles in complex systems, *SIAM J. Sci. Comp.* 26(2): 411–447.

Abramov, R. & Majda, A. (2007). Blended response algorithms for linear fluctuation-dissipation for complex nonlinear dynamical systems, *Nonlinearity* 20: 2793–2821.

Abramov, R. & Majda, A. (2008). New approximations and tests of linear fluctuation-response for chaotic nonlinear forced-dissipative dynamical systems, *J. Nonlin. Sci.* 18(3): 303–341.

Abramov, R. & Majda, A. (2009). New algorithms for low frequency climate response, *J. Atmos. Sci.* 66: 286–309.

Abramov, R. & Majda, A. (2011). Low frequency climate response of quasigeostrophic wind-driven ocean circulation, *J. Phys. Oceanogr.*, accepted.

Buizza, R., Miller, M. & Palmer, T. (1999). Stochastic representation of model uncertainty in the ECMWF Ensemble Prediction System, *Q. J. R. Meteor. Soc.* 125: 2887–2908.

Crommelin, D. & Vanden-Eijnden, E. (2008). Subgrid scale parameterization with conditional Markov chains, *J. Atmos. Sci.* 65: 2661–2675.

Crowley, T. (2000). Causes of climate change over the past 1000 years, *Science* 289: 270–277.

Delworth, T. & Knutson, T. (2000). Simulation of the early 20th century global warming, *Science* 287: 2246–2250.

Fatkullin, I. & Vanden-Eijnden, E. (2004). A computational strategy for multiscale systems with applications to Lorenz 96 model, *J. Comp. Phys.* 200: 605–638.

Franzke, C. (2002). Dynamics of low-frequency variability: Barotropic mode., *J. Atmos. Sci.* 59: 2909–2897.

Franzke, C., Majda, A. & Vanden-Eijnden, E. (2005). Low-order stochastic model reduction for a realistic barotropic model climate, *J. Atmos. Sci.* 62: 1722–1745.

Grote, M., Majda, A. & Grotta Ragazzo, C. (1999). Dynamic mean flow and small-scale interaction through topographic stress, *J. Nonlin. Sci.* 9: 89–130.

Hasselmann, K. (1976). Stochastic climate models, part I, theory, *Tellus* 28: 473–485.

Lions, J., Temam, R. & Wang, S. (1992a). New formulations of the primitive equations of the atmosphere and applications, *Nonlinearity* 5: 237–288.

Lions, J., Temam, R. & Wang, S. (1992b). On the equations of the large-scale ocean, *Nonlinearity* 5: 1007–1053.

Lions, J., Temam, R. & Wang, S. (1993a). Models of the coupled atmosphere and ocean, *Comp. Mech. Adv.* 1: 5–54.

Lions, J., Temam, R. & Wang, S. (1993b). Numerical analysis of the coupled models of atmosphere and ocean, *Comp. Mech. Adv.* 1: 55–119.

Lions, J., Temam, R. & Wang, S. (1995). Mathematical study of the coupled models of atmosphere and ocean, *Math. Pures Appl.* 74: 105–163.

Lorenz, E. (1963). Deterministic nonperiodic flow, *J. Atmos. Sci.* 20(2): 130–148.

Lorenz, E. (1996). Predictability: A problem partly solved, *Proceedings of the Seminar on Predictability*, ECMWF, Shinfield Park, Reading, England.

Lorenz, E. & Emanuel, K. (1998). Optimal sites for supplementary weather observations, *J. Atmos. Sci.* 55: 399–414.

Majda, A. (2003). *Introduction to PDEs and Waves for the Atmosphere and Ocean*, Vol. 9 of *Courant Lecture Notes*, American Mathematical Society and Courant Institute of Mathematical Sciences.

Majda, A., Abramov, R. & Grote, M. (2005). *Information Theory and Stochastics for Multiscale Nonlinear Systems*, Vol. 25 of *CRM Monograph Series of Centre de Recherches Mathématiques, Université de Montréal*, American Mathematical Society. ISBN 0-8218-3843-1.

Majda, A., Gershgorin, B. & Yuan, Y. (2010). Low frequency response and fluctuation-dissipation theorems: Theory and practice, *J. Atmos. Sci.* 67: 1186–1201.

Majda, A., Timofeyev, I. & Vanden-Eijnden, E. (1999). Models for stochastic climate prediction, *Proc. Natl. Acad. Sci.* 96: 14687–14691.

Majda, A., Timofeyev, I. & Vanden-Eijnden, E. (2001). A mathematical framework for stochastic climate models, *Comm. Pure Appl. Math.* 54: 891–974.

Majda, A., Timofeyev, I. & Vanden-Eijnden, E. (2002). A priori tests of a stochastic mode reduction strategy, *Physica D* 170: 206–252.

Majda, A., Timofeyev, I. & Vanden-Eijnden, E. (2003). Systematic strategies for stochastic mode reduction in climate, *J. Atmos. Sci.* 60: 1705–1722.

Palmer, T. (2001). A nonlinear dynamical perspective on model error: A proposal for nonlocal stochastic-dynamic parameterization in weather and climate prediction models, *Q. J. R. Meteor. Soc.* 127: 279–304.

Papanicolaou, G. (1977). Introduction to the asymptotic analysis of stochastic equations, in R. DiPrima (ed.), *Modern modeling of continuum phenomena*, Vol. 16 of *Lectures in Applied Mathematics*, American Mathematical Society.

Risken, H. (1989). *The Fokker-Planck Equation*, 2nd edn, Springer-Verlag, New York.

Selten, F. (1995). An efficient description of the dynamics of barotropic flow, *J. Atmos. Sci.* 52: 915–936.

Uhlenbeck, G. & Ornstein, L. (1930). On the theory of the Brownian motion, *Phys. Rev.* 36: 823–841.

Vanden-Eijnden, E. (2003). Numerical techniques for multiscale dynamical systems with stochastic effects, *Comm. Math. Sci.* 1: 385–391.

Volosov, V. (1962). Averaging in systems of ordinary differential equations, *Russian Math. Surveys* 17: 1–126.

Wilks, D. (2005). Effects of stochastic parameterizations in the Lorenz '96 system, *Q. J. R. Meteorol. Soc.* 131: 389–407.

Part 4

Applied Climatology

12

On the Relationship Between Boundary Layer Convergence and Cloud-to-Ground Lightning

Michael L. Gauthier*
United States Air Force Academy
USA

1. Introduction

It is generally accepted that significant electrification, and subsequent lightning generation, in clouds is attained via non-inductive charging (NIC) when sufficient numbers of ice crystals collide with graupel particles in the presence of supercooled liquid water [e.g. Saunders et al., 1991; Jayaratne et al., 1983; Takahashi, 1978]. As these particle scale interactions are driven by vertical motions it can be argued that, under appropriate thermodynamical and microphysical conditions, any process that enhances updraft strength should also enhance the storms ability to generate lightning.

Constrained by mass continuity, updrafts leading to deep moist convection are necessarily associated with sub-cloud horizontal mass convergence. Given that the Earth's surface is impermeable with respect to the wind, it is clear that horizontal convergence of boundary layer winds should result in compensating upward vertical motions with greater convergence over a given area resulting in greater vertical motions, possibly capable of initiating and/or intensifying convection. All else being equal (i.e., sufficient moisture and instability requisite for the development of deep moist convection), enhancements in boundary layer convergence (BLC) should deepen the planetary boundary layer (PBL), thereby enhancing the instability, with the end result being an increase in the number of updrafts capable of breaking the "cap" (capping inversion) allowing for more vigorous interactions between precipitation sized ice particles and ascending ice crystals within the charging zone, ultimately resulting in enhancements in thunderstorm electrification and lightning via NIC.

The effect of enhanced BLC, and the atmospheric response (i.e., rainfall and/or lightning) has been illustrated in previous observational studies from southern Florida [Ulanski and Garstang, 1978; Watson and Blanchard, 1984; Watson et al., 1987], highlighting the importance that boundary layer winds have on new thunderstorm growth and potential lightning production. In particular, Watson and Blanchard [1984], examined 121 convergence "events" during the 1975 Florida Area Cumulus Experiment (FACE) campaign and found a relationship between the change in total area divergence (the negative of which is used in this study; total area convergence (TAC), discussed in Section 3.3) and the amount

* The views expressed in this paper are those of the author and do not reflect the official policy or position of the U.S. Air Force, Department of Defense or the U.S. Government.

of radar-derived rainfall. Watson et al. [1987] coupled these findings with those that documented a proportionality between thunderstorm precipitation output and the total number of flashes [i.e., Workman and Reynolds, 1949; Battan, 1965], by relating cloud-to-ground (CG) lightning with surface wind convergence (both spatially and temporally) for 42 summer days in 1983 over central Florida. Their examination of divergence and lightning time series during this period showed that CG lightning began near the time of strongest convergence, peak lightning activity typically occurred when the total area divergence was near zero in the transition from convergence to divergence, and CG lightning ended just after peak divergence over the area. Further, through a set of scaling arguments Banacos and Schultz [2005] showed that horizontal mass convergence associated with smaller mesoscale boundaries such as lake/sea breezes, and active or remnant convective outflow boundaries, is at least an order of magnitude larger [$O(10^{-4}s^{-1})$] than convergence associated with typical frontal boundaries.

More recently, Gauthier [*in press*] found that, from a climatological sense, urban heat island (UHI) thermodynamics not only provided a more favorable environment for convection over the Houston area, but that the UHI acts to enhance the inland progression of the sea-breeze, thereby creating an enhanced area of localized convergence over the northern central portion of the city. The author then went on to assert that the spatial extent of the flash density features in and around the Houston area (to be described in Section 2.1) were "primarily the result of typical convective activity tied to the presence of a persistent thermal anomaly over the center of the city, giving rise to a preferential location of low-level convergence and convective enhancement."

Utilizing boundary layer convergence fields generated by the National Center for Atmospheric Research's (NCAR) Variational Doppler Radar Assimilation System (VDRAS; [Sun and Crook, 1997]), this study complements previous lightning [Gauthier et al., 2005], radar [Gauthier et al., 2006] and convergence [Gauthier, *in press*] climatologies presented in the literature, independently establishing the presence of an enhanced convergence zone located in the vicinity of the Houston metropolitan area, generally co-located with the observed climatological enhancement in cloud-to-ground lightning over and downwind of the city. Area averaged, and cell-scale analyses link boundary layer convergence (forcing #1) with enhancements in radar derived precipitation ice mass (response #1, forcing #2) and ultimately enhanced ground flash densities (response #2).

2. Background

2.1 Houston and the Houston lightning anomaly

An area lacking significant orographic features, Houston sits adjacent to the complex coastline of Galveston Bay (to the southeast) with over 50 km² of winding and stationary waterways dispersed throughout. In addition to the Houston shipping channel, a heavily industrialized link between the city and the Gulf of Mexico, there are four major bayous that transect the city, with Lake Houston, a significant ground-level water source located approximately 30 km to the northeast of city center. The climate of Houston is classified as humid subtropical, with its proximity to Galveston Bay and the Gulf of Mexico causing winters to be quite mild, while the summers are rather hot and humid. Summer air temperatures average 33° C (92° F) during the day and 23° C (73° F) at night with mean daily dewpoint temperatures of 22° C (72° F). Southerly winds prevail during the summer months with wind speeds averaging 4 m s^{-1}, with an average of 19 lightning days per summer [Gauthier et al., 2005].

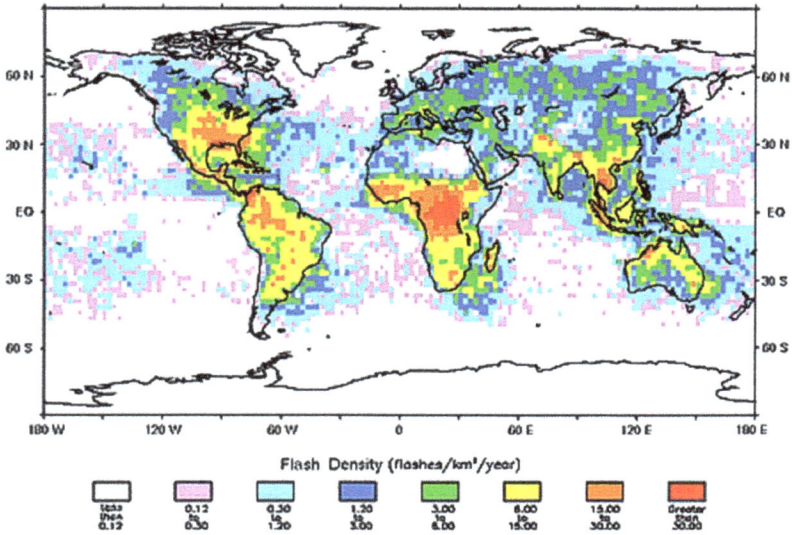

Fig. 1. Spatial distribution of annual global lightning activity between September 1, 1995 and August 31, 1996 as observed by NASA's OTD. Lightning flash densities (flashes km^{-2} year^{-1}) were calculated statistically using OTD data from more than 400 separate 3-minute observations of each location on the earth (SOURCE: http://thunder.msfc.nasa.gov/otd/).

Using 12-years of CG lightning data (1989-2000) gathered by the NLDN, Orville et al. [2001] documented a persistent, year-round, enhancement in CG lightning activity downwind of the Houston metropolitan area. Steiger et al. [2002], using the same data set (5 km spatial resolution), quantified the enhancement observed by Orville et al., reporting a 45% increase in annual CG lightning flash densities over and downwind of the Houston urban corridor relative to rural surroundings. Their findings are generally consistent with previous studies [i.e., Westcott, 1995], indicating that observed enhancements in CG flash densities can occur over, and downwind of urban corridors. To further study this documented lightning anomaly, Gauthier et al. [2005] performed an independent analysis on an extended subset of the CG lightning dataset (1995 – 2003) referenced in the literature [Orville et al., 2001; Steiger et al., 2002]. Their findings indicate that the local Houston CG lightning anomaly is a persistent summer-season feature (even when large lightning events were excluded from the analysis), with flash densities over and downwind of the Houston metropolitan area approaching 1.5-2 times that of immediate surroundings. However, when examined more regionally, they found the feature to be statistically non-unique, embedded with "anomalies" of similar scale within the more general enhancement of CG lightning along the Texas and Louisiana Gulf Coast. In fact, examination of annual lightning flash densities (flashes km^{-2} year^{-1}; Figure 1) observed by NASA's Optical Transient Detector (OTD) reveals a broad region of coastal enhancements, situated along the central Gulf of Mexico (including the Houston area), in *total*, not just CG lightning activity. This indicates that the Houston area likely falls victim to enhanced intra-cloud lightning activity as well.

2.2 Thunderstorm electrification – The physics of the problem

Although the exact processes by which thunderstorms become electrified and ultimately lower charge to the surface of the Earth is still debated, it is generally accepted that significant electrification, and subsequent lightning generation, in clouds is attained via NIC when sufficient numbers of ice crystals collide with graupel particles in the presence of supercooled liquid water [e.g. Saunders et al., 1991; Jayaratne et al., 1983; Takahashi, 1978]. Here, as graupel particles grow in the mixed-phase region of a developing storm, suspended by convective updrafts, smaller ice crystals are swept past the graupel and subsequently transported into the upper portions of the storm. Rebounding collisions between the graupel particles and ice crystals result in negative charge being deposited on the graupel particles, and positive charge being deposited on the ice crystals. This process leads to the development of an electrical dipole, with positive charge situated over negative charge. Once the graupel's terminal velocity exceeds the storm updraft speed, it will begin to fall towards the surface of the Earth. Graupel/ice collisions occurring below the charge reversal temperature, which is some threshold between –5°C and –10°C dependent on liquid water content (LWC), will result in the falling graupel particles becoming positively charged and the ascending ice crystals negatively charged. The rising ice crystals can contribute to the elevated negative charge region, while the descending graupel particles produce a small positive charge region below the main negative charge center. The result is the development of an electrical tripole (with the negative charge region sandwiched between two positive charge regions) in agreement with observed tripoles structure in typical thunderstorms [Williams, 1989].

While the tripole model just described is conveniently simple, it is important to note that there remains considerable complexity in the microphysical charge separation process, and not all thunderstorms ascribe to this simple model. However, the larger-scale physics associated with the NIC mechanism remain valid, with three key requirements for thunderstorm electrification and subsequent lightning production: (1) the existence of convective updrafts, capable of (2) suspending precipitation ice mass in the mixed-phase region of the storm, allowing (3) rebounding graupel-ice crystal collisions to occur. Although these requisite ingredients are critical to the thunderstorm electrification process, it is clear that (2) and (3) can not occur without the presence of a significant convective updraft, with stronger updrafts capable of producing more lightning with increased ground flash densities.

From the above discussions it is readily apparent that the occurrence of lightning should be correlated to updraft strength (which ultimately allows for mixed phase interactions between graupel particles and ice crystals). Therefore, under the appropriate thermodynamical and microphysical conditions, any process that enhances updraft strength should also enhance the storms ability to generate lightning. Given that the Earth's surface is impermeable with respect to the wind, it is clear that horizontal convergence of the winds occurring within the boundary layer will result in compensating vertical motions with greater convergence over a given area resulting in greater vertical motions. These convergence based vertical motions, if co-located with existing updrafts, can enhance updraft intensity which, given the underlying physics of the problem, should ultimately lead to enhancements in thunderstorm electrification.

As previously mentioned in Section 1, Gauthier [*in press*] found the spatial extent of the flash density features over the Houston area to be primarily the result of "typical" convective activity tied to the presence of a persistent thermal anomaly over the center of the city. This persistent thermal anomaly, coupled with the land surface heterogeneity of the surrounding area, gives rise to a preferential location of low-level convergence and convective enhancement. They concluded that, from a climatic viewpoint, the primary causative mechanisms responsible for the *intensity* of the Houston CG lightning anomaly (i.e., *those responsible for the predominant enhancements to the background, "typical", features just described*) to be those that serve to enhance low-level convergence; specifically that UHI thermodynamics contribute to an area of preferred convergence over, and to the east-northeast of the city, while also driving *mesoscale enhancements in sea breeze convergence*. Acting together, these two enhancements are argued to cause more frequent convection over the Houston urban area resulting in more lightning activity over the city.

2.3 VDRAS – Diagnosing convergence in the boundary layer

As VDRAS is an analysis tool developed, maintained and operated by NCAR, and the intent of this study it to evaluate the relationship between boundary layer convergence (as diagnosed by VDRAS), radar derived precipitation ice mass, and lightning, this section provides the reader with a cursory overview of VDRAS. For a detailed description of the system the reader is referred to Sun and Crook [1997], Sun and Crook [1998] and Crook and Sun [2001].

It has long been recognized that the success of numerical weather prediction (NWP) models depends strongly on the accuracy with which the atmosphere is represented at the time of model initialization. Unfortunately, by themselves, traditional weather observations provide a less than complete dataset from which to describe the initial state of the atmosphere in these models. Herein lies the need to incorporate, or assimilate, "unconventional" sources of weather information such as in-situ measurements and/or satellite observed/derived quantities into atmospheric NWP models to help "fill the gaps" in traditional observing networks, and provide a more complete (and hopefully more accurate) description of the atmosphere for use in the critical model initialization process. As its name implies, VDRAS utilizes a four-dimensional variational assimilation technique to incorporate a time series of radar observations (both radial velocity and reflectivity) from WSR-88D radars into a cloud-scale numerical model in order to better represent the evolution of flow (components of the horizontal wind) within the atmosphere. Through the minimization of a cost function, the model is fit to the observations over a specified period of time in order to develop a set of optimal initial conditions for use by the constraining numerical model.

VDRAS has been applied in both research and operational environments; the application of the system to different stages of convective development has demonstrated that the detailed structure of wind, thermodynamics and microphysics could be obtained with reasonable accuracy [Sun and Crook, 1998; Sun and Crook, 2001], thereby highlighting the utility of the system in diagnosing convergence within the boundary layer. Application of VDRAS generated fields to lightning studies such as this is a new endeavor; therefore, the work presented here can be considered a *"first of its kind"* [A. Crook, personal communications, 2005].

3. Data and method

This study complements previous datasets (CG lightning, radar reflectivity, and precipitation ice mass) used by the author to investigate the documented enhancement in cloud-to-ground lightning over, and downwind of, the Houston metropolitan area by incorporating BLC fields through the assimilation of radial velocity data contained within the radar reflectivity dataset used by Gauthier et al. [2006] (described in Section 3.1) into VDRAS. Data gathered were used to construct a spatial climatology of convergence throughout the domain, with accompanying area and cell-scale analyses used to investigate the relationship between BLC, precipitation ice mass (IM) and CG flash densities and counts (FD and FC, respectively).

Two separate analysis methods were used to compare warm-season statistics of VDRAS derived BLC, radar derived IM, and NLDN detected FDs and FCs, both of which mimic the approach used by Gauthier et al. [2006] where similar analyses were performed comparing IM with FDs and FCs. The first method compared time-integrated (TI) or cumulative means of BLC, IM and FD for each 4 km grid square within the domain yielding a total of 10,000 data points for comparison (one for each 16 km² pixel within the horizontal analysis domain). The second approach utilized the cell identification component of the Interactive Data Language (IDL) cell-tracking algorithm used by Gauthier et al. [2010] to compare storm integrated BLC values with storm IMs, FCs and FDs on a cell-by-cell basis (described in Section 3.4). Here, the original software was modified to incorporate the VDRAS BLC dataset. A total of 14,061 cells were identified and analyzed as part of this study.

3.1 Radar and CG lightning methodologies

Using the synoptically conditioned dataset from Gauthier [*in-press*], over 1,200 daytime (0900 – 1659 CST) convective volumes of archived WSR-88D Level II data [detailed in Gauthier et al., 2006] from 15 lightning days over Houston were selected for assimilation into VDRAS. In addition to the requirement for the occurrence of at least one CG lightning strike over the Houston area, included volumes occurred on days in which synoptic conditions were deemed favorable for the formation of a sea breeze circulation.

Radar data from each of these days were processed using two separate schemes, one for the VDRAS analysis performed at NCAR, and a second for the area and cell-scale analyses (conducted locally). In all instances the domain spanned 400 km x 400 km, centered on the location of the League City, TX radar site. For the VDRAS analyses, the Level II data were shipped to NCAR where they were interpolated to a 1 km Cartesian grid in the horizontal while an unfolding algorithm was applied to the radial velocity input data stream. Following gridding, the data underwent an automated quality control process, and was then further interpolated onto the 4 km model grid in the horizontal while remaining on the constant elevation angle levels in the vertical. The vertical resolution of the VDRAS output grids were 0.5 km, extending to an altitude of 7 km. Radar data was also locally gridded so that radar derived precipitation ice mass quantities could be calculated [as detailed by Gauthier et al., 2006] with horizontal resolutions matching that of the VDRAS grid; vertical resolution was 1 km, extending 20 km in the vertical to allow for computation of precipitation ice mass quantities throughout the domain.

Coincident with each radar volume, ground strike locations (i.e., flashes occurring from the beginning of one volume scan to the beginning of the subsequent volume scan) detected by the NLDN were gridded to match the horizontal dimensions of the Cartesian VDRAS and radar grids. For consistency, positive ground flashes with peak currents less than 10 kA have been disregarded [c.f., Cummins et al., 1998; Wacker and Orville, 1999a, b; Gauthier et al., 2005].

3.2 VDRAS methodology

The assimilation window for VDRAS is 10 minutes, meaning that a total of three consecutive volumes of radar data (constant elevation levels) were used in each window. Background wind fields, used to provide a first guess for the cold start cycle and boundary conditions for subsequent cycles, were generated using a velocity azimuth display (VAD; [Lhermitte and Atlas, 1961]), coupled with other available observations, as well as RUC model analyses. Once the background fields had been generated, the radar data were assimilated into the system.

Assimilation of the radar data was performed using a continuous cycling procedure [Sun and Crook, 2001]. As previously discussed in Section 2.3, an optimal fit between the model and the data was obtained by minimizing the difference between the model and observations (i.e., minimization of the cost function). This was accomplished by running (cycling) the numerical model forward and the adjoint of the model backward until an optimum solution was obtained. Model results (horizontal convergence at each pixel) were then saved as output grids. Mean BLC values at each horizontal grid-point were taken as the mean convergence within the lowest 1.5 km of the model domain (3 lowest model layers). Due to the length of the assimilation window, the VDRAS analysis resulted in an output grid for every other volume of radar data resulting in a total of 631 VDRAS output grids for use in this study.

3.3 Total Area Convergence (TAC) and Total Cell Convergence (TCC)

To quantify the overall magnitude of the boundary layer convergence occurring over a given area, the VDRAS convergence fields were used to compute the area-averaged convergence, or total-area convergence (TAC); given as the grid point mean convergence within an enclosed area [Cunning et al., 1982]. Equivalent to the negative of the line integral of the normal component of the wind around the area, this number provides a quantitative measure of the amount of horizontal mass flux into (positive values) or out of (negative values) an area, which can then be used to infer the relative magnitude and sign of compensating vertical motions.

Although TAC proves to be of utility when diagnosing horizontal mass flux in an *Eulerian* sense [e.g., Watson and Blanchard, 1984; Watson et al., 1987], and is an appropriate metric for the quantitative comparisons associated with the spatial climatologies to be presented in Section 4.1, this quantity is inappropriate for use in our cell-scale analyses where compensating areas of convergence/divergence within the BL associated with storm-scale up/down drafts routinely coexist; here we are most concerned with sub-cloud forcing that acts to enhance upward vertical motions within the cell. Therefore, a more appropriate method of determining that portion of the flow capable of forcing upward motions within a cell would be to integrate the positive mean BLC values contained within the cell; we call this quantity total cell convergence (TCC) and will use it as our primary metric when discussing differences in VDRAS derived mean BLC between convective cells in Section 4.2.

3.4 Cell identification methodology

The datasets described above were used as input parameters into an IDL variant [as modified by Gauthier et al., 2010] of the Thunderstorm Identification, Tracking, Analysis and Nowcasting (TITAN) software [Dixon and Wiener, 1993] where individual cells were identified throughout the domain for the entire period of interest.

For the purpose of this study, a storm cell is defined as a contiguous region (2 or more pixels) of low level (z=2 km above ground level) radar reflectivity with values greater than or equal to 30 dBZ. Using geometric logic regarding storm cell positions and shapes, the algorithm identifies cells that occured within the domain of interest. Since pixel relative BLC, IM and FC measurements had previously been computed, cell totals of each parameter were taken as the sum of positive BLCs (justified in Section 3.3), IMs and FCs associated with each pixel comprising the cell; cell FDs were computed by normalizing the cell FC by cell area. As pointed out by Gauthier et al. [2010], this approach effectively treated each cell as a vertical entity, accounting for neither vertical tilt, nor ground flashes coming to ground in regions of reflectivity less than 30 dBZ.

4. Results and discussion

As previously outlined, the data gathered for this investigation were used to construct a spatial climatology of convergence over and around the Houston metropolitan area. Herein, we present results in the form of time-integrated analyses (Section 4.1) with the results of the cell-scale analyses presented separately in Section 4.2.

4.1 Time integrated analysis

Figure 2a presents the spatial distribution of VDRAS derived BLC (mean convergence within the lowest 1.5 km of the atmosphere) averaged over all hours contained within the 15-day dataset (note that only positive convergent flow is contoured in this figure). Consistent with the wind and convergence climatologies presented in Gauthier [in-press], a definite area of localized convergence over the Houston area is apparent in the 15-day mean. Relating the forcing produced by the BLC to the atmospheric response, Figure 2b presents a spatial composite with ground flash densities (FDEN; blue contours) and shaded contours of precipitation ice mass (IM) overlaid upon the VDRAS derived BLC presented in Figure 2a. From this presentation, a new link in the forcing-response chain begins to emerge. Here, we see that persistently focused, low-level, mesoscale forcing (on the order of 4-5 cm s^{-1}, computed based on continuity) over the Houston area, may give rise to a preferential location of convective initiation (CI), leading to downwind enhancements of radar derived precipitation ice mass within the charging zone (i.e., displaced from peak forcing), and

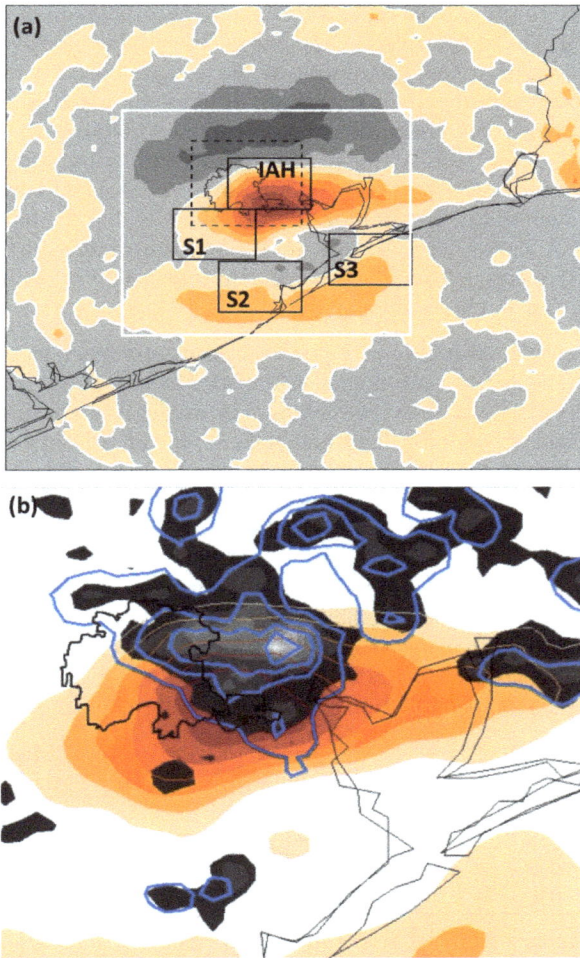

Fig. 2. Spatial distribution of the 15-day day time integrated mean (a) VDRAS derived boundary layer convergence (BLC) field with darkening shades of maroon associated with increasing positive convergence contoured at intervals of 1×10^{-5} sec^{-1}, relative to the zero value depicted by the white contour. Negative convergence (i.e., divergence) is represented by darkening shades of gray contoured at intervals of -1×10^{-5} sec^{-1}, relative to the same zero value contour, black reference boxes denote sub-domains discussed in the text, white box denotes that portion of the domain presented in Figure 2b; and (b) zoomed composite representation of normalized (relative to the domain maximum of each quantity) time integrated BLC (with shades of maroon increasing in darkness every 15%, beginning at 10% of the maximum BLC value within the domain, note divergent flow is not presented), radar derived precipitation ice mass (with shades of gray decreasing in intensity every 10%, beginning at 20%) and ground flash densities contoured in blue at the 20, 40, 60 and 80% levels, relative to the domain maximum flash density value associated with the 15-day mean.

finally enhancements in ground flash densities. Arguably, this sequence of events will, in and of itself, spawn localized (storm scale) regions of convergent and divergent flow serving to enhance and/or diminish the magnitude of the climatological convergence anomaly; however, the localized persistence of mesoscale BLC in the mean is consistent with the findings, and speculations contained in the aforementioned literature. Indeed, the persistent convergence located over the city is consistent with observations presented in Gauthier et al. [2010] where the authors found an area of enhanced composite cell merger occurrence located over and downwind of the city. The fact that the locations of enhanced convergence and cell merger activity are in general agreement further highlights the significance that cell mergers have with respect to convective intensity [Simpson, 1980; Simpson et al., 1980; Tao and Simpson, 1984].

Quantifying these spatial observations, Table 1 contains summary statistics of the spatial differences between the various sub-domains; statistics include daily mean values of TAC, IM and conditional FDEN (i.e., total flashes in each analysis box normalized by the area encompassing all pixels containing lightning) taken over all 15 days within our VDRAS subset. Clearly, relative to the other sub-domains, the Houston area (IAH) falls victim to enhanced forcing (TAC) leading to enhancements in PIM, and ultimately conditional FDENs. Comparison of the statistics associated with the remaining sub-domains (against one another) indicates that BLC alone, although an important ingredient, is not, in and of itself, sufficient to spawn deep moist convection and that meteorological conditions (i.e., the presence of sufficient moisture and instability) must be sufficient to allow this to occur. To illustrate this fact, we note that analysis boxes S2 and S3 have nearly equivalent values of TAC; however, the atmosphere responds more vigorously to the climatological forcing over S2 with greater enhancements in precipitation ice mass and resultant ground flash densities.

Analysis Box	Mean TAD $[10^{-5} \text{ s}^{-1}]$	Mean IM $[10^7 \text{ kg km}^{-2} \text{ day}^{-1}]$	Mean FDEN $[10^{-3} \text{ flashes km}^{-2} \text{ day}^{-1}]$
IAH	2.50	3.70	1.000
S1	0.94	1.80	0.200
S2	0.46	1.00	0.100
S3	0.45	0.51	0.027

Table 1. VDRAS summary statistics, see text for description.

Explanation for the dearth in lightning activity over the mouth of Galveston Bay (S3) can likely be found in the stabilizing effect that the cooler water temperatures have on the atmosphere. Although S2 (a coastal location) is also quite influenced by the Gulf waters, the stronger land-ocean contrasts likely serve to provide the necessary boost in instability enabling the enhanced response, relative to S3. Regardless of the reason for the subtle differences found within the climatological summary just presented, it is clear that the Houston area is more favorable for the development of deep moist convection resulting in enhanced ground flash densities relative to its immediate surroundings.

4.2 Cell scale analysis

Complementing the time integrated analyses just presented, all 631 convective radar volumes, along with accompanying VDRAS derived BLC fields, radar derived IMs and

NLDN detected CG lightning strikes were ingested into the modified cell tracking algorithm resulting in the identification of 1,928 lightning producing cells and 12,133 non-lightning producing cells throughout the domain; recall that the subjective definition of a cell is that of the area enclosed by the 30 dBZ radar reflectivity contour at 2 km above the ground. In this section we first discuss spatial differences between mean cell characteristics for cells geo-located in each of the sub-domains compared in Section 4.1 (IAH, S1, S2 and S3), we then go on to investigate differences between lightning and non-lightning cells in general, throughout the domain.

To that end, Table 2 provides a summary of the statistics to be discussed in this section. Focusing first on Table 2, columns a-d, we note that the IAH analysis box contained nearly 60% of the CG lightning producing cells (CG) observed in all analysis areas, and relative to the other sub-domains mean TCC associated with these cells were the greatest. Consistent with these findings we also note that the IAH CG cells contained larger quantities of precipitation ice mass located within the charging zone with higher CG flash rates (by a factor of 2, save S2). Comparing CG cells located within analysis boxes S1 and S2 (Table 2, columns b and c) , we note larger mean TCC values associated with S1 CG cells (mean values over twice that observed in S2); however, CG cells located within S2 contain 1.5 times as much IM and produce nearly 23% more lightning on a per cell basis. As in the previous section, the apparent discontinuity between the low-level forcing and the atmospheric response, lead us to conclude that relative to S1, analysis area S2 must be more favorable for the formation of deep moist convection, at least during the 15 days on which this analysis was based.

	(a) IAH		(b) S1		(c) S2		(d) S3		(e) DOMAIN	
	CG	noCG	CG	noCG	CG	noCG	CG	noCG	CG	noCG
COUNT	127	297	40	302	44	176	9	79	1,928	12,133
TCC [10^{-3} sec^{-1}]	1.80	0.29	1.10	0.30	0.50	0.30	0.36	0.12	1.60	0.23
IM [10^5 kg]	161	9	85	10	130	9	47	6	145	9
Flashes [per cell]	6.55		3.175		3.9		2.1		5.13	
FDEN [flashes km^{-2} cell^{-1}]	0.027		0.02		0.027		0.017		0.02	

Table 2. VDRAS summary statistics associated with 14,061 convective cells, see text for description.

Some insightful findings surrounding the characteristic differences between lightning and non-lightning (noCG) cells surface when comparing the mean statistics associated with all cells identified within the dataset (summarized in Table 2, column e). Here we see that, on average, CG cells have TCC values approximately 7 times greater than noCG cells with IM values being an order of magnitude larger. Not included in Table 2, we also note that the average peak radar reflectivity at tracking altitude (i.e., peak reflectivity at an altitude of 2 km) is on the order of 10 dBZ greater in the CG cells, nearly 50 dBZ.

The relationship between TCC, cell total IM and FDEN, is shown in Figure 3, where we present a three-parameter scatter plot of TCC and cell total IM, with varying pixel colors based on the flash density associated with the each particular cell. We first note a linear correlation (R = 0.7) between TCC and cell total IM with enhanced BL forcing associated with cells containing enhanced quantities of precipitation ice in the charging zone. Consistent with the findings of Gauthier et al. [2006], we further note that cells with more IM generally have larger ground flash densities associated with them (see dark gray and red data points in Figure 3). The fact that the most electrically intense storms (from a CG perspective) do not reside in the extreme upper-right hand portion of the parameter space is reassuring, as previous studies have highlighted an apparent anti-correlation between storms with extremely intense updrafts (> 30-35 m s^{-1}) and CG lightning production [e.g., MacGorman et al., 1989, Lang et al., 2000; Lang and Rutledge, 2002]. Hypothesized to be the result of an "elevated charge mechanism," it is argued that intense updrafts loft the main negative charge layer to greater altitudes than normal, thereby reducing the magnitude of the electric field between the main negative charge center and ground due to the increased spatial separation; this sequence of events would also likely delay the development of the lower positive charge center (LPCC) believed to be important in the development of ground flashes. It follows that this reduction in electric field may then favor intra-cloud (IC) lightning over CG lightning until such time as the storm begins to weaken, allowing the descent of precipitation ice to warmer regions of the cloud. This eventuality may then lead to the possible development of a the LPCC (recall discussions on the

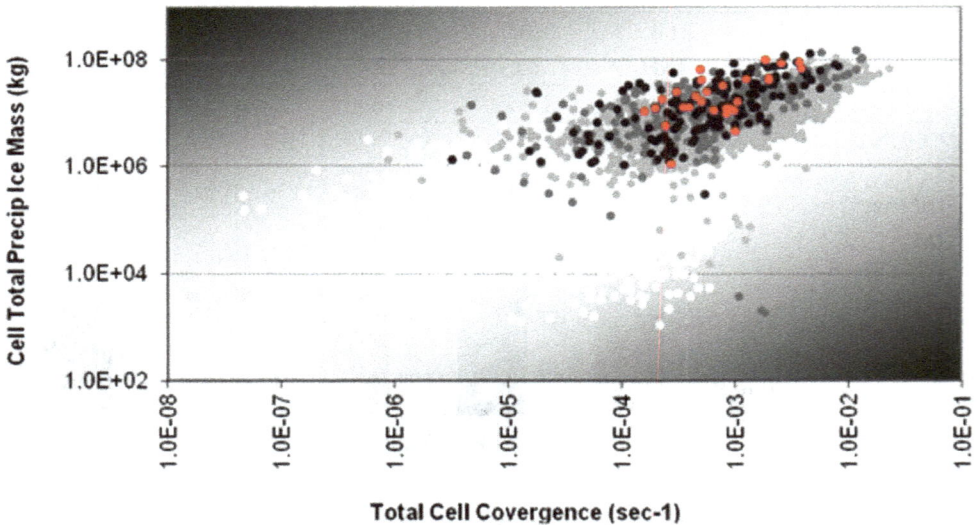

Total Cell Covergence (sec-1)

Fig. 3. A three-parameter scatter plot of cell total precipitation ice mass (IM; ordinate) as a function of total cell convergence (TCC; abscissa) and cell flash density (FDEN; no lightning = white, with increasing shades of gray associated with more intense flash densities, light gray = FDENs in the lower 3 quartiles of the conditional cumulative FDEN distribution, medium gray = FDEN within the 75-90th percentile range, dark gray = FDEN > 90th percentile with cells producing the most intense flash densities plotted in red).

development of the electrical tripole presented in Section 2.2) reinvigorating the occurrence of CG flashes. Although unable to quantify it, we suspect that this is what we are observing in Figure 3, where strong updraft enhancements (on the order of 15-25 m/s, based on convergence alone) are associated with a decrease in ground flash densities of storms that have been electrically "primed" (i.e., storms containing ample precipitation sized ice particles in the mixed-phase region).

5. Conclusion

Utilizing BL convergence fields generated by NCAR's Variational Doppler Radar Assimilation System, we investigated the relationship between VDRAS derived boundary layer convergence, radar derived precipitation ice mass and NLDN detected CG lightning over 15 non-consecutive summer days, independently establishing the presence of an enhanced convergence zone located in the vicinity of the Houston metropolitan area. Coupling *Eulerian* and cell-scale analyses over the Houston area, we clearly link the BL convergence zone with enhancements in radar derived precipitation ice mass and ultimately enhanced ground flash densities. Statistics associated with over 14,000 convective cells occurring throughout the domain highlight distinct differences between lightning and non-lightning producing cells, with the average lightning producing cells having convergence based updraft enhancements nearly 7 times that of non-lightning producing cells, with an order of magnitude increase in precipitation ice mass within the charging zone. Collectively, the findings presented in this chapter firmly establish a physical proportionality between the ability of the cloud ensemble to produce enhanced CG lightning activity, and the amount of low-level convergent flow available to feed the updraft(s).

6. Acknowledgment

This research was supported by funding from the U.S. Air Force Academy through the Air Force Institute of Technology. Special thanks go to Dr. Andrew Crook and his staff at NCAR for their assistance and expertise in running the VDRAS analyses used in this study, without their efforts this project would not have been possible. The views expressed in this paper are those of the author and do not reflect the official policy or position of the U.S. Air Force, Department of Defense, or the U.S. Government.

7. References

Banacos, P. C., and D. M. Schultz (2005). The use of moisture flux convergence in orecasting convective initiation: Historical and operational perspectives, *Weather and Forecasting*, 20(3), 351-366.

Battan L. J. (1965). Some factors governing precipitation and lightning from convective clouds, *Journal of Atmospheric Science*, 22, 79-84.

Crook N. A., and J. Sun (2004). Analysis and forecasting of the low-level wind during the Sydney 2000 Forecast Demonstration Project, *Weather and Forecasting.*, 19, 151-167.

Cummins, K. L., M. J. Murphy, E. A. Bardo, W. L. Hiscox, R. B. Pyle, and A. E. Pifer (1998). A combined TOA/MDF technology upgrade of the U.S. National

Lightning Detection Network, *Journal of Geophysical Research, 103*(D8), doi: 10.1029/98JD00153.

Cunning J. B., R. L. Holle, P. T. Gannon, and A. I. Watson (1982). Convective evolution and merger in the FACE experimental area: Mesoscale convection and boundary layer interactions, *Journal of Applied Meteorology, 21,* 953-977.

Dixon, M., and G. Wiener (1993). TITAN: thunderstorm identification, tracking, analysis, and nowcasting— A radar-based methodology, *Journal of Atmospheric and Oceanic Technology, 10,* 785-797.

Gauthier, M. L., W. A. Petersen, L. D. Carey, and R. E. Orville (2005). Dissecting the anomaly: A closer look at the documented urban enhancement in summer season ground flash densities in and around the Houston area, *Geophysical Research Letters, 32,* L10810, doi:10.1029/2005GL022725.

Gauthier, M.L., Petersen, W.A., Carey, L.D., Christian, H.J. (2006). Relationship between cloud-to-ground lightning and precipitation ice mass: a radar study over Houston. *Geophysical Research Letters, 33,* L20803. doi:10.1029/2006GL027244.

Gauthier, M. L., Petersen, W. A., Carey, L. D. (2010). Cell mergers and their impact on cloud to-ground lightning over the Houston area. *Journal of Atmospheric Research,* doi: 10.1016/j.atmosres.2010.02.010.

Gauthier, M. L. (*in-press*). Investigating possible causative mechanisms behind the Houston cloud-to-ground lightning anomaly, In: *Climatology: New Developments,* A. Herveoux, E. Suthland, Nova Science Publishers, Inc., ISBN: 978-1-62100-322-9, Hauppauge, New York.

Jayaratne, E. R., C. P. R. Saunders, and J. Hallet (1983). Laboratory studies of the charging of soft hail during ice crystal interactions, *Quarterly Journal of the Royal Meteorological Society, 109,* 609-630.

Lang, T. J., and S. A. Rutledge (2002). Relationships between convective storm kinematics, precipitation, and lightning, *Monthly Weather Review, 130,* 2492-2506.

Lang, T. J., S. A. Rutledge, J. E. Dye, M. Venticinque, P. Laroche, and E. Defer (2000). Anomalously low negative cloud-to-ground lightning flash rates in intense convective storms observed during STERAO-A. *Monthly Weather Review, 128,* 160-173.

Lhermitte, R. M., and D. Atlas (1961). Precipitation motion by pulse Doppler, *Proc. Ninth Weather Radar Conf.,* Kansas City, MO, American Meteorological Society, 218–223.

MacGorman, D. R., D. W. Burgess, V. Mazur, W. D. Rust, W. L. Taylor, and B. C. Johnson (1989). Lightning rates relative to tornadic storm evolution on 22 May 1981, *Journal of Atmospheric Science, 46,* 221-250.

Orville, R. E., G. Huffines, J. Nielsen-Gammon, R. Zhang, B. Ely, S. Steiger, S. Phillips, S. Allen, and W. Read (2001). Enhancement of cloud-to-ground lightning over Houston, Texas, *Geophysical Research Letters, 28*(13), doi:10.1029/2001GL012990.

OTD Global Lightning Image obtained from http://thunder.msfc.nasa.gov/otd/ maintained by NASA EOSDIS Global Hydrology Resource Center (GHRC) DAAC, Huntsville, AL. 2011. Data for the image were provided by the NASA EOSDIS GHRC DAAC.

Petersen, W. A., H. J. Christian, and S. A. Rutledge (2005). TRMM observations of the global relationship between ice water content and lightning, *Geophysical Researhc Letters, 32*, L14819, doi:10.1029/2005GL023236.

Saunders, C. P. R., W. D. Keith, and R. P. Mitzeva (1991). The effect of liquid water on thunderstorm charging, *Journal of Geophysical Reasearch, 96*, 11,007-11,017.

Simpson, J. (1980). Downdrafts as linkages in dynamic cumulus seeding effects, *Journal of Applied Meteorology, 19,*477-487.

Simpson, J. N. E. Westcott, R. J. Clerman, and R. A. Pielke (1980). On cumulus mergers, *Arch. Met. Geoph. Biokl., 29*, 1-40.

Steiger S. M., R. E. Orville, and G. Huffines (2002). Cloud-to-ground lightning characteristics over Houston, Texas: 1989-2000, *Journal of Geophyical Reserearch, 107*(D11), doi:10.1029/2001JD001142.

Sun J., and N. A. Crook (1997). Dynamical and microphysical retrieval from doppler radar observations using a cloud model and its adjoint. Part I: Model development and simulated data experiments, *Journal of Atmospheric Science, 54(12)*, 1642-1661.

Sun J., and N. A. Crook (1998). Dynamical and microphysical retrieval from doppler radar observations using a cloud model and its adjoint. Part II: Retrieval experiments of an observed Florida convective storm, *Journal of Atmospheric Science, 55(5)*, 835-852.

Sun, J., and N. A. Crook (2001). Real-time low-level wind and temperature analysis using single WSR-88D data, *Weather and Forecasting, 16(1)*, 117-132.

Takahashi, T. (1978), Rimin g electrification as a charge generation mechanism in thunderstorms, *Journal of Atmospheric Science, 35*, 1536-1548.

Tao, W.-K., and J. Simpson (1984). Cloud interactions and merging: Numerical simulations, *Journal of Atmospheric Science, 41*, 2901-2917.

Ulanski S. L., and M. Garstang (1978). The role of surface divergence and vorticity in the life cycle of convective rainfall. Part I: Observation and analysis, *Journal of Atmospheric Science, 35*, 1047-1062.

Wacker, R. S., and R. E. Orville (1999a). Changes in measured lightning flash count and return stroke peak current after the 1994 U.S. National Lightning Detection Network upgrade: 1. Observations, *Journal of Geophysical Research, 104*(D2), doi:10.1029/1998JD200060.

Wacker, R. S., and R. E. Orville (1999b). Changes in measured lightning flash count and return stroke peak current after the 1994 U.S. National Lightning Detection Network upgrade: 2. Theory, *Journal of Geophysical Research, 104*(D2), doi:10.1029/1998JD200059.

Watson A. I., and D. O. Blanchard (1984). The relationship between total area divergence and convective precipitation in South Florida, *Monthly Weather Review, 112*, 673-685.

Watson A. I., R. E. López, R. L. Holle, and J. R. Daugherty (1987). The relationship of lightning to surface convergence at Kennedy Space Center: A preliminary study, *Weather and Forecasting, 2*, 140-157.

Westcott, N. E. (1995). Summer-time cloud-to-ground lightning activity around major Midwestern urban areas, *Journal of Applied Meteorology, 34*, 133-1642.

Williams, E. R. (1989). The tripole structure of thunderstorms, *Journal of Geophysical Research,* *94,* 13,151-13,167.

Workman, E. J., and S. E. Reynolds (1949). Electrical activity as related to thunderstorm cell growth, *Buletin of the American Meteorological Society, 30,* 142-149.

13

Rainfall Prediction Using Teleconnection Patterns Through the Application of Artificial Neural Networks

Gholam Abbas Fallah-Ghalhari
Sabzevar Tarbiat Moallem University
I.R of Iran

1. Introduction

All aspects of human life are, directly or indirectly, affected by climatic processes. This effect is especially noticeable in such fields as agriculture, irrigation, economy, telecommunications, transportation, traffic, air pollution and military industries (Haltiner & Williams 1980).

A number of researchers have studied the possibility of forecasting rainfall several months in advance using climate indices such as SOI, PDOI and NPI (e.g. Silverman and Dracup 2000).

A well-known atmospheric phenomenon is the Southern Oscillation (SO). The SO is an atmospheric see-saw process in the tropical Pacific sea level pressure between the eastern and western hemispheres associated with the El Niño and La Niña oceanographic features. The oscillation can be characterized by a simple index, the Southern Oscillation Index (SOI). (Kawamura et al., 1998). The Pacific Decadal Oscillation index (PDOI) is the leading principal component of monthly sea surface temperature (SST) anomalies in the North Pacific Ocean north of 20°N (Zhang et al., 1997; Mantua et al., 1997). Trenberth and Hurrell (1994) have defined the North Pacific Index (NPI) as the area-weighted sea level pressure over the region 30°N to 65°N, 160°E to 140°W to measure the decadal variations of atmosphere and ocean in the north Pacific.

Furthermore, the existence of substantial databases of sea surface temperature anomalies (SST) opens the possibility of using these data to forecast rainfall several months in advance. Most of the research carried out in this area has used traditional statistical methods such as linear correlation or time series methods to identify the significant variables. These methods test for a linear relationship between the independent variables and rainfall, whereas the relationships are more likely to be non-linear as the underlying processes are themselves non-linear (Iseri et al. 2005).

Long-term rainfall prediction is very important to countries thriving on agro-based economy. In general, climate and rainfall are highly non-linear phenomena in nature giving rise to what is known as butterfly effect (Abraham et al. 2001). In their quest for new ways of predicting important meteorological factors, researchers have devised and developed techniques such as intelligent methods, which are viable and flexible means independent of system dynamic models.

Pongracz & Bartholy (2006) designed a model of monthly rainfall in Hungary using types of atmospheric circulation patterns and ENSO (El Niño - Southern Oscillation) index. To this end, they used a modeling technique based on fuzzy rules to establish a relationship between inputs and rainfall. Their results indicated that the model based on fuzzy rules provides an excellent means for the prediction of statistical features of rainfall using monthly occurrence of types of daily circulation pattern and delayed SOI.

Halid & Ridd (2002) used the fuzzy logic to design a model and predict local rainfall in January at Hasanuddin airport, Indonesia, which is the largest rice-producing area in the country. Their results indicated that, compared to other statistical models, the fuzzy logic model is more useful for the prediction of rainfall in January. Choi (1999) used neural networks and the Geographical Information System to forecast rainfall. The results indicated the efficiency of the Geographical Information System and neural networks in rainfall forecast. Cavazos (2000) used neural networks to forecast daily rainfall. The parameters which were used included the thickness between 500 and 1000 hecto-Pascal levels, the altitude of 500 hPa levels, and the humidity of 700 hPa levels. The outcomes indicated the efficiency of neural networks in the prediction of rainfall. Maria et al (2005) used neural networks and regression models to forecast rainfall in Sao Paulo, Brazil. The parameters they used included: potential temperature, vertical wind component, specific humidity, precipitable water, relative vorticity and humidity flux divergence. The results indicated the efficiency of both methods in forecasting rainfall.

One of the most crucial issues of global climatic variability is its effect on water resources. If more accurate predictions of rainfall were possible, this would enable more efficient utilization of water resources. However, long-term rainfall prediction models are still unsatisfactory, whereas short-term rainfall prediction models have undergone significant development. The probable reasons for the difficulties in conducting long-term rainfall prediction are the complexity of atmosphere-ocean interactions and the uncertainty of the relationship between rainfall and hydro meteorological variables. So far, long-term climate prediction using numerical models has not demonstrated useful performance, and statistical models have shown better performance than numerical models (Zwiers & Von Storch 2004). Consequently, in this study Artificial Neural Networks and linear regression models have been applied to nonlinear and linear statistical prediction.

Due to the significance of rainfall in many decision making processes such as water resources management and agriculture, the present study aims to find out the relationship between large-scale climatic synoptic patterns and regional rainfall using such synoptic patterns as sea level temperature and temperature difference, sea level pressure and pressure difference, precipitable water, air temperature at 700-hPa level, the thickness between 500 and 1000-hPa levels and the relative humidity at 300-hPa level.

2. Data and methods

2.1 Data

2.1.1 Study area

The region studied in this research is Khorasan Razavi Province. The time series studied is the average rainfall from April to June during 38 years. The data of spring rainfall for each year includes the rainfall in 34 visibility, climatology and rainfall measurement stations provided by the Weather Bureau and the Power Ministry. Of these, 24 stations are rainfall measurement stations of the Power Ministry and the rest belong to the Weather Bureau. Fig. 1 represents the map of the studied area and the name of the relevant stations. To compensate for some defects in rainfall data, subtractions and ratios method have used. Run test was also used to test the homogeneity of the data. The analysis of *runs* within a sequence is applied in statistics in many ways (for examples see Feller 1968 & Ducan 1974). The term *run* may in general be explained as a succession of items of the same class. Many concepts to analyze runs in a series of data have been studied. The main concepts are based on (i) the analysis of the total number of runs of a given class (see Guibas and Odlyzko 1980 and Knuth 1981) and (ii) examinations about the appearance of long runs (see Feller 1968, Guibas and Odlyzko 1980 and Wolfowitz 1944).

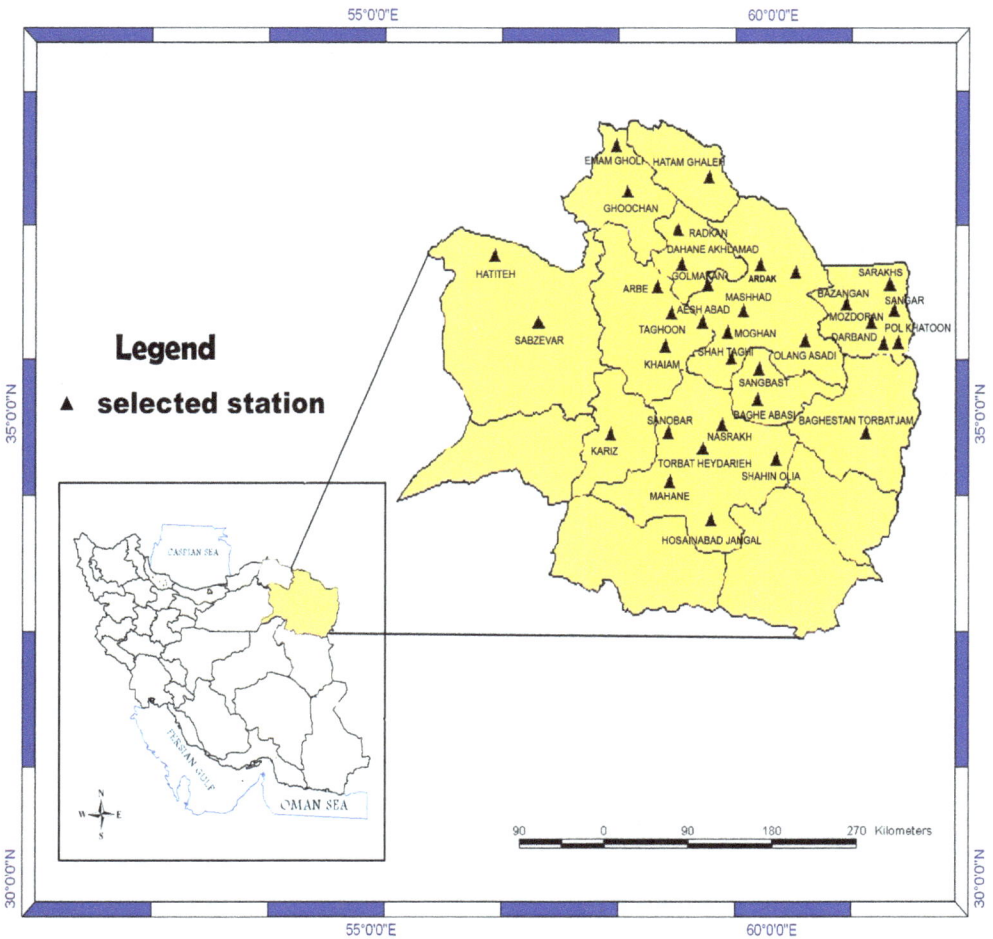

Fig. 1. Map of the region under study and selected stations (Fallah Ghalhary et al, 2010)

2.1.2 Climatological data

The data used in this study are:

1. 34 Rainfall station data for the seasonal rainfall (April – June) were obtained from Iranian Meteorological organization. All of these stations are in the north eastern region of Iran.
2. Large-scale ocean and atmospheric circulation variables such as Sea Surface Temperature (SST), Sea Level Pressure(SLP), the difference Sea Level Pressure, the difference Sea Surface Temperature between surface and 1000 hPa level, relative humid at 300 hPa level, geopotential height at 500 hPa level, air temperature at 850 hPa level during months (Oct-Mar). These data were obtained from NCEP/NCAR Re-analysis data. These data sets span the period of 1948 – current, covering the globe on a 2.5 × 2.5 grid and available at http://www.cdc.noaa.gov National Oceanic and Atmospheric Administration (NOAA) website. Table 1 summarizes the data used in this study.

Data	source	time	year
Rainfall station data	Iranian Meteorological organization	April – June	1970-2007
NCEP/NCAR Re-analysis data	Large-scale ocean and atmospheric circulation variables	Oct-Mar	1970-2007

Table 1. Summaries of the data used in this study

2.2 Methods

2.2.1 Spatial prediction and kriging

The need to obtain accurate predictions from observed data can be found in all scientific disciplines. Those that have embraced statistical notions of random variation are able to do this by exploiting the statistical dependence among the data and the variables(s) to be predicted. However, the statistical approach has not been without its detractors; for example, Philip and Waston (1986) argue that geostatistics is unhelpful for solving problems in mining and geology. Their article and the accompanying discussions are worth perusing.

It is helpful to explain first the terms used in the title of this chapter. Let $\{Z(s): s \in D \subset R^d\}$ be a random function (or process), from which n data $Z(s_1)$, ..., $Z(s_n)$ are collected. The data are used to perform inference on the process, here, to predict some known functional g ($\{Z(s): s \in D\}$) [or, more simply, g $(Z(.))$] of the random function $Z(.)$. For example, point prediction assumes g $(Z(.))=Z(S_0)$, where S_0 is a known spatial location. g is mostly real-valued. Sometimes interest is not in $Z(.)$, but in a "noiseless" version of it.

Suppose that

$$Z(s) = S(s) + \in (s), s \in D, \tag{1}$$

Where $\in (.)$ is a white-noise measurement-error process. In this case, one is interested in predicting a known functional $g(S(.))$ of the noiseless random function $S(.)$.

Spatial prediction refers to prediction either $g(Z(.))$ or $g(S(.))$ from data $Z(s_1)$, ... , $Z(s_n)$ observed at known spatial locations s_1 , ... , s_n.

Notice that my terminology encompasses the temporal notions of smoothing (or interpolation), filtering, and prediction (e.g., Lewis, 1986, PP. 36), which rely on time-ordering for their distinction. If temporal data are available from the past up to and including the present, smoothing refers to prediction of g(S (.)) at time points in the past, filtering refers to prediction of g(S (.)) at the present time and prediction refer to prediction of g(S (.)) at time points in the future. In this paper, the word "estimation" will be used exclusively for inference on fixed but unknown parameters; "prediction" is reserved for inference on random quantities.

Kriging is a minimum-mean-squared-error method of spatial prediction that (usually) depends on the second-order properties of the process Z (.).

Matheron (1963) named this method of optimal spatial linear prediction after D. G. Krige, a South African mining engineer who, in the 1950s, developed empirical methods for determining true ore-grade distributions from distributions based on sampled ore grades (e. g., Krige, 1951). However, the formulation of optimal spatial linear prediction did not come from krige's work. (See Matheron, 1971, pp. 117-119, and Cressie, 1990, for the extent of the early work of Krige.) The contributions of Wold (1938), Kolmogorov (1941), and Wiener (1949) all contain optimal linear prediction equations that reflect the notion that observations closer to the prediction point (for them, closer in time) should be given more weight in the predictor.

At the same time as geostatisstics was developing in mining engineering under G. Matheron in France, the very same ideas developed in meteorology under L. S. Gandin (Gandin, 1963) in the Soviet Union. The original (and simultaneous) contribution of these authors was to put optimal linear prediction (in terms of variogram) into a spatial setting. Gandins' name for his approach was *objective analysis*, and he used the terminology *optimum interpolation* instead of Kriging. Details of the origins of Kriging are set out in Cressie (1990) and Cressie (1993).

2.2.1.1 Observational and spatial scale

The following model is useful. Suppose that the data (Z (S₁)... Z (Sₙ)) represent Z values at points of $D \subset R^d$, and that they are modeled as a partial realization of the random process:

$$\{Z(s): S \in D \subset R^d\}, \tag{2}$$

Which satisfies the decomposition

$$Z(s) = \mu(s) + W(s) + \eta(s) + \epsilon(s), s \in D, \tag{3}$$

Where:

$\mu(.) \equiv E(Z(.))$ Is the deterministic mean structure that will be called *large-scale variation*.

W (.) is a zero-mean, L_2-continues [i.e., $E(W(s+h) - W(s))^2 \to 0$ as $\|h\| \to 0$], intrinsically stationarv process whose variogram rang (if it exist) is larger than min $\{\|s_i - s_j\| : 1 \le i < j \le n\}$. Call W (.) *smooth small-scale variation*.

η (.) Is a zero-mean, intrinsically stationarv process. independent of W, whose variogram rang exists and is smaller than min $\{\|s_i - s_j\| : 1 \le i < j \le n\}$. Call $\eta(.)$ *microscale variation*.

ϵ (.) is a zero-mean white-noise process, independent of W and η call ϵ (.) measurement error or noise, and denote var (ϵ (s))= C_{ME}. There are occasions when ϵ (.) may possess more structure than that of white noise (e.g., Laslett and McBratney, 1990).

In notation as follow,

$$2 y_Z(.) = 2 y_W(.) + 2 y_\eta(.) + 2 c_{ME}. \tag{4}$$

The quantities C_{ME} and $\gamma_Z(h)$, $\|h\|$ larger, are pertinent to the observational scale; the other quantities contain information on the spatial scale.

From the decomposition (3), write

$$Z(s) = S(s) + \in (s), s \in D, \tag{5}$$

Where the "signal" or smooth process S (.) is given by S (.) $\equiv \mu(.) + W(.) + \eta(.)$. The S process is often referred to as the noiseless version of the Z process or, in the engineering literature, as the *state* process. Also, write

$$Z(s) \equiv \mu(s) + \delta(s), \quad s \in D, \tag{6}$$

Where the correlated error process $\delta(.)$ is given by $\delta(.) \equiv W(.) + \eta(.) + \in (.)$. When the Correlation of $\delta(s)$ white $\delta(s+h)$ can be written as a function of $h / \alpha(h)$, where $0 < \alpha(h) < \infty$, then $\alpha(h)$ is sometimes called the *spatial correlation scale* of the process in direction h (Cressie, 1993).

2.2.1.2 Ordinary kriging

Ordinary Kriging produced better estimates than simple Kriging because of the non-stationarity of the data. The original data set had large areas where the values were low and large areas where the values were high. Simple Kriging requires the mean value of the data set to be provided, whereas ordinary Kriging calculates a mean for each individual block, based on the samples included in estimate. The local mean appears to be more meaningful in a situation where the global mean is not constant (Weber and Englund, 1992).

The word "Kriging" is synonymous with "optimal prediction" (as a noun) or with "optimally predicting" (as the present participle of a verb). In other words, it refers to making inferences on unobserved values of the random process Z (.) given by (2) or of S (.) given by (5) from data:

$$Z \equiv (Z(S_1), ..., Z(S_n)) \tag{7}$$

Observed at known spatial locations $\{S_1, ..., S_n\}$.

Denote the generic predictor of g (Z (.) or g(S (.)) by:

$$P(Z; g) \tag{8}$$

Choice of a good predictor will depend on the geometry and location of the region of space where prediction is desired and weather it is the Z process or the S process that is to be predicted. When g (Z (.)) = Z (B) $[\equiv \int_B Z(u)du / |B|]$ or g(S (.)) = S (B), write (3.2.3) as p (Z; B). In the special case of B = {s_0}, write (8) as p (Z; s_0).

Ordinary Kriging (Matheron, 1971; Journel and Huijbregts, 1978) refers to spatial prediction under the following two assumptions.

Model Assumption. In (6)

$$Z(s) = \mu + \delta(s), \quad s \in D, \mu \in \mathrm{R}, \text{ and } \mu \text{ unknown.} \tag{9}$$

Predictor Assumption.

$$p(Z;B) = \sum_{i=1}^{n} \lambda_i Z(s_i), \quad \sum_{i=1}^{n} \lambda_i = 1. \tag{10}$$

This latter condition, which the coefficients of the linear predictor sum to 1, guarantees uniform unbiasedness. E (p (Z; B)) = $\mu = E(Z(B))$, for all $\mu \in \mathrm{R}$.

There is a version of Kriging called simple Kriging, where μ in (9) is known and the coefficients are not constrained to sum to 1.

Hence, if

$$g(Z(.)) = Z(B) \equiv \begin{cases} \int_B Z(u)du \, / \, |B|, & |B| > 0 \\ ave\{Z(u) : u \in B\}, & |B| = 0 \end{cases} \tag{11}$$

Then the optimal p (. ; B) will minimize the mean-squared prediction error

$$\sigma_e^2 \equiv E(Z(B) - p(Z;B))^2 \tag{12}$$

Over the class of linear predictors $\sum_{i=1}^{n} \lambda_i Z(s_i)$ that satisfy $\sum_{i=1}^{n} \lambda_i = 1$ (Cressie, 1993).

2.2.1.3 Kriging and calculating the average regional rainfall

In this research, we have used from ordinary Kriging for obtaining the average regional rainfall in the area under study. Kriging is the estimation procedure used in geostatistics using known values and a semivariogram to determine unknown values. It was named after D. G. Krige from South Africa. The procedures involved in Kriging incorporate measures of error and uncertainty when determining estimations. Based on the semivariogram used, optimal weights are assigned to unknown values in order to calculate unknown ones. Since the variogram changes with distance, the weights depend on the known sample distribution (Davis 1990).

The final goal of studying spatial changes of rainfall is to simulate the changes in rainfall data in the spatial dimension in order to pave the way for attaining other goals such as forecasting rainfall and getting necessary information for the long-term analysis of rainfall in every region in the area under study. As mentioned before, Kriging method used in this study to calculate the average regional rainfall. The following steps were taken to obtain the time series of average regional rainfall:

1. Making input files for the Arc GIS 9.2 software
2. Obtaining the experimental variogram
3. Analyzing and drawing annual spatial changes of rainfall in the region
4. Obtaining the values of annual average rainfall in the region under study
5. Making time series of rainfall in the region under study.

Fig.2 for example shows the diagram of the Variogram using spherical model to estimate of the average regional rainfall for the year of 2007.

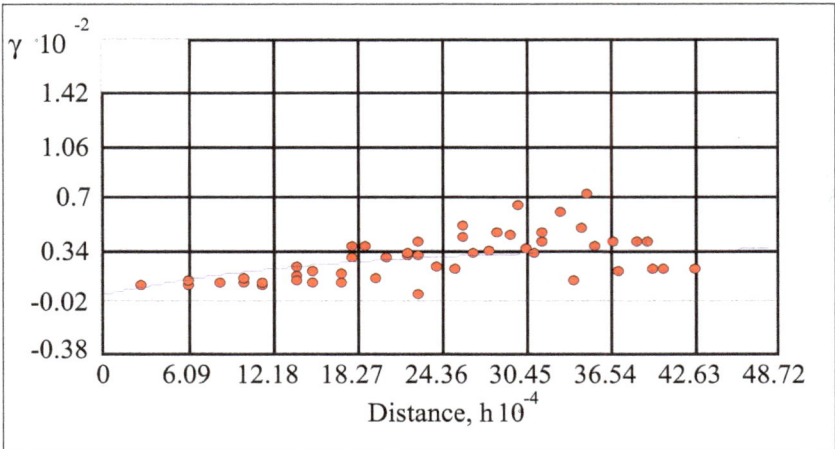

Fig. 2. Variogram diagram using spherical model to estimate of the average regional rainfall for year of 2007. Units on the horizontal and vertical axes are meter and square millimeter respectively.

2.2.2 Determining seasonal rainfall and predictors

The predictors studied in this research are classified into two groups: meteorological parameters at ground level and meteorological parameters at upper levels of the atmosphere. Table 2 shows these parameters.

Upper level of atmosphere	Ground level
Air temperature at 700-hPa level	Sea level pressure
the thickness between 500 and 1000hPa levels	Sea level pressure difference
Relative humidity of 300-hPa level	Sea level temperature
	Temperature difference between sea level and 1000-hPa level
	Zonal wind
	Meridional wind
	Precipitable water

Table 2. Meteorological Parameters used in this study

One distinctive scenario is considered in this study. This scenario uses input data with 6 months lags to investigate the possibility of forecasting more than 3 months in advance.

One of the objectives of this study is the identification of a possible relationship between rainfall in Iran and climatic predictors, using Pearson's correlation coefficient. The other objective is to verify the forecasts produced using the predictors identified with Pearson's correlation coefficient.

Seasonal rainfall and predictors have been determined using the average of the values of a predictor in order to predict the amount of seasonal rainfall. We made sure that the seasons of the predictor do not include months with rainfalls.

Since we aimed to investigate the relation between meteorological parameters and spring rainfall in this study (spring rainfall is very important in dry land cultivation and water recourses management), we used the average value of meteorological predictors in the period between October and March as the time series of the predictors and the average rainfall in the period between April and June as the rainfall time series.

To analyze the parameters of the upper levels of the atmosphere as well as three ground parameters, i.e. zonal wind, meridional wind and precipitable water in the present study, we used 5°×5° and 10°×10° degrees spatial resolution. The study areas, where meteorological parameters at ground level and upper levels of atmosphere have been factor-analyzed, are located between 0-80 °E and 10-50 °N in 5°×5° degrees spatial resolution and between 0-100 °E and 0-70 °N in 10°×10° degrees spatial resolution. The area includes regions where changes in the pattern of temperature, pressure, humidity, and wind speed affect Iranian rainfall (Fig. 3).

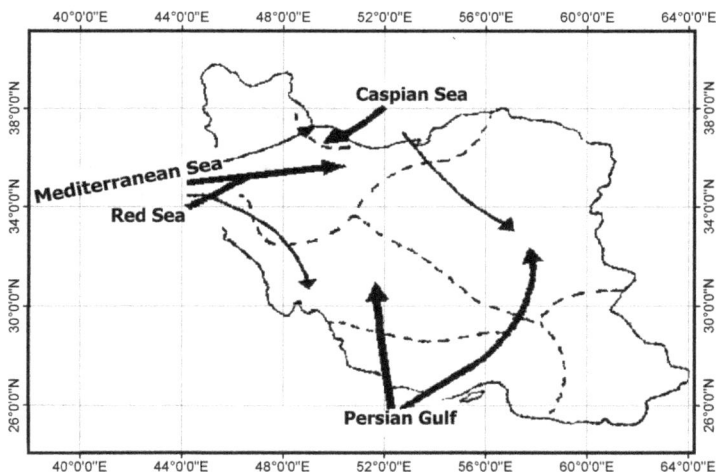

Fig. 3. Humidity sources of Iranian rainfalls in the spring season (Alijani, 2006). the wider arrows have more contribution.

For other parameters at ground level including pressure, temperature, pressure difference and temperature difference between sea level and 1000-hPa level, some points has been selected. That is, points were selected and analyzed in different parts of the seas, which were known to affect the climate of Iran from previous studies by other researchers (Nazemosadat & Cordery 2000, Alijani 2006).

We have used factor analysis to analyze the meteorological parameters at upper atmosphere and ground levels (zonal wind, meridional wind and precipitable water). The field of factor analysis involves the study of order and structure in multivariate data. The field includes both theory about the underlying constructs and dynamics which give rise to observed phenomena, and methodology for attempting to reveal those constructs and dynamics from observed data. Factor analysis is preferable to principal components analysis. Components analysis is only a data reduction method. It became common decades ago when computers were slow and expensive to use; it was a quicker, cheaper alternative to factor analysis (Gorsuch 1997). It is computed without regard to any underlying structure caused by latent variables; components are calculated using all of the variance of the manifest variables, and all of that variance appears in the solution (Ford et al. 1986).

In this statistical method designed to reduce the number of variables, the initial parameters are transformed into independent variables based on their correlation coefficients. These independent variables are called factors. The value of each of the observations in the new factors is calculated as factorial score. Hence, rather than true values of observations, their scores in new components are used as new criteria for clustering. The advantage of this method is that while it reduces the number of variables, it preserves the initial variance of the main data (Alijani 2006).

As mentioned above, to find out the relation between rainfall in the region under study and the changes in meteorological parameters of pressure, temperature, pressure difference and temperature difference, the points in different parts of the seas have been studied, which are supposed to affect the climate of Iran. These include points in the Mediterranean, Persian Gulf, Oman Sea, Aden Gulf, Arab Sea, Red Sea, Black Sea, the Adriatic, Aral Lake, Indian Ocean, The Atlantic, North Sea and Siberia (Fig. 4). Table 3 shows Time series of average regional rainfall (from April to June) and value of predictors.

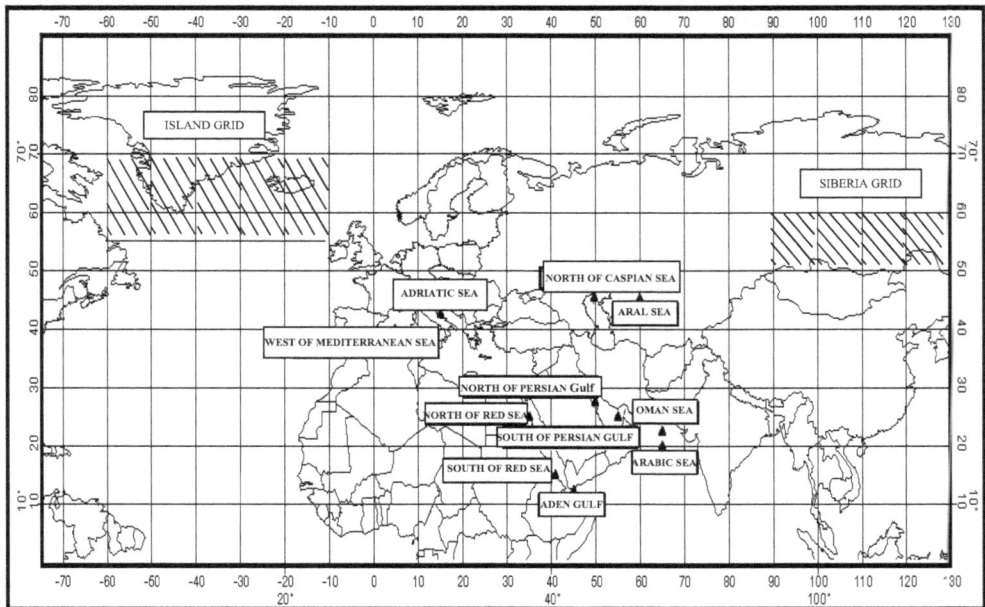

Fig. 4. Name and coordinates that have used for relation between rainfall and Remote Linkage Controlling (Fallah Ghalhary et al, 2010)

Year	Rainfall	x1	x2	x3	x4	x5	x6	x7	x8	x9	x10	x11	x12	x13	x14
1970	20.40	-1.15	0.27	0.17	-0.73	-1.65	-0.97	-1.63	-1.23	-1.28	0.20	-6.57	36.53	33.29	31.94
1971	17.35	-0.84	0.68	0.30	-0.19	-1.22	-1.30	-1.60	-1.46	0.02	0.07	-7.10	36.05	32.71	22.39
1972	27.93	-1.88	-0.46	0.18	1.81	-0.26	-0.03	-0.23	-0.30	-0.11	-1.80	-7.42	33.17	35.53	30.18
1973	30.96	-0.57	-0.45	-0.14	0.71	1.03	0.44	0.33	-0.22	0.29	-0.78	-7.35	36.44	36.31	27.58
1974	28.46	-0.65	-0.40	0.01	-0.58	-0.11	-1.03	-0.71	-1.04	0.46	-0.83	-7.12	34.52	32.60	25.40
1975	20.53	-0.67	-1.40	0.00	-0.08	0.51	-0.41	-0.06	-0.18	-0.69	-0.95	-7.04	34.65	34.97	24.05
1976	21.07	-1.80	-0.56	-0.26	0.84	-1.02	-1.54	-1.41	-0.83	0.16	-1.20	-7.83	34.08	35.95	28.07
1977	33.50	0.30	-1.02	0.12	-0.88	0.35	-1.02	-0.41	0.53	1.02	0.20	-5.94	36.75	34.26	34.78
1978	20.54	-0.56	0.38	0.23	0.49	0.91	0.49	-0.12	1.41	-1.49	0.11	-6.70	36.26	30.80	29.98
1979	20.88	-0.92	-0.18	0.32	-0.30	-0.12	0.16	-0.18	0.83	-0.51	0.61	-6.28	35.77	31.06	27.98
1980	17.18	0.28	-0.67	0.04	0.25	-0.14	0.10	0.18	-0.13	-1.44	0.18	-7.19	34.88	39.84	30.05
1981	22.40	0.28	-1.65	0.16	-0.02	-0.21	-0.20	-0.18	0.31	-1.41	1.10	-6.59	36.24	35.34	26.22
1982	28.08	-0.52	-0.06	0.02	-0.25	0.65	0.62	0.96	0.79	-0.91	-0.33	-7.10	35.36	34.97	26.45
1983	24.48	-0.27	-0.35	-0.38	-0.11	2.71	0.94	2.71	2.72	0.40	-1.33	-6.65	34.87	37.71	29.28
1984	19.19	-0.44	0.46	-0.02	0.15	0.21	0.02	0.54	0.53	-0.13	0.34	-6.98	39.57	31.10	22.57
1985	17.83	0.34	-1.00	0.04	-1.46	0.40	-1.21	-0.18	-0.09	1.01	0.07	-7.41	36.57	32.78	25.18
1986	16.11	-1.53	0.22	-0.09	0.10	0.26	-0.17	0.40	0.12	-0.05	0.14	-6.75	36.04	30.92	23.09
1987	20.41	-0.05	-0.21	-0.04	-0.46	0.95	1.44	1.60	1.60	-1.87	0.39	-7.66	36.27	28.92	20.10
1988	23.79	-0.37	1.09	-0.06	-0.27	-0.14	-1.12	-0.08	0.29	0.42	0.74	-6.64	39.42	32.70	24.55
1989	19.86	-0.46	0.00	-0.20	0.69	0.66	-0.07	0.35	0.73	1.77	-0.21	-6.91	35.09	25.91	20.53
1990	14.07	-0.13	1.26	-0.26	-0.54	0.75	0.66	0.51	0.48	1.00	0.01	-5.90	35.91	36.22	26.97
1991	17.73	0.53	-0.54	-0.07	-0.38	-0.59	0.59	-0.55	0.16	-0.11	0.48	-6.22	37.14	34.80	27.84
1992	36.57	-0.65	-1.25	-0.36	0.90	1.60	1.58	1.79	1.52	-0.72	-1.36	-7.63	33.74	37.07	26.62
1993	42.83	-0.72	-1.14	-0.33	0.57	1.18	1.31	1.34	0.92	0.27	-0.65	-7.45	34.89	35.34	28.89
1994	25.00	0.02	-2.17	-0.23	0.11	0.03	-0.26	0.09	0.31	0.81	0.56	-6.04	38.37	28.94	28.12
1995	16.99	1.54	0.21	0.05	-0.23	1.55	2.17	1.64	0.98	-1.79	-0.14	-6.58	35.94	33.07	20.51
1996	21.03	-1.05	0.71	0.13	-0.53	-1.32	-0.41	-1.11	-0.82	-0.25	-0.33	-6.73	35.61	30.50	25.18
1997	25.82	0.41	-0.70	-0.02	1.16	0.49	0.77	0.61	1.16	0.58	-0.33	-6.73	34.82	30.73	25.18
1998	32.48	0.04	1.84	-0.22	-0.09	-0.04	2.23	0.59	-0.15	-2.24	-0.36	-6.84	36.27	35.20	25.74
1999	19.65	1.69	-0.17	0.12	-0.49	-2.49	-1.34	-1.80	-1.57	1.27	0.67	-7.26	38.72	21.04	28.94
2000	9.18	1.42	0.47	0.16	-0.28	-1.00	-1.66	-1.24	-0.74	3.69	0.53	-5.14	37.10	22.23	18.58
2001	7.42	2.44	0.79	0.10	-0.71	-1.04	-1.56	-0.89	-1.63	2.08	0.53	-5.14	37.96	23.62	22.74
2002	20.14	0.92	0.98	0.27	0.05	-0.15	0.35	0.34	-0.81	-0.12	0.47	-7.25	36.17	26.29	23.25
2003	34.78	1.50	0.60	-0.06	0.04	-1.22	-0.01	-1.01	-0.52	-0.96	1.11	-6.39	37.06	29.85	25.70
2004	28.40	0.75	0.21	-0.07	0.69	-0.25	-0.10	-0.06	-0.43	0.10	0.14	-7.18	35.30	26.99	21.75
2005	21.59	1.03	0.76	0.44	0.72	-0.16	0.68	0.27	-0.82	-0.47	0.14	-7.18	34.53	24.89	25.40
2006	15.01	0.49	0.30	0.06	-0.88	-0.87	-0.13	-0.62	-1.20	0.20	1.06	-7.22	38.97	25.99	20.19
2007	10.09	1.26	3.14	-0.10	0.18	-0.22	-0.16	-0.21	-1.21	1.12	0.24	-5.42	35.94	22.85	20.74

Table 3. Time series of regional monthly average rainfall (from April to June) and value of predictors

2.2.3 Neural information processing systems

Artificial neural networks (ANNs) form a class of systems that is inspired by biological neural networks. They usually consist of a number of simple processing elements, called neurons that are interconnected to each other. In most cases one or more layers of neurons are considered that are connected in a feed forward or recurrent way (Zurada 1992, Grossberg 1988, Lippmann 1987). In trying to understand the emergence of the new discipline of neural networks it is useful to look at some historical milestones in Table 4 (Johan et al. 1997).

2.2.4 Basic neural network architectures

The best known neural network architecture is the multilayer feedforward neural network (multilayer perceptron). It is a static network that consists of a number of layers: input layer, output layer and one or more hidden layers connected in feedforward way (see e.g. Zurada 1992).

One signal neuron makes the simple operation of a weighted sum of the incoming signals and a bias term (or threshold), fed through an activation function σ and resulting in the output value of the neuron. A network with one hidden layer is described in matrix-vector notation as

$$y = W\sigma(V_x + \beta),\tag{13}$$

Or in elementwise notation:

$$y_i = \sum_{r=1}^{n_h} \omega_{ir}\sigma(\sum_{j=1}^{m} v_{rj}x_j + \beta_r) \quad i = 1,...,l.\tag{14}$$

Here $\chi \in$ Rm is the input and y ∈ Rl the output of the network and the nonlinear operation σ is taken elementwise. The intererconnection matrices are $W \in \Re^{l \times n_h}$ for the output layer $V \in \Re^{n_h \times m}$ for the hidden layer, $\beta \in \Re^{n_h}$ is the bias vector (thresholds of hidden neurons) with n_h the number of hidden neurons.

Year	Network	Inventor /Discoverer
1942	McCulloch-Pitts neuron	McCulloch, pitts
1957	Perceptron	Rosenblatt
1960	Madaline	Widrow
1969	Cerebellatron	Albus
1974	Back propagation network	Werbos, parker, Rumelhart
1977	Brain state in a box	Anderson
1978	Neocognitron	Fukushima
1978	Adaptive Resonance Theory	Carpenter, Grossberg
1980	Self-organizing map	Kohonen
1982	Hopfield	Hopfield
1985	Bidirectional assoc. mem.	Kosko
1985	Boltzmann machine	Hinton, Sejnowsky
1986	Counterpropation	Hecht- Nielsen
1988	Cellular neural network	Chua, yang

Table 4. The best known artificial neural network architectures together with their year of introduction and their inventor/ discoverer. See Hecht Nielsen (1988) for part of this table (Johan et al, 1997).

Fig. 5 shows a multilayer perceptron, which is a static nonlinear network that consists of a dummy input layer, an output layer and two hidden layer. A layer consists of a number of McCulloch-pitts neurons that perform the operation of a weighted sum of incoming signals, feeded through a saturation-like nonlinearity. One hidden layer is sufficient in order to be universal approximators for any continues nonlinear function (Johan et al. 1997).

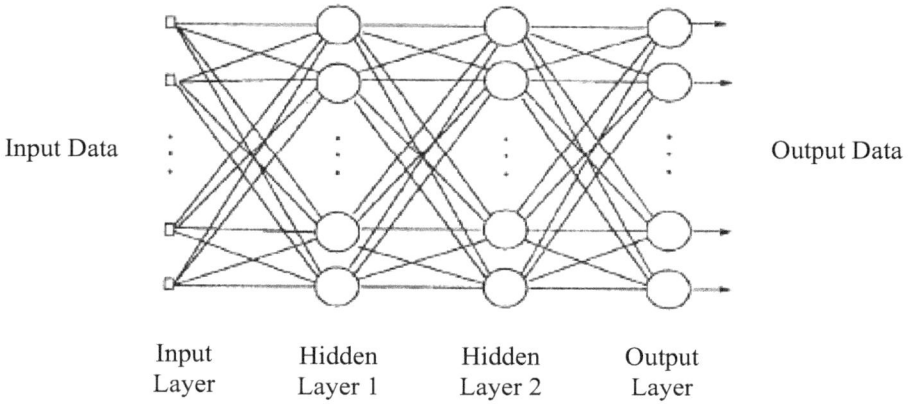

Fig. 5. Multilayer perceptron neural network (Johan et al, 1997).

For a network with two hidden layers (see Fig. 5):

$$y = W.\sigma(V_2.\sigma(V_1 x + \beta_1) + \beta_2) \tag{15}$$

Or:

$$y_i = \sum_{r=1}^{n_{h2}} w_{ir}\sigma(\sum_{s=1}^{n_{h1}} v_{rs}^{(2)} \sigma(\sum_{j=1}^{m} v_{sj}^{(1)} x_j + \beta_s^{(1)} + \beta_r^{(2)}), \quad i = 1,...,l. \tag{16}$$

The interconnection matrices are $W \in \Re^{l \times n_{h2}}$ for the output layer, $V_2 \in \Re^{n_{h2} \times n_{h1}}$ for the second hidden layer and $V_1 \in \Re^{n_{h1} \times m}$ for the first hidden layer. The bias vectors are $\beta_2 \in \Re^{n_{h2}}$, $\beta_1 \in \Re^{n_{h1}}$ for the second a first hidden layer respectively. In order to describe a network with L layer (L-1 hidden layers, because the input layer is a 'dummy' layer), the following notation will be used in the sequel.

$$x_i^l = \sigma(\xi_i^l), \tag{17}$$

$$\xi_i^l = \sum_{j=1}^{N_l} w_{ij}^l x_j^{l-1} \tag{18}$$

Where $l = 1,...,L$ is the layer index, N_l denotes the number of neurons in layer l and x_i^l is the output of the neurons at layer l. The thresholds are considered here to be part of the interconnection matrix, by defining additional constant inputs.

The choice of the activation function σ may depend on the application area. Typical activation functions are shown in Fig. 6.

Tanh function

Sigmoid function

Signum function

Step function

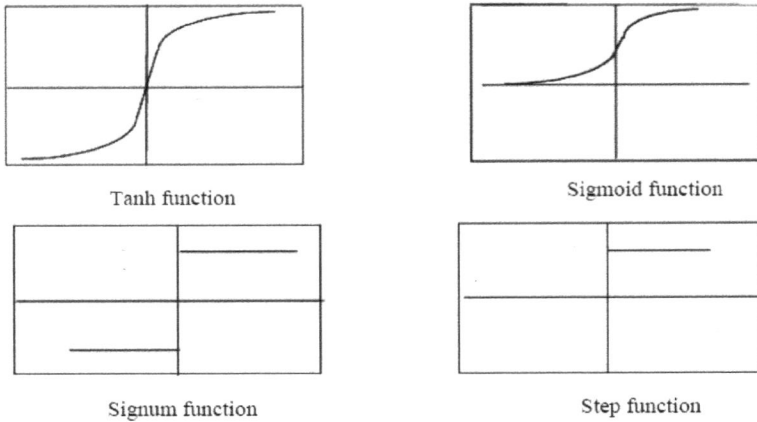

Fig. 6. Some possible activation functions for the neurons in the multilayer perceptron (Johan et al, 1997).

Fig. 6 shows some possible activation functions for the neurons in the multilayer precipitation. In this paper, we take the hyperbolic tangent function tanh. Note that this is a static nonlinearity that belongs to the sector [0, 1].

For applications in modeling and control the hyperbolic tangent functions:

$$\tanh(x) = (1 - \exp(-2x)) / (1 + \exp(-2x)) \tag{19}$$

Is normally used. In case of a 'tanh' the derivative of the activation function is:

$$\sigma' = 1 - \sigma^2 \tag{20}$$

The neurons of the input layer have a linear activation function.

A network that has received a lot of attention recently in the field of neural networks is the redial basic function network. This network can be described as:

$$y = \sum_{i=1}^{n_h} w_i g(\|x - c_i\|) \tag{21}$$

With $x \in \Re^m$ the input vector and $y \in \Re$ the output (models with multiple outputs are also possible). The network consists of one hidden layer with n_h hidden neurons. One of the basic differences with the multilayer perceptron is in the use of the activation function. In many cases one takes a Gaussian function for g, which is radially symmetric with respect to the input argument. The output layer has output weights $\omega \in \Re^{n_h}$. The parameters for the hidden layer are the centers $c_i \in \Re^m$ (Johan et al. 1997).

2.2.5 Recurrent neural network model

In recurrent neural network, some outputs of the nodes (output nodes or hidden nodes) are fed back to the previous layers. Most commonly used recurrent neural networks are external recurrent neural networks (Fig.7). In this scheme, the outputs of a neural network are fed back to form a part of its input layer. Another recurrent scheme is internal recurrent neural networks, in which the outputs of hidden nodes (instead of output nodes) are fed back to the input layer (Fig.8) (Aoyama et al, 1999).

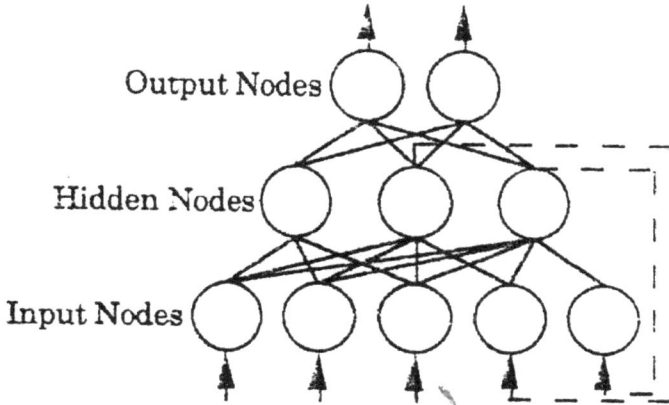

Fig. 7. External recurrent neural networks (Aoyama et al, 1999).

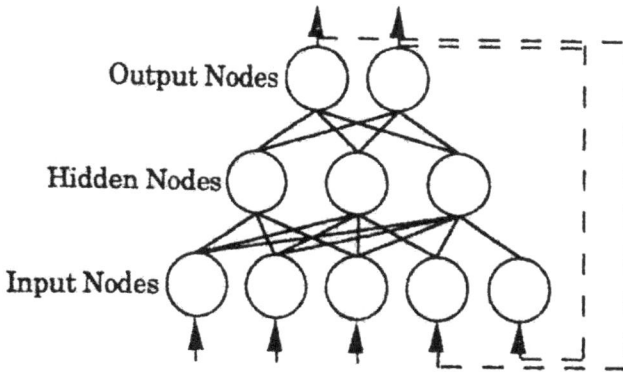

Fig. 8. Internal recurrent (state space) neural networks (Aoyama et al, 1999).

After various runs to test the network and the number of neurons of the hidden layer and different activation functions in the hidden and output layers, eventually found out that the final model with its one input layer, one hidden layer and one output layer (average regional rainfall) had the least error so in this research, used it as the main model. The numbers of the neurons in the input, hidden and output layers is fourteen, four and one respectively (14-4-1). The hidden layer activation function is a function of the hyperbolic tangent and the output layer activation function is a function of the linear hyperbolic tangent.

To assess the accuracy of the model, the index of Root Mean Square Error (RMSE) has been used which is calculated by the following formula:

$$RMSE = \sqrt{\frac{\sum_{i=1}^{n} (O_i - e_i)^2}{n}}$$ (22)

In the above formula, RMSE is Root Mean Square Error, O_i and e_i are the observed and predicted value of the variable respectively in the point i and n is number of network outputs.

3. Results and discussion

3.1 Predicting spring rainfall by means of artificial neural networks

In this study, Pearson Correlation Method has used to obtain meteorological predictors which affect regional rainfall. Thus, all the predictors which have shown a correlation with %5 level of significance in the period between October and March have been used as predictors in the structure of the rainfall forecast model. After numerous checking, it became clear that the optimum effect of predictors is when the period between October and March is used. Therefore, the following predictors in the period between October and March were used as predictors in rainfall forecast models: 1) SST Central Atlantic, 2) SST Western Mediterranean, 3) ΔSST Aral Lake, 4) ΔSST Labrador Sea, 5) SLP Northern Persian Gulf, 6) SLP Oman Sea, 7) SLP Southern Persian Gulf 8) SLP Southern Red Sea, 9) ΔSLP Between Eastern Mediterranean and Oman Sea, 10) air temperature at 700-hPa level in the region of index factor 2 in 5×5 degrees spatial resolution (Fig. 9), 11) air temperature at 700-hPa level in the index region of factor 3 in 5×5 degrees spatial resolution (Fig. 9), 12) precipitable water content in the index region of factor 10 in 10×10 degrees spatial resolution (Fig. 10), 13) relative humidity at 300-hPa level in the region of factor 2 in 5×5 degrees spatial resolution (Fig. 11), 14) relative humidity at 300-hPa level in the region of factor 4 in 5×5 degrees spatial resolution (Fig. 11). The variables which used as predictors in the rainfall forecast models show in the Table 5.

The above model divides the data into three different sections, namely, training data, validation data and testing data. The data belonging to 38 years were in turn divided into 19 years (1970– 1988) of training data, 9 years (1989–1997) of validation data and 10 years (1998–2007) of testing data. In the other words, from the whole set of historical data, two-thirds (1970-1997) were considered as calibration data, and one-third (1998 – 2007) as testing data. There is a clear analogy between the neural network weights and the parameters of other modeling approaches, and between the learning set and what we have before calles a period of calibration data. Work in neural networks often does not draw this analogy but it is a useful one in that just as an increase in the number of parameters gives a model more degrees of freedom in calibration but may result in over parameterization with respect to information in the data set, so in a neural network an increase in the number of layers, nodes and interconnections will also result in more degrees of freedom in fitting the learning set, also with the possibility of over parameterization (Beven 2001). Table 6 presents the results of the calibration period of the rainfall forecast model. As is shown, the minimum mean-square error after 1000 learning epochs is 0.169. Also, the maximum mean-square error is 0.169. In the other words, at this stage in the epoch of 1000, the network shows the maximum error. The minimum mean-square error of the validation epoch in the epoch of 3 is 0.234. The results of the prediction model are illustrated in Table 7 and Fig. 12. It is to be noted that these results were presented for the years 1998 to 2007, which was the testing epoch of the model. The features of this model have been presented in Table 8.

Symbol	The Name of Predictor	Time	Correlation Coefficient with the average regional rainfall	P-value
(X1)	SST Central Atlantic	Oct-Mar	-0.31	0.05
(X2)	SST Western Mediterranean	Oct-Mar	-0.38	0.01
(X3)	ΔSST Aral Lake	Oct-Mar	-0.35	0.03
(X4)	ΔSST Labrador Sea	Oct-Mar	0.36	0.02
(X5)	SLP Northern Persian Gulf	Oct-Mar	0.31	0.05
(X6)	SLP Oman Sea	Oct-Mar	0.38	0.01
(X7)	SLP Southern Persian Gulf	Oct-Mar	0.31	0.05
(X8)	SLP Southern Red Sea	Oct-Mar	0.34	0.03
(X9)	ΔSLP Between Eastern Mediterranean and Oman Sea	Oct-Mar	-0.32	0.04
(X10)	air temperature at 700-hPa level in the region of index factor 2 in 5°×5° degrees spatial resolution (Fig. 9)	Oct-Mar	-0.38	0.01
(X11)	air temperature at 700-hPa level in the region of index factor 3 in 5°×5° degrees spatial resolution (Fig. 9)	Oct-Mar	-0.45	0.004
(X12)	precipitable water in the region of index factor 10 in 10°×10° degrees spatial resolution (Fig. 10)	Oct-Mar	-0.31	0.05
(X13)	relative humidity at 300-hPa level in the region of index factor 2 in 5°×5° degrees spatial resolution (Fig. 11)	Oct-Mar	0.45	0.004
(X14)	relative humidity at 300-hPa level in the region of index factor 4 in 5°×5° degrees spatial resolution (Fig. 11)	Oct-Mar	0.46	0.003

Table 5. Selected predictor variables which used in the rainfall prediction model

Best Networks	Training	Cross Validation
Epoch	1000	3
Minimum Mean Squared Error	0.169	0.235
Final Mean Squared Error	0.169	0.426

Table 6. Minimum & Maximum Error in Training and Validation Epochs

As Table 8 shows, the mean square error is 46.5 and the normalized mean square error is 0.55. Also, the mean absolute error for this model was calculated to be 6.15 millimeters. The minimum absolute error is 0.13 millimeter and the maximum absolute error is 10.9 millimeters. Also, the correlation coefficient between observed and predicted rainfall for the model is 0.79. The root mean square error for this model was calculated to be 6.8 millimeters.

Year	Observed rainfall	Predicted rainfall
1998	32	27
1999	19	15
2000	9	15
2001	7	15
2002	20	15
2003	34	24
2004	28	17
2005	21	15
2006	15	15
2007	10	15

Table 7. Rainfall prediction in the region under study by means of neural network model

Performance	Y (mm)
Mean Squared Error	46.54
Normalized Mean Squared Error(MSE/variance desired output)	0.55
Mean Absolute Error	6.15
Minimum Absolute Error	0.13
Maximum Absolute Error	10.98
Linear correlation coefficient	0.79

Table 8. The features of artificial neural network model

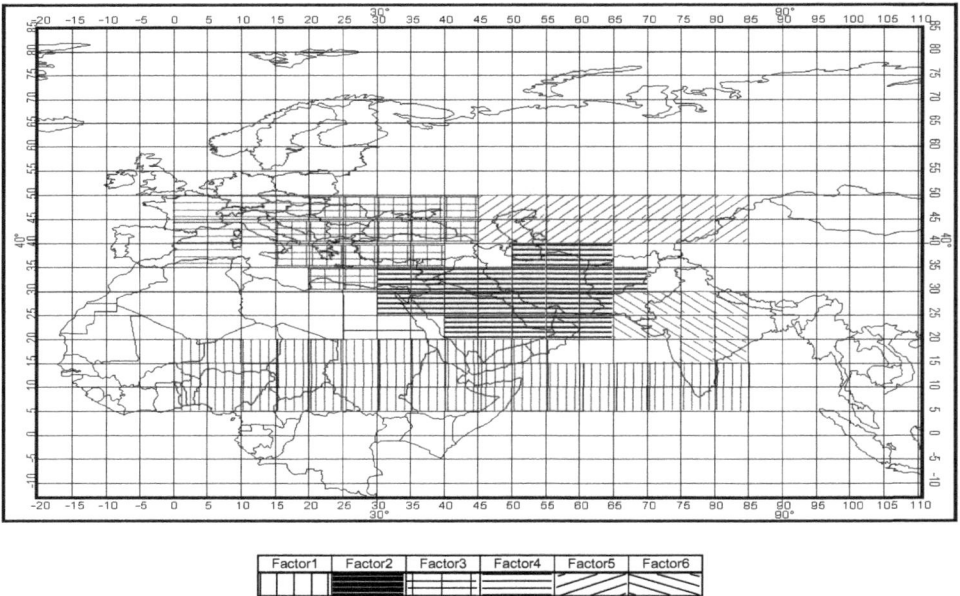

Fig. 9. The index areas detected by factor analysis at the temperature of 700-hPa level in the period October to March in networks of 5×5 degrees spatial resolution (Fallah Ghalhary et al, 2010)

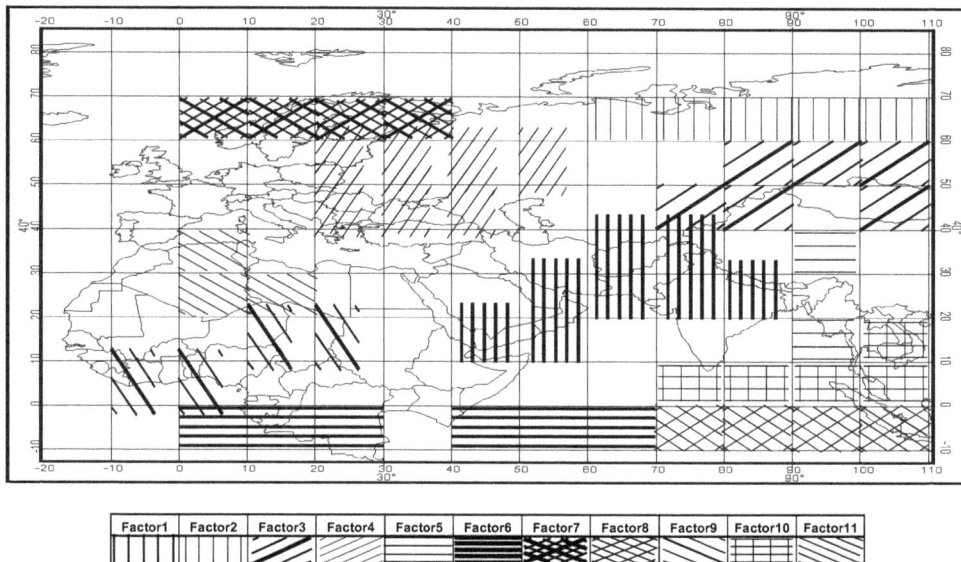

Fig. 10. The detected areas of precipitable water in the period October to March in networks of 10×10 degrees spatial resolution (Fallah Ghalhary et al, 2010)

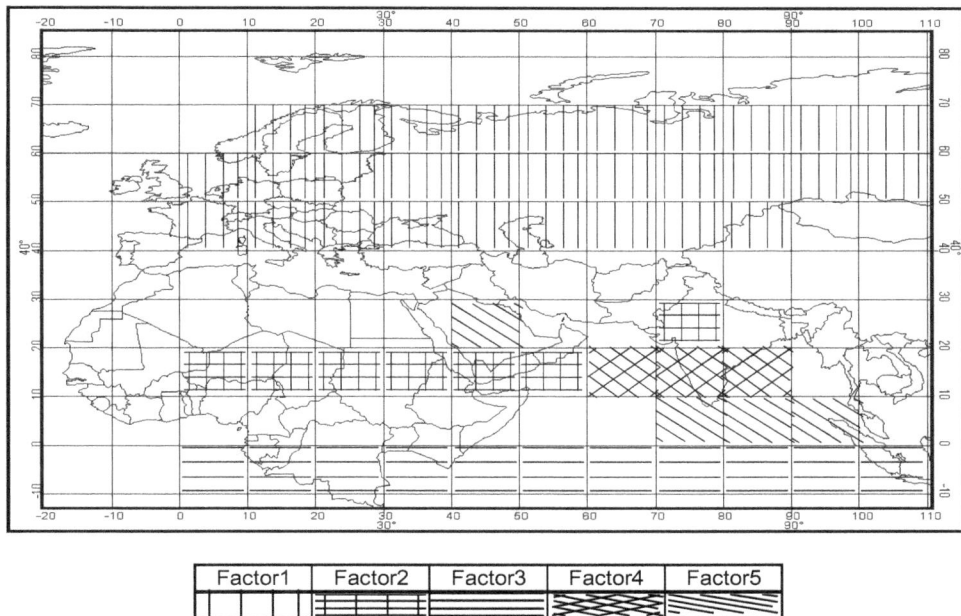

Fig. 11. The detected areas of relative humidity at 300-hPa level in networks of 5×5 degrees spatial resolution (Fallah Ghalhary et al, 2010)

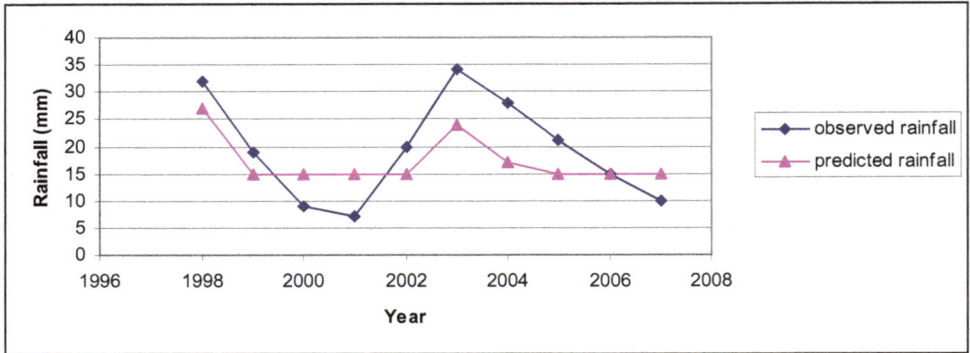

Fig. 12. Comparison of the observed and predicted rainfall in the region under study by means of artificial neural network model

The analysis of the results shows that the model is basically incapable of predicting rainfall in dry or wet extreme years. This is due to the fact that these extreme years have not been repeated in the calibration epoch of the prediction model and, for this reason, the model is not able to predict these extreme years. It must be noted that the minimum rainfall in the rainfall time series occurred in 2000 and 2001. To remove this problem, we must train the model with these extreme data. For this reason and in order to enable the neural network model to predict seasonal rainfall in such a way that it can be applied to all cases including dry, wet and normal years, we deleted the years 1998 and 2000 as the two extreme years in the test epoch of the model. One represents the dry extreme year and the other the wet extreme year. These two years were replaced by other data. They were taken out of the testing data and transferred into the training data and calibration epoch of the model. The results of the calibration epoch of the rainfall forecast model have been illustrated in Table 9. It is indicated that the minimum training error in the epoch of 17 is 0.108. The ultimate training error is 0.175. Also, the minimum validation error in the epoch of 1 is 0.087 and the ultimate validation error is 0.399. As the data show the accuracy of the model in detecting all dry, wet and normal years has amazingly increased. We can see that the accuracy of the model is higher than the previous model, as shown in Fig. 13, changes in the type of training data have affected the results of rainfall forecast model and that this model can estimate rainfall with higher accuracy.

Best Networks	Training	Cross Validation
Epoch	17	1
Minimum Mean Squared Error	0.108	0.087
Final Mean Squared Error	0.175	0.399

Table 9. Minimum and maximum error in training and validation epochs after modifying the network with proper historical data.

The observed and predicted rainfall by the model after modifying the input vector of the model in the training epoch has been presented in Table 10. the accuracy of the model has highly increased in this case and the root mean-square error has reached 2.5 millimeter, which indicates the high efficiency of the model in predicting rainfall in the area under study. The features of this model have been presented in Table 11. The mean-square error is 6.7 millimeter and the normalized mean-square error is 0.12. Furthermore, the mean absolute error for this model was equal to 2.12 millimeters. The minimum absolute error was 0.03 millimeter and the maximum absolute error was 4.97 millimeters. The correlation coefficient between observed and predicted rainfall was 0.95, which is very reasonable. Fig. 14 shows the observed values of rainfall versus predicted values. The equation of the regression line for the changes in the values of observed versus predicted rainfall is as follows:

Observed Rainfall (mm) = 0.05 + 0.93 Predicted Rainfall

Year	Observed rainfall	Predicted rainfall
1996	21	22
1997	25	27
1998	32	35
1999	19	18
2001	7	7,7
2002	20	22
2004	28	25
2005	21	26
2006	15	15
2007	10	14

Table 10. Rainfall forecast in the region under study after modifying the network with historical data

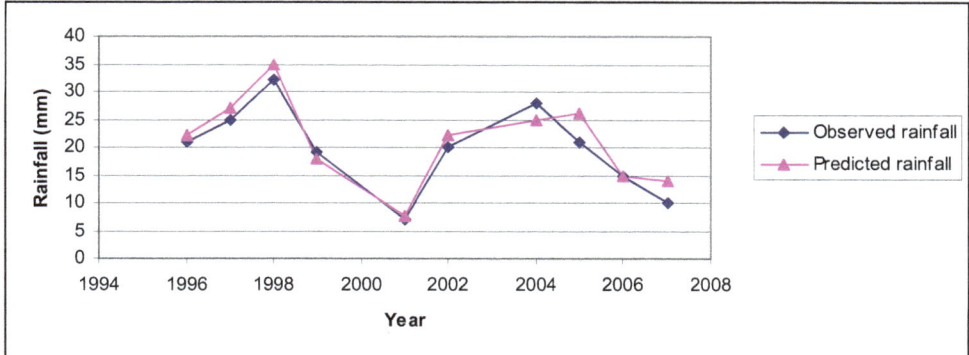

Fig. 13. Comparison of observed and predicted rainfall in the region under study by artificial neural network model after modifying the network with historical data.

the observations of the linear regression between observed values of rainfall and predicted values and the results of the variance analysis of the linear regression between the observed values of rainfall and predicted values have been summarized in Table 12 and Table 13 respectively. As we can see in Table 12, considering the confidence range of %99 of the linear regression between the observed values of rainfall and predicted values, the root mean-square error was calculated to be 2.46 millimeters. Table 13 shows that the F ratio is significant at %1 level, which is indicative of a strong relation between the changes in the observed values of rainfall versus predicted values. The significance test of the gradient of the regression line between the true values of rainfall and the values predicted by the model has also been done (Table 14). As observed here, the gradient of the regression line is also significant at %1 level. The P-value for the significance test of the gradient of the regression equation was smaller than 0.0001.

Performance	Y (mm)
Mean Squared Error	6.73
Normalized Mean Squared Error(MSE/variance desired output)	0.12
Mean Absolute Error	2.12
Minimum Absolute Error	0.03
Maximum Absolute Error	4.97
Linear correlation coefficient	0.95

Table 11. Features of artificial neural network after modifying the network with historical data

Variable	Value
RSquare	0.908079
RSquare Adj	0.896589
Root Mean Square Error	2.466808
Mean of Response	19.8
Observations (or Sum Wgts)	10

Table 12. Summary of the observations of the linear regression between the true values of rainfall and predicted values

Source	DF	Sum of Squares	Mean Square	F Ratio
Model	1	480.91887	480.919	79.0317
Error	8	48.68113	6.085	Prob > F
C. Total	9	529.60000		<.0001

Table 13. The variance analysis of the linear regression between the true values of rainfall and predicted values

Term	Estimate	Std Error	t Ratio	Prob>|t|
Intercept	-0.075552	2.367907	-0.03	0.9753
Predicted rainfall (mm)	0.9388546	0.105608	8.89	<.0001

Table 14. Summary of the statistical observations of the estimation of the parameters of the model

Model	RMSE	Maximum Absolute Error	Minimum Absolute Error
ANN Model	6.82	10.98	0.13
Modified ANN Model	2.5	4.97	0.03

Table 15. Statistical comparative of two models

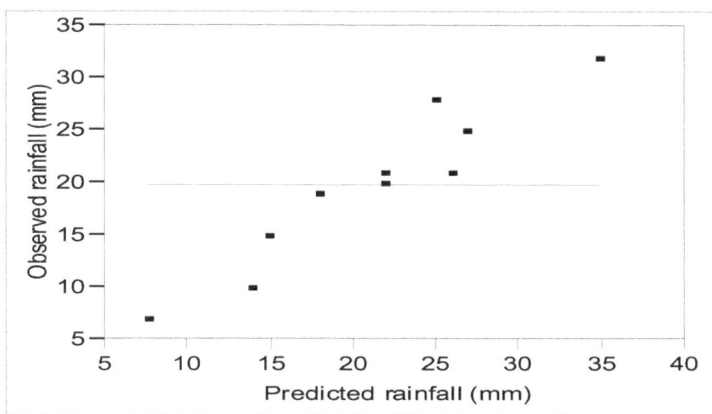

Fig. 14. Changes in the true values of rainfall versus predicted values. The slanted line is the regression line

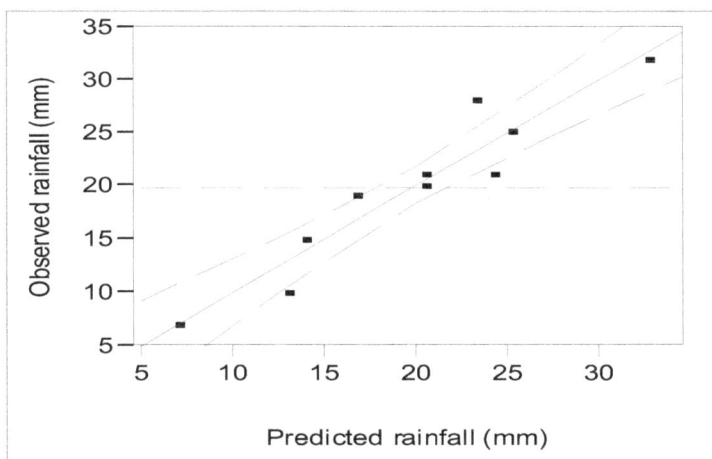

Fig. 15. The confidence range of %99 for changes of the observed rainfall versus predicted values

The confidence range of %99 for the changes of the observed rainfall versus the predicted values has been illustrated (Fig. 15). Here again, the changes of the observed rainfall and the predicted rainfall shows a close correspondence, being significant at 0.01 level. Fig. 16 Displays the changes in the values of residuals versus predicted values of rainfall. Here again, the changes in the values of residuals versus predicted values of rainfall are completely accidental and normal, indicating the high accuracy of the model in predicting rainfall. In sum, the analysis of the results indicates that the difference between the observed and predicted rainfall is within a reasonable range; moreover, the model has been able to predict rainfall in all years with an acceptable error. The root mean-square error for this model was 2.5 millimeter, which is very small, indicating the accuracy of the model in predicting rainfall. We can conclude from the above discussion that the variables used in the model have been able to detect the distribution pattern of rainfall in the region with great ease and accuracy. We can also decide that the model can be successfully used for predicting rainfall in spring.

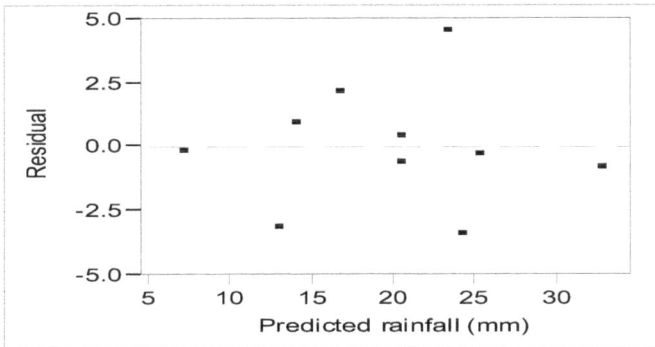

Fig. 16. Changes of the values of residuals versus predicted values of rainfall in terms of millimeter

4. Conclusion

Based on the obtained results, we can conclude that ANN models were successful in the prediction of spring rainfall, but the ANN model has higher accuracy after the revision of training data with a root mean square error of 2.5 milliliters. This is clearly observable in Fig.13, ANN model after the revision of training data has been more successful than the before. Table 15 shows the results of two ANN models to predicting the amount of the rainfall in the area under study. At the end, we can result that the variables entering rainfall prediction models have been well able to detect the rainfall distribution patterns in the region and can be used in rainfall prediction patterns in the region. This plays a vital role in the management and planning of drink and agriculture water resources. Considering these predictions, we can plan future policies for optimizing the costs and possibilities for maximum efficiency.

5. Acknowledgement

This paper presents part of the findings of the research project "Predicting spring rainfall in Khorasan Razavi Province based on meteorological predictors by means of fuzzy logic, artificial neural networks and adaptive neuron-fuzzy networks" by the same author. The author greatly appreciates the cooperation of the directors and managers of the "Climatological Research Institute" who provided the possibilities required for the completion of this project.

6. References

Abraham A., Sajith N., Joseph B. (2001). Will We Have a Wet Summer? Long-term Rain Forecasting Using Soft Computing Models: Modelling and Simulation, Publication of the Society for Computer Simulation International, Prague, Czech Republic, 1044-1048

Alijani B., 2006. Synoptic Climatology: 2nd ed. Samt Publication, Tehran, Iran, 258 pp.

Aoyama A., Doyle F. J., Vankat V. (1999). Fuzzy neural network systems techniques and their applications to nonlinear chemical process control systems, Academic press, 2:485-493.

Beven K., 2001. Rainfall-Runoff Modelling, John Wiley & Sons, LTD, First Ed, ISBN: 978-0-470-86671-9, PP.360.

Cavazos T., 2000. Using Self-Organizing Maps to Investigate Extreme Climate Event: An Application to wintertime Rainfall in the Balkans, J. Climatol. 13: 1718-1732.

Choi L., 1999. An application hydroinformatic tools for rainfall forecasting: Thesis (PhD). University of New South Wales (Australia), 752 pp.

Cressie, N., 1990. The origins of Kriging. Mathematical geology. 22. 239-252.

Cressie, Noel A.C., 1993. Statistics for Spatial Data, Revised Edition. New York: John Wiley & Sons, Inc. 900 p.

Davis J.C., 1990. Statistics and Data Analysis in Geology, John Wiley & Sons Inc, ISBN-10: 0471198951, PP. 564.

Ducan A.J., 1974. Quality Control and Industrial Statistics. Richard D. IRWIN, INC., 4 editions.

Fallah Ghalhary G. A., Habibi Nokhandan M., Mousavi Baygi M., Khoshhal J. and Shaemi Barzoki A. (2010). Spring rainfall prediction based on remote linkage controlling using adaptive neuro-fuzzy inference system (ANFIS), Theor Appl Climatol, 101:217–233.

Feller W., 1968. An Introduction to Probability Theory and its Applications, 1. John Wiley & Sons, Inc., 3 rd edition.

Ford J.K., MacCallum R.C. and Tait M. (1986). The Application of Exploratory Factor-Analysis in Applied- Psychology - a Critical-Review and Analysis. Personnel Psychology, 39(2):.291-314.

Gandin L.S., 1963. Objective analysis of meteorological fields. Gidrometeorologicheskoe Izdatel'stvo (GIMIZ), Leningrad (translated by Israel program for scientific translations, Jerusalem, 1965)

Gorsuch R.L., 1997. Common Factor-Analysis versus Component Analysis - Some Well and Little Known Facts, Multivar Behav Res, 25(1): 33-39.

Grossberg S., 1988. Nonlinear neural networks: principles, mechanisms and architectures. Neural Networks, 1: 17-61.

Guibas L.J. and Odlyzko A.M. (1980). Long Repetive Patterns in Random Sequences. Z. Wahrscheinlichkeitstheorie, 53: 241-262.

Halid H. and Ridd P. (2002). Modelling Inter-Annual Variation of Local Rainfall Data: Using a Fuzzy Logic Technique. Proceedings of International Forum on Climate Prediction, James Cook University, Australia. First Proof. Pages, 166-170.

Haltiner G.J. and Williams R.T. (1980). Numerical Prediction and Dynamic Meteorology. 2nd Edition. New York, Wiley & Sons, PP. 447.

Hecht Nielsen R., 1988. Neurocomputer applications in NATO ASI Series, NEURAL COMPUT, 41, Eds. R. Eckmiller, Ch.v.d.Malsburg, Springer-Verlag.

Iseri Y., Dandy G.C., Maier H.R., Kawamura A. and Jinno K. (2005). Medium Term Forecasting of Rainfall using Artificial Neural Networks, Neural Networks, 1834-1840.

Johan A.K.S., Joos P.L.V. and Bart L.R.D.M. (1997). Artificial Neural Networks for Modelling and Control of Non-Linear Systems, Kluwer Academic Publishers, second Ed, PP.235. ISBN: 0-7923-9678-2.

Journel A.G. and Huijbregts C.H.J. (1978). Mining geostatistics, Academic press, London.

Kawamura A., McKerchar A.I., Spigel R. H. and Jinno K. (1998). Chaotic characteristics of the Southern Oscillation Index time series, Journal of Hydrology. 204: 168-181.

Knuth D.E., 1981. The Art of Computer Programming, 2: Semi numerical Algorithms. Addison-Wesley, Reading, MA, 2nd edition.

Kolmogorov A.N., 1941. Interpolation and extrapolation of stationary random sequences. Izvestiia Akademii Nauk SSSR, Seriia Matematicheskiia, 5, 3-14. [Translated, 1962, Memo Rm-3090-PR, Rand Corp, Santa Monica, CA.,].

Kridge D.G., 1951. A statistical approaches to some basic mine valuation problems on the Witwatersrand. Journal of the chemical, Metallurgical and Mining society of South Africa, 52, 119-139.

Laslett G.M. and McBratney A.B. (1990). Further comparison of spatial methods for predicting soil pH, Soil Sci. Soc. Am. J. 54, 1553-1558.

Lewis F.L., 1986. Optimal estimation. Wiley, New York.

Lippmann R.P., 1987. Introduction to computing with neural nets, IEEE ASSP magazine, April: 4-22.

Mantua N.J. Hare S.R., Zhang Y., Wallace J.M. and Francis R.C. (1997). A Pacific interdecadal climate oscillation with impacts on salmon production, Bulletin of the American Meteorological Society 78: 1069-1079.

Maria C., Haroldo F. and Ferreira N. (2005). Artificial neural network technique for rainfall forecasting applied to the Sao Paulo region. J. Hydrol , 301:146-162.

Matheron G., 1963. Principle of geostatistics, Economic Geology, 58, 1246-1266.

Matheron G., 1971. The theory of regionalized variables and its application. Cahiers du Centre de Morphologie Mathematique, No.5. Fontainebleau, France.

Nazemosadat M.J. and Cordery I. (2000). On the Relationship between ENSO and Autumn Rainfall in Iran. J. Climatol. 1: 47-62.

Philip G.M. and Watson D.F. (1986). Matheronian geostatistics-Quo vadis? Mathematical geology, 18, 93-117.

Pongracz R. and Bartholy J. (2006). Regional Effects of ENSO in Central/Eastern Europe, j. ADGEO. 6:133-137.

Silverman D. and Dracup J.A. (2000). Artificial neural networks and long-range precipitation prediction in California, J Appl Meteorol, 39: 57-66.

Trenberth K.E. and Hurrel J.W. (1994). Decadal atmosphere-ocean variations in the Pacific. Climate Dynamics 9: 303-319.

Weber D. and Englund E. (1992). Evaluation and comparison of spatial interpolators. Mathematical Geology 24:4, 381-391.

Wold H., 1938. A study in the analysis of stationary time series. Almqvist and Wiksells, Uppsala.

Wolfowitz J., 1944. Asymptotic distribution of runs up and down. Annals Math. Stat., 15: 163-165

Zhang Y., Wallace J.M. and Battisi D.S. (1997). ENSO-like interdecadal variability 1900-1993, Journal of Climate 10: 1004-1020.

Zurada J.M., 1992. Introduction to Artificial Neural Systems, West publishing company.

Zwiers F.W. and von Storch H. (2004). On the role of statistics in climate research. Int. J Climatol. 24: 665–680.

14

Synthesizing High-Resolution Climatological Wind Fields with a Mesoscale Atmospheric Boundary Layer Model Forced with Local Weather Observations

Guillermo J. Berri

National Meteorological Service, Buenos Aires
Argentina

1. Introduction

The region of the La Plata River in southeastern South America (see Fig. 1) concentrates important economic and social activities since one third of the population of Argentina and more than one half the population of Uruguay live there. Large urban complexes, different commercial activities and important industries are located along its shores. In addition, the La Plata River and its tributary the Paraná River are main ship tracks with some of the largest ports of the southern cone of South America. Therefore, the region is of vital importance for the two countries. The La Plata River is a large water surface that projects into the continent, conditioning the local weather and climate. Thus, other related environmental aspects are strongly influenced by the local weather and climate conditions, such as environmental pollution, water currents and tidal regime, commercial fishing at the oceanic front, port operations, navigation and tourism.

Fig. 1. Location of La Plata River region in South America

The La Plata River is 300 km long, with a variable width between 40 km and 200 km. The region creates a considerable surface temperature contrast with the continent that sets up the stage for the development of a low-level circulation, with sea-land breeze characteristics.

During the daytime hours the lower layers over land are warmer than over the river, creating a land-river surface temperature gradient that establishes a river to land wind component known as sea breeze. Over the northern shore of the La Plata River the surface winds increase the southerly component, while over the southern shore they increase the northerly wind component.

The daytime inland surface wind components create horizontal divergence and subsidence over the river and convergence and upward motion over land near the river shores. During the nighttime hours the land is cooler than the river, the land-river surface temperature gradient reverses and the winds tend to blow from land to river establishing the land breeze. The nighttime land breeze is not so well developed as the daytime sea breeze basically because of weaker low level winds due to the nocturnal stability. The daily cycle of the land-river surface temperature contrast gives rise to significant changes of the predominant wind direction across the region throughout the day. This can be appreciated in Fig. 2 that shows the observed 1959-1984 mean winds at five weather stations in the

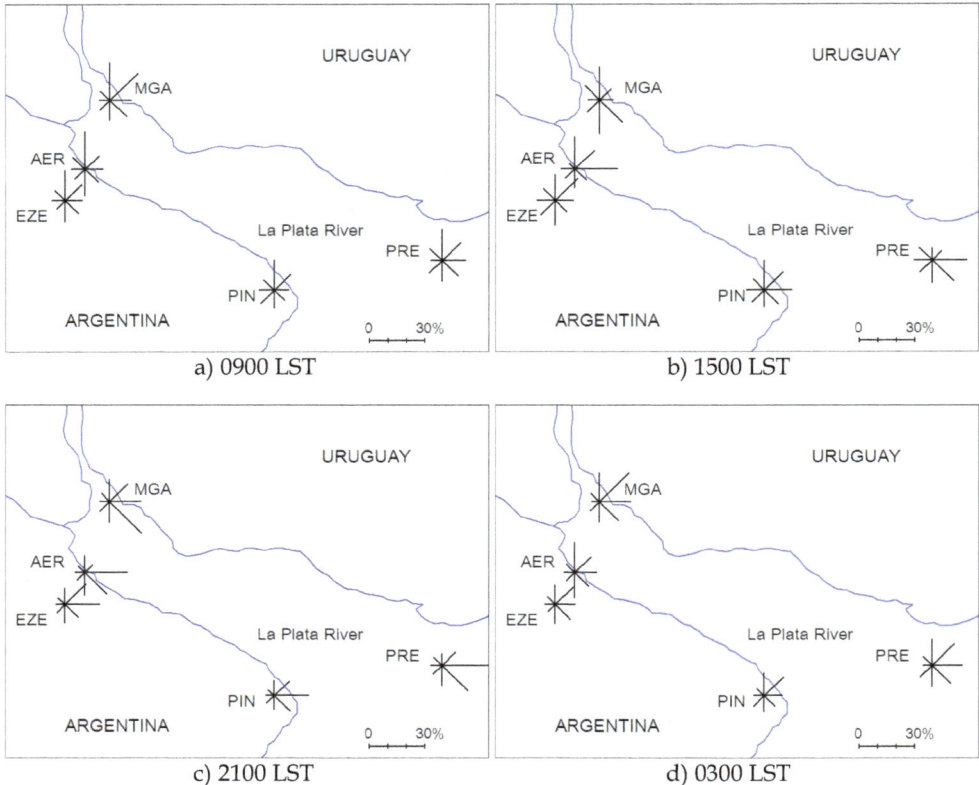

a) 0900 LST

b) 1500 LST

c) 2100 LST

d) 0300 LST

Fig. 2. Observed 1959-1984 mean wind direction frequencies (in percentage) at four local standard times (LST): a) 0900 LST, b) 1500 LST, c) 2100 LST and d) 0300 LST. The weather stations are Ezeiza (EZE), Aeroparque (AER), Martín García (MGA), Punta Indio (PIN) and Pontón Recalada (PRE).

region. The weather stations used in the study are Ezeiza (EZE), Aeroparque (AER), Martín García (MGA), Punta Indio (PIN) and Pontón Recalada (PRE). At 0900 LST (local standard time) (Fig. 2a), all weather stations show predominant N and NE wind sectors that together amount between 30%-40% of the time. At 1500 LST (Fig. 2b) the weather stations over land in Argentina display the N, NE and E wind sectors as the dominant ones, totalizing 40%-60% of the time (in particular AER has an E wind frequency of 23%). In contrast, the weather stations over the river display as dominant sectors the E, SE and S wind directions.

At 2100 LST (Fig. 2c) the dominant wind sector over land is E, followed by SE, as well as over PRE at the river mouth, whereas over MGA is SE followed by E. The wind direction frequency distribution at 0300 LST (Fig. 2d) looks quite similar to that of 0900 LST. Table 1 shows the mean wind speed by wind sector at the four different times of the day. Throughout the day the weather stations over the river display a significant change of the dominant wind sectors of more than one quadrant (90° to 130°), while in the stations over land the daily change is less than one quadrant.

Atmospheric mesoscale models have the capability of reproducing the major aspects of the sea-land breeze circulation when the horizontal thermal contrast is properly defined. In some regions where the sea breeze circulation is important, it is common to adapt existent numerical models to study this local circulation. For example Pielke et al. (1992) describe the use of the RAMS model in sea-land breeze studies, employing telescopic and nested grids; Case et al. (2002) use the RAMS model coupled to the Eta model to simulate the sea breeze over the eastern coast of the Florida peninsula, and Colby (2004) employs the MM5 model with different resolutions in the study of the sea breeze over the New England coast. Sea-breeze cases over northwestern Hawaii were simulated using the NCEP (National Centers for Environmental Prediction) mesoscale spectral model coupled with an advanced land surface model with 3 km horizontal resolution (Zhang, 2005). The inland penetration of the lake breeze on the western shore of Lake Michigan was studied by Roebber & Gehring (2000) using the MM5 mesoscale model with 5 km grid spacing. Some boundary layer forecast models have been developed to study the breeze in complex terrains, for example Daggupaty (2001) that simulates the 3–dimensional circulation associated with lake breezes in southwestern Ontario. One of the main problems with regional models is the resolution required to represent the short to mid–term regional–scale processes, as discussed by Colby (2004). An important aspect in modelling local scale circulations is the verification of predictions, as done for example by Case (2004) over east central Florida.

Berri & Nuñez (1993) show that the sea-land breeze circulation over the La Plata River region can be simulated with a mesoscale boundary layer model (BLM) especially developed for the region. The results obtained by the authors in a case study show good agreement between observed and modeled surface wind direction changes throughout the day. The daily cycle of the sea-land breeze responds to the atmospheric pressure anomaly field induced by the cyclic thermal contrast at the surface. Thus, the driving mechanism of the abovementioned model is the daily variation of the horizontal temperature difference across the river shores.

A recent study (Sraibman & Berri 2009) finds that operational low level wind forecast for the La Plata River region can be improved by running the BLM model forced by the Eta model operational forecasts. This study concludes that the improvement obtained with the BLM model is a consequence of the appropriate definition of the land-river surface temperature contrast that is fundamental for resolving the small scale details of the low-level circulation over the region.

0300 LST	N	NE	E	SE	S	SW	W	NW
EZE	3.8	3.5	2.9	3.7	4.0	3.5	3.2	3.2
AER	5.2	5.2	3.9	5.1	5.7	4.7	4.2	4.3
MGA	4.1	4.6	4.5	5.3	6.1	4.8	4.3	4.3
PIN	4.6	4.6	4.7	4.8	5.0	4.6	3.9	4.4
PRE	6.4	7.0	6.5	7.1	8.3	9.0	6.8	6.1
0900 LST	N	NE	E	SE	S	SW	W	NW
EZE	4.1	4.2	3.7	4.4	4.7	4.1	3.6	3.7
AER	4.7	4.4	3.7	5.3	5.7	5.1	4.8	4.4
MGA	4.4	4.9	4.7	5.8	6.1	4.8	4.2	4.3
PIN	5.1	5.0	5.3	5.5	5.1	4.8	4.4	5.0
PRE	6.4	6.0	5.6	7.5	7.6	7.8	7.1	6.2
1500 LST	N	NE	E	SE	S	SW	W	NW
EZE	4.7	4.6	4.7	5.0	5.3	5.4	5.0	4.6
AER	3.9	2.9	3.5	5.3	6.6	6.7	6.6	5.7
MGA	3.7	3.9	4.2	4.7	4.2	4.4	4.3	3.5
PIN	5.4	4.5	5.6	6.0	6.0	5.9	5.7	6.2
PRE	3.9	4.8	5.5	6.4	6.7	7.0	5.8	5.4
2100 LST	N	NE	E	SE	S	SW	W	NW
EZE	3.1	2.9	3.6	4.5	4.1	3.5	3.1	2.8
AER	3.8	3.2	4.1	5.4	5.9	5.1	4.1	3.6
MGA	3.5	4.3	4.5	5.6	5.9	5.2	4.2	3.6
PIN	3.7	4.1	4.8	4.9	5.3	4.5	3.7	3.8
PRE	4.4	5.3	6.8	7.5	7.7	8.3	6.3	5.1

Table 1. Mean wind speed by wind sector (m s⁻¹), at the four main observing times of the day, at the five weather stations of the study Ezeiza (EZE), Aeroparque (AER), Martín García (MGA), Punta Indio (PIN) and Pontón Recalada (PRE). The averaging period is 1959-1984.

2. BLM model formulation

The BLM is a hydrostatic and incompressible model that has been developed by Berri (1987). The model equations are based on the three principles that govern the atmospheric motion, i.e. conservation of momentum, mass and energy. Since the model is formulated for studying the atmospheric circulation in the boundary layer, the vertical component of the equation of motion becomes the hydrostatic equation and the mass conservation principle is approximated by the continuity equation for an incompressible fluid. Since the model is dry, all energy sources have been neglected, except the surface heating, so that the energy equation reduces to the conservation of potential temperature. The BLM model equations are:

$$\frac{\partial u}{\partial t} = -u\frac{\partial u}{\partial x} - v\frac{\partial u}{\partial y} - w\frac{\partial u}{\partial z} - \alpha_0\frac{\partial p}{\partial x} + fv + \frac{\partial}{\partial x}\left(K_{mh}\frac{\partial u}{\partial x}\right) + \frac{\partial}{\partial y}\left(K_{mh}\frac{\partial u}{\partial y}\right) + \frac{\partial}{\partial z}\left(K_{mz}\frac{\partial u}{\partial z}\right) \tag{1}$$

$$\frac{\partial v}{\partial t} = -u\frac{\partial v}{\partial x} - v\frac{\partial v}{\partial y} - w\frac{\partial v}{\partial z} - \alpha_0\frac{\partial p}{\partial y} - fu + \frac{\partial}{\partial x}\left(K_{mh}\frac{\partial v}{\partial x}\right) + \frac{\partial}{\partial y}\left(K_{mh}\frac{\partial v}{\partial y}\right) + \frac{\partial}{\partial z}\left(K_{mz}\frac{\partial v}{\partial z}\right) \tag{2}$$

$$\frac{\partial \theta}{\partial t} = -u\frac{\partial \theta}{\partial x} - v\frac{\partial \theta}{\partial y} - w\frac{\partial \theta}{\partial z} + \frac{\partial}{\partial x}\left(K_{\theta h}\frac{\partial \theta}{\partial x}\right) + \frac{\partial}{\partial y}\left(K_{\theta h}\frac{\partial \theta}{\partial y}\right) + \frac{\partial}{\partial z}\left(K_{\theta z}\frac{\partial \theta}{\partial z}\right) \tag{3}$$

$$\frac{\partial w}{\partial z} = -\frac{\partial u}{\partial x} - \frac{\partial v}{\partial y} \tag{4}$$

$$\frac{\partial p_0}{\partial z} = -\frac{g}{\alpha_0} \tag{5}$$

$$\frac{\partial p'}{\partial z} = \frac{g}{\alpha_0}\frac{\theta'}{\theta_0} \tag{6}$$

$$p = p_0 + p' \tag{7}$$

$$\theta' = \theta - \theta_0 \tag{8}$$

All symbols in the equations have the usual meteorological meaning, subscript o refers to a horizontal mean value over the entire domain, and superscript ´ refers to a local departure from the horizontal mean value. Equations (1) to (3) are the forecast equations for the u and v wind components and potential temperature θ, respectively. Equations (4) to (8) are the diagnostic equations for the vertical motion w, standard pressure p_o, pressure perturbation p', total pressure p, and potential temperature perturbation θ', respectively. Equations (1) to (3) are solved from the top of the surface layer ($z = 40$ m) to the material top of the model by a semi–implicit numerical scheme. Within the constant flux layer, the similarity theory is applied and the forecast equations become the well-known diagnostic equations, i.e. the logarithmic vertical profiles of wind and potential temperature, as a function of stability. The boundary conditions at the top of the model are: $u = u_g$, $v = v_g$, $w = p' = \theta' = 0$ where u_g and v_g are the geostrophic wind components. At the lower boundary the conditions are: $u = v = w = 0$, whereas θ is defined at every time step (see below Section 3.1). At the lateral boundaries, all variables are allowed to change in order to provide a zero gradient across the boundaries at each time step, except the pressure since its gradient provide the geostrophic wind. The model is initialized under conditions of horizontal homogeneity for all the variables except pressure, since its gradient supports the geostrophic wind at the initial state. Thus, Equations (1) and (2) become the well–known Ekman layer equations:

$$\frac{\partial u}{\partial t} = f(v - v_g) + \frac{\partial}{\partial z}\left(K_{mz}\frac{\partial u}{\partial z}\right) \tag{9}$$

$$\frac{\partial v}{\partial t} = -f(u - u_g) + \frac{\partial}{\partial z}\left(K_{mz} \frac{\partial v}{\partial z} \right) \tag{10}$$

in which the pressure force terms are replaced by the expressions of the geostrophic equilibrium with constant u_g and v_g. Equation (5) is integrated once and Equations (9) and (10) are integrated during three inertial periods (about 60 hours at this latitude), in order to remove any possible inertial oscillations from the solution. Thus, the steady state solution achieved for u and v at every grid point of the domain depends only on the particular values adopted at the upper and lower boundaries. These are defined from the 0900 LST observations and are kept constant during the initialization process. For more details about the model formulation and the numerical method of solution, please refer to Berri & Nuñez (1993).

The inner rectangle of Fig. 1 depicts the domain of the BLM forecasts that consists of 79 points in the x direction (350 km) and 58 points in the y direction (316 km). The horizontal resolution is 0.05° that corresponds to an average of 5 km. The vertical domain has 12 levels distributed according to a log–linear spacing. The first level is the roughness length z_o (equal to 0.001 m over water and 0.01 m over land), and the last one is the material top of the model at 2000 m. The intermediate levels are located at the following heights: 10, 40, 80, 140, 220, 350, 550, 800, 1100 and 1500 m.

3. Methodology for the model climatology

The low-level wind field climatology of the model is defined as the ensemble result obtained by running a series of 18-hour forecasts. Each ensemble member is a forecast obtained by forcing the model with a different upper and lower boundary condition defined from the local observations. The upper boundary condition consists of a given value of wind direction and wind speed at the top of the model, defined from the Ezeiza radiosonde observations (EZE in Fig. 2). The lower boundary condition consists of a surface heating function (see below Section 3.1), defined from the temperature observations at the surface weather stations in the region. The ensemble result is obtained by averaging the wind direction and wind speed of all the ensemble members. The data corresponds to 1959-1984, the only extended period with complete observations available in a suitable manner for the study. The model results are validated at 0300, 0900, 1500 and 2100 LST, which are the times of the day when the observations are available in the historical database. Since the model is initialized at 0900 LST, the 18-hour forecast runs until 0300 LST of the following day that is the last time of validation.

3.1 Boundary conditions

The model ensemble consists of 192 members, each one characterized by a wind direction and a wind speed at the top of the model. The 192 members correspond to 16 wind direction classes (N, NNE, NE,…, NNW), and 12 wind speed classes with the following upper bounds in ms^{-1}: 2, 3, 4, 5, 6, 7, 8, 9, 10, 12, 14, with the last class representing wind speeds greater than 14 ms^{-1}. Each ensemble member has a probability of occurrence p_j (j=1 to 192), that is determined from the mean wind frequency distribution of the 0900 LST Ezeiza radiosonde data of the period 1959-1984.

For each ensemble member, the surface heating function defines the temperature at the surface as follows: $T(x,y,t) = T_o + F_1(t)F_2(x,y)$, where T_o is the daily mean temperature of the ensemble, $F_1(t)$ defines the daily cycle of the maximum river-land temperature difference, and $F_2(x,y)$ defines the river-land temperature difference as a function of the distance between every (x, y) point and the coast. Except nearby the coasts, the horizontal air temperature gradients over the land and over the river are much smaller than the river-land air temperature gradient. Thus, the main forcing that drives the model at the surface is the daily variation of the horizontal air temperature difference across the coasts. Two weather stations are chosen for determining this forcing, one in the land –Ezeiza (EZE)- and the other one in the river –Pontón Recalada (PRE)-. The temperature difference $T_{EZE} - T_{PRE}$ is calculated from the four daily observations at 0300, 0900, 1500 and 2100 LST. Then, the mean value is obtained as the average of all days in which the Ezeiza radiosonde observation corresponds to every wind direction and wind speed class at the top of the model. Thus, for each ensemble member there are four daily temperature differences, which are interpolated by means of a harmonic analysis in order to obtain $F_1(t)$. Fig. 3 depicts typical $F_1(t)=T_{EZE} - T_{PRE}$ curves, as monthly mean values for four different months of the year. The land-river temperature difference is defined as follows: $F_2(x,y)=\{1 + \tanh[s(x,y)/B]\}/2$, where $s(x,y)$ is the minimum distance from every grid point to the coast (positive over the land and negative over the river).

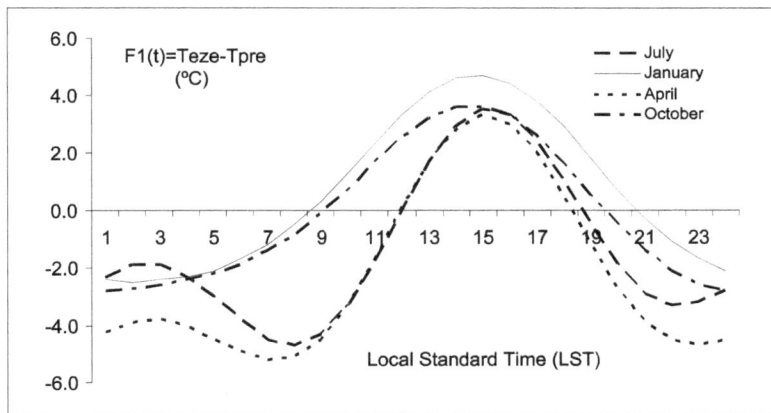

Fig. 3. Typical hourly interpolated $F_1(t)$ curves, in °C, from the temperature difference Teze minus Tpre observed at 0300 LST, 0900 LST, 1500 LST and 2100 LST. The curves are monthly mean values for four different months of the year.

The hyperbolic tangent distributes the land-river temperature difference symmetrically with respect to the coasts. In the present study the parameter B is set equal to 1000 meters, which provides 75% (90%) of the temperature change over a distance of $2B$ ($3B$) across the coasts. Different B values were tested and the adopted one minimized the averaged error of the wind distribution. Fig. 4 shows an example of $F_2(x,y)$ as a function of the perpendicular distance to the coast, across a narrow band centered at the coastline. Over the river and away from the coast, the surface temperature $T(x,y,t) = T_o$ remains constant, since $F_2(x,y) = 0$. Over land and away from the coast the surface temperature develops a full daily cycle given by $T(x,y,t) = T_o + F_1(t)$, since $F_2(x,y) = 1$. At the lateral boundaries all forecast variables, except pressure, are allowed to change in order to provide a zero gradient across the boundaries at each time step.

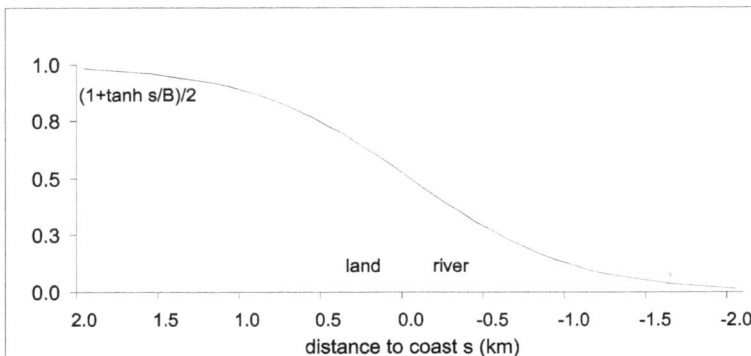

Fig. 4. Example of $F_2(x,y)$ as a function of the perpendicular distance to the coast, across a narrow band centered at the coastline, with parameter B=1000 m.

3.2 Model validation

Each ensemble member provides a forecast of the horizontal wind components u and v at 10 m, and contributes to the wind field climatology with a probability p_j with $j=1$ to n ($n = 192$). The u and v forecast is expressed as a wind direction d (degrees from the north), and a wind speed $v = (u^2 + v^2)^{1/2}$ in m s^{-1}. The wind direction d defines the wind sector identified by index $i=1$ to 9 (clockwise from N $-i=1-$ until NW $-i=8-$), with $i=9$ indicating calm, i.e. wind speeds smaller than a given threshold discussed later on. Matrices $D_{i,j}^k$ and $V_{i,j}^k$ store the wind direction and wind speed forecasts, respectively, in which index $k=1$ to 4 represents the local time 0300, 0900, 1500 and 2100 LST, respectively.

Thus, the wind direction forecast simply consists of the occurrence of a given wind sector and matrix $D_{i,j}^k$ counts the number of cases. Once the jth ensemble member forecast is completed, the quantity p_j is added to the (i,j,k) component of matrix $D_{i,j}^k$, and the quantity $v.p_j$ is added to the (i,j,k) component of matrix $V_{i,j}^k$. After completing the series of forecasts, the elements of matrix $D_{i,j}^k$ are either zero or p_j, and $\sum_{i=1}^{9}\sum_{j=1}^{n} D_{i,j}^k = 1$, since $\sum_{j=1}^{n} p_j = 1$. At every jth and kth element of matrices $D_{i,j}^k$ and $V_{i,j}^k$, only one of the nine ith elements is not zero, i.e. the one that corresponds to the occurring wind direction sector.

At every grid point, the wind direction frequency distribution f_i^k obtained with the model (in percentage) is:

$$f_i^k = 100 \cdot \sum_{j=1}^{n} D_{i,j}^k \tag{11}$$

and the corresponding mean wind speed per wind sector v_i^k (in ms^{-1}) is:

$$v_i^k = \sum_{j=1}^{n} V_{i,j}^k \Bigg/ \sum_{j=1}^{n} D_{i,j}^k \tag{12}$$

since the $V_{i,j}^k$ elements are of the form $v.p_j$, and the $D_{i,j}^k$ elements are of the form p_j.

The relative error in wind direction is defined as $ed_i^k = (f_i^k - fo_i^k) / fo_i^k$ and in wind speed as $ev_i^k = (v_i^k - vo_i^k) / vo_i^k$. In these expressions fo_i^k and vo_i^k are the mean observed wind direction frequency distribution and mean wind speed per wind sector, respectively, at the five surface weather stations of the study. The model distributions f_i^k and v_i^k are calculated with the averaged value of the four grid points that surround every weather station.

The averaged model errors are expressed as the root mean squared value of the relative error (from now on referred to as RMS), in wind direction ErD^k, from Equation (11), and wind speed ErV^k, from Equation (12), both weighted by the mean observed wind direction frequency fo_i^k, as follows:

$$ErD^k = \left[\sum_{i=1}^{9} fo_i^k (ed_i^k)^2 \Bigg/ \sum_{i=1}^{9} fo_i^k \right]^{1/2} \tag{13}$$

$$ErV^k = \left[\sum_{i=1}^{9} fo_i^k (ev_i^k)^2 \Bigg/ \sum_{i=1}^{9} fo_i^k \right]^{1/2} \tag{14}$$

4. BLM low-level wind climatology

4.1 Annual mean

Fig. 5 presents the wind direction frequency distribution and averaged wind speed by sector, obtained with the model according to Equations (11) and (12) (one every six grid points are plotted). It represents the averaged value of the four times of the day, when the observations are available in the database of the period 1959-1984. Since the model is initialized with the observations at 0900 LST, the 6-hour forecast is validated at 1500 LST, the 12-hour forecast at 2100 LST and the 18-hour forecast at 0300 LST. The 0900 LST forecast is taken after 30-minute integration in order to allow for the model spin-up.

The weather stations report the observed wind direction in eight compass sectors, and a ninth category that corresponds to calm wind, which means that the wind speed is below the instrument threshold. For the purpose of model validation, a calm wind observation represents a problem since the model never predicts a zero wind speed. Calm wind observations over the region are variable and depend on the weather station and time of the day, and in occasions they exceed a 30% of the observations. Thus, the inappropriate handling of the calm wind predictions may have a significant impact in the model errors, so it was necessary to adopt a criterion in order to overcome the problem. Test runs were conducted with the model with the purpose of determining the wind speed value below which the resultant percentage of calm winds would match the observations at the nearby weather station. This matching value was then adopted as the wind speed threshold below which the model result would be considered as calm wind. The thresholds varied depending on the time of the day and the position of the weather station in the domain. The nearby grid points to the weather stations over land displayed similar results, although with values always greater than those of the grid points nearby the river weather stations. The set of wind speed thresholds adopted for the grid points over land (in m s⁻¹) is: 1.6, 2.6, 1.4 and 1.4 at 0300, 0900, 1500 and 2100 LST, respectively. For the grid points over the river the wind speed thresholds (in m s⁻¹) are: 0.6, 0.8, 0.8 and 0.6, at 0300, 0900, 1500 and 2100 LST, respectively.

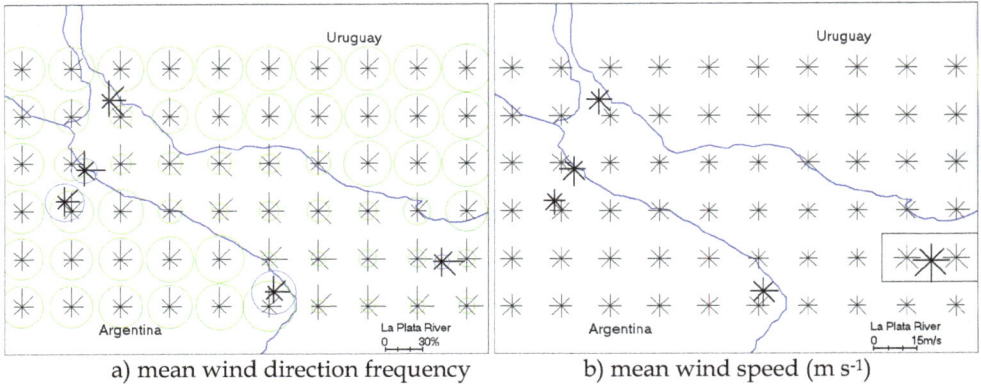

a) mean wind direction frequency   b) mean wind speed (m s⁻¹)

Fig. 5. a) mean wind direction frequency in percentage, and b) mean wind speed by sector in m s⁻¹ at 10 m. Model results are plotted in thin lines and observations of the five weather stations of the study are plotted in thick lines. Circles of panel a) represent the frequency of calm wind (see text for details). The rectangular box on the right hand side of panel b) surrounds the model grid points that are plotted at 21 m. (see text for details). The averaging period is 1959-1984.

The wind sectors with higher frequencies obtained with the model are N, NE, E, SE and S, in coincidence with the observations (Fig. 5a). However, the dominant wind sectors vary according to the position over the domain and, in general; there is more contrast between points closer to the coasts and over the river. Over land and far away from the river, N and NE are, in this order, the wind sectors that display the higher frequencies. Over the river the wind direction distributions are different than over land and the E and SE wind sectors become, depending on the position, the dominant ones. The only two weather stations in the river are located at both ends, MGA at the river spring and PRE at the river mouth, so it is not possible to verify the model results along the river. For example, the wind direction distribution obtained with the model over the river centerline resembles the PRE observations, i.e. relatively high frequencies of the E and SE wind directions. However, the model results over the river mouth, with N and NE dominant wind directions, differ from the PRE observations with predominantly E and SE wind directions. This may be due to the fact that the grid points over this area are far away from the coast and have, therefore, less influence of the river-land surface thermal contrast. At the other end of the river the wind directions obtained with the model have a better agreement with the MGA observations. Over land there are only three weather stations with available observations for the study. The wind direction distributions obtained with the model agree better with the observations at EZE and PIN, than at AER. The first two weather stations are located a few tens of kilometers away from the coast, whereas AER is so close to it (about five hundred meters), that the horizontal resolution of the model may be a limiting factor.

Fig. 5b) shows the averaged wind speed by sector obtained with the model, as well as the observed wind speed at the five weather stations of the study. The wind instrument at PRE is at 21 meters above the surface so that the model grid points within the rectangular box on the right hand side of Fig. 5b) are plotted at that height. Throughout the domain, the model wind speeds are in general slightly greater in the S, SW and W wind sectors, in comparison with the others. The observations display a similar pattern, with the exception of PIN that shows less difference among wind sectors. There is also a general agreement between the magnitude of the observed and the modeled wind speeds, with the exception of the neighboring points to PRE where the observed wind speeds are clearly greater.

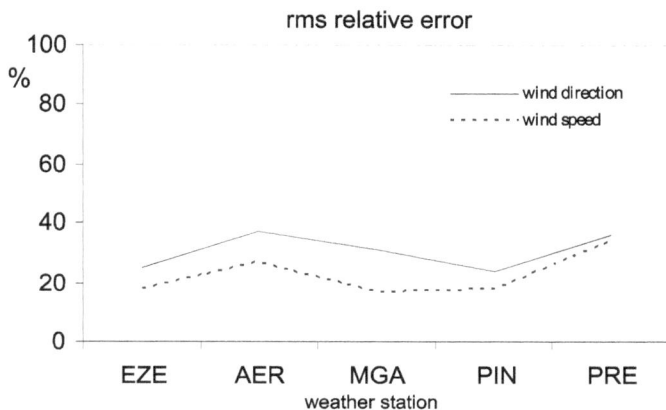

Fig. 6. Percentage RMS of the model relative errors in wind direction (Equation 13) and wind speed (Equation 14), averaged over the four grid points that surround every weather station: Ezeiza (EZE), Aeroparque (AER), Martín García (MGA), Punta Indio (PIN) and Pontón Recalada (PRE). The averaging period is 1959-1984.

Fig. 6 shows the percentage RMS of relative errors produced by the model in wind direction and wind speed, according to Equations (13) and (14) respectively, averaged over the four grid points that surround every weather station. At the five weather stations the error in wind speed is smaller than the error in wind direction. As discussed in the previous paragraph, the largest errors occur at PRE, while the smallest ones are at EZE and PIN. AER and MGA display errors with intermediate values in comparison with the other weather stations. The smallest errors are obtained at the weather stations located over land and away from the coast; because AER, also over land but very close to the coast, shows the second greatest error of the five weather stations. As mentioned above, the model resolution of 5 km may be the limiting factor responsible for the relatively large wind errors at AER.

The surface wind direction frequency distribution obtained with the model at four times of the day is shown in Fig. 7, together with the local observations. The regions along the river and over land near the coasts display more variability of the dominant wind sectors throughout the day. For example at 0900 LST (Fig. 7b), the dominant wind sectors are N and NE, while at 1500 LST (Fig. 7c) they are E and SE. This change is well represented by the model, particularly the shift towards the E sector, while the SE sector is more dominant over the central part of the river. Another particular aspect at the river mouth is the significant change in the dominant wind direction throughout the day, since the N and NE sectors at 0300 LST (Fig. 7a) and 0900 LST (Fig. 7b) become E, SE and S at 1500 LST (Fig. 7c) and at 2100 LST (Fig. 7d), which is reasonably well reproduced by the model. At the river spring, MGA shows predominance of the N and NE sectors at 0900 LST (Fig. 7b), in coincidence with the model results. At 1500 LST (Fig. 7c), MGA displays the S, N and SE as the dominant sectors, while the model shows the N, NE and SE sectors around MGA, with increasing frequency of the SE and S wind sectors towards the centerline of the river.

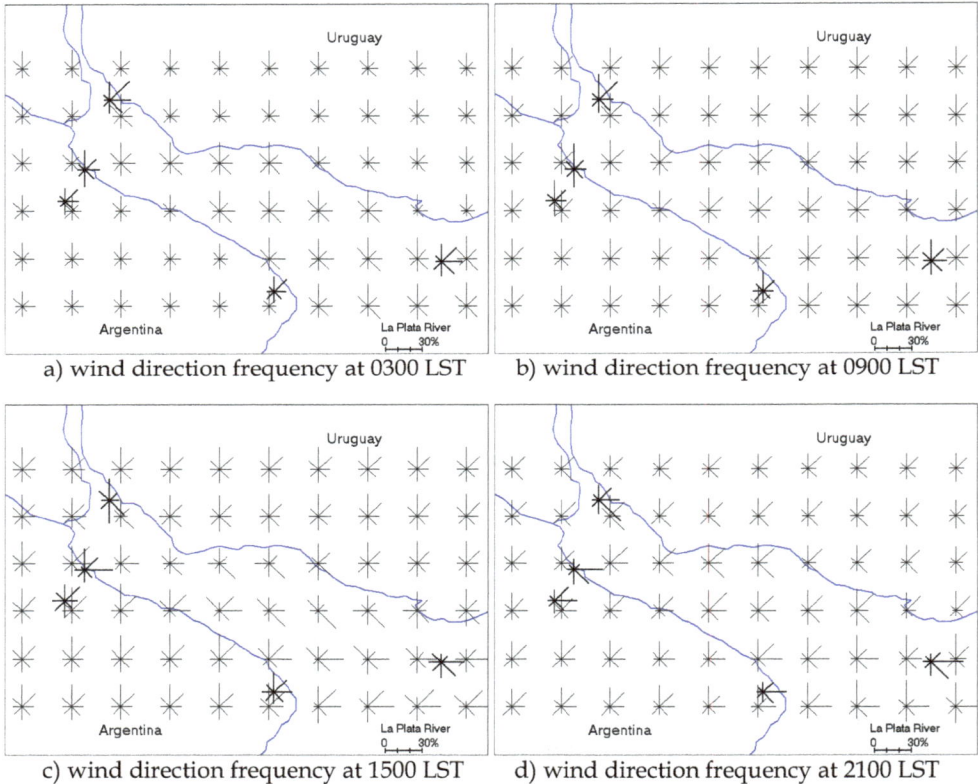

a) wind direction frequency at 0300 LST   b) wind direction frequency at 0900 LST

c) wind direction frequency at 1500 LST   d) wind direction frequency at 2100 LST

Fig. 7. Mean wind direction frequency by sector obtained with the model –thin lines-, and observed at the five weather stations of the study -thick lines- at 10 m. at a) 0300 LST, b) 0900 LST, c) 1500 LST and d) 2100 LST. The north wind direction points upward and each tick mark of the scale in lower right corner represents a 10% frequency. The averaging period is 1959-1984.

At 2100 LST (Fig. 7d) the MGA observations display the predominance of the SE, E and NE sectors, and the model agrees in the case of the SE and NE sectors, while the presence of E winds becomes more evident at the centerline and towards the river mouth. Over land at 0900 LST (Fig. 7b) the dominant sectors are N and NE, in agreement with the model results. At 1500 LST (Fig. 7c) the AER observations show the highest frequency in the E sector, but the model indicates the NE sector. There is also good agreement of the model results with the observations at EZE in the N and NE wind sectors, and at PIN in the N, NE and E wind sectors. At 2100 LST (Fig. 7d) the observations show an overall predominance of the E and SE sectors, in agreement with the model results, although the latter displays a significant N frequency, particularly towards the river mouth, that is not revealed by the observations. At 0300 LST (Fig. 7d) the observations show the N, NE, E and SE sectors as the dominant ones, while the model agrees with the N and E sectors, in general, although the NE frequency is relatively small in particular over the river and over land near the coast.

Fig. 8 shows the mean surface wind speed by sector obtained with the model, together with the observations of the five weather stations of the study at four times of the day. As indicated before, the grid points that surround PRE (rectangular box on the right hand side of each panel) are plotted at 21 meters above the surface in coincidence with the height of the wind instrument. In general, and throughout the domain, the observed and modeled mean wind speeds of all sectors are quite similar.

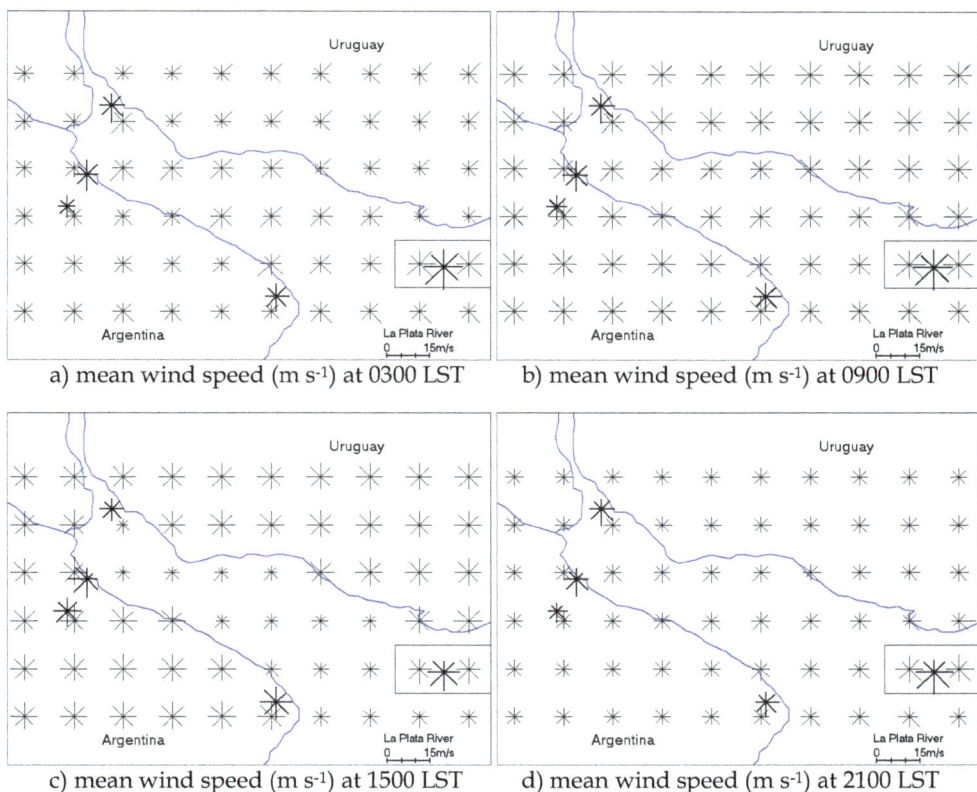

a) mean wind speed (m s⁻¹) at 0300 LST   b) mean wind speed (m s⁻¹) at 0900 LST

c) mean wind speed (m s⁻¹) at 1500 LST   d) mean wind speed (m s⁻¹) at 2100 LST

Fig. 8. Mean wind speed by sector obtained with the model –thin lines-, and observed at the five weather stations of the study -thick lines- at 10 m. at a) 0300 LST, b) 0900 LST, c) 1500 LST and d) 2100 LST. The rectangular box on the right hand side of each panel surrounds the model grid points that are plotted at 21 m. (see text for details). Each tick mark of the scale in lower right corner represents 5 m s⁻¹. The averaging period is 1959-1984.

The major observed contrast is between the wind speeds over the river and over land. For instance, at 0300 LST the wind speed over the river is clearly larger than over land, but at 1500 LST the situation is the opposite and the wind speed over land is significantly greater than over the river. At 0900 LST and 2100 LST the wind speed contrast between land and river is minimum. This spatial pattern and the changes that take place throughout the day is confirmed by the observations.

		0300 LST			0900 LST			1500 LST			2100 LST		
EZE	model	N	E	NE	N	NE	S	N	NE	E	NE	E	N
	observed	NE	SE	E	N	NE	S	NE	N	S	E	NE	SE
AER	model	N	E	NE	N	NE	S	N	NE	E	N	NE	E
	observed	N	S	NE	N	S	NE	E	N	NE	E	SE	S
MGA	model	N	NW	NE	N	NE	S	N	NE	SE	N	NE	SE
	observed	NE	E	N	NE	N	SE	S	N	SE	SE	E	NE
PIN	model	N	E	NW	N	NE	S	N	NE	E	NE	N	E
	observed	NE	N	E	N	NE	SW	N	E	NE	E	SE	NE
PRE	model	N	E	SE	N	NE	S	E	NE	S	N	NE	S
	observed	NE	E	N	N	NE	E	E	SE	S	E	SE	S

Table 2. First three observed and modeled wind sectors with highest frequency, in decreasing order, at the five weather stations of the study as a function of time of the day. In bold face are shown the matching wind direction sectors, regardless of their order.

Another way of comparing the model results with the observations is by analyzing the coincidence of the wind direction sectors with highest frequency. Table 2 lists the first three observed and modeled wind sectors with highest frequency, in decreasing order, that together amount 40% to 60% of the observations. For example, the upper-left box (EZE at 0300 LST) displays the NE, SE and E as the first, second and third wind sectors with highest frequency, whereas the model indicates N, E and NE, respectively.

The modeled and observed wind sectors that match are shown in bold face, regardless of the order in which they agree, since in some cases the ranking is defined by a few percent points. The best situation is when the three sectors match in the same order, as for example at 0900 LST at EZE, whereas the worst case is at 2100 LST at AER, when only the highest observed frequency is captured by the model but in third place. There is better agreement at the five points at 0900 LST and 1500 LST, since the average number of hits is 2.4, whereas at 0300 LST and 2100 LST the average number of hits is 1.6. At PRE the model shows less overall agreement with the observed predominant wind directions, since the average number of hits is 1.75. At the other four locations the model presents a similar behavior, at EZE, AER and PIN the average number of is 2.25 and at MGA is 2.0. In particular AER shows contrasting results because is the only place with twice three hits, i.e. at 0900 LST and 1500 LST, while at 2100 LST shows only one hit.

The model errors are not constant throughout the day, as can be seen in Fig. 9 that shows the percentage RMS of relative error, averaged of the five weather stations. Except at 0900 LST and 1500 LST, the wind speed error is much smaller than the wind direction error, and shows a minor dependence on the time of the day. The error in wind direction varies more throughout

Synthesizing High-Resolution Climatological Wind Fields with a Mesoscale Atmospheric Boundary
Layer Model Forced with Local Weather Observations

367

the day; it is minimum at 0900 LST and maximum at 2100 LST. The time evolution of the model errors does not follow a straightforward deterioration with time since the 13-hour (2100 LST) forecast error is larger than the 19-hour (0300 LST) forecast error.

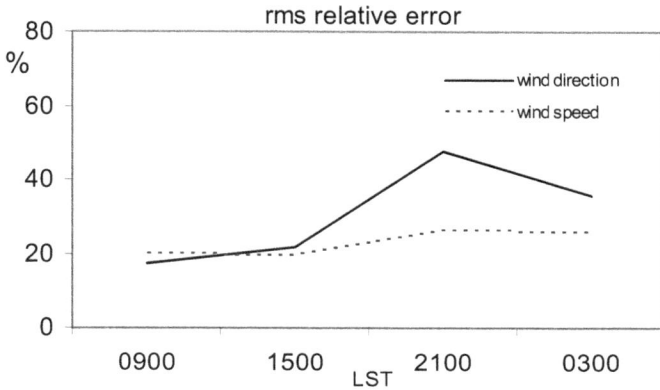

Fig. 9. Percentage RMS of the model relative errors in wind direction (Equation 13) and wind speed (Equation 14), as a function of the local standard time, averaged of the five weather stations. The averaging period is 1959-1984.

Fig. 10 shows the model errors in wind direction (top panel) and wind speed (bottom panel) as a function of weather station and time of the day. As mentioned before, wind direction errors (Fig. 10a) are more variable with time and space than wind speed errors (Fig. 10b). At 0900 LST and 1500 LST wind direction errors are minima and very small everywhere, while at 0300 LST and 2100 LST they are generally at their maxima, particularly at AER and PRE. At MGA the wind direction error shows a small variation throughout the day. On the other hand PRE, that has the greatest combined error in wind direction and wind speed (see Fig. 6) and displays strong daily variations, presents the minimum wind direction error at 0900 LST. The wind speed errors (Fig. 10b) are more homogeneous throughout the domain and present a similar variation with time of the day at the five weather stations.

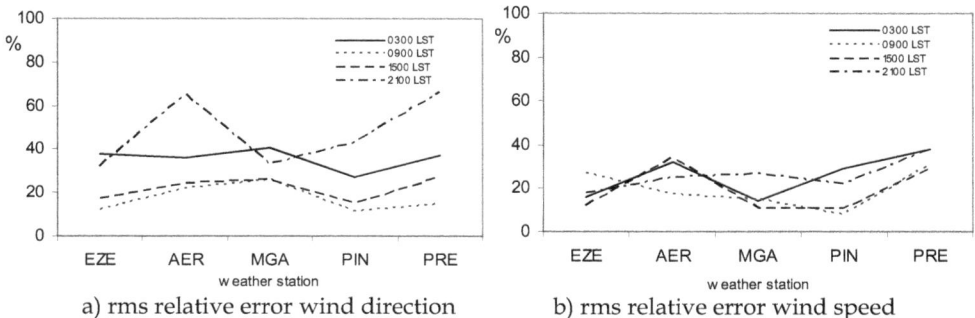

a) rms relative error wind direction   b) rms relative error wind speed

Fig. 10. Percentage RMS of the model relative errors in a) wind direction (Equation 13) and b) wind speed (Equation 14), averaged of the four grid points that surround every weather station: Ezeiza (EZE), Aeroparque (AER), Martín García (MGA), Punta Indio (PIN) and Pontón Recalada (PRE). The averaging period is 1959-1984.

4.2 Seasonal mean

In order to assess the seasonal performance of the model, the same methodology as described in Section 3.1 is applied to the four seasons of the year defined as summer -DJF-, autumn –MAM, winter –JJA- and spring SON. Fig. 11 shows the model errors at the five weather stations as a function of the season of the year for wind direction (top panel) and wind speed (bottom panel). In the case of wind direction (Fig. 11a), the model error is minimum in winter and maximum in summer at the five locations. In spring the model errors are smaller that in autumn and the annual error is very similar to that of winter. The only exception is PRE where the autumn wind direction error is smaller than the annual one. The spread of the wind direction errors among the different seasons is similar at the five weather stations. The wind speed errors (Fig. 11b) display no seasonality, except in a minor degree in the case of EZE that has the minimum in summer and the maximum in winter and autumn, in contrast with the case of the wind direction error.

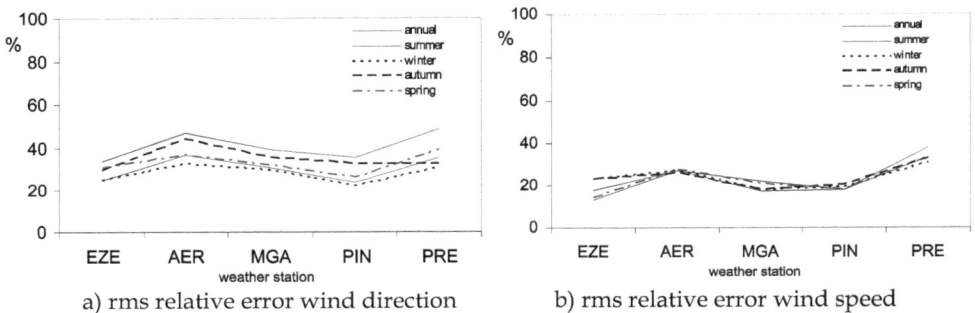

a) rms relative error wind direction	b) rms relative error wind speed

Fig. 11. Percentage RMS of the model relative errors as a function of season of the year, in a) wind direction (Equation 13) and b) wind speed (Equation 14). Values are the average of the four grid points that surround every weather station: Ezeiza (EZE), Aeroparque (AER), Martín García (MGA), Punta Indio (PIN) and Pontón Recalada (PRE). The averaging period is 1959-1984.

5. Validation of the synthesized method

Normally, the climatological mean value of any meteorological variable is calculated by averaging all available observations. In the case of a model climatology the same concept applies, so that the climatological mean value should be the result of averaging a long series of individual realizations. The question is: how good is the "ensemble method" for calculating the low level climatological wind field in comparison with a conventional method based on individual daily forecasts?. Berri et al. (2011) address this question by calculating the climatological wind field in two different ways and comparing results with the observations. For this purpose, the same model version and data set period are employed. In one case, namely the "ensemble method", the climatological wind field is calculated, as described in Section 3, and in detail in Berri et al. (2010); as the average of 192 members and their associated probabilities. In the other case, namely the "daily method", the climatological wind field is calculated as the average of 3248 days with available data during the same period 1959-84. The resulting wind frequency distributions, as well as the observed one, are calculated in a similar manner. The objective is to evaluate the goodness

of the ensemble method to synthesizing low level climatological wind fields, based on a
reduced number of realizations, in comparison with the conventional method that employs
a long series of individual realizations.

5.1 Results of the validation

Fig. 12 compares the averaged daily RMSEs of the low level wind climatology obtained with
both methods, at the five meteorological stations of the study. Except at PRE, at the other
stations the wind direction RMSE (panel a) of the daily method (dashed line) is a few
percent points greater than that of the ensemble method (solid line). In the case of wind
speed, panel b) shows almost no difference between the two methods.

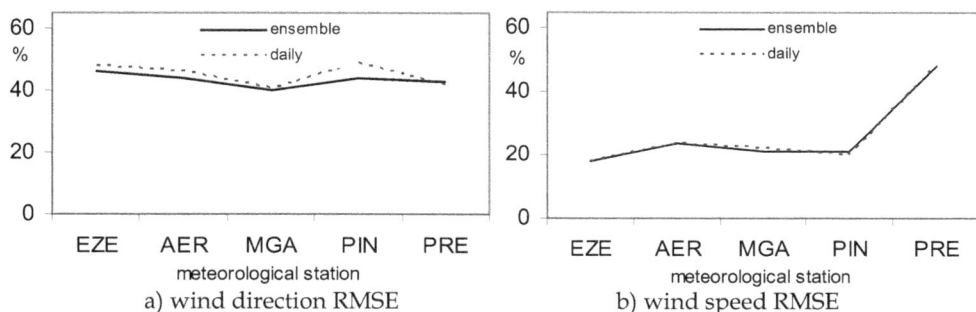

a) wind direction RMSE   b) wind speed RMSE

Fig. 12. Ensemble method and daily method RMSEs, in percentage, at the meteorological
stations of Ezeiza (EZE), Aeroparque (AER), Martín García (MGA), Punta Indio (PIN) and
Pontón Recalada (PRE), averaged during the period 1959-1984. Panel a) corresponds to wind
direction and panel b) corresponds to wind speed.

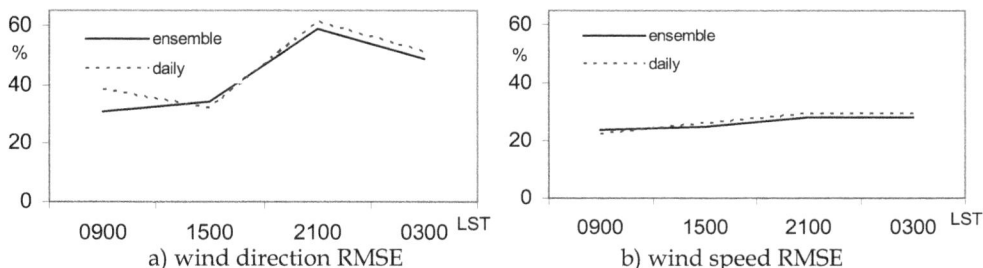

a) wind direction RMSE   b) wind speed RMSE

Fig. 13. Ensemble method and daily method RMSEs, in percentage, as a function of the local
standard time (LST), averaged over the five meteorological stations during the period 1959-
1984. Panel a) corresponds to wind direction and panel b) corresponds to wind speed.

The RMSE of the two methods, averaged over the five meteorological stations and as a
function of the local standard time, is compared in Fig. 13. In the case of wind direction
(panel a), the ensemble method (solid line) has smaller RMSE in all cases except at 1500 LST.
The largest difference in RMSE between the two methods is at 0900 LST, being smaller that
of the ensemble method (solid line). In the case of wind speed (panel b), the ensemble
method has smaller RMSE than the daily method, except at 0900 LST, although the
differences are always of a few percent points.

The comparison between the RMSE of both methods is shown in more detail in Fig. 14, by time of the day and meteorological station. In the case of wind direction (black line), EZE (Fig. 14a) and PIN (Fig. 14b) are the sites where at all times the ensemble method (solid line) has smaller RMSE. The largest differences in favour of the daily method (dashed line) are obtained for AER (Fig. 14b) at 1500 LST and MGA (Fig. 14c) at 0300 LST, i.e. 13 and 9 percent points, respectively. In the case of PRE (Fig. 14e), the daily method (dashed line) gives smaller RMSE at 1500 and 2100 LST.

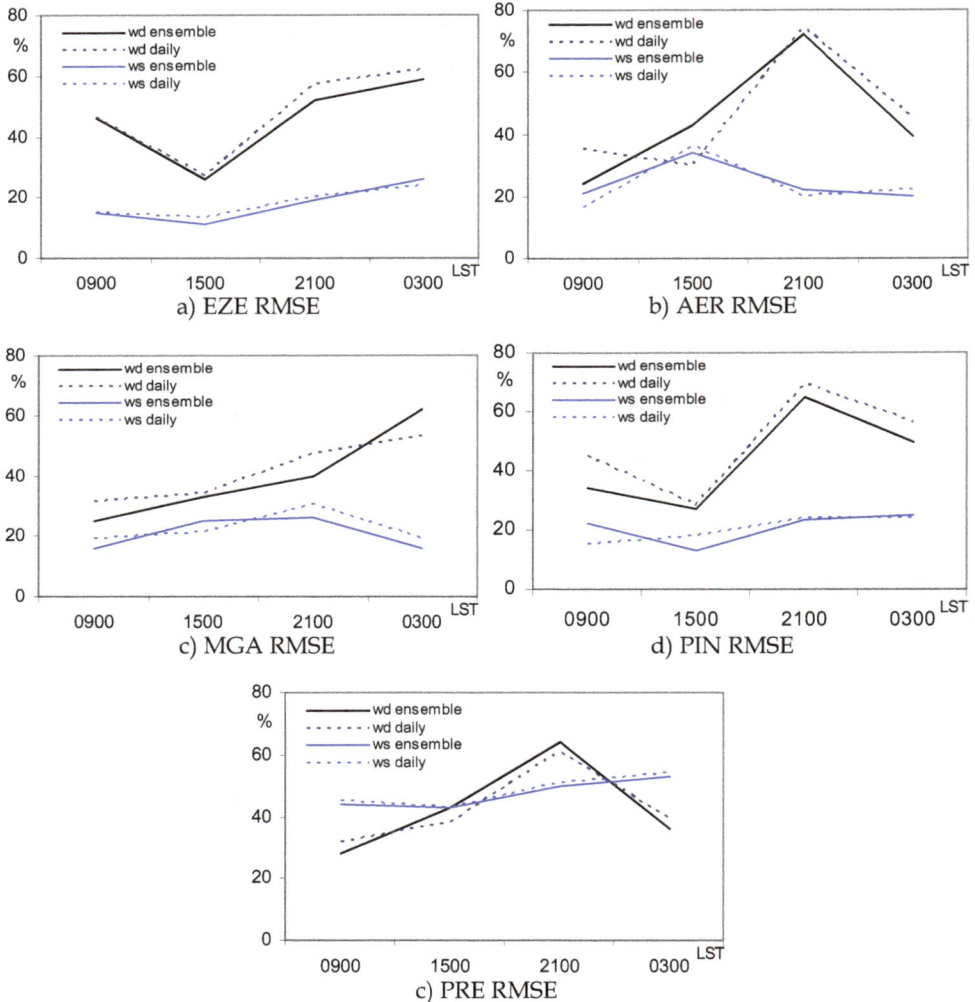

a) EZE RMSE

b) AER RMSE

c) MGA RMSE

d) PIN RMSE

c) PRE RMSE

Fig. 14. Ensemble method and daily method RMSEs, in percentage, as a function of the local standard time (LST), at the five meteorological stations during the period 1959-1984. Wind direction (speed) RMSEs are shown in black (blue).

In the case of wind speed (blue line), PRE (Fig. 14e) is the only site where at all times the ensemble method (solid line) has smaller RMSE, although by one percent point. PIN (Fig. 14d) and AER (Fig. 14b) show two times of the day with smaller RMSE for the daily method (dashed line), reaching 7 percent points in the case of PIN at 0900 LST. Finally, MGA (Fig. 14c) and EZE (Fig. 14a) present only one time of the day with smaller RMSE for the daily method (dashed line).

A qualitative summary of results is shown in Table 3 that presents the RMSE difference daily minus ensemble method. Considering individual boxes, 75% of them have greater or equal than zero values, meaning equal or better results with the ensemble method. When the wind direction and wind speed are considered together, the ensemble method shows smaller RMSE in 50% of the cases; while the daily method shows no cases like those.

Despite the fact that the ensemble method has, in general, smaller error; in some cases the wind direction RMSE of the daily method is small enough to overcome a larger RMSE in wind speed. In these situations the daily method has a smaller combined error; outperforming the ensemble method. The most notable cases are MGA at 0300 LST and AER at 1500 LST.

	EZE		AER		MGA		PIN		PRE	
LST	wd	ws	wd	ws	wd	ws	wd	Ws	wd	ws
0300	3	-2	6	2	-9	3	6	-1	3	1
0900	0	0	11	-5	6	3	11	-7	4	1
1500	1	2	-13	2	1	-4	1	5	-5	0
2100	5	1	2	-2	7	4	4	1	-3	1

Table 3. Absolute difference daily method minus ensemble method RMSE (in percent points) for wind direction (wd) and wind speed (ws), as a function of the local standard time (LST) at the meteorological stations of Ezeiza (EZE), Aeroparque (AER), Martín García (MGA), Punta Indio (PIN) and Pontón Recalada (PRE); averaged during the period 1959-1984. Positive numbers indicate smaller RMSE for the ensemble method.

The combined wind direction and wind speed error is largest at the evening (2100 LST). Berri et al. (2010) argued that the maximum error of the ensemble method at 2100 LST could be due to the fact that the transition from unstable daytime conditions to stable night time conditions takes place around that time of the day. In summer, 2100 LST is just after sunset; while in winter is about three hours after sunset. Since the ensemble method averages the surface conditions of different days with the same upper boundary condition, there could be an inherent limitation for appropriately resolving the transition from unstable to stable conditions. Therefore, the daily method would offer the possibility of overcoming such limitation by considering the individuality of each day of the data base. However, the present study indicates that this is not the case since the results of the daily method are not better than those of the ensemble method at 2100 LST. In the average, the results of the ensemble method are not outperformed by the daily method.

6. Operational low-level wind forecast

In order to test the BLM capability to produce operational low–level wind field forecasts over the La Plata River region, with higher temporal and spatial resolution than the presently available one, three experiments were conducted using different boundary conditions (see Sraibman & Berri, 2009 for the details). In Experiment I the BLM model is forced in both the upper and lower boundaries with the forecasts of the Eta/CPTEC (Centro de Previsao de Tempo e Estudos Climáticos, Brazil) model (Mesinger & Black, 1992). In Experiment II the BLM is forced in the upper boundary with the Eta model forecasts and in the lower boundary with the observed surface temperature. Experiment III employs only observations for defining the forcing conditions of the upper boundary (radiosonde data) and the lower boundary (surface temperature). In each experiment the BLM wind forecasts are compared to the available observational data and to the Eta model wind forecasts.

Fig. 15. Mean observed wind roses for the period November 2003-April 2004 at the meteorological stations Aeroparque, Don Torcuato, Ezeiza, Colonia and Carrasco. The scale in the lower left corner indicates the wind direction frequency in percentage

6.1 Experiment design

Each experiment consists of 142 diurnal forecasts obtained with the BLM model during the summer semester period of November 2003 to April 2004. The number of cases is less than the 182 days of that period of time, since the set of forecasts is restricted to those days with complete hourly observations. Each set member is a 12–hour forecast starting at 0900 local standard time (LST), approximately 2 to 3 hours after sunrise, when there is a minimum land–river temperature contrast. In this first study of BLM operational forecast verification, the forecast horizon is limited to the diurnal cycle, since the results of Berri & Nuñez (1993) show that the most remarkable changes of the wind pattern take place during that part of the day. The forecast wind is compared to the observed wind in five weather stations in the region, Aeroparque, Don Torcuato and Ezeiza in Argentina and Colonia and Carrasco in Uruguay (see Fig.15 for their location). The 0900 LST BLM forecast is after 30 minutes of model integration, in order to facilitate the model spin–up. Fig. 15 shows the observed mean wind direction frequencies at the five weather stations of the study for the period November 2003–April 2004. The figure displays significant differences in the predominant wind direction sectors across the La Plata River region, as a consequence of the dominant sea–land breeze circulation.

6.2 Evaluation method

The BLM and Eta model 10-m wind forecasts are validated with observations at the weatherstations, by means of two accuracy measures. One is the hit rate (HR) that counts the percentage of cases that the forecast wind direction matches the observed wind direction (Wilks, 1995), and the other one is the root-mean-squared error (RMSE) of the horizontal wind components.

In order to calculate the HR, the horizontal wind components are transformed to a wind direction expressed as one of the eight standard compass sectors (north, north-east, east, etc.). The RMSE is a usual accuracy measure employed in forecast verification (for example Hanna & Yang 2001; White et al. 1999; Zhong & Fast 2003), and is calculated as

$$RMSE = \sqrt{\frac{1}{N}\left(\sum_{j=1}^{N}\left(u_o - u_f\right)_j^2 + \sum_{j=1}^{N}\left(v_o - v_f\right)_j^2\right)} \qquad (15)$$

where N is the number of available observations, u and v are the zonal and meridional wind components, respectively, subscripts f and o refer to forecast and observation, respectively, and subscript j identifies the observation to be verified. The BLM forecasts are verified with each available observation between 0900 and 2100 LST, while the Eta forecasts are verified with the 0900, 1500 and 2100 LST observations, since these are the only model outputs available for the study. Since the terrain is flat, the verification of both models is made with the interpolated value of the four surrounding grid points to each weather station (the interpolation uses the same technique that is described in Section 5.3.1).

6.3 Experiment I

In this set of forecasts the BLM is forced at the upper and lower boundaries with the Eta forecasts; the Eta model produces twice daily forecasts initialized at 1200 UTC and 0000 UTC (corresponding to 0900 LST and 2100 LST, respectively). In the present study we use the 1200 UTC forecast of the previous day, since, according to Mesinger & Black (1992), Seluchi & Chou (2001) and Bustamante et al. (1999), the Eta forecasts for the second 24 h are better than for the first 24 h. Thus, the BLM is initialized at 0900 LST with a 24-h Eta forecast. Since the Eta forecasts are available at 6-h intervals (i.e. 1200 UTC, 1800 UTC and 0000 UTC), it is necessary to interpolate the data in time in order to define the boundary conditions at the 1-min BLM timestep. For this purpose a cubic spline interpolating polynomial (Cormier and Marsh 2001) is implemented, which has the property of being continuous and having first-order and second-order continuous derivatives. Additionally, it is necessary to interpolate the data in space from the 40-km resolution of the Eta model to the 10-km BLM resolution, with the technique described in the following subsection.

6.3.1 Upper boundary interpolation

The 850 hPa Eta wind forecasts are interpolated using the Cressman method (Cressman, 1959). The first step consists of calculating the distance $D_{ij}^{\ k}$ from each BLM grid point, identified by subscripts i and j, to every Eta grid point, identified with the superscript k. This distance is given by $D_{ij}^{k2} = \left(X_k - X_{ij}\right)^2 + \left(Y_k - Y_{ij}\right)^2$, where X_{ij} and Y_{ij} are the longitude and latitude, respectively, of the BLM grid points, and X_k and Y_k are the longitude and latitude, respectively, of the Eta grid points. The second step consists of calculating the weighting factor of each grid point as follows: $W_{ij}^{\ k} = N^2 - D_{ij}^{\ k2} \big/ N^2 + D_{ij}^{\ k2}$, if $N \geq D_{ij}^{\ k}$, and $W_{ij}^{k} = 0$, if $N < D_{ij}^{\ k}$. N is a fixed value that, after testing the results of the interpolation, it was set equal to 0.4 degrees. Finally, the corresponding value of any variable at each BLM grid point is given by $V_{ij} = \sum_k V^k W_{ij}^{\ k} \big/ \sum_k W_{ij}^k$, where V^k is the corresponding value at the Eta grid point.

6.3.2 Lower boundary interpolation

At the lower boundary a similar interpolating routine is implemented, although it is applied separately to the grid points over the land and over the river, respectively. The reason for this procedure is because the temperature over the river is quite homogeneous and changes in a very small amount during the day, while over land the temperature daily cycle is large, reaching values of up to 10°C or more. Thus, a 1–km transition band centred at the coast is established, in which the land–river surface temperature difference is linearly interpolated. This provides a smooth transition between the two regions and concentrates the land–river temperature contrast at the coast.

6.3.3 Results of experiment I

Table 4 shows that the BLM RMSE is between 2.4 and 3.3 m s⁻¹ smaller than the Eta RMSE, while the BLM HR value is almost three times the corresponding Eta value in all cases. The comparison of the HR obtained with both models at every weather station is shown in Fig. 16 at three different times of the day, as well as the average value of the three times. In all the weather stations the wind direction forecasts obtained with the BLM are better than those of the Eta model, and in particular at 0900 LST they provide the largest HR values. The improvement of the BLM forecasts over the Eta forecasts is greatest in Aeroparque, the closest weather station to the river shore.

The analysis of the BLM wind field forecast (not shown) revealed an excessive predominance of the inland wind component at 1500 LST, indicating a stronger than normal sea breeze. The BLM surface forcing is the land–river temperature difference, so that a large thermal contrast will induce a strong sea–land breeze circulation.

	0900 LST		1500 LST		2100 LST		Average 0900, 1500, 2100 LST		Average from 0900 to 2100 LST	
	HR (%)	RMSE (m s⁻¹)	HR (%)	RMSE (m s⁻¹)	HR (%)	RMSE (m s⁻¹)	HR (%)	RMSE (m s⁻¹)	HR (%)	RMSE (m s⁻¹)
BLM	68	3.2	50	3.5	58	3.6	59	3.4	58	3.5
Eta	21	5.6	18	6.8	15	6.5	18	6.4		

Table 4. HR and RMSE for Experiment I

In order to compare the magnitude of the Eta model thermal contrast with observations, we calculate the temperature difference between Ezeiza (inland) and Pontón Recalada (river). We chose Ezeiza as the most representative inland weather station, since it is located several kilometres away from the coast, while Pontón Recalada is the only one weather station over the river (see Fig. 15). The observed temperature difference is compared to the temperature difference between the nearest Eta grid points to each of the above mentioned weather stations.

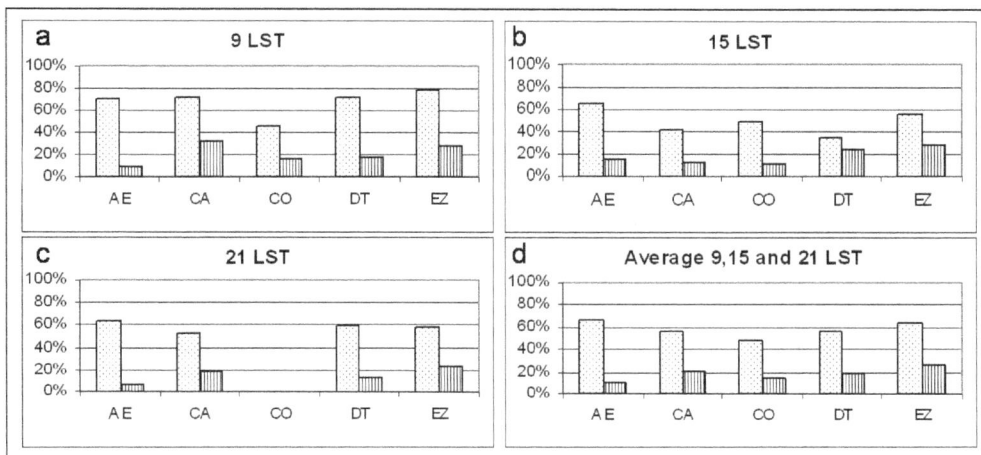

Fig. 16. HR rates for Experiment I at a) 0900 LST, b) 1500 LST, c) 2100 LST and d) average of 0900, 1500 and 2100 LST, for the meteorological stations Aeroparque (AE), Carrasco (CA), Colonia (CO), Don Torcuato (DT) and Ezeiza (EZ). The dotted (dashed) bar corresponds to the BLM (Eta) model.

The mean value of the observed temperature difference is 1°C, 4°C and 1°C (at 0900, 1500 and 2100 LST, respectively), whereas the Eta forecast provides, respectively, the following values 1°C, 10°C and −2°C. The largest disagreement between observations and Eta forecasts of the land–river thermal contrast is in the afternoon, which is responsible for the stronger than normal sea breeze component. Therefore, with the purpose of improving the BLM forecast, a new experiment is designed in which the surface forcing is defined from the observations.

	0900 LST		1500 LST		2100 LST		Average 0900, 1500, 2100 LST		Average from 0900 to 2100 LST	
	HR (%)	RMSE (m s^{-1})	HR (%)	RMSE (m s^{-1})	HR (%)	RMSE (m s^{-1})	HR (%)	RMSE (m s^{-1})	HR (%)	RMSE (m s^{-1})
BLM	64	3.7	60	4.2	45	3.7	56	3.8	58	4.3
Eta	21	5.6	18	6.8	15	6.5	18	6.4		

Table 5. HR and RMSE for Experiment II

6.4 Experiment II

In this case the BLM is forced by a surface potential temperature given by the expression:
$\theta(x,y,t) = \theta_0 + A(t)[1 + \tanh(s(x,y)/B)]$, where θ_o is the mean value of the surface potential temperature over the entire domain at the initial state. The land–river temperature difference, $A(t)$, is calculated as the harmonic analysis of the observed temperature difference between

Ezeiza and Pontón Recalada at 0300, 0900, 1500 and 2100 LST. Finally, $s(x,y)$ is the shortest distance from every grid point to the coast. The hyperbolic tangent distributes the land–river temperature difference symmetrically with respect to the coast and the parameter $B=1000$ m provides 75% (90%) of the temperature change in a distance of $2B$ ($3B$) across the coast. The upper boundary condition of the BLM model is the same as Experiment I.

6.4.1 Results of experiment II

Table 5 compares the RMSE and HR values of both models, and again the BLM forecasts result more accurate than the Eta forecasts. However, the average results are not as good as those of Experiment I, since the average HR of 56% is slightly smaller than the 59% of Experiment I (see column 4), and the RMSE equal to 3.8 m s^{-1} is slightly larger than the 3.4 m s^{-1} of Experiment I. However, when looking at different times of the day we can see that at 1500 LST the HR equal to 60% represents an improvement over the 50% of Experiment I, although the RMSE equal to 4.2 m s^{-1} is slightly worse than the 3.5 m s^{-1} of Experiment I. On the other hand, the BLM forecasts at 0900 and 2100 LST are degraded in comparison with Experiment I. For example the HR value drops from 68% to 64% at 0900 LST and from 58% to 45% at 2100 LST. Similarly, the RMSE increases from 3.2 m s^{-1} to 3.7 m s^{-1} at 0900 LST and from 3.6 m s^{-1} to 3.7 m s^{-1} at 2100 LST. The improvement of the BLM afternoon forecasts is the result of a more appropriate definition of the land–river thermal contrast, since the excessive forcing obtained from the Eta forecast develops a much stronger inland sea breeze component. The last column of Tables 4 and 5 contains the average values of BLM HR and RMSE of all the hours between 0900 LST and 2100 LST, and Experiment I shows a slight improvement over Experiment II in the case of the RMSE, but not in the HR which has the same value. Clearly, a better definition of the surface forcing improves the BLM forecasts at 1500 LST.

Fig. 17 shows the HR value at each weather station and time of the day. It can be clearly seen that at 1500 LST (Fig. 17b), Experiment II has a better forecast performance than Experiment I, except at the Ezeiza weather station.

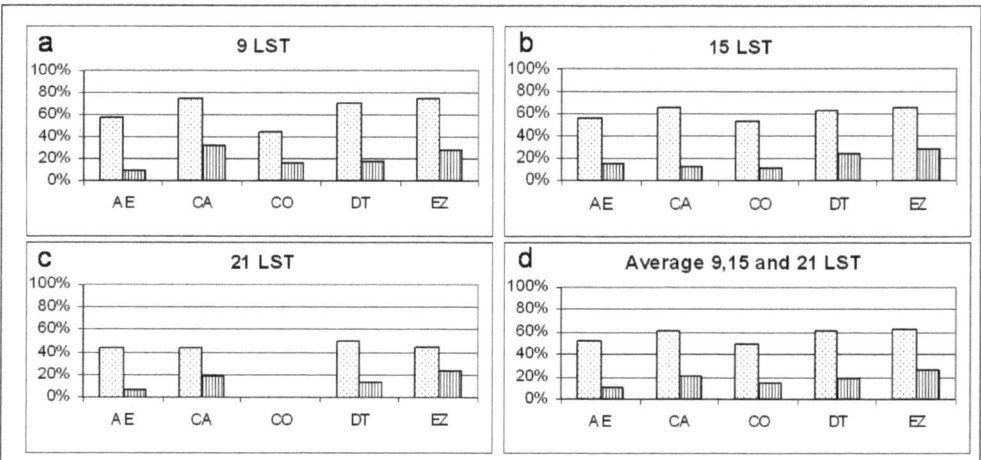

Fig. 17. same as Fig. 16, but for Experiment II.

	BLM boundary conditions from Eta forecasts		BLM boundary conditions from observations			
Upper boundary	850 hPa Exp I	1000 hPa	850 hPa	925 hPa	1000 hPa Exp III	Eta
Lower boundary	Eta temperature	Eta temperature	temperature function	temperature function	temperature function	
RMSE(m s⁻¹)	3.4	3.4	3.8	3.4	3.4	6.4
HR (%)	59	53	41	48	64	18

Table 6. Averaged 0900, 1500 and 2100 LST RMSE and HR values at the five meteorological stations, obtained with the Eta and the BLM models forecasts under different boundary conditions. Eta temperature means that the interpolation from Eta temperature is used, and temperature function means that the surface potential temperature function is used.

6.5 Experiment III, model forced with observations

In this case, both the upper and lower boundary conditions for the BLM forecasts are defined exclusively from the local observations. The surface forcing condition is the same as Experiment II, but the upper boundary condition is taken from the 1200 UTC (0900 LST) 850 hPa level Ezeiza radiosonde sounding. Since there is only one sounding a day, the upper boundary condition remains constant during the integration period.

6.5.1 Results of experiment III

Table 6 compares the averaged RMSE and HR values (0900, 1500 and 2100 LST) at the five weather stations obtained with the BLM forecasts performed with different boundary conditions. In the first two columns the boundary conditions are taken from the Eta forecasts (first column corresponds to Experiment I); while in the following three columns the boundary conditions are taken from the observations. The last column shows the averaged RMSE and HR values of the Eta forecasts. In all the cases the BLM forecasts are more accurate than the Eta forecasts, although there are remarkable differences among the different cases.

	0900 LST		1500 LST		2100 LST		Average 0900, 1500, 2100 LST		Average from 0900 to 2100 LST	
	HR (%)	RMSE (m s⁻¹)	HR (%)	RMSE (m s⁻¹)	HR (%)	RMSE (m s⁻¹)	HR (%)	RMSE (m s⁻¹)	HR (%)	RMSE (m s⁻¹)
BLM	73	3.0	64	3.7	54	3.7	64	3.4	62	3.3
Eta	21	5.6	18	6.8	15	6.5	18	6.4		

Table 7. HR and RMSE for Experiment III

When the 850 hPa level of the radiosonde sounding is used as the upper boundary condition (same level of the Eta forecasts used in the previous case); the BLM forecasts degrade with respect to Experiment I, since the RMSE increases from 3.4 m s⁻¹ to 3.8 m s⁻¹ and the HR drops from 59% to 41% (Table 6, column 3). In view of this, we decided to run the BLM forecasts using other levels of the radiosonde sounding for defining the upper boundary condition. The other two standard levels available in the observations are 925 hPa and 1000 hPa, and the results shown in columns 4 and 5 of Table 6 indicate that the forecast performance improves as the level of the boundary condition approaches the surface. The best result is obtained with the 1000 hPa condition, which represents an improvement, although small, with respect to the case in which the Eta forecasts are used to define both boundary conditions, i.e. the RMSE remains unchanged but the HR increases from 59% to 64% (Table 6, column 5). The La Plata River region is under the influence of the South Atlantic anticyclone whose temperature inversion defines the top of the boundary layer. The analysis of the Ezeiza radiosonde data shows that in 72% of the observations during the validation period, the base of the temperature inversion is below 850 hPa, and in 54% of the cases it is below 925 hPa. Therefore, most of the time those two levels are above the top of the boundary layer, which may be reason why the best results are obtained when forcing the model with an upper boundary condition taken from the 1000 hPa level.

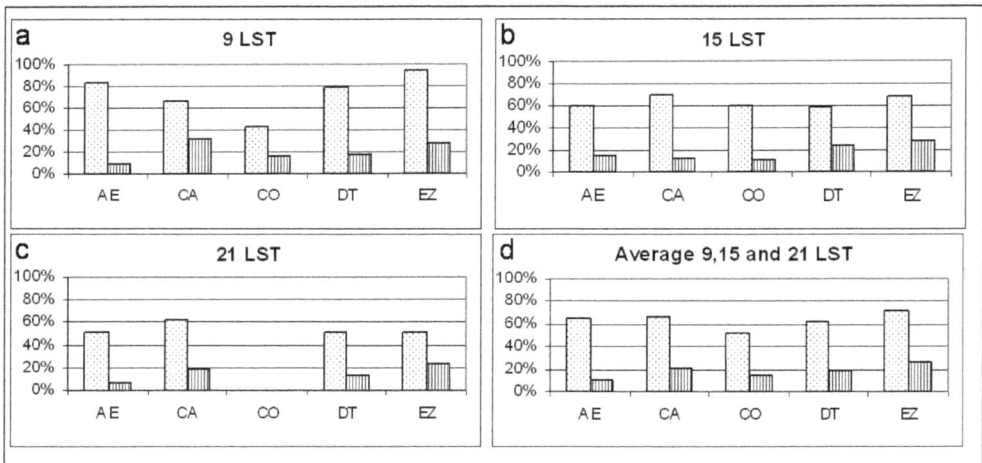

Fig. 18. same as Fig. 16, but for Experiment III.

Table 7 details the results of Experiment III and shows forecast degradation with time as in the two previous experiments. Fig. 18 shows that the HR obtained with the BLM at every weather station is, in the average, better than that of the Eta forecasts. When comparing the results of this experiment with the other two, the HR is always equal or better at every weather station, with the exception of Colonia. For comparison, the BLM is run with the upper boundary condition taken from the 1000 hPa Eta forecasts (Table 6, column 2), but now the results worsen since the HR drops from 59% to 53%, although the RMSE remains unchanged and equal to 3.4 m s^{-1}.

Table 8 compares the results of the three experiments by time of the day. In the average (rightmost two columns), the results of Experiment III are better than those of the other two experiments in terms of both the HR and the RMSE. This is also the case at 0900 LST and 1500 LST. However, at 2100 LST Experiment I provides the best result, indicating that for longer lead times the use of forecast boundary conditions is better than the persistence of observations. Also, in all cases the BLM wind forecasts are more accurate than the Eta forecasts, for example the BLM RMSE is between 1.9 m s^{-1} and 3.3 m s^{-1} smaller than the Eta RMSE and the HR value is more than three times greater than that of the Eta forecasts. Table 8 also shows systematic forecast degradation with time in Experiment III, in which only observations are used for defining the boundary conditions. On the other hand, when only the Eta forecasts are used for defining the boundary conditions, i.e. Experiment I, the 12–hour (2100 LST) forecast results are not only slightly more accurate than the 6–hour forecast, but they are also unmatched. These results indicate that for longer lead forecasts there is a clear advantage of using the Eta model outputs for defining the boundary conditions.

6.6 Example of forecast for 18 January 2004

Since point verification may not be well suited for quantifying forecast models, and only standard meteorological observations are available in the region, a qualitative measure of verification is implemented. The BLM vertical velocity forecast is compared to the cumulus cloud distribution of a satellite image, in order to give a regional scale assessment of the model ability. As an example, this section presents the forecast for 18 January 2004, a date that was chosen with the following consideration. A first screening identified the days with an 850-hPa Eta forecast of light winds from the east or north-east throughout the daytime hours. This is the appropriate regional condition for the development of the typical low clouds that are induced by the sea-breeze circulation in the region. Out of the initially identified 7 days, the date of the example was the only one with a clear satellite image available around noon.

Firstly, the 10-m wind field forecasts of the BLM and Eta models are compared to the observations. Figure 19a shows the BLM forecast at 0900 LST along with the wind observations (thick arrows) at the five weather stations of the study (left panel), and the Eta forecast (right panel). The wind field for both models is mainly from the east and north-east, in agreement with the observations. However, there are discrepancies in the wind speeds since the Eta forecast around the weather stations in Argentina shows smaller values than the BLM forecast, which in turn agrees better with the observations. Besides, the BLM winds are weaker over the river and stronger over the land, whereas the Eta forecast shows systematically decreasing wind speeds from east to west.

Figure 19b corresponds to 1500 LST, the time of the day when the sea breeze is well developed. The Eta forecast shows south-easterly winds across the river and neighbouring regions, whereas the observations reveal different wind directions. Over the coastal region of Uruguay the BLM and the Eta wind direction forecasts are from the south-east, in agreement with the observations. Over the coastal region of Argentina the BLM wind direction forecast is east, which agrees with the observations, whereas the Eta wind direction forecast is south-east. The observed wind pattern over the region of the river springs displays the inland sea-breeze component, which is very well represented by the BLM forecast but clearly ignored by the Eta forecast. Figure 19c corresponds to 2100 LST, when the observed wind direction field over the coast of Uruguay does not show changes with respect to the 1500 LST forecast. The BLM wind direction forecast agrees with the observations, whereas the Eta forecast shows a wind direction shift to the eastern sector over the entire northern half of the region. Over the coast of Argentina the BLM forecast shows easterly and east-south-easterly winds, mostly in agreement with the observations. Instead, the counter-clockwise wind direction change from panel (b) to (c) of the Eta forecast is opposed to the observed one, between 1500 LST and 2100 LST.

The comparison of the wind fields at these three times clearly reveals the significant variability in time and space of the sea–land breeze circulation across the region. On the other hand, these changes are reasonably well represented by the BLM forecast, particularly over the river springs, but completely ignored by the Eta forecast. Figure 20a presents the forecast vertical velocity field at 1,200 m for 1100 LST, 18 January 2004. The regions identified with letters A and E display downward motion, whereas the rest of the domain displays upward motion. Since there are no vertical velocity observations available, it is possible to perform a qualitative verification of the vertical motion by analyzing the cumulus cloud distribution of a satellite image. Figure 20b shows the 1344 UTC (1044 LST) NOAA-17 satellite image, in which the low-level cumulus clouds can be clearly identified. The region identified with letter A in Fig. 20a matches the cloud free region over most of the river and the neighbouring inland region of Argentina of Fig. 20b. The region identified with letter B in Fig. 20a displays upward motion over land and up to the river shore, in coincidence with the cumulus cloud distribution in Fig. 20b, whereas in region C the clouds are located further inland (Fig. 20b); in agreement with the vertical motion field of Fig. 20a. Over the river mouth (Fig. 20b) there are scattered clouds, again in agreement with the upward motion of region D in Fig. 20a. Region E (Fig. 20a), inland over Uruguay and far from the river, is characterized by downward motion, coinciding with a cloud free region in Fig. 20b. Region F is clearly the one without any qualitative agreement between cloudiness and vertical motion. There, the model results indicate upward motion, with a maximum near the coast, whereas the satellite image is free of clouds. One of the tributaries of the La Plata River is the Uruguay River, which can be clearly seen in the satellite image coming from the north. The other tributary is the Paraná River, which drains from the north-west (upper left corner of Fig. 20b), and despite having flow that double the flow of the Uruguay River, is not clearly visible in the image. This is because the Paraná River runs along multiple streams that form a river delta, merging the La Plata River immediately to the north-west of the city of Buenos Aires. Therefore, a significant part of region F is a very humid flatland where the river–land temperature contrast is not confined to a narrow region as in the rest of the La Plata River shores. This singularity modifies the convergence/divergence pattern of the horizontal motion and, consequently, the vertical motion as well. Since the BLM model does not consider different soil types, this particular characteristic of region F is not represented, which in turn may explain the disagreement between the observed cloud distribution and the predicted vertical motion. Figure 21 presents a vertical cross-section at 58.4° W of the BLM vertical velocity forecast for 1500LST.

Synthesizing High-Resolution Climatological Wind Fields with a Mesoscale Atmospheric Boundary
Layer Model Forced with Local Weather Observations

381

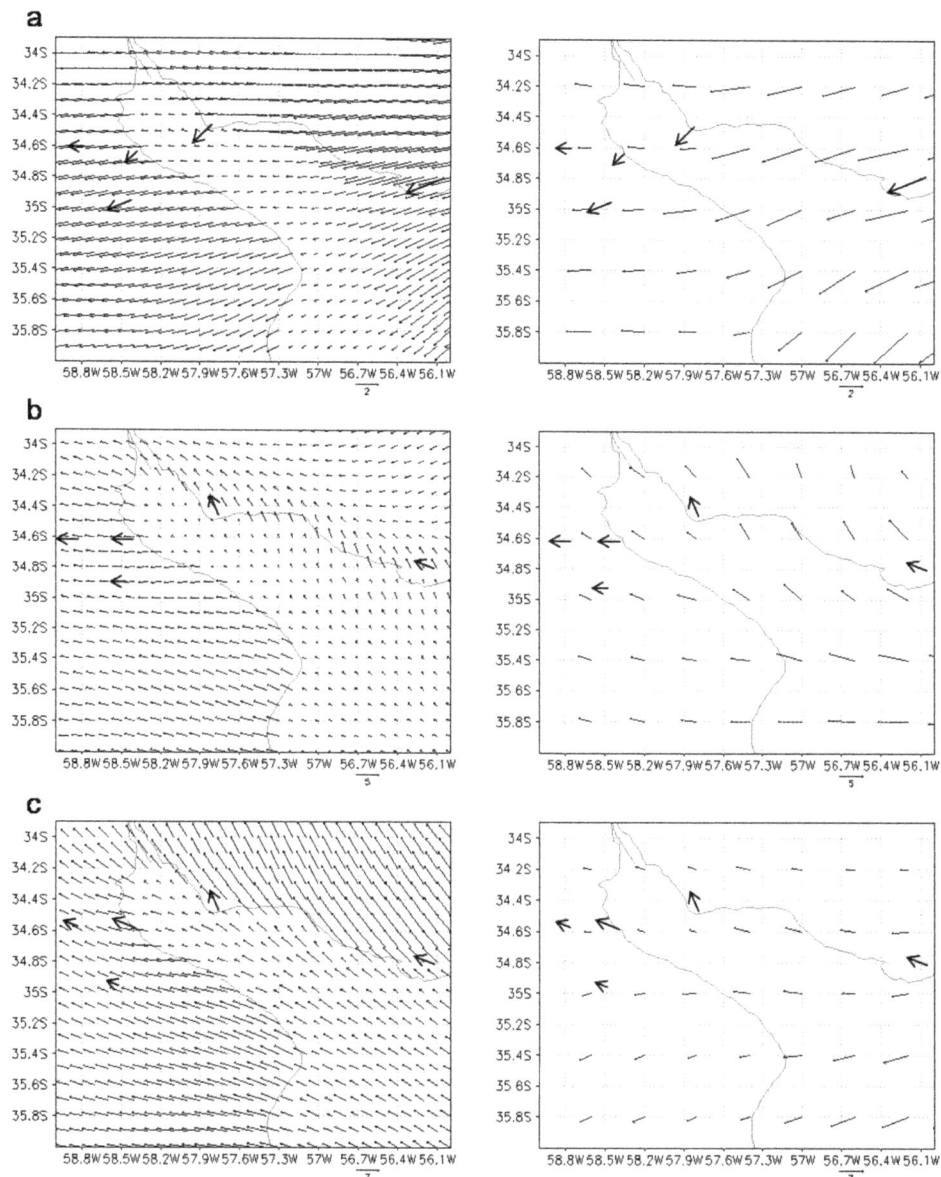

Fig. 19. Example of 10 meters wind field forecast of BLM model (left panels), and Eta model (right panels). The bold arrows represent the standard wind observations at the five weather stations. The date is 18 January 2004 and panel (a) corresponds to 0900 LST, panel (b) corresponds to 1500 LST and panel (c) corresponds to 2100 LST.

At this time of the day the sea breeze is at its peak, so that the land regions display upward motion and the river region, at the centre of the figure, displays downward motion. This vertical velocity pattern is typical of a well-developed mid-afternoon sea-breeze circulation that results from the low-level wind convergence/divergence fields. A similar vertical cross-section of the Eta forecast (not shown) does not reveal any detail since the magnitude of the vertical motion is of the order of 0.1 mm s^{-1}, with no clearly defined spatial pattern.

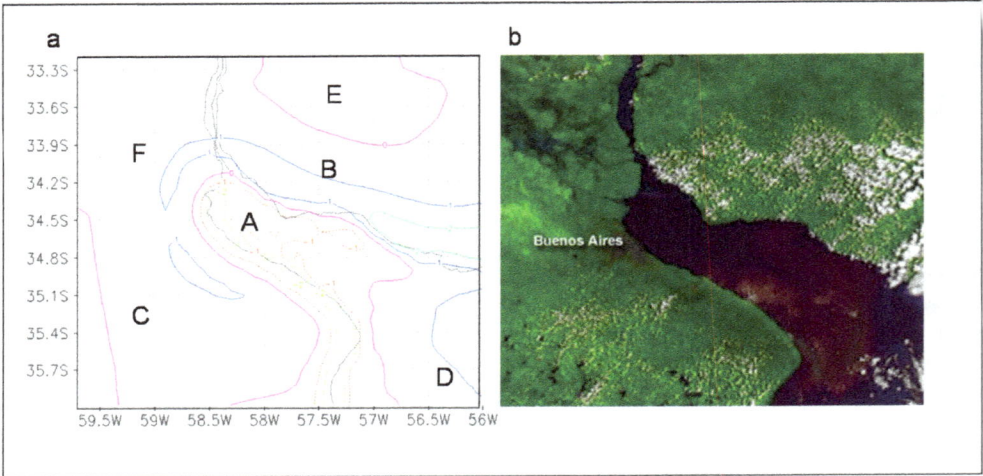

Fig. 20. (a) BLM vertical velocity forecast at 1200 m corresponding to 1100 LST of 18 January 2004; (b) NOAA-17 satellite image at 13:44 UTC (10:44 LST) provided by the National Meteorological Service of Argentina.

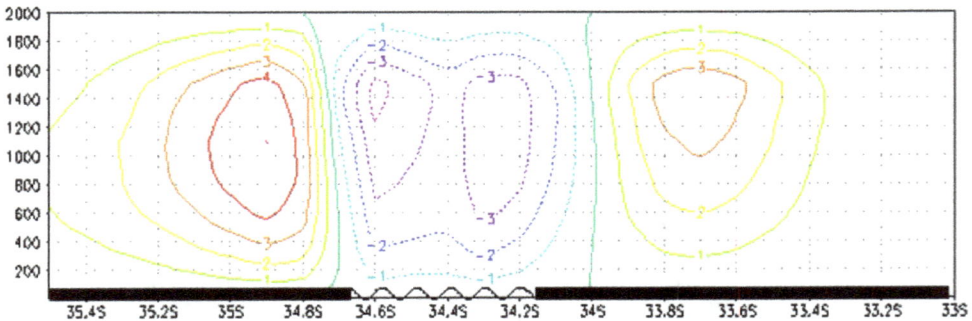

Fig. 21. Vertical cross section at 58.8°W of the BLM vertical velocity forecast (cm s^{-1}) at 1500 LST.

6.7 Discussion of operational forecast validation

In Experiment I the BLM is forced with the Eta 850-hPa wind forecasts at the upper boundary and the Eta surface temperature forecasts at the lower boundary. The BLM wind forecasts are substantially better than the Eta wind forecasts. The BLM RMSE is about 50% smaller, and the HR is more than three times greater, than the corresponding Eta forecasts. The river–land temperature differences forecast by the Eta model are always greater than the observed ones, the consequence of which is a degradation of the wind direction forecast, since the excessive thermal forcing creates a stronger than normal inland wind component during the afternoon. In view of this problem, Experiment II defines the surface forcing as a function of the observed temperature difference between Ezeiza (land) and Pontón Recalada (river) weather stations. The wind direction forecast is significantly better in the afternoon, although it is accompanied by a small degradation of the wind-speed forecast. On the other hand, the degradation of the wind direction forecast by the evening is larger than in Experiment I, indicating that the forecast improvement achieved during the first hours cannot be sustained with time.

Experiment III is designed to validate the BLM forecasts when forced exclusively with observations. The lower boundary condition is the same as in Experiment II, while the upper boundary condition is taken from the 0900 LST Ezeiza radiosonde sounding at the three standard levels within the boundary layer, i.e. 850, 925 and 1,000hPa. When the 850 and 925 hPa levels are used, the forecast results are not better than those of Experiment I. Instead, the 1,000-hPa level provides the best result since the error measures are the minimum of all experiments. The South Atlantic anticyclone has a strong influence over the La Plata River region and its temperature inversion defines the top of the boundary layer. The morning sounding reveals that most of the time the temperature inversion is below 925 hPa, and this may be the reason why the best result is obtained when forcing the model at the upper boundary condition with the 1,000-hPa data.

The BLM forecasts show greater degradation with time when the boundary conditions are defined exclusively from the observations. On the other hand, when they are defined exclusively from the Eta forecasts, the 12-h forecasts are more accurate than the 6-h forecasts. This result clearly indicates the advantage of using the Eta model outputs for defining the boundary conditions for longer lead BLM forecasts.

Although the short lead forecasts improve when forcing the BLM with observations, this case has operational disadvantages. Since the model is forced with the only daily radiosonde sounding at 0900 LST, the synoptic scale changes that may take place during the forecast period are ignored. Also, the necessity of specifying the lower boundary forcing as a function of time would require a forecast of the land–river temperature difference, so that an additional uncertainty may affect the final result. In any case this method would allow for a 12-h forecast for the rest of the day, but, on the other hand, the use of 48-h Eta model outputs for forcing the BLM allows for a forecast with 1 day in advance, i.e. 36-h lead forecast.

As an additional measure of verification, the vertical velocity field forecast by the model is compared to the cumulus cloud distribution of a satellite image, in order to have a regional scale assessment of the model simulation. The result shows a good agreement between the spatial patterns inmost parts of the region, i.e. cumulus cloud areas (cloud free areas) coincide with the areas where the model predicts upward (downward) motion. The improvement of the low-level wind forecast obtained with the BLM model, in comparison to the Eta model forecast, is not a straightforward consequence of the higher horizontal resolution of the former, in terms of grid spacing. An additional experiment is performed in which the BLM is forced at the lower boundary with the Eta model surface temperature forecasts, with the following consideration. The Eta model outputs are simply interpolated to the BLM resolution, i.e. ignoring the coastal geometry as in all other experiments. The result, although representing a minor improvement over the Eta forecasts, is far from achieving the performance of Experiment I. It is clear that the advantage of using the BLM model for forecasting the low-level wind field over the La Plata River region is the result of a more appropriate definition of the land–river surface temperature contrast. The particular formulation that the BLM model has for the geometry of the river coasts is fundamental for resolving the smaller scale details of the low-level local circulation.

Despite the large errors in the surface winds displayed by the Eta forecasts, its 850-hPa wind and surface temperature forecasts are able to drive the BLM model to obtain surface wind forecasts with smaller errors than the Eta model.

7. Conclusion

The conclusion is that the ensemble method is an appropriate methodology for determining high resolution, low level climatological wind fields, with the BLM model applied to a region with strong diurnal cycle of surface thermal contrast. The proposed methodology is of particular utility for synthesizing wind fields over regions with limited meteorological observations, since the 192-member matrix can be reasonably defined with few observing points, and even in the case of incomplete records. Also, operational low-level wind forecasts for the La Plata River region can be improved by forcing the BLM model with the Eta operational forecasts.

8. Acknowledgements

Partial support for this research was provided by research grants PIPs 5575 and 0772 from Consejo Nacional de Investigaciones Científicas y Técnicas (CONICET) of Argentina, and PICTs 2005/38193 and 2008/1417, from Agencia Nacional de Promoción Científica y Tecnológica of Argentina. Also acknowledged is the Servicio Meteorológico Nacional of Argentina for providing the meteorological station data, with special thanks to Jose Ares for preparing the data base.

9. References

Berri, G.J. (1987). *Thermo-hydrodynamic study of the atmospheric boundary layer over the La Plata River region with a numerical simulation model*, Doctoral Thesis, available at Department of Atmospheric and Oceanic Sciences, University of Buenos Aires.

Berri G.J. & Nuñez M.N. (1993). *Transformed shoreline–following horizontal coordinates in a mesoescale model: A sea–land breeze case study*, J. Appl. Meteorol., 5, 918–928pp.

Berri, G.J., L. Sraibman, R.A. Tanco & G. Bertossa (2010). *Low-level wind field climatology over the La Plata River region obtained with a mesoscale atmospheric boundary layer model forced with local weather observations.* J. Appl. Meteorol., 49, 1293-1305

Berri, G.J., J. Galli Nuin, L. Sraibman & G. Bertossa (2011). *Verification of a synthesized method for the calculation of low-level climatological wind fields using a mesoscale boundary-layer model.* In press Bound.-Layer Meteor.

Black T.L. (1994). *The new NMC mesoscale Eta model: Description and forecast examples,* Wea. Forecasting, 9, 265–278pp.

Bustamante, J.; Gomes, J. L.; Chou, S.C.; & Rozante, J. R. (1999). *Evaluation of April 1999 rainfall forecasts over South America using the Eta Model.* Climanálise, N° 5, Cachoeira Paulista, SP, Brazil.

Case J.L., Manobianco J., Lane J.E., Immer C.D., & Merceret F.J. (2004). *An Objective Technique for Verifying Sea Breezes in High–Resolution Numerical Weather Prediction Models,* Wea. Forecasting , 19, 690–705pp.

Colby F.P. Jr. (2004). *Simulation of the New England Sea Breeze: The effect of Grid Spacing,* Wea. Forecasting, 19, 277–285pp.

Cormier D.R. & Marsh L. (2001). *Spline Regression Models. Sage Publications,* Thousand Oaks, California, 80 pp.

Cressman G.P. (1959). *An operative objective analysis scheme.* Mon. Wea. Rev., 87, 367–374pp.

Daggupaty S.M. (2001). *A Case Study of the simultaneous Development of Multiple Lake–Breeze Fronts with a Boundary Layer Forecast Model,* J. Appl. Meteorol., 40, 289–311pp.

Hanna S. R. & Yang R. (2001). *Evaluations of mesoscale models' simulations of near–surface winds, temperature gradients, and mixing depths,* J. Appl. Meteorol., 40, 1095–1104pp.

Mesinger F. & Black T.L. (1992). *On the impact on forecast accuracy of the step–mountain (eta) vs. sigma coordinate,* Meteorol. Atmos. Phys., 50, 47–60pp.

Pielke R.A. (1974). *A Three Dimensional Numerical Model of the Sea Breezes Over the South Florida,* Mon. Wea. Rev. 102, 115–139pp.

Pielke R.A., Cotton W.R., Walko R.L., Tremback C.J., Lyons W.A., Grasso L.D., Nicholls M.E., Moran M.D., Wesley D.A., Lee T.J. & Copeland J.H. (1992). *A Comprehensive Meteorological Modeling System RAMS,* Meteorol. Atmos. Phys., 49, 69–91pp.

Roebber P.J. & Gehring Mark G. (2000). *Real–Time Prediction of the Lake Breeze on the Western Shore of Lake Michigan,* Wea. Forecasting, 15, 298–312.

Seluchi M.E. & Chou S.C. (2001). *Evaluation of two Eta model versions over South America,* Rev. Geofis., 40, 219–238.

Sraibman L., & G.J. Berri, (2009). *Low level wind forecast over La Plata River region with a mesoscale boundary layer model forced by regional operational forecasts,* Boundary Layer Meteorology, 130, 407-422.

White B.G., Paegle J., Steenburgh J.W., Horel J.D., Swanson R.T., Cook L.K, Onton D.J. & Miles J.G. (1999). *Short–term forecast validation of six models,* Wea. Forecasting, 14, 84–108.

Wilks, Daniel S. (1995). *Statistical Methods in the Atmospheric Sciences,* Academic Press, 240 pp.

Zhang Y., Chen Y., Schroeder T.H. & Kodama K. (2005). *Numerical Simulations of Sea-Breeze Circulations over Northwest Hawaii*, Wea. Forecasting, 20, 827–846.

Zhong S. & Fast J. (2003). *An Evaluation of the MM5, RAMS, and Meso–Eta Models at Subkilometer Resolution Using VTMX Field Campaign Data in the Salt Lake Valley*, Mon. Wea. Rev, 131, 1301–1322.

Permissions

The contributors of this book come from diverse backgrounds, making this book a truly international effort. This book will bring forth new frontiers with its revolutionizing research information and detailed analysis of the nascent developments around the world.

We would like to thank Dr. Shih-Yu (Simon) Wang and Dr. Robert R. Gillies, for lending their expertise to make the book truly unique. They have played a crucial role in the development of this book. Without their invaluable contribution this book wouldn't have been possible. They have made vital efforts to compile up to date information on the varied aspects of this subject to make this book a valuable addition to the collection of many professionals and students.

This book was conceptualized with the vision of imparting up-to-date information and advanced data in this field. To ensure the same, a matchless editorial board was set up. Every individual on the board went through rigorous rounds of assessment to prove their worth. After which they invested a large part of their time researching and compiling the most relevant data for our readers. Conferences and sessions were held from time to time between the editorial board and the contributing authors to present the data in the most comprehensible form. The editorial team has worked tirelessly to provide valuable and valid information to help people across the globe.

Every chapter published in this book has been scrutinized by our experts. Their significance has been extensively debated. The topics covered herein carry significant findings which will fuel the growth of the discipline. They may even be implemented as practical applications or may be referred to as a beginning point for another development. Chapters in this book were first published by InTech; hereby published with permission under the Creative Commons Attribution License or equivalent.

The editorial board has been involved in producing this book since its inception. They have spent rigorous hours researching and exploring the diverse topics which have resulted in the successful publishing of this book. They have passed on their knowledge of decades through this book. To expedite this challenging task, the publisher supported the team at every step. A small team of assistant editors was also appointed to further simplify the editing procedure and attain best results for the readers.

Our editorial team has been hand-picked from every corner of the world. Their multi-ethnicity adds dynamic inputs to the discussions which result in innovative outcomes. These outcomes are then further discussed with the researchers and contributors who give their valuable feedback and opinion regarding the same. The feedback is then collaborated with the researches and they are edited in a comprehensive manner to aid the understanding of the subject.

Apart from the editorial board, the designing team has also invested a significant amount of their time in understanding the subject and creating the most relevant covers. They scrutinized every image to scout for the most suitable representation of the subject and create an appropriate cover for the book.

The publishing team has been involved in this book since its early stages. They were actively engaged in every process, be it collecting the data, connecting with the contributors or procuring relevant information. The team has been an ardent support to the editorial, designing and production team. Their endless efforts to recruit the best for this project, has resulted in the accomplishment of this book. They are a veteran in the field of academics and their pool of knowledge is as vast as their experience in printing. Their expertise and guidance has proved useful at every step. Their uncompromising quality standards have made this book an exceptional effort. Their encouragement from time to time has been an inspiration for everyone.

The publisher and the editorial board hope that this book will prove to be a valuable piece of knowledge for researchers, students, practitioners and scholars across the globe.

List of Contributors

Jin-Ho Yoon
Pacific Northwest National Laboratory, Richland, WA, USA

Wan-Ru (Judy) Huang
Guy Carpenter Asia-Pacific Climate Impact Centre, School of Energy and Environment, City University of Hong Kong, Hong Kong, China

Yuriy Kuleshov
National Climate Centre, Australian Bureau of Meteorology, Melbourne School of Mathematical and Geospatial Sciences, Royal Melbourne Institute of Technology (RMIT) University, Melbourne, Australia

Siddarth Shankar Das
Space Physics Laboratory, Vikram Sarabhai Space Centre ISRO-PO, Trivandrum, India

Viviane B. S. Silva and Vernon E. Kousky
NOAA/National Weather Service, Climate Services Division, Climate Prediction Center, USA

Aniello Russo
Marche Polytechnic University-DISVA, Ancona, Italy
The National Research Council, Institute of Marine Sciences, Venice, Italy

Sandro Carniel and Mauro Sclavo
The National Research Council, Institute of Marine Sciences, Venice, Italy

Maja Krzelj
University of Split, Center of Marine Studies, Split, Croatia

S. Y. Simon Wang and Robert R. Gillies
Utah Climate Center / Department of Plants, Soil and Climate, Utah State University, Logan UT, USA

Daji Huang, Xiaohua Zhu and Dongfeng Xu
State Key Laboratory of Satellite Ocean Environment Dynamics, Second Institute of Oceanography, State Oceanic Administration, China
Department of Ocean Science and Engineering, Zhejiang University, China

Xiaobo Ni
State Key Laboratory of Satellite Ocean Environment Dynamics, Second Institute of Oceanography, State Oceanic Administration, China

Qisheng Tang
Yellow Sea Fisheries Research Institute, Chinese Academy of Fishery Sciences, China

Mohan Kuppusamy and Prosenjit Ghosh
Centre for Earth Science, IISc, Bangalore, India

Jill S. M. Coleman and Steven A. LaVoie
Ball State University, USA

Rafail V. Abramov
Dept. of Mathematics, Statistics and Computer Science, University of Illinois at Chicago, USA

Michael L. Gauthier
United States Air Force Academy, USA

Gholam Abbas Fallah-Ghalhari
Sabzevar Tarbiat Moallem University, I.R of Iran

Guillermo J. Berri
National Meteorological Service, Buenos Aires, Argentina

www.ingramcontent.com/pod-product-compliance
Lightning Source LLC
Chambersburg PA
CBHW070712190326
41458CB00004B/957